T0319462

PROBABILITY AND STOCHASTIC PROCESSES

PROBABILITY AND STOCHASTIC PROCESSES

Ionuţ Florescu
Stevens Institute of Technology

Published by John Wiley & Sons, Inc., Hoboken, New Jersey.
Published simultaneously in Canada.

For general information on our other products and services please contact our Customer Care
Department with the U.S. at 877-762-2974, outside the U.S. at 317-572-3993 or fax 317-572-4002.

Wiley also publishes its books in a variety of electronic formats. Some content that appears in print,
however, may not be available in electronic format.

Library of Congress Cataloging-in-Publication Data:

Florescu, Ionuţ, 1973–
  Probability and stochastic processes / Ionuţ Florescu, Stevens Institute of Technology, Hoboken, NJ.
      pages cm
  Includes bibliographical references and index.
  ISBN 978-0-470-62455-5 (cloth)
  1. Probabilities. 2. Stochastic processes. I. Title.
  QA273.F655 2015
  519.2–dc23
                                                              2015018455
Printed in the United States of America.

10 9 8 7 6 5 4 3 2 1

To M.G. and C.

CONTENTS IN BRIEF

CONTENTS

LIST OF FIGURES

LIST OF TABLES

PREFACE

This book originated as a series of lecture notes for a first year graduate class taught during two semesters at Stevens Institute of Technology. It covers probability, which is taught during the first semester, and stochastic processes, taught in the second semester. Thus the book is structured to cover both subject in a wide enough manner to allow applications to many domains.

Probability is an old subject. Stochastic processes is a new subject that is quickly becoming old. So why write a book on an old subject? In my opinion, this book is necessary at the current day and age. The fundamental textbooks are becoming too complex for the new students and, in an effort to make the material more accessible to students, new applied probability books discard the rigor of the old books and the painstaking details that is put forward in these old textbooks. At times, reading these new books feels like the authors are inventing new notions to be able to skip the old reasoning. I believe that this is not needed. I believe that it is possible to have a mathematically rigorous textbook which is at the same time accessible to students. The result is this work. This book does not try to reinvent the concepts only to put them into an accessible format. Throughout, I have tried to explain complex notions with as many details as possible. For this reason to a versed reader in the subject, many of the derivations will seem to contain unnecessary details. Let me assure you that for a student seeing the concepts for the first time these derivations are vital.

This textbook is not a replacement for the fundamental textbooks. Many results are not proven and, for a deeper understanding of each of the subjects, the reader is advised to delve deeper into these fundamental textbooks. However, in my opinion this textbook contains all the material needed to start research in probability, complete a qualifying exam in probability and stochastic processes, or make sound probability reasoning for applied problems.

I. FLORESCU

Hoboken, New Jersey
April, 2014

ACKNOWLEDGMENTS

I want to acknowledge all my teachers of Mathematics and Probability. They are the ones who made this book possible. These great people that taught me probability are listed in the order in which they taught me: *Nicolae Gheorghilã, Valentin Nicula, Gabriela Licea, Ioan Cuculescu, Constantin Tudor, Vasile Preda, Thomas Sellke, Ioannis Kontoyiannis, Rebecca Doerge, Burgess Davis, Tony Cai, Herman Rubin, Phillip Protter,* and of course my advisor *Frederi Viens*.

I would also like to acknowledge all my students who helped improve this book by providing feedback and corrections, among them Lazer Teran, Pojen Huan, Chris Flynn, and Brad Warren.

I. F.

INTRODUCTION

What is Probability? In essence:

> Mathematical modeling of random events and phenomena. It is fundamentally
> different from modeling deterministic events and functions, which constitutes the
> traditional study of Mathematics.

However, the study of probability uses concepts and notions straight from Mathematics; in fact Measure Theory and Potential Theory are expressions of abstract mathematics generalizing the Theory of Probability.

Like so many other branches of mathematics, the development of probability theory has been stimulated by the variety of its applications. In turn, each advance in the theory has enlarged the scope of its influence. Mathematical statistics is one important branch of applied probability; other applications occur in such widely different fields as genetics, biology, psychology, economics, finance, engineering, mechanics, optics, thermodynamics, quantum mechanics, computer vision, geophysics,etc. In fact I compel the reader to find one area in today's science where no applications of probability theory can be found.

Early history

In the XVII-th century the first notions of Probability Theory appeared. More precisely, in 1654 Antoine Gombaud Chevalier de Méré, a French nobleman with an interest in gaming and gambling questions, was puzzled by an apparent contradiction

Probability and Stochastic Processes, First Edition. Ionuţ Florescu

concerning a popular dice game. The game consisted of throwing a pair of dice 24 times; the problem was to decide whether or not to bet even money on the occurrence of at least one "double six" during the 24 throws. A seemingly well-established gambling rule led de Méré to believe that betting on a double six in 24 throws would be profitable (based on the payoff of the game). However, his own calculations based on many repetitions of the 24 throws indicated just the opposite. Using modern probability language de Méré was trying to establish if such an event has probability greater than 0.5 (we are looking at this question in example 1.7). Puzzled by this and other similar gambling problems he called on the famous mathematician Blaise Pascal. This, in turn led to an exchange of letters between Pascal and another famous French mathematician Pierre de Fermat. This is the first known documentation of the fundamental principles of the theory of probability. Before this famous exchange of letters, a few other simple problems on games of chance had been solved in the XV-th and XVI-th centuries by Italian mathematicians; however, no general principles had been formulated before this famous correspondence.

In 1655 during his first visit to Paris, the Dutch scientist Christian Huygens learned of the work on probability carried out in this correspondence. On his return to Holland in 1657, Huygens wrote a small work *De Ratiociniis in Ludo Aleae*, the first printed work on the calculus of probabilities. It was a treatise on problems associated with gambling. Because of the inherent appeal of games of chance, probability theory soon became popular, and the subject developed rapidly during the XVIII-th century.

The XVIII-th century

The major contributors during this period were Jacob Bernoulli (1654–1705) and Abraham de Moivre (1667-1754). Jacob (Jacques) Bernoulli was a Swiss mathematician who was the first to use the term integral. He was the first mathematician in the Bernoulli family, a family of famous scientists of the XVIII-th century. Jacob Bernoulli's most original work was *Ars Conjectandi* published in Basel in 1713, eight years after his death. The work was incomplete at the time of his death but it still was a work of the greatest significance in the development of the Theory of Probability. De Moivre was a French mathematician who lived most of his life in England[1]. De Moivre pioneered the modern approach to the Theory of Probability, in his work *The Doctrine of Chance: A Method of Calculating the Probabilities of Events in Play* in the year 1718. A Latin version of the book had been presented to the Royal Society and published in the *Philosophical Transactions* in 1711. The definition of statistical independence appears in this book for the first time. *The Doctrine of Chance* appeared in new expanded editions in 1718, 1738 and 1756. The birthday problem (example 1.12) appeared in the 1738 edition, the gambler's ruin problem (example 1.11) in the 1756 edition. The 1756 edition of *The Doctrine of Chance* contained what is probably de Moivre's most significant contribution to probability, namely the approximation of the binomial distribution with the normal distribution in the case of a large number of trials - which is now known by most probability textbooks as "The First Central Limit Theorem" (we will discuss this theorem in Chapter 4). He understood the notion of

---

[1] A protestant, he was pushed to leave France after Louis XIV revoked the Edict of Nantes in 1685, leading to the expulsion of the Huguenots

standard deviation and is the first to write the normal integral (and the distribution density). In *Miscellanea Analytica* (1730) he derives Stirling's formula (wrongly attributed to Stirling) which he uses in his proof of the central limit theorem. In the second edition of the book in 1738 de Moivre gives credit to Stirling for an improvement to the formula. De Moivre wrote:

> "I desisted in proceeding farther till my worthy and learned friend Mr James Stirling, who had applied after me to that inquiry, [discovered that $c = \sqrt{2}$]."

De Moivre also investigated mortality statistics and the foundation of the theory of annuities. In 1724 he published one of the first statistical applications to finance *Annuities on Lives*, based on population data for the city of Breslau. In fact, in *A History of the Mathematical Theory of Probability* (London, 1865), Isaac Todhunter says that probability:

> ... owes more to [de Moivre] than any other mathematician, with the single exception of Laplace.

De Moivre died in poverty. He did not hold a university position despite his influential friends Leibnitz, Newton, and Halley, and his main income came from tutoring.

De Moivre, like Cardan (Girolamo Cardano), predicted the day of his own death. He discovered that he was sleeping 15 minutes longer each night and summing the arithmetic progression, calculated that he would die on the day when he slept for 24 hours. He was right!

The XIX-th century

This century saw the development and generalization of the early Probability Theory. Pierre-Simon de Laplace (1749–1827) published *Théorie Analytique des Probabilités* in 1812. This is the first fundamental book in probability ever published (the second being Kolmogorov's 1933 monograph). Before Laplace, probability theory was solely concerned with developing a mathematical analysis of games of chance. The first edition was dedicated to Napoleon-le-Grand, but the dedication was removed in later editions![2]

The work consisted of two books and a second edition two years later saw an increase in the material by about 30 per cent. The work studies generating functions, Laplace's definition of probability, Bayes rule (so named by Poincaré many years later), the notion of mathematical expectation, probability approximations, a discussion of the method of least squares, Buffon's needle problem, and inverse Laplace transform. Later editions of the *Théorie Analytique des Probabilités* also contains supplements which consider applications of probability to determine errors in observations arising in astronomy, the other passion of Laplace.

On the morning of Monday 5 March 1827, Laplace died. Few events would cause the Academy to cancel a meeting but they did so on that day as a mark of respect for one of the greatest scientists of all time.

---

[2]The close relationship between Laplace and Napoleon is well documented and he became Count of the Empire in 1806. However, when it was clear that royalists were coming back he offered his services to the Bourbons and in 1817 he was rewarded with the title of marquis.

Century XX and modern times

Many scientists have contributed to the theory since Laplace's time; among the most important are Chebyshev, Markov, von Mises, and Kolmogorov.

One of the difficulties in developing a mathematical theory of probability has been to arrive at a definition of probability that is precise enough for use in mathematics, yet comprehensive enough to be applicable to a wide range of phenomena. The search for a widely acceptable definition took nearly three centuries and was marked by much controversy. The matter was finally resolved in the 20th century by treating probability theory on an axiomatic basis. In 1933, a monograph by the Russian *giant mathematician* Andrey Nikolaevich Kolmogorov (1903–1987) outlined an axiomatic approach that forms the basis for the modern theory. In 1925, the year he started his doctoral studies, Kolmogorov published his first paper with Khinchin on the probability theory. The paper contains, among other inequalities about partial series of random variables, the three series theorem which provides important tools for stochastic calculus. In 1929, when he finished his doctorate, he already had published 18 papers. Among them were versions of the strong law of large numbers and the law of iterated logarithm.

In 1933, two years after his appointment as a professor at Moscow University, Kolmogorov published *Grundbegriffe der Wahrscheinlichkeitsrechnung* his most fundamental book. In it he builds up probability theory in a rigorous way from fundamental axioms in a way comparable with Euclid's treatment of geometry. He gives a rigorous definition of the conditional expectation which later became fundamental for the definition of Brownian motion, stochastic integration, and Mathematics of Finance. (Kolmogorov's monograph is available in English translation as *Foundations of Probability Theory*, Chelsea, New York, 1950). In 1938 he publishes the paper *Analytic methods in probability theory* which lay the foundation for the Markov processes, leading toward a more rigorous approach to the Markov chains.

Kolmogorov later extended his work to study the motion of the planets and the turbulent flow of air from a jet engine. In 1941 he published two papers on turbulence which are of fundamental importance in the field of fluid mechanics. In 1953–54 two papers by Kolmogorov, each of four pages in length, appeared. These are on the theory of dynamical systems with applications to Hamiltonian dynamics. These papers mark the beginning of KAM-theory, which is named after Kolmogorov, Arnold and Moser. Kolmogorov addressed the International Congress of Mathematicians in Amsterdam in 1954 on this topic with his important talk *General Theory of Dynamical Systems and Classical Mechanics*. He thus demonstrated the vital role of probability theory in physics. His contribution in the topology theory is also of outmost importance[3].

Closer to the modern era, I have to mention Joseph Leo Doob (1910-2004), who was one of the pioneers in the modern treatment of stochastic processes. His book *Stochastic Processes* (Doob, 1953) is one of the most influential in the treatment of modern stochastic processes (specifically martingales). Paul-André Meyer

---

[3]Kolmogorov had many interests outside mathematics, for example he was interested in the form and structure of the poetry of the greatest Russian poet Alexander Sergeyevich Pushkin (1799-1837).

(1934–2003) is another big probabilist influential in later development of Stochastic Processes. He was instrumental in developing the theory of continuous time Markov processes, and in his treatment of continuous time martingales (Doob-Meyer decomposition). Later on, when applications to Fiance were developed this particular result was so influential that the entire theory started with it. The whole French probability school owe him a great deal of gratitude.

For my own development, the most influential books were Karatzas and Shreve (1991) which laid the foundations and my understanding of probability and the first edition of Protter (2003), which took my understanding to a whole other level by looking at everything from the perspective of semimartingales.

I would also like to conclude this introduction by mentioning the Romanian probabilists who directly influenced my development, not by the books they published but by their amazing lecture notes and their exposition of the material in lectures. I include here Professors Gabriela Licea, Ioan Cuculescu and the much regretted Constantin Tudor (1950–2011) who are directly responsible for most of the Romanian probabilists existing today.

For its immense success and wide variety of applications, the Theory of Probability can arguably be viewed as the most important area in Mathematics.

The book you have in your hands is divided into two parts. This is done for historical reasons as well as logistic since the book was designed to serve as material for two separate courses. The first part contains the fundamental treatment of Probability, and this part lays the foundation of everything that follows including material not covered in this book (such as statistics, time series etc.). The advanced topics on Stochastic Processes are contained in the second part. Rather than going directly into advanced topics (continuous time processes) I am presenting first fundamental processes, like the renewal process and discrete time Markov Chains, because they provide some of the most powerful results and techniques of the whole stochastic processes. The book concludes with more advanced continuous time processes.

Throughout the book I have added applications of the theory, which in my mind, is aligned to the original spirit of the creators of this area of mathematics who believed in the applicability of this domain. It is from my personal experience that a person with strong theoretical background can apply the theory easily, while a person with only applied background needs to resort to "quick and dirty fixes" to make algorithms work, and then only working in their particular context.

I am going to mention here one of my first truly applied research episodes, working with an applied collaborator. After spending a week designing an algorithm based on sequential Bayes we separated for the weekend. On Monday he met me very excited, telling me that he could not wait and implemented the algorithm by himself over the weekend, and that the algorithm really, really works. To my obviously confused figure he proceed to add that in his experience of 20 years programming no algorithm works from the first time and usually one tweaks parameters to make it work for the particular sequence analyzed. Of course, to me the fact that the algorithm works was no surprise since it is based on a theoretically sound technique. If the algorithm would not have worked that would have been the real surprise pointing most probably to a programming error or toward an error when judging the assumptions of the problem.

I will conclude this introduction with my hope that this book will prove to be as useful for you as it was for many generations of students learning probability theory at Stevens Institute of Technology.

PART I

PROBABILITY

CHAPTER 1

ELEMENTS OF PROBABILITY MEASURE

The axiomatic approach of Kolmogorov is followed by most books on probability theory. This is the approach of choice for most graduate level probability courses. However, the immediate applicability of the theory learned as such is questionable, and many years of study is required to understand and unleash its full power.

On the other hand, the books on applied probability completely disregard this approach, and they go more or less directly into presenting applications, thus leaving gaps in the reader's knowledge. On a cursory glance, this approach appears to be very useful (the presented problems are all very real and most are difficult). However, I question the utility of this approach when confronted with problems that are slightly different from the ones presented in such books.

I believe no present textbook strikes the right balance between these two approaches. This book is an attempt in this direction. I will start with the axiomatic approach and present as much as I feel will be be necessary for a complete understanding of the theory of probability. I will skip proofs which I consider will not bring something new to the development of the student's understanding.

Probability and Stochastic Processes, First Edition. Ionuţ Florescu
© 2015 John Wiley & Sons, Inc. Published 2015 by John Wiley & Sons, Inc.

1.1 Probability Spaces

Let Ω be an abstract set containing all possible outcomes or results of a random experiment or phenomenon. This space is sometimes denoted with S and is named the *sample space*. I call it "an abstract set" because it could contain anything. For example, if the experiment consists in tossing a coin once, the space Ω could be represented as $\{Head, Tail\}$. However, it could just as well be represented as $\{Cap, Pajura\}$, these being the Romanian equivalents of $Head$ and $Tail$. The space Ω could just as well contain an infinite number of elements. For example, measuring the diameter of a doughnut could result in all possible numbers inside a whole range. Furthermore, measuring in inches or in centimeters would produce different albeit equivalent spaces.

We will use ω, where $\omega \in \Omega$ to denote a generic outcome or a sample point.

Any collection of outcomes is called an event. That is, any subset of Ω is an event. We shall use capital letters from the beginning of the alphabet $(A, B, C, \ldots)$ to denote events.

So far so good. The proper definition of Ω is one of the most important issues when treating a problem probabilistically. However, this is not enough. We have to make sure that we can calculate the probability of all the items of interest.

Think of the following possible situation: Poles of various sizes are painted in all possible colors. In other words, the poles have two characteristics of interest: size and color. Suppose that in this model we have to calculate the probability that the next pole would be shorter than 15 in. and painted either red or blue. In order to answer such questions, we have to properly define the sample space Ω and, furthermore, give a definition of probability such that the calculations are consistent. Specifically, we need to describe the elements of Ω which **can be** measured.

To this end, we have to group these events in some way that would allow us to say: yes, we can calculate the probability of all the events in this group. In other words, we need to talk about the notion of a collection of events.

We will introduce the notion of σ-algebra (or σ-field) to deal with the problem of the proper domain of definition for the probability. Before we do that, we introduce a special collection of events:

$$\mathscr{P}(\Omega) = \text{The collection of all possible subsets of } \Omega \qquad (1.1)$$

We could define probability on this very large set. However, this would mean that we would have to define probability for every single element of $\mathscr{P}(\Omega)$. This will prove impossible except in the case when Ω is finite. In this case, the collection $\mathscr{P}(\Omega)$ is called *the power set of* Ω. However, even in this case we have to do it consistently. For example, if, say, the set $\{1, 2, 3\}$ is in Ω and has probability 0.2, how do we define the probability of $\{1, 2\}$? How about the probability of $\{1, 2, 5\}$? A much better approach would be to define probability for a smaller set of important elements which generate the collection $\mathscr{P}(\Omega)$. For example, if we define the probability of each of the generators $1, 2, \ldots, 5$, perhaps we can then say something about the probabilities of the bigger events.

How do we do this? Fortunately, algebra comes to our rescue. The elements of a collection of events are the events. So, first we define operations with them: *union, intersection, complement* and slightly less important *difference and symmetric difference.*

$$
\begin{cases}
A \cup B & = \text{set of elements that are \textbf{either} in } A \textbf{ or } \text{in } B \\
A \cap B & = AB = \text{set of elements that are \textbf{both} in } A \textbf{ and } \text{in } B \\
A^c & = \bar{A} = \text{set of elements that are in } \Omega \text{ but \textbf{not} in } A
\end{cases}
\tag{1.2}
$$

$$
\begin{cases}
A \setminus B = & \text{set of elements that are in } A \text{ but \textbf{not} in } B \\
A \triangle B = & (A \setminus B) \cup (B \setminus A)
\end{cases}
$$

We will also use the notation $A \subseteq B$ to denote the case when all the elements in A are also in B, and we say A is a subset of B. $A \subset B$ will denote the case when A is a proper subset of B, that is, B contains at least one other element besides those in A.

We can express every such operation in terms of union and intersection, or basically reduce to only two operations. For example, $A \setminus B = A \cap B^c$. Union may be expressed in terms of intersection, and vice versa. In fact, since these relations are very important, let us state them separately.

De Morgan laws

$$
\begin{cases}
(A \cup B)^c & = A^c \cap B^c \\
(A \cap B)^c & = A^c \cup B^c
\end{cases}
\tag{1.3}
$$

We will mention one more property, which is the distributivity property, that is, union and intersection distribute to each other:

Distributivity property of union/intersection

$$
A \cup (B \cap C) = (A \cup B) \cap (A \cup C)
\tag{1.4}
$$
$$
A \cap (B \cup C) = (A \cap B) \cup (A \cap C)
$$

There is much more to be found about set operations, but for our purpose this is enough. We recommend you to look up Billingsley (1995) or Chung (2000) for a wealth of details.

Definition 1.1 (Algebra on Ω) *A collection $\mathcal{F}$ of events in Ω is called an algebra (or field) on Ω iff*

1. $\Omega \in \mathcal{F}$;

2. Closed under complementarity: If $A \in \mathcal{F}$ then $A^c \in \mathcal{F}$;

3. Closed under finite union: If $A, B \in \mathcal{F}$ then $A \cup B \in \mathcal{F}$.

Remark 1.2 *The first two properties imply that $\varnothing \in \mathscr{F}$. The third is equivalent to $A \cap B \in \mathscr{F}$ by the second property and the de Morgan laws (1.3).*

Definition 1.3 (σ-Algebra on Ω) *If $\mathscr{F}$ is an algebra on Ω and, in addition, it is closed under countable unions, then it is a σ-algebra (or σ-field) on Ω.*

Note   Closed under countable unions means that the third property in Definition 1.1 is replaced with the following: If $n \in \mathbb{N}$ is a natural number and $A_n \in \mathscr{F}$ for all n, then

$$\bigcup_{n \in \mathbb{N}} A_n \in \mathscr{F} .$$

The σ-algebra provides an appropriate domain of definition for the probability function. However, it is such an abstract thing that it will be hard to work with it. This is the reason for the next definition. It will be much easier to work with the generators of a *sigma*-algebra. *This will be a recurring theme in probability. In order to show a property for a certain category of objects, we show that the property holds for a small set of objects which generate the entire class. This will be enough – standard arguments will allow us to extend the property to the entire category of objects.*

■   **EXAMPLE 1.1   A simple example of σ-algebra**

Suppose a set $A \subset \Omega$. Let us calculate $\sigma(A)$. Clearly, by definition Ω is in $\sigma(A)$. Using the complementarity property, we clearly see that A^c and $\emptyset$ are also in $\sigma(A)$. We only need to take unions of these sets and see that there are no more new sets. Thus

$$\sigma(A) = \{\Omega, \emptyset, A, A^c\}.$$

■

Definition 1.4 (σ algebra generated by a class $\mathscr{C}$ of sets in Ω) *Let $\mathscr{C}$ be a collection (class) of sets in Ω. Then $\sigma(\mathscr{C})$ is the smallest σ-algebra on Ω that contains $\mathscr{C}$.*

Mathematically

1. $\mathscr{C} \subseteq \sigma(\mathscr{C})$,

2. $\sigma(\mathscr{C})$ is a σ-field,

3. If $\mathscr{C} \subseteq \mathscr{G}$ and $\mathscr{G}$ is a σ-field, then $\sigma(\mathscr{C}) \subseteq \mathscr{G}$.

As we have mentioned, introducing generators of the large collection of sets $\sigma(\mathscr{C})$ is fundamental. Suppose that we want to show that a particular statement is true for every set in the large collection. If we know the generating subset $\mathscr{C}$, we can verify whether the statement is true for the elements of $\mathscr{C}$, which is a much easier task. Then, because of the properties that would be presented later, the particular statement will be valid for all the sets in the larger collection $\sigma(\mathscr{C})$.

Proposition 1.5 *Properties of σ-algebras:*

- $\mathscr{P}(\Omega)$ *is the largest possible σ-algebra defined on Ω;*

- *If $\mathscr{C}$ is already a σ-algebra, then $\sigma(\mathscr{C}) = \mathscr{C}$;*

- *If $\mathscr{C} = \{\varnothing\}$ or $\mathscr{C} = \{\Omega\}$, then $\sigma(\mathscr{C}) = \{\varnothing, \Omega\}$, the smallest possible σ-algebra on Ω;*

- *If $\mathscr{C} \subseteq \mathscr{C}'$, then $\sigma(\mathscr{C}) \subseteq \sigma(\mathscr{C}')$;*

- *If $\mathscr{C} \subseteq \mathscr{C}' \subseteq \sigma(\mathscr{C})$, then $\sigma(\mathscr{C}') = \sigma(\mathscr{C})$.*

■ EXAMPLE 1.2

Suppose that $\mathscr{C} = \{A, B\}$, where A and B are two sets in Ω such that $A \subset B$. Let us list the sets in $\sigma(\mathscr{C})$. A common mistake made by students is the following argument:

$A \subset B$, therefore using the fourth property in the proposition above $\sigma(A) \subseteq \sigma(B)$ and therefore the sigma algebra asked is $\sigma(A, B) = \{\Omega, \emptyset, B, B^c\}$, done.

This argument is wrong on several levels. First, the quoted property refers to collections of sets and not to the sets themselves. While it is true that $A \subset B$, it is not true that $\{A\} \subset \{B\}$ as collections of sets. Instead, $\{A\} \subset \{A, B\}$ and, indeed, this implies $\sigma(A) \subseteq \sigma(A, B)$. But this just means that the result should contain all the sets in $\sigma(A)$ (the sets in the previous example). Furthermore, as the example will show and as the Proposition 1.6 says, it isn't true either that $\sigma(A) \cup \sigma(B) = \sigma(A, B)$. The only way to solve this problem is the hard way.

Clearly, $\sigma(A, B)$ should contain the basic sets and their complements, thus $\sigma(A, B) \supset \{\Omega, \emptyset, A, B, A^c, B^c\}$. It should also contain all their unions according to the definition. Therefore, it must contain

$$A \cup B = B$$
$$A \cup B^c$$
$$A^c \cup B = \Omega$$
$$A^c \cup B^c = A^c$$

where the equalities are obtained using that $A \subset B$. So the only new set to be added is $A \cup B^c$ and thus its complement as well: $(A \cup B^c)^c = A^c \cap B$. Now we need to verify that by taking unions we do not obtain any new sets, a task left for the reader. In conclusion, when $A \subset B$,

$$\sigma(A, B) = \{\Omega, \emptyset, A, B, A^c, B^c, A \cup B^c, A^c \cap B\}$$

Proposition 1.6 (Intersection and union of σ-algebras) *Suppose that $\mathscr{F}_1$ and $\mathscr{F}_2$ are two σ-algebras on Ω. Then*

1. *$\mathscr{F}_1 \cap \mathscr{F}_2$ is a sigma algebra.*

2. *$\mathscr{F}_1 \cup \mathscr{F}_2$ is **not** a sigma algebra. The smallest σ algebra that contains both of them is $\sigma(\mathscr{F}_1 \cup \mathscr{F}_2)$ and is denoted $\mathscr{F}_1 \vee \mathscr{F}_2$.*

Proof: For part 2, there is nothing to show. A counterexample is provided by the example 1.2. Take $\mathscr{F}_1 = \sigma(A)$ and $\mathscr{F}_2 = \sigma(B)$. The example above calculates $\sigma(A, B)$ and it is simple to see that $\mathscr{F}_1 \cup \mathscr{F}_2$ needs more sets to become a sigma algebra (e.g., $A \cup B^c$).

For part 1, we just need to verify the definition of the sigma algebra. Take a set A in $\mathscr{F}_1 \cap \mathscr{F}_2$. So A belongs to both collections of sets. Since $\mathscr{F}_1$ is a sigma algebra by definition, $A^c \in \mathscr{F}_1$. Similarly, $A^c \in \mathscr{F}_2$. Therefore, $A^c \in \mathscr{F}_1 \cap \mathscr{F}_2$. The rest of the definition is verified in a similar manner.   ∎

Finite, countable, uncountable (infinite) sets

In general, listing the elements of a sigma algebra explicitly is hard. It is only in simple cases that this can be done. Specifically, we can do this in the case of finite and countable spaces Ω. But what exactly is a countable space and how does it differ from uncountable spaces? Let us present some simple definitions to this end.

Definition 1.7 (Injective, surjective, bijective functions) *Let $f : \Omega \longrightarrow \Gamma$ be some function. We say that the function f is*

i) ***Injective** or **one to one** if $\forall x \neq y$ both in Ω we have $f(x) \neq f(y)$. Injectivity is usually checked by the equivalent statement "$f(x) = f(y) \Rightarrow x = y$".*

ii) ***Surjective** or **onto** if for all $y \in \Gamma$, there exist at least an $x \in \Omega$ such that $f(x) = y$. This is normally checked by picking an arbitrary y in Γ and actually finding the x that goes into it. This procedure gives the set $f^{-1}(\{y\})$.*

iii) ***Bijective** if it is both one to one and onto. The function is a bijection if the inverse set $f^{-1}(\{y\})$ has exactly one element for all y. In this case, f^{-1} is a proper function defined on Γ with values in Ω called the inverse function.*

Definition 1.8 (Cardinality of a set) *Suppose that we can construct a bijection between two sets Ω and Γ. Then we say that the cardinality of the two sets is the same (denoted $|\Omega| = |\Gamma|$).*

Important sets:

$$\mathbb{N} = \{0, 1, 2, 3, \ldots\} = \text{natural numbers}$$
$$\mathbb{Z} = \{\ldots, -2, -1, 0, 1, 2, \ldots\} = \text{integer numbers}$$
$$\mathbb{Q} = \left\{ \frac{m}{n} \mid m, n \in \mathbb{Z} \right\} = \text{rational numbers}$$
$$\mathbb{R} = \text{real numbers} \ (\sigma\text{-field on } \mathbb{Q} \text{ under addition and multiplication})$$
$$\mathbb{C} = \{a + bi \mid a, b \in \mathbb{R}\} = \text{complex numbers}$$

We shall denote with a star any set not containing 0. For example, $\mathbb{Z}^* = \mathbb{Z} \backslash \{0\}$.

Suppose $n \in \mathbb{N}$ is some natural number. We define the cardinality of the set $A = \{1, 2, \ldots, n\}$ as $|A| = n$. We define the cardinality of $\mathbb{N}$ with $\aleph_0$. I will not get into details about these *aleph* numbers. For details, refer to an introductory textbook such as Halmos (1998).

Definition 1.9 *Any set with cardinality equal to a finite number $n \in \mathbb{N}$ is called a* finite set. *Any set with cardinality $\aleph_0$ is called a* countable set. *Any set which is not one of the first two is called* uncountable *(infinite).*

The definition of infinite is a bit strange ($\mathbb{N}$ has an infinite number of elements), but it is important to us because it helps us to make the distinction between say $\mathbb{Q}$ and any interval, say $(0.00000000000000001, 0.00000000000000002)$. This tiny interval has a lot more elements than the entire $\mathbb{Q}$. I will give two more results then we return to probability.

Proposition 1.10 *The following results hold:*

1) $|\mathbb{N}| = |\mathbb{Z}|$

2) $|\mathbb{N}| = |\mathbb{Q}|$

3) $|\mathbb{R}| = |(0, 1)|$

4) $\mathbb{R}$ *is uncountable (infinite).*

Proof: The proof is very simple and amusing, so I am giving it here.

1) Let $f : \mathbb{N} \to \mathbb{Z}$, which takes odd numbers in negative integers and even numbers in positive integers, that is

$$f(x) = \begin{cases} x/2, & \text{if } x \text{ is even} \\ -(x + 1)/2, & \text{if } x \text{ is odd} \end{cases}$$

It is easy to show that this function is a bijection between $\mathbb{N}$ and $\mathbb{Z}$; thus, according to the definition, the two have the same cardinality ($\mathbb{Z}$ is countable).

2) From definition

$$\mathbb{Q} = \bigcup \left\{ \frac{m}{n} \,\middle|\, m, n \in \mathbb{Z} \right\} = \bigcup_{n \in \mathbb{Z}} \left\{ \frac{m}{n} \,\middle|\, m \in \mathbb{Z} \right\} = \bigcup_{n \in \mathbb{Z}} Q_n,$$

where we use the notation $Q_n = \{ \frac{m}{n} \mid m \in \mathbb{Z} \}$. For any fixed n, there clearly exists a bijection between $\mathbb{Z}$ and Q_n $(g(x) = x/n)$. Thus Q_n is countable. But then $\mathbb{Q}$ is a countable union of countable sets. Lemma 1.11 proved next shows that the result is always a countable set (the proof is very similar to the proof above of $\mathbb{Z} = \mathbb{N} \cup (-\mathbb{N})$).

3) Let $f : \mathbb{R} \to (0, 1)$,

$$f(x) = \frac{e^x}{1 + e^x}.$$

This function is a bijection (exercise).

4) From the previous point 3, it is enough to show that $(0, 1)$ is uncountable. Assume by absurd that the interval is countable. Then there must exist a bijection with $\mathbb{N}$ and therefore its elements may be written as a sequence:

$$x_1 = 0.x_{11}x_{12}x_{13} \ldots$$
$$x_2 = 0.x_{21}x_{22}x_{23} \ldots$$
$$x_3 = 0.x_{31}x_{32}x_{33} \ldots$$
$$\vdots \quad \vdots$$

where x_{ij} are the digits of the number x_i. To finish the reduction to absurd, we construct a number which is not on the list. This number is going to be different than each of the numbers listed in at least one digit. Specifically, construct the number $y = 0.y_{11}y_{22}y_{33} \ldots$, such that the digit i is

$$y_{ii} = \begin{cases} 2, & \text{if } x_{ii} = 1 \\ 1, & \text{if } x_{ii} \neq 1 \end{cases}$$

The point is that the digit i of the number y is different from the digit i of the number x_i, and therefore the two numbers cannot be equal. This is happening for all the i's, so y is a new number not on the list. This contradicts the assumption that we may list all the numbers in the interval. Thus $(0, 1)$ is not countable.

■

In the above proof, we need the following lemma which is very useful by itself.

Lemma 1.11 *Suppose* $A_1, A_2, \ldots$ *are countable sets. Then* $\cup_{n \geq 1} A_n$ *is a countable set.*

Proof: The proof basically uses the fact that a composition of two bijections is a bijection. Suppose we look at one of these sets say A_n. Since this set is countable, we may list its elements as $A_n = \{a_{m1}, a_{m2}, a_{m3}, \ldots \}$, and similarly for all the sets. Now consider the function $f : \cup_{n \geq 1} A_n \to \mathbb{N}$, with

$$f(a_{mn}) = \frac{(m + n - 1)(m + n - 2)}{2} + m$$

This function basically assigns a_{11} to 1, a_{12} to 2, a_{21} to 3, a_{22} to 5, and so on.

The function f is clearly an injection because no two elements are taken at the same point, but it may not be surjective (for example, some set A_i may be finite and thus those missing elements a_{ij} may not be taken into integers $\frac{(i+j-1)(i+j-2)}{2} + i$ and thus these integers have $f^{-1}(\{y\}) = \emptyset$). However, we can restrict the codomain to the image of the function, and thus the set $\cup_{n \geq 1} A_n$ has the same cardinality as a set included in $\mathbb{N}$ but any subset of $\mathbb{N}$ is countable (easy to show), done. ∎

Finally, the following result is relevant for the next chapter. The lemma may be applied to the distribution function (a nondecreasing bounded function).

Lemma 1.12 *Let $f : \mathbb{R} \to \mathbb{R}$ be a nondecreasing function. Then the set of points where f has a discontinuity is countable.*

Proof: Let A the set of discontinuity points of f. We need to show that A is countable. Let $x \in A$. The function is discontinuous at x, so the left and right limits are different: $f(x-) = \lim_{y \uparrow x} f(y) \neq \lim_{y \downarrow x} f(y) = f(x+)$. Since f is increasing and the two values are different, there must exist some number $q_x \in \mathbb{Q}$ such that

$$f(x-) < q_x < f(x+).$$

This allows us to construct a function $g : A \to \mathbb{Q}$, by $g(x) = q_x$. This function is injective. Why? Let $x \neq y \in A$. Then we have, say, $x < y$. Since f is increasing, we have $f(x) < f(y)$. But then we must also have

$$q_x < f(x+) < f(y-) < q_y$$

Thus $q_x \neq q_y$, and so the function g is injective. Again, like we did in the previous lemma, we restrict the codomain to the image of A, which makes the function a bijection. Therefore, the set A is countable. ∎

Back to σ-algebras.

Remark 1.13 (Very important. The case of countable space Ω) *When the sample space is finite, we can and typically will take the sigma algebra to be $\mathscr{P}(\Omega)$. Indeed, any event of a finite space can be trivially expressed in terms of individual outcomes. In fact, if the space Ω is finite, say it contains m possible outcomes, then the number of possible events is finite and is equal to 2^m.*

Remark 1.13 is important. It basically says that if you are looking to understand the concept and usefulness of sigma algebras, looking at coin tosses or rolling of dies is irrelevant. For these simple experiments, the notion of σ-algebra is useless because we can just take the events that can be measured as all the possible events. σ-algebras are useful even in these simple examples if we start thinking about tossing the coin or rolling the die ad infinitum. That sample space is not finite. We next present the Borel σ-algebra, which is going to be essential for understanding random variables.

An example: Borel σ-algebra

Let Ω be a topological space (think geometry is defined in this space and this assures us that the open subsets exist in this space)[1].

Definition 1.14 *We define*

$$\mathscr{B}(\Omega) = \text{The Borel } \sigma\text{-algebra} \tag{1.5}$$
$$= \sigma\text{-algebra generated by the class of open subsets of } \Omega$$

In the special case when $\Omega = \mathbb{R}$, we denote $\mathscr{B} = \mathscr{B}(\mathbb{R})$, the Borel sets of $\mathbb{R}$. This $\mathscr{B}$ is the most important σ-algebra. The reason for this is that most experiments can be brought to equivalence with $\mathbb{R}$ (as we shall see when we will talk about random variables). Thus, if we define a probability measure on $\mathscr{B}$, we have a way to calculate probabilities for most experiments. ∎

Most subsets of $\mathbb{R}$ are in $\mathscr{B}$. However, it is possible (though very difficult) to explicitly construct a subset of $\mathbb{R}$ which is not in $\mathscr{B}$. See (Billingsley, 1995, page 45) for such a construction in the case $\Omega = (0, 1]$.

There is nothing special about the open sets, except for the fact that they can be defined in any topological space (thus they always exist given a topology). In $\mathbb{R}$, we can generate the Borel σ-algebra using many classes of generators. In the end, the same σ-algebra is reached (see problem 1.8).

Probability measure

We are finally in a position to give the domain for the probability measure.

Definition 1.15 (Measurable space) *A pair $(\Omega, \mathscr{F})$, where Ω is a set and $\mathscr{F}$ is a σ-algebra on Ω, is called a* measurable space.

Definition 1.16 (Probability measure. Probability space) *Given a measurable space $(\Omega, \mathscr{F})$, a probability measure is any function $\mathbf{P} : \mathscr{F} \to [0, 1]$ with the following properties:*

i) $\mathbf{P}(\Omega) = 1$

---

[1]For completion, we present the definition of a topological space even though it is secondary to our purpose. A topological space is a set Ω together with a collection of subsets of Ω, called open sets, and satisfying the following three axioms:

1. The empty set and Ω itself are open;

2. Any union of open sets is open;

3. The intersection of any finite number of open sets is open.

This collection of open sets (denoted $\mathscr{T}$) is called a *topology on Ω* or an *open set topology*. The sets in $\mathscr{T}$ are called *open sets*, and their complements in Ω are called *closed sets*. A subset of Ω may be neither closed nor open, either closed or open, or both.

ii) *(countable additivity) For any sequence* $\{A_n\}_{n\in\mathbb{N}}$ *of disjoint events in* $\mathscr{F}$ *(i.e.,* $A_i \cap A_j = \varnothing$*, for all* $i \neq j$*):*

$$\mathbf{P}\left(\bigcup_{n=1}^{\infty} A_n\right) = \sum_{n=1}^{\infty} \mathbf{P}(A_n).$$

The triple $(\Omega, \mathscr{F}, \mathbf{P})$ *is called a* probability space.

Note that the probability measure is a set function (i.e., a function defined on sets).

The next two definitions define slightly more general measures than the probability measure. We will use these notions later in this book in the hypotheses of some theorems to show that the results apply to more general measures.

Definition 1.17 (Finite measure) *Given a measurable space* $(\Omega, \mathscr{F})$*, a finite measure is a set function* $\mu : \mathscr{F} \to [0, 1]$ *with the same countable additivity property as in Definition 1.16, but the measure of the space is a finite number (not necessarily 1). Mathematically, the first property in Definition 1.16 is replaced with*

$$\mu(\Omega) < \infty$$

Note that we can always construct a probability measure from a finite measure μ. For any set $A \in \mathscr{F}$, define a new measure by

$$\mathbf{P}(A) = \frac{\mu(A)}{\mu(\Omega)}.$$

So this notion isn't more general than the probability measure. However, the next notion is.

Definition 1.18 (σ-finite measure) *A measure* μ *defined on a measurable space* $(\Omega, \mathscr{F})$ *is called* σ-finite *if it is countably additive and there exists a partition of the space* Ω *with sets* $\{\Omega_i\}_{i\in I}$*, and* $\mu(\Omega_i) < \infty$ *for all* $i \in I$*. Note that the index set* I *is allowed to be countable.*

To be precise, a partition of any set Ω is any collection of sets $\{\Omega_i\}_{i\in I}$, which are disjoint (i.e., $\Omega_i \cap \Omega_j = \emptyset$, if $i \neq j$) such that their union recreates the original set: $\cup_{i\in I}\Omega_i = \Omega$. The partition may be finite or infinite depending on whether the number of sets in the partition is finite or countable.

As an example of the difference between finite and sigma finite measure spaces, consider the interval $[0, 1)$ and assume that a probability measure is defined on that space (we will talk later about the Lebesgue measure generated by the length of intervals). Typically, such a measure may be extended to any interval of length 1: $[a, a+1)$. But note that the measure will become infinite when extended to $\mathbb{R}$. However, since we can write $\mathbb{R} = \cup_{n\in\mathbb{Z}}[n, n + 1)$, we can see that the measure can be extended to a sigma-finite measure on $\mathbb{R}$. Typically, sigma-finite measures have the same nice properties of probability measures.

■ EXAMPLE 1.3   Discrete probability space

Let Ω be a countable space. Let $\mathscr{F} = \mathscr{P}(\Omega)$. Let $p : \Omega \to [0, N]$ be a function on Ω such that $\sum_{\omega \in \Omega} p(\omega) = N < \infty$, where N is a finite constant. Define a probability measure by

$$\mathbf{P}(A) = \frac{1}{N} \sum_{\omega \in A} p(\omega) \tag{1.6}$$

We can show that $(\Omega, \mathscr{F}, \mathbf{P})$ is a probability space. Indeed, from the definition (1.6),

$$\mathbf{P}(\Omega) = \frac{1}{N} \sum_{\omega \in \Omega} p(\omega) = \frac{1}{N} N = 1.$$

To show the countable additivity property, let A be a set in Ω such that $A = \bigcup_{i=1}^{\infty} A_i$, with A_i disjoint sets in Ω. Since the space is countable, we may write $A_i = \{\omega_1^i, \omega_2^i, \ldots\}$, where any of the sets may be finite, but $\omega_j^i \neq \omega_l^k$ for all i, j, k, l where either $i \neq k$ or $j \neq l$. Then using the definition (1.6) we have

$$\mathbf{P}(A) = \frac{1}{N} \sum_{\omega \in \bigcup_{i=1}^{\infty} A_i} p(\omega) = \frac{1}{N} \sum_{i \geq 1, j \geq 1} p(\omega_j^i)$$

$$= \frac{1}{N} \sum_{i \geq 1} \left(p(\omega_1^i) + p(\omega_2^i) + \ldots \right) = \sum_{i \geq 1} \mathbf{P}(A_i)$$

■

This is a very simple example, but it shows the basic probability reasoning.

Remark 1.19 *Example 1.3 gives a way to construct discrete probability measures (distributions). For example, take $\Omega = \mathbb{N}$ (the natural numbers) and take $N = 1$ in the definition of the probability of an event. Then using various probability assignments, $p(\omega)$ produces distributions with which you may already be familiar.*

- $p(\omega) = \begin{cases} 1 - p & , \text{if } \omega = 0 \\ p & , \text{if } \omega = 1 \\ 0 & , \text{otherwise} \end{cases}$ *, gives the Bernoulli(p) distribution.*

- $p(\omega) = \begin{cases} \binom{n}{\omega} p^\omega (1-p)^{n-\omega} & , \text{if } \omega \leq n \\ 0 & , \text{otherwise} \end{cases}$ *, gives the binomial(n,p) distribution.*

- $p(\omega) = \begin{cases} \binom{\omega-1}{r-1} p^r (1-p)^{\omega-r} & , \text{if } \omega \geq r \\ 0 & , \text{otherwise} \end{cases}$ *, gives the negative binomial(r,p) distribution.*

- $p(\omega) = \frac{\lambda^\omega}{\omega!} e^{-\lambda}$*, gives the Poisson ($\lambda$) distribution.*

■ **EXAMPLE 1.4   Uniform distribution on (0,1)**

Let $\Omega = (0,1)$ and $\mathscr{F} = \mathscr{B}((0,1))$ be the Borel sigma algebra generated by the open sets. Define a probability measure U as follows: for any open interval $(a,b) \subseteq (0,1)$, let $U((a,b)) = b - a$ the length of the interval. For any other open interval O, define $U(O) = U(O \cap (0,1))$.

Note that we did not specify the measure $U(A)$ for all of the Borel sets $A \in \mathscr{B}$, but rather only for the generators of the Borel σ-field. This illustrates the probabilistic concept discussed after Definition 1.4 where we define or prove a statement using only the generators of a sigma algebra.

In our specific situation, under very mild conditions on the generators of the σ-algebra, any probability measure defined only on the generators can be uniquely extended to a probability measure on the whole σ-algebra (Carathèodory extension theorem). In particular, when the generators are open sets, these conditions are true and we can restrict the definition to the open sets alone. This example is going to be expanded further in Section 1.5.

Proposition 1.20 (Elementary properties of probability measure) *Let $(\Omega, \mathscr{F}, \mathbf{P})$ be a probability space. Then*

1. $\forall A, B \in \mathscr{F}$ with $A \subseteq B$, then $\mathbf{P}(A) \leq \mathbf{P}(B)$

2. $\mathbf{P}(A \cup B) = \mathbf{P}(A) + \mathbf{P}(B) - \mathbf{P}(A \cap B)$, $\forall A, B \in \mathscr{F}$.

3. (General inclusion-exclusion formula, also named Poincaré formula):

$$\mathbf{P}(A_1 \cup A_2 \cup \cdots \cup A_n) = \sum_{i=1}^{n} \mathbf{P}(A_i) - \sum_{i<j\leq n} \mathbf{P}(A_i \cap A_j) \qquad (1.7)$$
$$+ \sum_{i<j<k\leq n} \mathbf{P}(A_i \cap A_j \cap A_k) - \cdots + (-1)^{n-1}\mathbf{P}(A_1 \cap A_2 \cap \cdots \cap A_n)$$

Note that successive partial sums are alternating between over- and underestimating.

4. (Finite subadditivity, sometimes called Boole's inequality):

$$\mathbf{P}\left(\bigcup_{i=1}^{n} A_i\right) \leq \sum_{i=1}^{n} \mathbf{P}(A_i), \quad \forall A_1, A_2, \ldots, A_n \in \mathscr{F}.$$

1.1.1   Null element of $\mathscr{F}$. Almost sure (a.s.) statements. Indicator of a set

An event $N \in \mathscr{F}$ is called a *null event* if $P(N) = 0$.

Definition 1.21 *A statement S about points $\omega \in \Omega$ is said to be true almost surely (a.s.), almost everywhere (a.e.), or with probability 1 (w.p.1) if the set M, defined as*

$$M := \{\omega \in \Omega \,|\, S(\omega) \text{ is true}\} ,$$

is in $\mathscr{F}$ and $\mathbf{P}(M) = 1$, (or, equivalently M^c is a null set).

We will use the notions a.s., a.e., and w.p.1. to denote the same thing – the definition above. For example, we will say $X \geq 0$ a.s. and mean that $\mathbf{P}\{\omega | X(\omega) \geq 0\} = 1$, or equivalently $\mathbf{P}\{\omega | X(\omega) < 0\} = 0$. The notion of almost sure is a fundamental one in probability. Unlike in deterministic cases where something has to always be true no matter what, in probability we care about "the majority of the truth." In other words, probability recognizes that some phenomena may have extreme outcomes, but if they are extremely improbable then we do not care about them. Note that for this notion to make any sense, once again you have to think outside the realm of finite dimensional spaces. If the space Ω has a finite number of outcomes, say $10^{10000000000}$ all with a very small but strict positive probability, then neglecting any one of them makes the probability of any resulting set less than 1. To get what we are talking about with the null sets, you need once again to think about spaces with an infinite number of elements.

Definition 1.22 *We define the indicator function of an event A as the (simple) function $\mathbf{1}_A : \Omega \rightarrow \{0,1\}$,*

$$\mathbf{1}_A(\omega) = \begin{cases} 1 &, \quad \text{if } \omega \in A \\ 0 &, \quad \text{if } \omega \notin A \end{cases}$$

Sometimes, this function is denoted as I_A.

Note that the indicator function is a regular function (not a set function). Indicator functions are very useful in probability theory. Here are some useful relationships:

$$\mathbf{1}_{A \cap B}(\omega) = \mathbf{1}_A(\omega)\mathbf{1}_B(\omega), \quad \forall \, \omega \in \Omega$$

If the family of sets $\{A_i\}$ forms a partition of Ω (i.e., the sets A_i are disjoint and $\Omega = \bigcup_{i=1}^n A_i$), then

$$\mathbf{1}_B(\omega) = \sum_i \mathbf{1}_{B \cap A_i}(\omega), \text{ for any set } B \in \Omega$$

Remember this very simple function. We shall use it over and over throughout this book.

1.2 Conditional Probability

Let $(\Omega, \mathscr{F}, \mathbf{P})$ be a probability space.

Definition 1.23 *For $A, B \in \mathcal{F}$, with $\mathbf{P}(B) \neq 0$, we define the conditional probability of A given B by*

$$\mathbf{P}(A|B) = \frac{\mathbf{P}(A \cap B)}{\mathbf{P}(B)}.$$

We can immediately rewrite the formula above to obtain the *multiplicative rule*:

$$\mathbf{P}(A \cap B) = \mathbf{P}(A|B)\mathbf{P}(B),$$
$$\mathbf{P}(A \cap B \cap C) = \mathbf{P}(A|B \cap C)\mathbf{P}(B|C)\mathbf{P}(C)$$
$$\mathbf{P}(A \cap B \cap C \cap D) = \mathbf{P}(A|B \cap C \cap D)\mathbf{P}(B|C \cap D)\mathbf{P}(C|D)\mathbf{P}(D), \quad \text{and so on.}$$

This multiplicative rule is very useful for stochastic processes (part 2 of the book) and for estimation of parameters of a distribution.

Total probability formula: Given $A_1, A_2, \ldots, A_n$, a partition of Ω (i.e., the sets A_i are disjoint and $\Omega = \bigcup_{i=1}^{n} A_i$), then

$$\mathbf{P}(B) = \sum_{i=1}^{n} \mathbf{P}(B|A_i)\mathbf{P}(A_i), \quad \forall B \in \mathcal{F} \tag{1.8}$$

Bayes Formula: If $A_1, A_2, \ldots, A_n$ form a partition of Ω

$$\mathbf{P}(A_j|B) = \frac{\mathbf{P}(B|A_j)\mathbf{P}(A_j)}{\sum_{i=1}^{n} \mathbf{P}(B|A_i)\mathbf{P}(A_i)}, \quad \forall B \in \mathcal{F}. \tag{1.9}$$

◼ EXAMPLE 1.5

A biker leaves the point O in the figure below. At each crossroad, the biker chooses a road at random. What is the probability that he arrives at point A ?

Let B_k, $k = 1, 2, 3, 4$ be the event that the biker passes through point B_k. These four events are mutually exclusive and they form a partition of the space. Moreover, they are equiprobable ($\mathbf{P}(B_k) = 1/4, \forall k \in \{1, 2, 3, 4\}$). Let A denote the event "the biker reaches the destination point A." Conditioned on each of the possible points B_1-B_4 of passing, we have

$$\mathbf{P}(A|B_1) = 1/4$$
$$\mathbf{P}(A|B_2) = 1/2$$
$$\mathbf{P}(A|B_3) = 1$$

At B_4 is slightly more complex. We have to use the multiplicative rule:

$$\mathbf{P}(A|B_4) = 1/4 + \mathbf{P}(A \cap B_5|B_4) + \mathbf{P}(A \cap B_6 \cap B_5|B_4)$$
$$= 1/4 + \mathbf{P}(A|B_5 \cap B_4)\mathbf{P}(B_5|B_4)$$
$$\quad + \mathbf{P}(A|B_6 \cap B_5 \cap B_4)\mathbf{P}(B_6|B_5 \cap B_4)\mathbf{P}(B_5|B_4)$$
$$= 1/4 + 1/3(1/4) + 1(1/3)(1/4) = 3/12 + 2/12 = 5/12$$

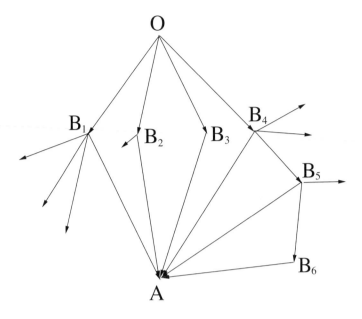

Figure 1.1   The possible trajectories of the biker. O is the origin point and A is the arrival point. B_k's are intermediate points.

Finally, by the law of total probability, we have

$$\mathbf{P}(A) = \mathbf{P}(A|B_1)\mathbf{P}(B_1) + \mathbf{P}(A|B_2)\mathbf{P}(B_2) + \mathbf{P}(A|B_3)\mathbf{P}(B_3) + \mathbf{P}(A|B_4)\mathbf{P}(B_4)$$
$$= 1/4(1/4) + 1/2(1/4) + 1/4(1) + 5/12(1/4) = 13/24$$

∎

■ **EXAMPLE 1.6   De Mére's paradox**

As a result of extensive observation of dice games, the French gambler Chevaliér De Mére noticed that the total number of spots showing on three dice thrown simultaneously turn out to be 11 more often than 12. However, from his point of view this is not possible since 11 occurs in six ways :
  $(6 : 4 : 1); (6 : 3 : 2); (5 : 5 : 1); (5 : 4 : 2); (5 : 3 : 3); (4 : 4 : 3)$,
while 12 also occurs in six ways:
  $(6 : 5 : 1); (6 : 4 : 2); (6 : 3 : 3); (5 : 5 : 2); (5 : 4 : 3); (4 : 4 : 4)$
What is the fallacy in the argument?

Proof (Solution due to Pascal): The argument would be correct if these "ways" would have the same probability. However, this is not true. For example, (6:4:1) occurs in 3! ways, (5:5:1) occurs in 3 ways, and (4:4:4) occurs in 1 way.

As a result, we can calculate $\mathbf{P}(11) = 27/216$; $\mathbf{P}(12) = 25/216$, and indeed his observation is correct and he should bet on 11 rather than on 12 if they have the same game payoff. ∎

■ **EXAMPLE 1.7   Another De Mére's paradox**

Which one of the following is more probable?

1. Throw four dice and obtain at least one 6;

2. Throw two dice 24 time and obtain at least once a double 6.

Proof (Solution): For option 1: $1 - \mathbf{P}(\text{No } 6) = 1 - (5/6)^4 = 0.517747$.
  For option 2: $1 - \mathbf{P}(\text{None of the 24 trials has a double 6}) = 1 - (35/36)^{24} = 0.491404$ ∎

■ **EXAMPLE 1.8   Monty Hall problem**

This is a problem named after the host of the American television show "Let's make a deal." Simply put, at the end of a game you are left to choose between three closed doors. Two of them have nothing behind, and one contains a prize. You chose one door but the door is not opened automatically. Instead, the presenter opens another door that contains nothing. He then gives you the choice of changing the door or sticking with the initial choice.

Most people would say that it does not matter what you do at this time, but that is not true. In fact everything depends on the host's behavior. For example, if the host knows in advance where the prize is and always reveals at random some other door that does not contain anything, then it is always better to switch.

Proof (Solution): This problem generated a lot of controversy since its publication (in the 1970s) since the solution seems so counterintuitive. We mention articles talking about this problem in more detail (Morgan et al., 1991; Mueser and Granberg, 1991). We are presenting the problem here since it exemplifies the conditional probability reasoning. The key in any such problem is the sample space which has to be complete enough to be able to answer the questions asked.

Let D_i be the event that the prize is behind door i. Let SW be the event that switching wins the prize[2]

It does not matter which door we chose initially. The reasoning is identical with all three doors. So, we assume that initially we pick door 1.

Events D_i $i = 1, 2, 3$ are mutually exclusive, and we can write

$$\mathbf{P}(SW) = \mathbf{P}(SW|D_1)\mathbf{P}(D_1) + \mathbf{P}(SW|D_2)\mathbf{P}(D_2) + \mathbf{P}(SW|D_3)\mathbf{P}(D_3).$$

When the prize is behind door 1, since we chose door 1, the presenter has two choices for the door to show us. However, neither would contain the prize, and in

---

[2]As a side note, this event is the same as the event "not switching loses."

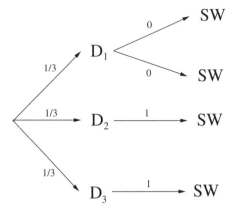

Figure 1.2   The tree diagram of conditional probabilities. Note that the presenter has two choices in case D_1, neither of which results in winning if switching the door.

either case switching does not result in winning the prize, therefore $\mathbf{P}(SW|D_1) = 0$. If the prize is behind door 2, since our choice is door 1, the presenter has no alternative but to show us the other door (3) which contains nothing. Thus, switching in this case results in winning the price. The same reasoning works if the prize is behind door 3. Therefore

$$\mathbf{P}(SW) = 1\frac{1}{3} + 1\frac{1}{3} + 0\frac{1}{3} = \frac{2}{3}$$

Thus switching has a higher probability of winning than not switching.

A generalization to n doors shows that it still is advantageous to switch, but the advantage decreases as $n \to \infty$. Specifically, in this case $\mathbf{P}(D_i) = 1/n$; $\mathbf{P}(SW|D_1) = 0$ still, but $\mathbf{P}(SW|D_i) = 1/(n-2)$ if $i \neq 1$, which gives

$$\mathbf{P}(SW) = \sum_{i=2}^{n} \frac{1}{n}\frac{1}{n-2} = \frac{n-1}{n-2}\frac{1}{n} > \frac{1}{n}$$

Furthermore, different presenter strategies produce different answers. For example, if the presenter offers the option to switch only when the player chooses the right door, then switching is always bad. If the presenter offers switching only when the player has chosen incorrectly, then switching always wins. These and other cases are analyzed in Rosenthal (2008).   ∎

◨ EXAMPLE 1.9   Bertrand's box paradox

This problem was first formulated by Joseph Louis François Bertrand in his Calcul de Probabilités (Bertrand, 1889). In some sense this problem is related to the previous problem but it does not depend on any presenter strategy and the solution is much clearer. Solving this problem is an exercise in Bayes formula.

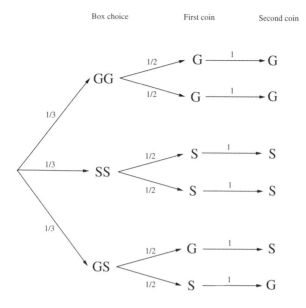

Box choice   First coin   Second coin

Figure 1.3   The tree diagram of conditional probabilities.

Suppose that we have three boxes. One box contains two gold coins, a second box has two silver coins, and a third box has one of each. We choose a box at random and from that box we choose a coin also at random. Then we look at the coin chosen. Given that the coin chosen was gold, what is the probability that the other coin in the box chosen is also gold. At first glance it may seem that this probability is $1/2$, but after calculation this probability turns out to be $2/3$.

Proof (Solution):

We plot the sample space in Figure 1.3. Using this tree we can calculate the probability:

$$\mathbf{P}(\text{Second coin is } G | \text{First coin is } G) = \frac{\mathbf{P}(\text{Second coin is } G \text{ and First coin is } G)}{\mathbf{P}(\text{First coin is } G)}.$$

Now, using the probabilities from the tree we continue:

$$= \frac{\frac{1}{3}\frac{1}{2}1 + \frac{1}{3}\frac{1}{2}1}{\frac{1}{3}\frac{1}{2}1 + \frac{1}{3}\frac{1}{2}1 + \frac{1}{3}\frac{1}{2}1} = \frac{2}{3}.$$

Now that we have seen the solution, we can recognize a logical solution to the problem as well. Given that the coin seen is gold, we can throw away the middle box. If this would be box 1, then we have two possibilities that the other coin is gold (depending on which one we have chosen in the first place). If this is box 2, then there is one possibility (the remaining coin is silver). Thus the probability should be 2/3 since we have two out of three chances. Of course, this argument does not work if we do not choose the boxes with the same probability.   ∎

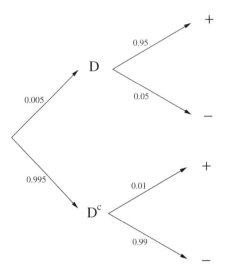

Figure 1.4   Blood test probability diagram

◧ EXAMPLE 1.10

A blood test is 95% effective in detecting a certain disease when it is in fact present. However, the test yields a false positive result for 1% of the people who do not have the disease. If 0.5% of the population actually have the disease, what is the probability that a randomly chosen person is diseased, given that his test is positive?

Proof (Solution): This problem illustrates once again the application of the Bayes rule. I do not like to use the formula literally, instead I like to work from first principles and obtain the Bayes rule without memorizing anything. We start by describing the sample space. Refer to Figure 1.4 for this purpose.

So, given that the test is positive means that we have to calculate a conditional probability. We may write

$$\mathbf{P}(D|+) = \frac{\mathbf{P}(D \cap +)}{\mathbf{P}(+)} = \frac{\mathbf{P}(+|D)\mathbf{P}(D)}{\mathbf{P}(+)} = \frac{0.95(0.005)}{0.95(0.005) + 0.01(0.995)} = 0.323$$

How about if only 0.05% (i.e., 0.0005) of the population has the disease?

$$\mathbf{P}(D|+) = \frac{0.95(0.0005)}{0.95(0.0005) + 0.01(0.9995)} = 0.0454$$

This problem is an exercise in thinking. A good rate of correctly identifying the disease does not necessarily translate into a good rate of a person having the disease if the test is positive. The latter strongly depends on the actual proportion of the population having the disease. ∎

▣ EXAMPLE 1.11   Gambler's ruin problem

We conclude this section with an example which we will see many times through-out this book. This problem appeared in De Moivre's doctrine of chance, but an earlier version was also published by Huygens (1629–1695).

The formulation is simple: a game of heads or tails with a fair coin. Player wins 1 dollar if he successfully calls the side of the coin which lands upwards and loses \$1 otherwise. Suppose the initial capital is X dollars and he intends to play until he wins m dollars but no longer. What is the probability that the gambler will be ruined (loses all his/her money)?

Proof (Solution): We will display what is called a *first-step analysis*.

Let $p(x)$ denote the probability that the player is going to be eventually ruined if he starts with x dollars.

If he wins the next game, then he will have \$ $x + 1$ and he will be ruined from this position with prob $p(x + 1)$.

If he loses the next game, then he will have \$ $x - 1$, so he is ruined from this position with prob $p(x - 1)$.

Let R be the event in which he is eventually ruined. Let W be the event in which he wins the next trial. Let L be the event in which he loses this trial. Using the total probability formula, we get

$$\mathbf{P}(R) = \mathbf{P}(R|W)\mathbf{P}(W) + \mathbf{P}(R|L)\mathbf{P}(L) \Rightarrow p(x) = p(x+1)(1/2) + p(x-1)(1/2)$$

Is this true for all x? No. This is true for $x \geq 1$ and $x \leq m - 1$. In the rest of the cases, we obviously have $p(0) = 1$ and $p(m) = 0$, which give the boundary conditions for the equation above.

This is a linear difference equation with constant coefficients. You can find a re-fresher of the general methodology in Appendix A.1.

Applying the method in our case gives the characteristic equation

$$y = \frac{1}{2}y^2 + \frac{1}{2} \Rightarrow y^2 - 2y + 1 = 0 \Rightarrow (y - 1)^2 = 0 \Rightarrow y_1 = y_2 = 1$$

In our case the two solutions are equal, thus we seek a solution of the form $p(x) = (C + Dx)1^n = C + Dx$. Using the initial conditions, we get $p(0) = 1 \Rightarrow C = 1$ and $p(m) = 0 \Rightarrow C + Dm = 0 \Rightarrow D = -C/m = -1/m$; thus the general probability of ruin starting with wealth x is

$$p(x) = 1 - x/m.$$

■

1.3 Independence

Definition 1.24 *Two events, A and B, are independent if and only if*

$$\mathbf{P}(A \cap B) = \mathbf{P}(A)\mathbf{P}(B).$$

The events $A_1, A_2, A_3, \ldots$ are called mutually independent *(or sometimes* simply independent*) if for every subset J of $\{1, 2, 3, \ldots\}$ we have*

$$\mathbf{P}\left(\bigcap_{j \in J} A_j\right) = \prod_{j \in J} \mathbf{P}(A_j)$$

The events $A_1, A_2, A_3, \ldots$ are called pairwise independent *(sometimes* jointly independent*) if*

$$\mathbf{P}(A_i \cap A_j) = \mathbf{P}(A_i)\mathbf{P}(A_j), \quad \forall i, j \in J.$$

Note that jointly independent does not imply independence.
Two sigma fields $\mathscr{G}, \mathscr{H} \in \mathscr{F}$ are $\mathbf{P}$-independent if

$$\mathbf{P}(G \cap H) = \mathbf{P}(G)\mathbf{P}(H), \quad \forall G \in \mathscr{G}, \forall H \in \mathscr{H}.$$

See Billingsley (1995) for the definition of independence of $k \geq 2$ sigma algebras.

■ **EXAMPLE 1.12   The birthday problem**

This is one of the oldest published probability problems. Suppose that there are n people meeting in a room. Each person is born on a certain day of the year and we assume that each day is equally likely for each individual (for simplicity we neglect leap years) and the probability of being born on a specific day is $1/365$. What is the probability that two or more people in the room share a birthday? What is the minimum number n such that this probability is greater than $1/2$?

This problem appeared for the first time in De Moivre's *The Doctrine of Chance: A method of calculating the probabilities of events in play* (1718).

The key is to calculate the complement of the probability requested. That is

$\mathbf{P}\{\text{at least 2 people share a birthday}\} = 1 - \mathbf{P}\{\text{Nobody shares a birthday}\}$

The probability that no one shares a birthday is easy to calculate using, for example, a conditional argument. To formalize using mathematical notation, let us denote the birthdays of the n individuals with $B_1, B_2, \ldots, B_n$. Let us denote the number of distinct elements in a set with $|\{B_1, B_2, \ldots, B_n\}|$. So, for example, $|\{B_1, B_2, \ldots, B_n\}| = n - 1$ means that there are $n - 1$ distinct elements in the set. Obviously, $|\{B_1, B_2, \ldots, B_n\}| = n$ means that there are no shared birthdays. Clearly, this probability is 0 if $n > 365$, so we assume that $n \leq 365$. We can write

$$\mathbf{P}\{|\{B_1, \ldots, B_n\}| = n\} = \mathbf{P}\{|\{B_1, B_2, \ldots, B_n\}| = n \mid |\{B_1, B_2, \ldots, B_{n-1}\}| = n - 1\}$$
$$\mathbf{P}\{|\{B_1, B_2, \ldots, B_{n-1}\}| = n - 1\}$$
$$= \mathbf{P}\{B_n \notin \{B_1, \ldots, B_{n-1}\} \mid |\{B_1, B_2, \ldots, B_{n-1}\}| = n - 1\}$$
$$\mathbf{P}\{|\{B_1, B_2, \ldots, B_{n-1}\}| = n - 1\}$$

and we can follow the argument further until we get to B_1. The conditional probability is simply the probability that the nth individual is born on a different day than the first $n - 1$ given that the $n - 1$ individuals all have different birthdays. This probability is $(365 - (n - 1))/365$ since there are $n - 1$ choices which are not good. If we continue the argument until we reach the last individual, we finally obtain the desired probability:

$$\mathbf{P}\{\text{at least 2 people share the birthday}\} = 1 - \frac{365 - (n - 1)}{365} \cdots \frac{364}{365} \frac{365}{365}$$

$$= 1 - \frac{365!}{(365 - n)!} \frac{1}{365^n}$$

To answer the second question (what should n be to have the above probability over 0.5) we can use Stirling's approximation of the factorials appearing in the formula. This approximation will be given later, but to obtain the numbers at this time we are just going to run a search algorithm using a computer (n is at most 365).

Using a computer when $n = 22$, we obtain the probability 0.475695308, and when $n = 23$ we obtain the probability 0.507297234. So we need about 23 people in the room to have the probability of two or more people sharing the birthday greater than 0.5. However, once programmed properly it is easy to play around with a computer program. By changing the parameters, it is easy to see that with 30 people in the room the probability is over 0.7; with 35 people the probability is over 0.8; and if we have 41 people in a room we are 90% certain that at least two of them have a shared birthday.

1.4 Monotone Convergence Properties of Probability

Let us take a step back for a minute and comment on what we have seen thus far. σ-algebra differs from regular algebra in that it allows us to deal with a countable (not finite) number of sets. Again, this is a recurrent theme in probability: learning to deal with infinity. On finite spaces, things are more or less simple. One has to define the probability of each individual outcome and everything proceeds from there. However, even in these simple cases, imagine that one repeats an experiment (such as a coin toss) over and over. Again, we are forced to cope with infinity. This section introduces a way to deal with this infinity problem.

Let $(\Omega, \mathcal{F}, \mathcal{P})$ be a probability space.

Lemma 1.25 *The following are true:*

1. If $A_n, A \in \mathcal{F}$ and $A_n \uparrow A$ (i.e., $A_1 \subseteq A_2 \subseteq \ldots A_n \subseteq \ldots$ and $A = \bigcup_{n \geq 1} A_n$), then $\mathbf{P}(A_n) \uparrow \mathbf{P}(A)$ as a sequence of numbers.

2. If $A_n, A \in \mathcal{F}$ and $A_n \downarrow A$ (i.e., $A_1 \supseteq A_2 \supseteq \ldots A_n \supseteq \ldots$ and $A = \bigcap_{n \geq 1} A_n$), then $\mathbf{P}(A_n) \downarrow \mathbf{P}(A)$ as a sequence of numbers.

3. *(Countable subadditivity)* If $A_1, A_2, \ldots,$ and $\bigcup_{i=1}^{\infty} A_n \in \mathscr{F}$, with A_i's not necessarily disjoint, then

$$\mathbf{P}\left(\bigcup_{n=1}^{\infty} A_n\right) \leq \sum_{n=1}^{\infty} \mathbf{P}(A_n)$$

Proof: 1. Let $B_1 = A_1, B_2 = A_2 \setminus A_1, \ldots, B_n = A_n \setminus A_{n-1}$. Because the sequence is increasing, we have that the B_i's are disjoint; thus

$$\mathbf{P}(A_n) = \mathbf{P}(B_1 \cup B_2 \cup \cdots \cup B_n) = \sum_{i=1}^{n} \mathbf{P}(B_i).$$

Thus, using countable additivity

$$\mathbf{P}\left(\bigcup_{n \geq 1} A_n\right) = \mathbf{P}\left(\bigcup_{n \geq 1} B_n\right) = \sum_{i=1}^{\infty} \mathbf{P}(B_i) = \lim_{n \to \infty} \sum_{i=1}^{n} \mathbf{P}(B_i) = \lim_{n \to \infty} \mathbf{P}(A_n).$$

2. Note that $A_n \downarrow A \Leftrightarrow A_n^c \uparrow A^c$ and from part 1 this means $1 - \mathbf{P}(A_n) \uparrow 1 - \mathbf{P}(A)$.

3. Let $B_1 = A_1, B_2 = A_1 \cup A_2, \ldots, B_n = A_1 \cup \cdots \cup A_n, \ldots$ From the finite subadditivity property in Proposition 1.20, we have that $\mathbf{P}(B_n) = \mathbf{P}(A_1 \cup \cdots \cup A_n) \leq \mathbf{P}(A_1) + \cdots + \mathbf{P}(A_n)$.

$\{B_n\}_{n \geq 1}$ is an increasing sequence of events, thus from part 1 we get that $\mathbf{P}(\bigcup_{n=1}^{\infty} B_n) = \lim_{n \to \infty} \mathbf{P}(B_n)$. Combining the two relations above, we obtain

$$\mathbf{P}(\bigcup_{n=1}^{\infty} A_n) = \mathbf{P}(\bigcup_{n=1}^{\infty} B_n) \leq \lim_{n \to \infty} (\mathbf{P}(A_1) + \cdots + \mathbf{P}(A_n)) = \sum_{n=1}^{\infty} \mathbf{P}(A_n)$$

∎

Lemma 1.26 *The union of a countable number of $\mathbf{P}$-null sets is a $\mathbf{P}$-null set.*

This Lemma is a direct consequence of the countable subadditivity.

Recall from analysis: For a sequence of numbers $\{x_n\}_n$ lim sup and lim inf are defined

$$\limsup x_n = \inf_m \{\sup_{n \geq m} x_n\} = \lim_{m \to \infty} (\sup_{n \geq m} x_n),$$

$$\liminf x_n = \sup_m \{\inf_{n \geq m} x_n\} = \lim_{m \to \infty} (\inf_{n \geq m} x_n),$$

and they represent the highest (respectively lowest) limiting point of a subsequence included in $\{x_n\}_n$.

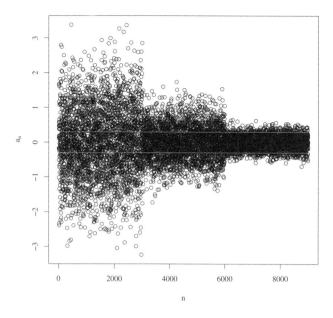

Figure 1.5 A sequence which has no limit. However, there exist sub-sequences that are convergent. The highest and lowest limiting points (gray lines) give the lim sup and the lim inf, respectively

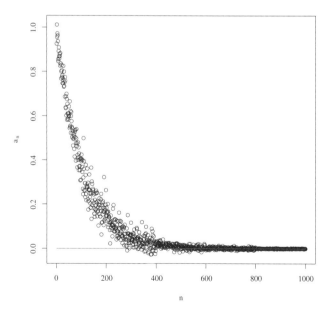

Figure 1.6 A sequence which has a limit (gray line). The limiting point is equal to both the lim sup and the lim inf of the sequence

Note that if z is a number such that $z > \limsup x_n$, then $x_n < z$ eventually[3]. Likewise, if $z < \limsup x_n$, then $x_n > z$ infinitely often[4].

These notions are translated to probability in the following way.

Definition 1.27 *Let $A_1, A_2, \ldots$ be an infinite sequence of events in some probability space $(\Omega, \mathscr{F}, \mathbf{P})$. We define the following events:*

$$
\limsup_{n \to \infty} A_n = \bigcap_{n \geq 1} \bigcup_{m=n}^{\infty} A_m = \{\omega : \omega \in A_n \text{ for infinitely many } n\}
$$

$$
= \{A_n \text{ infinitely often}\}
$$

$$
\liminf_{n \to \infty} A_n = \bigcup_{n \geq 1} \bigcap_{m=n}^{\infty} A_m = \{\omega : \omega \in A_n \text{ for all } n \text{ large enough}\}
$$

$$
= \{A_n \text{ eventually}\}
$$

Let us clarify the notions of "infinitely often" and "eventually" a bit more. We say that an outcome ω happens infinitely often for the sequence $A_1, A_2, \ldots, A_n, \ldots$ if ω is in the set $\bigcap_{n=1}^{\infty} \bigcup_{m \geq n} A_m$. This means that for any n (no matter how big) there exist an $m \geq n$ and $\omega \in A_m$.

We say that an outcome ω happens eventually for the sequence $A_1, A_2, \ldots, A_n, \ldots$ if ω is in the set $\bigcup_{n=1}^{\infty} \bigcap_{m \geq n} A_m$. This means that there exists an n such that for all $m \geq n$, $\omega \in A_m$, so from this particular n and up, ω is in all the sets.

Why do we give such complicated definitions? The basic intuition is the following: say, you roll a die infinitely many times, then it is obvious what it means for the outcome 1 to appear infinitely often. Also, we can say that the average of the rolls will eventually be arbitrarily close to 3.5 (this will be shown later). It is not very clear how to put this in terms of events happening infinitely often. The framework above provides a generalization to these notions.

The Borel Cantelli lemmas

With these definitions, we are now capable of giving two important lemmas.

Lemma 1.28 (First Borel–Cantelli) *If $A_1, A_2, \ldots$ is any infinite sequence of events with the property $\sum_{n \geq 1} \mathbf{P}(A_n) < \infty$, then*

$$
\mathbf{P}\left(\bigcap_{n=1}^{\infty} \bigcup_{m \geq n} A_m\right) = \mathbf{P}\left(A_n \text{ events are true infinitely often}\right) = 0
$$

This lemma essentially says that if the probabilities of events go to zero and the sum is convergent, then necessarily A_n will stop occurring. However, the reverse of

---

[3]i.e., there is some n_0 very large so that $x_n < z$, for all $n \geq n_0$

[4]That is, for any n there exists an $m \geq n$ such that $x_m > z$.

the statement is not true. To make it hold, we need a very strong condition (independence).

Lemma 1.29 (Second Borel–Cantelli) *If $A_1, A_2, \ldots$ is an infinite sequence of **independent** events, then*

$$\sum_{n \geq 1} \mathbf{P}(A_n) = \infty \quad \Leftrightarrow \quad \mathbf{P}(A_n \text{ infinitely often}) = 1.$$

Proof:

First Borel–Cantelli

$$\mathbf{P}(A_n \text{ i.o.}) = \mathbf{P}\left(\bigcap_{n \geq 1} \bigcup_{m=n}^{\infty} A_m\right) \leq \mathbf{P}\left(\bigcup_{n=m}^{\infty} A_m\right) \leq \sum_{m=n}^{\infty} \mathbf{P}(A_m), \forall n,$$

where we used the definition and countable subadditivity. By the hypothesis, the sum on the right is the tail end of a convergent series and therefore converges to zero as $n \to \infty$. Thus we are done. ∎

Proof (Second Borel–Cantelli:): The "$\Rightarrow$" part. Clearly, showing that $\mathbf{P}(A_n \text{ i.o.}) = \mathbf{P}(\limsup A_n) = 1$ is the same as showing that $\mathbf{P}((\limsup A_n)^c) = 0$.

By the definition of $\limsup$ and the DeMorgan's laws,

$$(\limsup A_n)^c = \left(\bigcap_{n \geq 1} \bigcup_{m=n}^{\infty} A_m\right)^c = \bigcup_{n \geq 1} \bigcap_{m=n}^{\infty} A_m^c.$$

Therefore, it is enough to show that $\mathbf{P}(\bigcap_{m=n}^{\infty} A_m^c) = 0$ for all n (recall that a countable union of null sets is a null set). However

$$\mathbf{P}\left(\bigcap_{m=n}^{\infty} A_m^c\right) = \lim_{r \to \infty} \mathbf{P}\left(\bigcap_{m=n}^{r} A_m^c\right) = \lim_{r \to \infty} \underbrace{\prod_{m=n}^{\infty} \mathbf{P}(A_m^c)}_{\text{by independence}}$$

$$= \lim_{r \to \infty} \prod_{m=n}^{r} (1 - \mathbf{P}(A_m)) \leq \lim_{r \to \infty} \underbrace{\prod_{m=n}^{r} e^{-\mathbf{P}(A_m)}}_{1-x \leq e^{-x} \text{ if } x \geq 0}$$

$$= \lim_{r \to \infty} e^{-\sum_{m=n}^{r} \mathbf{P}(A_m)} = e^{-\sum_{m=n}^{\infty} \mathbf{P}(A_m)} = 0$$

The last equality follows since $\sum \mathbf{P}(A_n) = \infty$.

Note that we have used the inequality $1 - x \leq e^{-x}$, which is true if $x \in [0, \infty)$. One can prove this inequality with elementary analysis.

The "$\Leftarrow$" part. This implication is the same as the first lemma. Indeed, assume by absurd that $\sum \mathbf{P}(A_n) < \infty$. By the first Borel–Cantelli lemma this implies that $\mathbf{P}(A_n \text{ i.o.}) = 0$, a contradiction with the hypothesis of the implication. ∎

The Fatou lemmas

Again, assume that $A_1, A_2, \ldots$ is a sequence of events.

Lemma 1.30 (Fatou lemma for sets) *Given any measure (not necessarily finite) μ, we have*

$$\mu(A_n \text{ eventually}) = \mu(\liminf_{n \to \infty} A_n) \leq \liminf_{n \to \infty} \mu(A_n)$$

Proof: Recall that $\liminf_{n \to \infty} A_n = \bigcup_{n \geq 1} \bigcap_{m=n}^{\infty} A_m$, and denote this set with A. Let $B_n = \bigcap_{m=n}^{\infty} A_m$, which is an increasing sequence (less intersections as n increases) and $B_n \uparrow A =$. By the monotone convergence property of measure (Lemma 1.25), $\mu(B_n) \to \mu(A)$. However,

$$\mu(B_n) = \mu(\bigcap_{m=n}^{\infty} A_m) \leq \mu(A_m), \forall m \geq n,$$

thus $\mu(B_n) \leq \inf_{m \geq n} \mu(A_m)$. Therefore

$$\mu(A) \leq \lim_{n \to \infty} \inf_{m \geq n} \mu(A_m) = \liminf_{n \to \infty} \mu(A_n)$$

■

Lemma 1.31 (The reverse of the Fatou lemma) *If $\mathbf{P}$ is a finite measure (e.g., probability measure), then*

$$\mathbf{P}(A_n \text{ i.o.}) = \mathbf{P}(\limsup_{n \to \infty} A_n) \geq \limsup_{n \to \infty} \mathbf{P}(A_n).$$

Proof: This proof is entirely similar. Recall that $\limsup_{n \to \infty} A_n = \bigcap_{n \geq 1} \bigcup_{m=n}^{\infty} A_m$, and denote this set with A. Let $B_n = \bigcup_{m=n}^{\infty} A_m$. Then clearly B_n is a decreasing sequence and $B_n \downarrow A$. By the monotone convergence property of measure (Lemma 1.25) and since the measure is finite, $\mathbf{P}(B_1) < \infty$ so $\mathbf{P}(B_n) \to \mathbf{P}(A)$. However,

$$\mathbf{P}(B_n) = \mathbf{P}(\bigcup_{m=n}^{\infty} A_m) \geq \mathbf{P}(A_m), \forall m \geq n,$$

thus $\mathbf{P}(B_n) \geq \sup_{m \geq n} \mathbf{P}(A_m)$, again since the measure is finite. Therefore

$$\mathbf{P}(A) \geq \lim_{n \to \infty} \sup_{m \geq n} \mathbf{P}(A_m) = \limsup_{n \to \infty} \mathbf{P}(A_n)$$

■

Kolmogorov zero-one law

We like to present this theorem since it introduces the concept of *a sequence of σ-algebras*, a notion essential for stochastic processes.

For a sequence $A_1, A_2, \ldots$ of events in the probability space $(\Omega, \mathscr{F}, \mathscr{P})$, consider the generated sigma algebras $\mathscr{T}_n = \sigma(A_n, A_{n+1}, \ldots)$ and their intersection

$$\mathscr{T} = \bigcap_{n=1}^{\infty} \mathscr{T}_n = \bigcap_{n=1}^{\infty} \sigma(A_n, A_{n+1}, \ldots),$$

called the tail σ-field.

Theorem 1.32 (Kolmogorov's 0–1 law) *If the sets $A_1, A_2, \ldots$ are independent in the sense of definition 1.24, then every event A in the tail σ field $\mathscr{T}$ defined above has probability $\mathbf{P}(A)$ either 0 or 1.*

Remark 1.33 *This theorem says that any event in the tail sigma algebra either happens all the time or it does not happen at all. As any Kolmogorov result, this one is very useful in practice. In many applications, we want to show that, despite the random nature of the phenomenon, eventually something will happen for sure. One needs to show that the event desired is in the tail sigma algebra and give an example (a sample path) where the limiting behavior is observed. However, this is an old result. A practical limitation today is the (very strong) assumption that the sets are independent.*

Proof: We skip this proof and only give the steps of the theorem. The idea is to show that A is independent of itself, thus $\mathbf{P}(A \cap A) = \mathbf{P}(A)\mathbf{P}(A) \Rightarrow \mathbf{P}(A) = \mathbf{P}(A)^2 \Rightarrow \mathbf{P}(A)$ is either 0 or 1. The steps of this proof are as follows:

1. First define $\mathscr{A}_n = \sigma(A_1, \ldots, A_n)$ and show that it is independent of $\mathscr{T}_{n+1}$ for all n.

2. Since $\mathscr{T} \subseteq \mathscr{T}_{n+1}$ and $\mathscr{A}_n$ is independent of $\mathscr{T}_{n+1}$, then $\mathscr{A}_n$ and $\mathscr{T}$ are independent for all n.

3. Define $\mathscr{A}_\infty = \sigma(A_1, A_2, \ldots)$. Then from the previous step we deduce that $\mathscr{A}_\infty$ and $\mathscr{T}$ are independent.

4 Finally, since $\mathscr{T} \subseteq \mathscr{A}_\infty$ by the previous step, $\mathscr{T}$ is independent of itself and the result follows.

$\blacksquare$

Note that $\limsup A_n$ and $\liminf A_n$ are tail events. However, it is only in the case when the original events are independent that we can apply Kolmogorov's theorem. Thus in that case $\mathbf{P}\{A_n \text{ i.o.}\}$ is either 0 or 1.

1.5   Lebesgue Measure on the Unit Interval (0,1]

We conclude this chapter with the most important measure available. This is the unique measure that makes things behave in a normal way (e.g., the interval $(0.2, 0.5)$ has measure 0.3).

Let $\Omega = (0, 1]$. Let $\mathscr{F}_0 =$ class of semiopen subintervals (a,b] of Ω. For an interval $I = (a, b] \in \mathscr{F}_0$, define $\lambda(I) = |I| = b - a$. Let $\varnothing \in \mathscr{F}_0$ the element of length 0. Let $\mathscr{B}_0 =$ the algebra of finite disjoint unions of intervals in (0,1]. This algebra is not a σ-algebra. The proof is in Problem 1.4 at the end of this chapter.

If $A = \bigcup_{i=1}^{n} I_n \in \mathscr{B}_0$ with I_n disjoint $\mathscr{F}_0$ sets, then

$$\lambda(A) = \sum_{i=1}^{n} \lambda(I_i) = \sum_{i=1}^{n} |I_i|$$

The goal is to show that λ is countably additive on the algebra $\mathscr{B}_0$. This will allow us to construct a measure (actually a probability measure since we are working on (0,1]) using the Caratheodory's theorem (Theorem 1.36). The resulting measure is well defined and is called the *Lebesgue measure*.

Theorem 1.34 (Theorem for the length of intervals) *Let $I = (a, b] \subseteq (0, 1]$ and I_k of the form $(a_k, b_k]$ bounded but not necessarily in $(0, 1]$.*

(i) If $\bigcup_k I_k \subseteq I$ and I_k are disjoint, then $\sum_k |I_k| \leq |I|$.

(ii) If $I \subseteq \bigcup_k I_k$ (with the I_k not necessarily disjoint), then $|I| \leq \sum_k |I_k|$.

(iii) If $I = \bigcup_k I_k$ and I_k disjoint, then $|I| = \sum_k |I_k|$.

Proof: Exercise (*Hint:* use induction) ∎

Note: Part (iii) shows that the function λ is well defined.

Theorem 1.35 λ *is a (countably additive) probability measure on the field $\mathscr{B}_0$. λ is called the Lebesgue measure restricted to the algebra $\mathscr{B}_0$*

Proof: Let $A = \bigcup_{k=1}^{\infty} A_k$, where A_k are disjoint $\mathscr{B}_0$ sets. By definition of $\mathscr{B}_0$,

$$A_k = \bigcup_{j=1}^{m_k} J_{k_j}, \quad A = \bigcup_{i=1}^{n} I_i,$$

where the J_{k_j} are disjoint. Then,

$$\lambda(A) = \sum_{i=1}^{n} |I_i| = \sum_{i=1}^{n} \left(\sum_{k=1}^{\infty} \sum_{j=1}^{m_k} |I_i \cap J_{k_j}|\right) = \sum_{k=1}^{\infty} \sum_{j=1}^{m_k} \left(\sum_{i=1}^{n} |I_i \cap J_{k_j}|\right)$$

and since $A \cap J_{k_j} = J_{k_j} \Rightarrow |A \cap J_{k_j}| = \sum_{i=1}^{n} |I_i \cap J_{k_j}| = |J_{k_j}|$, the above is continued:

$$= \underbrace{\sum_{k=1}^{\infty} \sum_{j=1}^{m_k} |J_{k_j}|}_{=|A_k|} = \sum_{k=1}^{\infty} \lambda(A_k)$$

∎

The next theorem will extend the Lebesgue measure to the whole $(0, 1]$, thus we define the probability space $((0, 1], \mathscr{B}((0, 1]), \lambda)$. The same construction with minor modifications works in $(\mathbb{R}, \mathscr{B}(\mathbb{R}), \lambda)$.

Theorem 1.36 (Caratheodory's extension theorem) *A probability measure on an algebra has a unique extension to the generated σ-algebra.*

Note: The Caratheodory theorem practically constructs all the interesting probability models. However, once we construct our models we have no further need of the theorem. It also reminds us of the central idea in the theory of probabilities: If one wants to prove something for a big set, one needs to look first at the generators of that set.

Proof: (skipped), in the problems 1.11,1.12. ∎

Definition 1.37 (Monotone class) *A class $\mathscr{M}$ of subsets in Ω is monotone if it is closed under the formation of monotone unions and intersections, that is:*

(i) $A_1, A_2, \cdots \in \mathscr{M}$ and $A_n \subset A_{n+1}, \bigcup_n A_n = A \Rightarrow A \in \mathscr{M}$,

(ii) $A_1, A_2, \cdots \in \mathscr{M}$ and $A_n \supset A_{n+1} \Rightarrow \bigcap_n A_n \in \mathscr{M}$.

The next theorem is only needed for the proof of the Caratheodory theorem. However, the proof is interesting and thus is presented here.

Theorem 1.38 *If $\mathscr{F}_0$ is an algebra and $\mathscr{M}$ is a monotone class, then $\mathscr{F}_0 \subseteq \mathscr{M} \Rightarrow \sigma(\mathscr{F}_0) \subseteq \mathscr{M}$.*

Proof: Let $m(\mathscr{F}_0) = $ minimal monotone class over $\mathscr{F}_0 = $ the intersection of all monotone classes containing $\mathscr{F}_0$

We will prove that $\sigma(\mathscr{F}_0) \subseteq m(\mathscr{F}_0)$.

To show this, it is enough to prove that $m(\mathscr{F}_0)$ is an algebra. Then Exercise 1.12 will show that $m(\mathscr{F}_0)$ is a σ algebra. Since $\sigma(\mathscr{F}_0)$ is the smallest, the conclusion follows.

To this end, let $\mathscr{G} = \{A : A^c \in m(\mathscr{F}_0)\}$.

(i) Since $m(\mathscr{F}_0)$ is a monotone class, so is $\mathscr{G}$.

(ii) Since $\mathscr{F}_0$ is an algebra, its elements are in $\mathscr{G} \Rightarrow \mathscr{F}_0 \subset \mathscr{G}$.

(i) and (ii) $\Rightarrow m(\mathscr{F}_0) \subseteq \mathscr{G}$. Thus $m(\mathscr{F}_0)$ is closed under complementarity.

Now define $\mathscr{G}_1 = \{A : A \cup B \in m(\mathscr{F}_0), \forall B \in \mathscr{F}_0\}$.

We show that $\mathscr{G}_1$ is a monotone class:

Let $A_n \nearrow$ an increasing sequence of sets, $A_n \in \mathscr{G}_1$. By definition of $\mathscr{G}_1$, for all n $A_n \cup B \in m(\mathscr{F}_0), \forall B \in \mathscr{F}_0$.

But $A_n \cup B \supseteq A_{n-1} \cup B$, and thus the definition of $m(\mathscr{F}_0)$ implies

$$\bigcup_n (A_n \cup B) \in m(\mathscr{F}_0), \forall B \in \mathscr{F}_0 \Rightarrow \left(\bigcup_n A_n\right) \cup B \in m(\mathscr{F}_0), \forall B,$$

and thus $\bigcup_n A_n \in \mathscr{G}_1$.

This shows that $\mathscr{G}_1$ is a monotone class. But since $\mathscr{F}_0$ is an algebra, its elements (the contained sets) are in $\mathscr{G}_1{}^5$, thus $\mathscr{F}_0 \subset \mathscr{G}_1$. Since $m(\mathscr{F}_0)$ is the smallest monotone class containing $\mathscr{F}_0$, we immediately have $m(\mathscr{F}_0) \subseteq \mathscr{G}_1$.

Let $\mathscr{G}_2 = \{B : A \cup B \in m(\mathscr{F}_0), \forall A \in m(\mathscr{F}_0)\}$
$\mathscr{G}_2$ **is a monotone class.** (identical proof – see problem 1.11)

Let $B \in \mathscr{F}_0$. Since $m(\mathscr{F}_0) \subseteq \mathscr{G}_1$ for any set $A \in m(\mathscr{F}_0) \Rightarrow A \cup B \in m(\mathscr{F}_0)$. Thus, by the definition of $\mathscr{G}_2 \Rightarrow B \in \mathscr{G}_2 \Rightarrow \mathscr{F}_0 \subseteq \mathscr{G}_2$.

The previous implication and the fact that $\mathscr{G}_2$ is a monotone class imply that $m(\mathscr{F}_0) \subseteq \mathscr{G}_2$.

Therefore, $\forall A, B \in m(\mathscr{F}_0) \Rightarrow A \cup B \in m(\mathscr{F}_0) \Rightarrow m(\mathscr{F}_0)$ is an algebra.   ∎

Problems

1.1   Roll a die. Then $\Omega = \{1, 2, 3, 4, 5, 6\}$. An example of an event is $A = \{$ Roll an even number$\} = \{2, 4, 6\}$. Find the cardinality (number of elements) of $\mathscr{P}(\Omega)$ in this case.

1.2   Suppose $\Omega = \{a, b, c\}$ is a probability space, and a discrete probability function is introduced such that $\mathbf{P}(\{a\}) = 1/3$ and $\mathbf{P}(\{b\}) = 1/2$. List all events in the maximal sigma algebra $\mathcal{P}(\Omega)$ and calculate the probability of each such event.

1.3   Suppose two events A and B are in some space Ω. List the elements of the generated σ algebra $\sigma(A, B)$ in the following cases:

> a) $A \cap B = \emptyset$ b) $A \subset B$

c) $A \cap B \neq \emptyset; A \setminus B \neq \emptyset$ and $B \setminus A \neq \emptyset$.

1.4   **An algebra which is not a σ-algebra**
Let $\mathscr{B}_0$ be the collection of sets of the form $(a_1, a_1'] \cup (a_2, a_2'] \cup \cdots \cup (a_m, a_m']$, for any $m \in \mathbb{N}^* = \{1, 2 \ldots\}$ and all $a_1 < a_1' < a_2 < a_2' < \cdots < a_m < a_m'$ in $\Omega = (0, 1]$.
Verify that $\mathscr{B}_0$ is an algebra. Show that $\mathscr{B}_0$ is not a σ-algebra.

1.5   Let $\mathscr{F} = \{A \subseteq \Omega | A$ finite **or** A^c is finite$\}$.
> a) Show that $\mathscr{F}$ is an algebra.
> b) Show that, if Ω is finite, then $\mathscr{F}$ is a σ-algebra.
> c) Show that, if Ω is infinite, then $\mathscr{F}$ is **not** a σ-algebra.

---

[5]One can just verify the definition of $\mathscr{G}_1$ for this.

1.6 A σ-Algebra does not necessarily contain all the events in Ω

Let $\mathscr{F} = \{A \subseteq \Omega | \, A \text{ countable } \textbf{or } A^c \text{ is countable}\}$. Show that $\mathscr{F}$ is a σ-algebra. Note that, if Ω is uncountable, it implies that it contains a set A such that both A and A^c are uncountable and thus $A \notin \mathscr{F}$.

1.7 Show that the Borel sets of $\mathbb{R}$ $\mathscr{B} = \sigma \left(\{(-\infty, x] | \, x \in \mathbb{R}\}\right)$.

Hint: show that the generating set is the same, that is, show that any set of the form $(-\infty, x]$ can be written as countable union (or intersection) of open intervals, and, vice versa, that any open interval in $\mathbb{R}$ can be written as a countable union (or intersection) of sets of the form $(-\infty, x]$.

1.8 Show that the following classes all generate the Borel σ-algebra, or, put differently, show the equality of the following collections of sets:

$$\sigma \left((a, b) : a < b \in \mathbb{R}\right) = \sigma \left([a, b] : a < b \in \mathbb{R}\right) = \sigma \left((-\infty, b) : b \in \mathbb{R}\right)$$
$$= \sigma \left((-\infty, b) : b \in \mathbb{Q}\right),$$

where $\mathbb{Q}$ is the set of rational numbers.

1.9 Properties of probability measures

Prove properties 1–4 in the Proposition 1.20 on page 21.

Hint: You only have to use the definition of probability. The only thing nontrivial in the definition is the countable additivity property.

1.10 No matter how many zeros do not add to more than zero

Prove the Lemma 1.26 on page 32.

Hint: You may use countable subadditivity.

1.11 If $\mathscr{F}_0$ is an algebra, $m(\mathscr{F}_0)$ is the minimal monotone class over $\mathscr{F}_0$, and $\mathscr{G}_2$ is defined as

$$\mathscr{G}_2 = \{B : A \cup B \in m(\mathscr{F}_0), \forall A \in m(\mathscr{F}_0)\}$$

Then show that $\mathscr{G}_2$ is a monotone class.

Hint: Look at the proof of theorem 1.38 on page 39, and repeat the arguments therein.

1.12 A monotone algebra is a σ-algebra

Let $\mathscr{F}$ be an algebra that is also a monotone class. Show that $\mathscr{F}$ is a σ-algebra.

1.13 Prove the *total probability formula* equation (1.8) and the *Bayes Formula* equation 1.9.

1.14 If two events are such that $A \cap B = \emptyset$, are A and B independent? Justify.

1.15 Show that $\mathbf{P}(A|B) = \mathbf{P}(A)$ is the same as independence of the events A and B.

1.16 Prove that, if two events A and B are independent, then so are their complements.

1.17 Generalize the previous problem to n sets using induction.

1.18   Calculating Odds. If we know the probabilities of a certain event, another way we can express these probabilities is through the odds – that is, the return given to betting on the event. To give an example, we can easily calculate the probability of the next card in a deck of 52 cards being a red card. So the odds offered on the next card being a red card are 1:1; that is, for every dollar bet, I give back 1 more dollar if the bet is a win. The probability is $0.5 = \frac{1}{1+1}$. In the same spirit, the odds given to draw a club are 3:1 or 3\$ for every 1\$ bet. (In fact, the return is 4 if you count the dollar initially bet.) The probability of course is $0.25 = \frac{1}{1+3}$. This idea will be seen in much more detail later when we talk about martingales. What happens if the probabilities are not of the form $1/n$. For example, it is easy to calculate that the probability of drawing a black ball from an urn containing three red and four black balls is $4/7$. No problem, the correct odds for betting on black are 3:4; that is, for every four dollars bet, you receive back three more. That is easy to see because the event has probability greater than $1/2$. So the next questions are all asking you to formulate the correct odds (in a fair game) of the following events:

   a) A card chosen at random from a deck of 52 is an ace.
   b) A card chosen at random is either a Queen or a Club.
   c) In 4 cards drawn from a deck of 52, there are at least 2 Hearts.
   d) A hand of bridge (13 cards) is missing a suit (i.e., does not contain all four clubs, hearts, diamond, and spades).
   e) You see two heads when tossing a fair coin twice.
   f) You see either a sum of 7 or a sum of 11 when rolling two six-sided dies.

   Of course, you can also reconstruct the probabilities from the odds. What are the probabilities of the following events?

   g) You are offered 3:1 odds that the Lakers will win the next game with the Pacers.
   h) Odds 5:2 that the last shot is taken by player X from Lakers.
   i) Odds 30:1 that the game is won on a last shot from player Y from the Pacers.
   You should see by now from this very simple example that, in order that the probabilities to be consistent, one must have some sort of a model to avoid arbitrage (i.e., making money with no risk).

1.19   One urn contains w_1 white balls and b_1 black balls. Another urn contains w_2 white balls and b_2 black balls. A ball is drawn at random from each urn, then one of the two such chosen is selected at random.

   a) What is the probability that the final ball selected is white?
   b) Given that the final ball selected was white, what is the probability that in fact it came from the first urn (with w_1 and b_1 balls).

1.20   At the end of a well-known course, the final grade is decided with the help of an oral examination. There are a total of m possible subjects listed on some pieces of paper. Of them, n are generally considered "easy."

Each student enrolled in the class, one after another, draws a subject at random, and then presents it. Of the first two students, who has the better chance of drawing a "favorable" subject?

1.21   Andre Agassi and Pete Sampras decide to play a number of games together. They play nonstop and at the end it turns out that Sampras won n games while Agassi m, where $n > m$. Assume that in fact any possible sequence of games was possible to reach this result. Let $P_{n,m}$ denote the probability that from the first game until the last Sampras is always in the lead. Find

1. $P_{2,1}$; $P_{3,1}$; $P_{n,1}$

2. $P_{3,2}$; $P_{4,2}$; $P_{n,2}$

3. $P_{4,3}$; $P_{5,3}$; $P_{5,4}$

4. Make a conjecture about a formula for $P_{n,m}$.

1.22   My friend Andrei has designed a system to win at the roulette. He likes to bet on red, but he waits until there have been six previous black spins and only then he bets on red. He reasons that the chance of winning is quite large since the probability of seven consecutive back spins is quite small. What do you think of his system. Calculate the probability that he wins using this strategy.

Actually, Andrei plays his strategy four times and he actually wins three times out of the four he played. Calculate the probability of the event that just occurred.

1.23   Ali Baba is caught by the sultan while stealing his daughter. The sultan is being gentle with him and he offers Ali Baba a chance to regain his liberty.
There are two urns and m white balls and n black balls. Ali Baba has to put the balls in the two urns; however, he likes the only condition that no urn is empty. After that, the sultan will chose an urn at random, and then pick a ball from that urn. If the chosen ball is white Ali Baba is free to go, otherwise Ali Baba's head will be at the same level as his legs.
How should Ali Baba divide the balls to maximize his chance of survival?

CHAPTER 2

RANDOM VARIABLES

All the definitions with sets presented in Chapter 1 are consistent, but if we wish to calculate and compute numerical values related to abstract spaces we need to standardize the spaces. The first step is to give the following definition.

Definition 2.1 (Measurable function (m.f.)) *Let $(\Omega_1, \mathscr{F}_1)$, $(\Omega_2, \mathscr{F}_2)$ be two measurable spaces. Let $f : \Omega_1 \longrightarrow \Omega_2$ be a function. f is called a measurable function if and only if for any set $B \in \mathscr{F}_2$ we have $f^{-1}(B) \in \mathscr{F}_1$. The inverse function is a set function defined in terms of the pre-image. Explicitly, for a given set $B \in \mathscr{F}_2$,*

$$f^{-1}(B) = \{\omega_1 \in \Omega_1 : f(\omega_1) \in B\}$$

Note: This definition makes it possible to extend probability measures to other spaces. For instance, let f be a measurable function and assume that there exists a probability measure P_1 on the first space $(\Omega_1, \mathscr{F}_1)$. Then we can construct a probability measure on the second space $(\Omega_2, \mathscr{F}_2)$ by $(\Omega_2, \mathscr{F}_2, P_1 \circ f^{-1})$. Note that since f is measurable, $f^{-1}(B)$ is in $\mathscr{F}_1$, thus $P_1 \circ f^{-1}(B) = P_1(f^{-1}(B))$ is well defined. In fact, this $P_1 \circ f^{-1}$ is a new probability measure induced by the measurable function f. We shall see just how important this measure is in the next section.

Probability and Stochastic Processes, First Edition. Ionuţ Florescu

Reduction to $\mathbb{R}$. Random variables

Definition 2.2 *Any measurable function with codomain* $(\Omega_2, \mathscr{F}_2) = (\mathbb{R}, \mathscr{B}(\mathbb{R}))$ *is called a* random variable.

Consequence: Since the Borel sets in $\mathbb{R}$ are generated by $(-\infty, x]$, we can have the definition of a random variable directly by

$$f : \Omega_1 \longrightarrow \mathbb{R} \text{ such that } f^{-1}(-\infty, x] \in \mathscr{F} \text{ or } \{\omega : f(\omega) \le x\} \in \mathscr{F}, \forall x \in \mathbb{R}.$$

We shall sometimes use $\{f \le x\}$ as a shortcut to denote the set $\{\omega : f(\omega) \le x\} = f^{-1}(-\infty, x]$. Traditionally, the random variables are denoted with capital letters from the end of the alphabet $X, Y, Z, \ldots$, and their values are denoted with the corresponding small letters $x, y, z, \ldots$.

Definition 2.3 (Distribution of random variable) *Assume that on the measurable space* $(\Omega, \mathscr{F})$ *we define a probability measure* $\mathbf{P}$ *so that it becomes a probability space* $(\Omega, \mathscr{F}, \mathbf{P})$. *If a random variable* $X : \Omega \to \mathbb{R}$ *is defined, then we call its distribution the set function* μ *defined on the Borel sets of* $\mathbb{R}$: $\mathscr{B}(\mathbb{R})$, *with values in* $[0, 1]$:

$$\mu(B) = \mathbf{P}\left(\{\omega : X(\omega) \in B\}\right) = \mathbf{P}\left(X^{-1}(B)\right) = \mathbf{P} \circ X^{-1}(B).$$

Remark 2.4 *First note that the measure* μ *is defined on sets in* $\mathbb{R}$ *and takes values in the interval* $[0, 1]$. *Therefore, the random variable* X *allows us to apparently eliminate the abstract space* Ω. *However, this is not the case since we still have to calculate probabilities using* $\mathbf{P}$ *in the definition of* μ *above.*

However, there is one simplification we can make. If we recall the result of Exercises 1.7 and 1.8, we know that all Borel sets are generated by the same type of sets. Using the same idea as before, it is enough to describe how to calculate μ for the generators. We could, of course, specify any type of generating sets we wish (open sets, closed sets, etc.) but it turns out the simplest way is to use sets of the form $(-\infty, x]$, since we only need to specify one end of the interval (the other is always $-\infty$). With this observation, we only need to specify the measure $\mu = P \circ X^{-1}$ directly on the generators to completely characterize the probability measure.

Definition 2.5 *[The distribution function of a random variable] The distribution function of a random variable* X *is a function* $F : \mathbb{R} \to [0, 1]$ *with*

$$F(x) = \mu(-\infty, x] = \mathbf{P}\left(\{\omega : X(\omega) \in (-\infty, x]\}\right) = \mathbf{P}\left(\{\omega : X(\omega) \le x\}\right)$$

But, this is exactly the definition of the cumulative distribution function (cdf) which you can find in any lower level probability classes. It is exactly the same thing except that in the effort to dumb down (in whosoever opinion it was to teach the class that way) the meaning is lost and we cannot proceed with more complicated things. From the definition above, we can deduce all the elementary properties of the cdf that you have learned (right continuity, increasing, taking values between 0 and 1). In fact, let me ask you to prove these properties in the next Proposition.

Proposition 2.6 *The distribution function for any random variable X has the following properties:*

(i) *F is increasing (i.e., if $x \leq y$, then $F(x) \leq F(y)$)*[1]

(ii) *F is right continuous (i.e., $\lim_{h \downarrow 0} F(x + h) = F(x)$)*

(iii) *$\lim_{x \to -\infty} F(x) = 0$ and $\lim_{x \to \infty} F(x) = 1$*

◼ EXAMPLE 2.1   Indicator random variable

Recall the indicator function from Definition 1.22. Let $\mathbf{1}_A$ be the indicator function of a set $A \subseteq \Omega$. This is a function defined on Ω with values in $\mathbb{R}$. Therefore, it could also be viewed as a possible random variable. According to the definition, the said function is a random variable if and only if the function is measurable. It is simple to show that this happens if and only if $A \in \mathcal{F}$ the σ-algebra associated with the probability space.

Now suppose that $A \in \mathcal{F}$, so the indicator is a random variable; according to the construction we had, this variable must have a distribution. Let us calculate its distribution function.

According to the definition, we have to calculate $\mathbf{P} \circ \mathbf{1}_A^{-1}((-\infty, x])$ for any x. However, the function $\mathbf{1}_A$ only takes two values: 0 and 1. We can calculate immediately

$$\mathbf{1}_A^{-1}((-\infty, x]) = \begin{cases} \emptyset & , \text{ if } x < 0 \\ A^c & , \text{ if } x \in [0, 1) \, . \\ \Omega & , \text{ if } x > 1 \end{cases}$$

Therefore,

$$F(x) = \begin{cases} 0 & , \text{ if } x < 0 \\ \mathbf{P}(A^c) & , \text{ if } x \in [0, 1) \, . \\ 1 & , \text{ if } x \geq 1 \end{cases}$$

Proving the following lemma 2.7 is elementary using the properties of the probability measure (Proposition 1.20) and is left as an exercise.

---

[1] In other math books a function with this property is called nondecreasing. I do not like the negation, so I prefer to call a function like this increasing with the distinction that a function with the following property $x < y$ implies $F(x) < F(y)$ is going to be called a **strictly increasing** function.

Lemma 2.7 (Probability calculations using the distribution function) *Let F be the distribution function of X. Denote by $F(x-) = \lim_{y \nearrow x} F(y)$ the left limit of F at x. We have*

(i) $\mathbf{P}(X \geq x) = 1 - F(x-)$

(ii) $\mathbf{P}(x < X \leq y) = F(y) - F(x)$

(iii) $\mathbf{P}(X = x) = F(x) - F(x-)$.

Recall that, in fact, $\{X \leq x\}$ is a notation for $\{\omega \in \Omega : X(\omega) \leq x\}$. The probabilities on the left are well defined in the abstract set Ω, but the distribution function effectively transforms their calculation to knowing the function F.

We defined a random variable as any measurable function with codomain $(\mathbb{R}, \mathscr{B}(\mathbb{R}))$. A even more specific case is obtained when the random variable has the domain also equal to $(\mathbb{R}, \mathscr{B}(\mathbb{R}))$. In this case, the random variable is called a *Borel function*.

Definition 2.8 (Borel measurable function) *A function $g : \mathbb{R} \to \mathbb{R}$ is called Borel (measurable) function if g is a measurable function from $(\mathbb{R}, \mathscr{B}(\mathbb{R}))$ into $(\mathbb{R}, \mathscr{B}(\mathbb{R}))$.*

■ EXAMPLE 2.2

Show that any continuous function $g : \mathbb{R} \to \mathbb{R}$ is Borel-measurable.

Proof (Solution): This is very simple. Recall that the Borel sets are generated by open sets. So it is enough to see what happens to the pre-image of an open set B. But g is a continuous function, and therefore $g^{-1}(B)$ is an open set and thus $g^{-1}(B) \in \mathscr{B}(\mathbb{R})$. Therefore, by definition, g is Borel measurable. ■

2.1 Discrete and Continuous Random Variables

Definition 2.9 (pdf pmf and all that) *Note that the distribution function F always exists. In general, the distribution function F is not necessarily derivable. However, if it is, we call its derivative $f(x)$ the* probability density function *(pdf):*

$$F(x) = \int_{-\infty}^{x} f(z)dz$$

Traditionally, a variable X with this property is called a continuous random variable.

Furthermore, if F is piecewise constant (i.e., constant almost everywhere), or in other words there exist a countable sequence $\{a_1, a_2, \dots\}$ such that the function F is constant for every point except these a_i's and we denote $p_i = F(a_i) - F(a_i-)$, then *the collection of p_i's is the traditional* probability mass function *(pmf) that characterizes a* discrete random variable[2].

Remark 2.10 *Traditional undergraduate textbooks segregate between discrete and continuous random variables. Because of this segregation, they are the only variables presented and it appears that all the random variables are either discrete or continuous. In reality, these are the only types that can be presented without following the general approach we take here. The definitions we presented here cover any random variable. In this book, we do not segregate the random variables by type and the treatment of random variables is the same.*

Important   So what is the point of all this? What did we just accomplish here?

The answer is, we successfully moved from the abstract space $(\Omega, \mathcal{F}, P)$ to something perfectly equivalent but defined on $(\mathbb{R}, \mathcal{B}(\mathbb{R}))$. Because of this, we only need to define probability measures on $\mathbb{R}$ and show that anything coming from the original abstract space is equivalent to one of these distributions on $\mathbb{R}$. We have just constructed our first model.

What's in the name? Continuous random variable. Absolutely continuous measure. Radon–Nikodym theorem

So where do these names "discrete" and "continuous" random variable comes from? What is so special about them?

For the discrete random variable it is easy to understand. The variable only takes at most a countable number of outcomes and it is not possible to take values within an interval, thus the outcomes are discrete. What about continuous? To understand this, we need to go back to 1913 and understand the concept of the absolute continuity of measure.

Definition 2.11 *Suppose we have two measures μ and ν defined on the same measurable space $(\Omega, \mathcal{F})$. Then ν is said to be absolutely continuous with respect to μ, or dominated by μ, if $\nu(A) = 0$ for every set A for which $\mu(A)=0$. We denote this by $\nu \ll \mu$. Mathematically*

$$\nu \ll \mu \quad \text{if and only if} \quad \mu(A) = 0 \Rightarrow \nu(A) = 0$$

The Radon–Nikodym theorem completely characterizes the absolute continuity for σ-finite measures. The theorem is named after Johann Radon, who proved the theorem for the special case when the underlying space is $\mathbb{R}^n$ in 1913, and Otton Nikodym, who proved the general case of any measurable space in 1930 (Nikodym, 1930).

---

[2]We used the notation $F(x-)$ for the left limit of function F at x or, in a more traditional notation,

$$F(x-) = \lim_{z \to x, z < x} F(z) = \lim_{z \downarrow x} F(z).$$

Theorem 2.12 (Radon–Nikodym) *Given a measurable space $(\Omega, \mathscr{F})$ and two σ-finite measures μ and ν, if $\nu \ll \mu$, then there exists a measurable function $f : \Omega \to [0, \infty)$ such that*

$$\nu(A) = \int_A f d\mu.$$

The function f is called the Radon–Nikodym derivative *for reasons that will become evident in a minute and is usually denoted with $\frac{d\nu}{d\mu}$.*

We still need to define properly the integral on the right, and we shall do so in the next chapter. However, for the purposes of this section the Riemann integral will suffice.

Connection with continuous random variables   Do you recall the Lebesgue measure on the unit interval λ which we introduced in Section 1.5? We stated at the time that this is the most important measure. Well, here is the reason why. First of all, the Lebesgue measure may be extended uniquely to the whole $\mathbb{R}$ in such a way that the measure of an interval is the length of the interval. Clearly, this measure is not finite anymore, but it is σ-finite. This is easy to see by writing $\mathbb{R} = \bigcup_{n \in \mathbb{Z}} [n, n+1)$. Clearly, any probability measure is σ-finite as well.

Consider a random variable X on some probability space. As we saw, $\mathbf{P} \circ X^{-1}$ is a Borel probability measure. So using the Radon–Nikodym theorem above, if this probability measure is absolutely continuous with respect to the Lebesgue measure, then there exists a function f such that

$$\mathbf{P} \circ X^{-1}(A) = \int_A f d\lambda \quad \forall A \in \mathscr{B}(\mathbb{R}).$$

However, recall that the Borel sets are generated by sets of the form $(-\infty, x]$ and that Lebesgue measure on the interval is the length of the interval (i.e., $\lambda(A) = \int_A dx$), so the distribution function of a random variable absolutely continuous with the Lebesgue measure may be written as

$$F(x) = \mathbf{P} \circ X^{-1}((-\infty, x]) = \int_{(-\infty, x]} f d\lambda = \int_{-\infty}^{x} f(x) dx,$$

where the last term is a simple Riemann integral. It is easy to see now that f is by definition the derivative of F and thus in this case (i.e., absolute continuity with respect to the Lebesgue measure) the Radon–Nikodym derivative is simply the usual derivative of the function. This is exactly the definition of continuous random variable and here is where the name is coming from: a continuous random variable is a random variable whose distribution is absolutely continuous with respect to the Lebesgue measure. Of course, this is too complex to remember, and hence the simplified version "continuous random variable".

Examples of random variables

⬛ EXAMPLE 2.3 Indicator r.v. (continued)

This indicator variable is also called the *Bernoulli random variable*. Notice that the variable only takes values 0 and 1, and the probability that the variable takes the value 1 may be easily calculated using the previous definitions:

$$\mathbf{P} \circ \mathbf{1}_A^{-1}(\{1\}) = \mathbf{P}\{\omega : \mathbf{1}_A(\omega) = 1\} = \mathbf{P}(A).$$

Therefore, the variable is distributed as a Bernoulli random variable with parameter $p = \mathbf{P}(A)$. Alternately, we may obtain this probability using the previously computed distribution function

$$\mathbf{P}\{\omega : \mathbf{1}_A(\omega) = 1\} = F(1) - F(1-) = 1 - \mathbf{P}(A^c) = \mathbf{P}(A).$$

⬛ EXAMPLE 2.4

Roll a six-sided fair die. Say, $X(\omega) = 1$ if the die shows 1 ($\omega = 1$), $X = 2$ if the die shows 2, and so on. Find $F(x) = \mathbf{P}(X \leq x)$.

Proof (Solution):

If $x < 1$, then $\mathbf{P}(X \leq x) = 0$.

If $x \in [1, 2)$, then $\mathbf{P}(X \leq x) = \mathbf{P}(X = 1) = 1/6$.

If $x \in [2, 3)$, then $\mathbf{P}(X \leq x) = \mathbf{P}(X(\omega) \in \{1, 2\}) = 2/6$.

We continue this way to get

$$\mathbf{F}(x) = \begin{cases} 0 & \text{if } x < 1 \\ i/6 & \text{if } x \in [i, i+1) \text{ with } i = 1, \cdots, 5 \\ 1 & \text{if } x \geq 6 \end{cases}$$

■

Exercise 1 (Mixture of continuous and discrete random variable) *Say, a game asks you to toss a coin. If the coin lands Tail, you lose 1\$, and if Head then you draw a number from $[1, 2]$ at random and gain that number. Furthermore, suppose that the coin lands a Head with probability p. Let X be the amount of money won or lost after one game. Find the distribution of X.*

Proof (Solution): Let $\omega = (\omega_1, \omega_2)$, where $\omega_1 \in \{\text{Head}, \text{Tail}\}$ and ω_2 in the defining experimental space for the uniform distribution. Define a random variable $Y(\omega_2)$ on the uniform $[1, 2]$ space. Then the random variable X is defined as

$$X(\omega) = \begin{cases} -1 & \text{, if } \omega_1 = \text{Tail} \\ Y(\omega_2) & \text{if } \omega_1 = \text{Head} \end{cases}.$$

If $x \in [-1, 1)$ we get

$$\mathbf{P}(X \le x) = \mathbf{P}(X = -1) = \mathbf{P}(\omega_1 = \text{Tail}) = 1 - p.$$

If $x \in [1, 2)$, we get

$$\mathbf{P}(X \le x) = \underbrace{\mathbf{P}(X = -1 \text{ or } X \in [1, x])}_{\text{the two events are disjoint}} = 1 - p + \mathbf{P}(\omega_1 = \text{ heads}, Y \le x)$$

$$= 1 - p + p \underbrace{\mathbf{P}(Y \in [1, x])}_{\text{Uniform}[1,2]}$$

$$= 1 - p + p \int_1^x 1 dy = 1 - p + p(x - 1)$$

$$= 1 - 2p + px.$$

Note that, if the two parts of the game are not independent of each other, we cannot calculate this distribution.

Finally, we obtain

$$\mathbf{F}(x) = \begin{cases} 0 & \text{if } x < -1 \\ 1 - p & \text{if } x \in [-1, 1) \\ 1 - 2p + px & \text{if } x \in [1, 2) \\ 1 & \text{if } x \ge 2 \end{cases}$$

Checking that our calculation is correct It is always a good idea to check the result. We can verify the distribution function's properties and plot the function to confirm this.   ∎

In the previous example, we used a uniformly distributed random variable. The next section introduces formally this and many other distributions.

2.2   Examples of Commonly Encountered Random Variables

In this section we present a few important distributions. This is by no means an extensive list, but it is a list of distributions which any probabilist needs to be familiar with.

Discrete random variables

For discrete random variables we give the probability mass function, and it will describe completely the distribution (recall that the distribution function is piecewise linear).

(i) *Bernoulli distribution*, the random variable only takes two values:

$$\mathbf{X} = \begin{cases} 1 & \text{with } \mathbf{P}(X = 1) = p \\ 0 & \text{with } \mathbf{P}(X = 0) = 1 - p \end{cases}$$

We denote a random variable X with this distribution as $X \sim Bernoulli(p)$.

This distribution models experiments, which may result in one of two possible outcomes: for example, a coin toss, or rolling a die and looking for outcome 6, or rolling two dies and looking for the sum of them to equal 7, and so on.

(ii) *Binomial(n, p) distribution*, the random variable takes values in $\mathbb{N}$ with

$$\mathbf{P}(X = k) = \begin{cases} \binom{n}{k} p^k (1 - p)^{n-k} & \text{for any } k \in \{0, 1, 2, \dots, n\} \\ 0 & \text{otherwise} \end{cases}$$

We denote a random variable X with this distribution as $X \sim Binom(n, p)$.

Note that X has the same distribution as $Y_1 + \cdots Y_n$, where $Y_i \sim Bernoulli(p)$. Because of this, X may be thought of as the number of successes in n repetitions of a single phenomenon which may result in two possible outcomes (generally labeled success and failure). We also note that the probabilities are the terms in the binomial expansion: $(p + (1 - p))^n$, so they obviously sum to 1.

(iii) *Geometric (p) distribution*:

$$\mathbf{P}(X = k) = \begin{cases} (1 - p)^{k-1} p & \text{for any } k \in \{1, 2 \cdots\} \\ 0 & \text{otherwise} \end{cases}$$

This random variable is sometimes called the geometric "number of trials" distribution. We may also talk about a geometric "number of failures distribution", defined as

$$\mathbf{P}(Y = k - 1) = \begin{cases} (1 - p)^{k-1} p & \text{for any } k \in \{1, 2 \cdots\} \\ 0 & \text{otherwise} \end{cases}$$

Most of the time, when we write $X \sim Geometric(p)$ we mean that X has a geometric number of trials distribution. In the rare cases when we use the other one, we will specify very clearly.

(iv) *Negative binomial (r, p) distribution*

$$\mathbf{P}(X = k) = \begin{cases} \binom{k-1}{r-1} (1 - p)^{r-k} p^r & \text{for any } k \in \{r, r + 1, \dots\} \\ 0 & \text{otherwise} \end{cases}$$

Similarly, with the $Geometric(p)$ distribution we can talk about a "number of failures" distribution:

$$\mathbf{P}(X = k - 1) = \begin{cases} \binom{k-1}{r-1}(1 - p)^{r-k}p^r & \text{for any } k \in \{r, r+1, \ldots\} \\ 0 & \text{otherwise} \end{cases}$$

Unless specified otherwise, we always mean the number of trials when we write $X \sim NegBinom(r, p)$. In the same way as a binomial may be expressed as a sum of Bernoulli random variables, any negative binomial with parameters r and p may be expressed as a sum of r $Geometric(p)$ random variables.

Let us stop for a moment and see where these distributions are coming from. Suppose we repeat a simple experiment many times. This experiment only has two possible outcomes: "success" with probability p, and "failure" with probability $1 - p$. Each repetition is independent of any of the previous ones.

- The variable X, which takes value 1 if the experiment is a success and 0 otherwise, has a $Bernoulli(p)$ distribution.

- Repeat the experiment n times in such a way that no experiment influences the outcome of any other experiment[3] and we count how many of the n repetition actually resulted in success. Let Y be the variable denoting this number. Then $Y \sim Binom(n, p)$.

- If instead of repeating the experiment a fixed number of times we repeat the experiment as many times as are needed to see the first success, then the number of trials needed is going to be distributed as a $Geometric(p)$ random variable. If we count failures until the first success, we obtain the $Geometric(p)$ "number of failures" distribution.

- If we repeat the experiment until we see r successes, the number of trials needed is a $negativebinomial(r, p)$.

- Because of their relationship, there is a simple way to relate the probability of a binomial and negative binomial. Specifically, suppose that $Y \sim NegBinom(r, p)$; then

$$\begin{aligned} \mathbf{P}(Y \leq k) &= \mathbf{P}(\text{the } r\text{th success is before trial } k) = \\ &= \mathbf{P}(\text{in } k \text{ trials there must be at least } r \text{ successes}) \\ &= \mathbf{P}(X \geq r), \end{aligned}$$

where $X \sim Binom(k, p)$.

---

[3]This is the idea of independence, which we will discuss a bit later.

(v) *Hypergeometric distribution(N,m,n,p)*,

$$\mathbf{P}(X = k) = \frac{\binom{m}{k}\binom{N-m}{n-k}}{\binom{N}{n}} \quad k \in \{0, 1 \cdots m\}.$$

This may be thought of as drawing n balls from an urn containing m white balls and $N - m$ black balls, where X represents the number of white balls in the sample.

(vi) *Poisson distribution*, the random variable takes values in $\mathbb{N}$,

$$\mathbf{P}(X = k) = \frac{\lambda^k}{k!}e^{-k}, \quad k = 0, 1, 2, \ldots.$$

Continuous random variables

In this case, every random variable has a pdf and we will specify this function directly.

(i) *Uniform distribution[a,b]*, the random variable represents the position of a point taken at random (without any preference) within the interval $[a, b]$.

$$f(x) = \begin{cases} \frac{1}{b-a} & , \text{if } x \in [a, b] \\ 0 & , \text{otherwise} \end{cases}$$

Figure 2.1 presents the density and the distribution function for a $Uniform(0, 1)$ random variable.

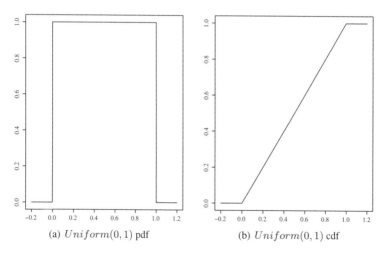

(a) $Uniform(0, 1)$ pdf   (b) $Uniform(0, 1)$ cdf

Figure 2.1   The pdf and cdf of the uniform distribution in the interval $[0, 1]$.

(ii) *Exponential distribution(θ)*

$$f(x) = \frac{1}{\theta} e^{-x/\theta}, \quad x \geq 0.$$

Proposition 2.13 *Suppose $X \sim Exp(\lambda)$, $\lambda > 0$. Then its distribution function is*

$$F_X(x) = 0 \text{ if } x < 0$$

and

$$F_X(x) = 1 - e^{-\lambda x}, \text{ if } x \geq 0.$$

Proof: The first part is obvious because the support of the density of X is the set of positive real numbers. If $x \geq 0$, then

$$F_X(x) = \int_0^x \lambda e^{-\lambda t} dt = 1 - e^{-\lambda x}.$$

■

Figure 2.2 presents the density and the distribution function for an *Exponential* ($\lambda = 0.2$) random variable.

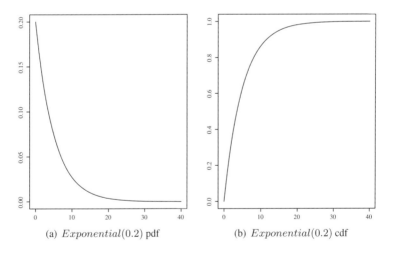

(a) *Exponential*(0.2) pdf   (b) *Exponential*(0.2) cdf

Figure 2.2   *Exponential* distribution

The exponential distribution is often used to model the lifespan of organisms and objects: for example, the lifetime of bacteria living in a culture, lifetimes of light bulbs, and so on. We will encounter this distribution many times throughout this book. It has a special relation with the Poisson distribution, which will be explained in Chapter 10.

(iii) *Normal distribution*(μ, σ)

$$f(x) = \frac{1}{\sqrt{2\pi\sigma^2}} e^{\frac{-(x-\mu)^2}{2\sigma^2}}, \quad x \in \mathbb{R}.$$

Figure 2.3 presents the density and distribution for a $Normal(0, 1)$ random variable.

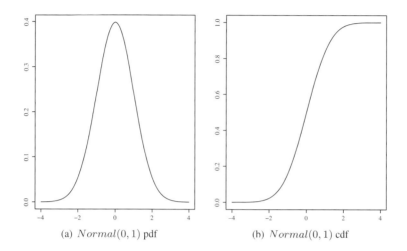

(a) $Normal(0, 1)$ pdf   (b) $Normal(0, 1)$ cdf

Figure 2.3 $Normal$ distribution

The normal distribution is commonly encountered in practice, and is used throughout statistics, natural sciences, and social sciences as a simple model for complex phenomena. For example, the observational error in an experiment is usually assumed to follow a normal distribution, and the propagation of un-certainty is computed using this assumption. Note that a normally distributed variable has a symmetric distribution about its mean. Quantities that grow expo-nentially, such as prices, incomes, or populations, are often skewed to the right, and hence may be better described by other distributions, such as the log-normal distribution or the Pareto distribution. In addition, the probability of seeing a nor-mally distributed value that is far from the mean (i.e., extreme values) drops off extremely rapidly. As a result, statistical inference using a normal distribution is not robust to the presence of outliers (data that is unexpectedly far from the mean). When the data contains outliers a heavy-tailed distribution such as the Student t-distribution may be a better fit for the data.

(iv) *Gamma distribution* (a, λ)

$$f_{a,\lambda}(x) = \frac{\lambda^a}{\Gamma(a)} e^{-\lambda x} x^{a-1} 1_{(0,\infty)}(x) \tag{2.1}$$

with $a, \lambda > 0$. We will denote $X \sim \Gamma(a, \lambda)$, meaning that the random variables X follow a Gamma distribution with density (2.1).

The Gamma function in the definition is defined by

$$\Gamma(x) = \int_0^\infty t^{x-1} e^{-t} dt. \tag{2.2}$$

Proposition 2.14 *The Gamma function satisfies the following properties:*

a. For every $x > 0$, we have $\Gamma(x + 1) = x\Gamma(x)$.

b. For every $n \in \mathbb{N}^\star$, we have $\Gamma(n + 1) = n!$.

c. $\Gamma\left(\frac{1}{2}\right) = \sqrt{\pi}$.

Figure 2.4 presents the density and distribution for a $Gamma(10, 0.2)$ random variable.

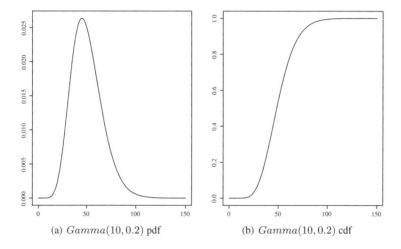

(a) $Gamma(10, 0.2)$ pdf   (b) $Gamma(10, 0.2)$ cdf

Figure 2.4 $Gamma$ distribution

Remark 2.15 *If $a = 1$, then the law $\Gamma(1, \lambda)$ is the exponential law with parameter λ (with expectation $\frac{1}{\lambda}$).*

Another special case is when $a = \lambda = \frac{1}{2}$. In this case, we obtain the law of a squared normal. Specifically, suppose that X is a standard normal random variable, that is, $X \sim N(0, 1)$. Then

$$X^2 \sim \Gamma\left(\frac{1}{2}, \frac{1}{2}\right).$$

This distribution $\Gamma\left(\frac{1}{2}, \frac{1}{2}\right)$ is also called a *chi-squared distribution* with one degree of freedom. In general, the sum of squares of n normal random variables with mean 0 and variance 1 is a chi-squared distribution with $n-1$ degree of freedom. Let us formalize this distribution in a definition (which is also a $\Gamma\left(\frac{n}{2}, \frac{1}{2}\right)$).

Definition 2.16 *We say that a random variables X follows a chi-squared distribution with n degrees of freedom if its density function is*

$$\frac{1}{2^{n/2}\Gamma\left(\frac{n}{2}\right)} x^{\frac{n}{2}-1} e^{-\frac{x}{2}}.$$

We will use the notation $X \sim \chi_n^2$.

Figure 2.5 presents the density and distribution for a χ_{10}^2 random variable.

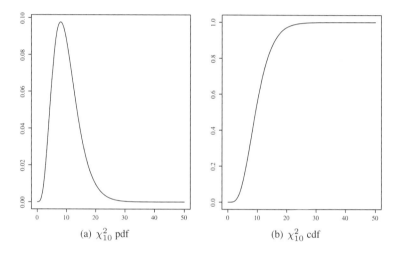

(a) χ_{10}^2 pdf   (b) χ_{10}^2 cdf

Figure 2.5   The χ_n^2 distribution

If $X \sim \Gamma(a, \lambda)$, then the random variable

$$Y = \frac{1}{X}$$

follows an inverse Gamma distribution. This is a probability law appearing in several practical applications. The density of the inverse Gamma distribution with parameters $a, \lambda > 0$ is defined by

$$f(x) = \frac{\lambda^a}{\Gamma(a)} \left(\frac{1}{x}\right)^{a+1} e^{-\lambda/x}. \tag{2.3}$$

Figure 2.6 presents the density and distribution for an $InverseGamma(10, 0.2)$ random variable.

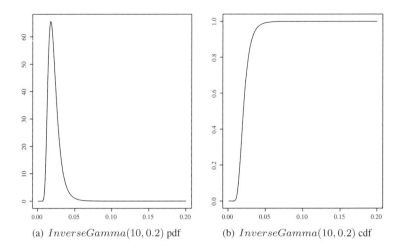

(a) $InverseGamma(10, 0.2)$ pdf   (b) $InverseGamma(10, 0.2)$ cdf

Figure 2.6   Inverse Gamma distribution

(v) *Beta Distribution (a,b)*

$$f_{a,b}(x) = \frac{x^{a-1}(1-x)^{b-1}}{\beta(a,b)} 1_{[0,1]}(x) \tag{2.4}$$

with $a, b > 0$. In the density, the *beta function* is defined as

$$\beta(a,b) = \int_0^1 t^{a-1}(1-x)^{b-1} dx.$$

We will denote a random variable X with this distribution with $X \sim Beta(a,b)$ to distinguish from the beta function.

Figure 2.7 presents the density and distributions for various *Beta* random variables.

The Beta distribution is typically used to model probabilities of events. It is heavily used in Bayesian statistics due to its properties and relationship with the binomial distribution. We also note that $Beta(1,1)$ is exactly the uniform distribution in the interval $[0, 1]$.

(vi) *Student or t distribution*

$$f(x) = \frac{\Gamma(\frac{n+1}{2})}{\sqrt{n\pi}\Gamma(\frac{n}{2})} \left(1 + \frac{x^2}{n}\right)^{-\frac{n+1}{2}}, \tag{2.5}$$

where $n > 0$ is a parameter named the degree of freedom and Γ is the gamma function previously introduced. We use n to denote the degrees of freedom

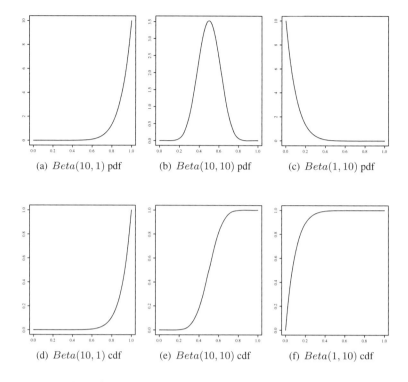

(a) $Beta(10, 1)$ pdf   (b) $Beta(10, 10)$ pdf   (c) $Beta(1, 10)$ pdf

(d) $Beta(10, 1)$ cdf   (e) $Beta(10, 10)$ cdf   (f) $Beta(1, 10)$ cdf

Figure 2.7   Beta distributions for various parameter values

because in most applications n is going to be an integer. It does not need to be, however.

This distribution has a very interesting history. To start, we remark that is one takes $X_1, \ldots, X_n$ standard normal random variables and forms the expression

$$\frac{\bar{X}}{\sqrt{S^2(X)}},$$

then this expression has a t-distribution with $n - 1$ degrees of freedom.

In 1908, William Sealy Gosset published a derivation of the distribution of this expression ($\frac{\bar{X}}{\sqrt{S^2(X)}}$), where the original variables $X_1, \ldots, X_n$ were standard normals Student (1908). He was working for the Guinness Brewery in Dublin, Ireland, at that time. Guinness forbade members of its staff from publishing scientific papers to not allow the competition to acquire secrets of its famous brew manufacturing. Gosset realized the importance of his discovery and decided such a result deserved to be known even under a pseudonym. The distribution became popular when applied by Sir Ronald Aylmer Fisher (Fisher, 1925) who calls it "Student's t" distribution.

Figure 2.8 presents the density and distribution for a t_{10} random variable.

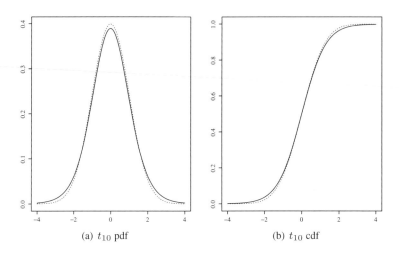

(a) t_{10} pdf   (b) t_{10} cdf

Figure 2.8 t_{10} distribution (continuous line) overlapping a standard normal (dashed line) to show the heavier tails

The shape of this distribution resembles the bell shape of the normal, but it has fatter tails. In fact, a t_n distribution has no moments greater than n. One more interesting fact is that the t_n pdf converges to the normal pdf as $n \to \infty$.

(vii) *Log-normal*(μ, σ^2)

$$f(x) = \frac{1}{x\sigma\sqrt{2\pi}} e^{-\frac{(\log x - \mu)^2}{2\sigma^2}} 1_{(0,\infty)}. \qquad (2.6)$$

The parameters are $\mu \in \mathbb{R}$ and $\sigma > 0$. We shall denote $X \sim LogN(\mu, \sigma^2)$ a distribution with these parameters.

The name comes from the fact that the logarithm of a random variable is normally distributed. Specifically, if Y is log-normally distributed, then $X = log(Y)$ is normally distributed. Vice versa, if X is a random variable with a normal distribution, then $Y = e^X$ has a log-normal distribution.

Figure 2.9 presents the density and distribution for a $LogNormal(0, 1)$ random variable.

(viii) *Normal mixture distribution*

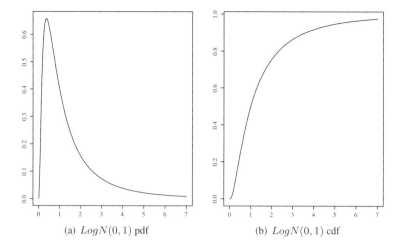

(a) $LogN(0,1)$ pdf   (b) $LogN(0,1)$ cdf

Figure 2.9 $Log - normal$ distribution

Definition 2.17 *A random variable X is a mixture of m normal random variables (or **mix-norm**) if it takes the values*

$$X = \begin{cases} N_1, & \text{with prob. } p_1 \\ N_2, & \text{with prob. } p_2 \\ N_3, & \text{with prob. } p_3 \\ \vdots \\ N_m, & \text{with prob. } p_m, \end{cases}$$

where $N_i \sim N(\mu_i, \sigma_i^2)$, $i = 1, \ldots, m$, with $\sum_{j=1}^{m} p_j = 1$. We will use the notation

$$MN_m((p_1, \mu_1, \sigma_1^2), (p_2, \mu_2, \sigma_2^2), \ldots, (p_m, \mu_m, \sigma_m^2))$$

to denote this distribution. The density of X has the formula

$$f_X(x) = \sum_{i=1}^{m} p_i \frac{1}{\sqrt{2\pi\sigma_i^2}} e^{\frac{-(x-\mu_i)^2}{2\sigma_i^2}}$$

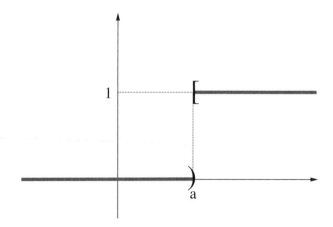

Figure 2.10   A simple distribution function, yet with no density

A special random variable: Diraç Delta distribution

For a fixed a real number, consider the following distribution function:

$$F_\delta(x) = \begin{cases} 0 & \text{if } x < a \\ 1 & \text{if } x \geq a \end{cases}$$

This function is plotted in Figure 2.10. Note that the function has all the properties of a distribution function (increasing, right continuous, and limited by 0 and 1). However, the function is not derivable (the distribution does not have a pdf).

The random variable with this distribution is called a Diraç impulse function at a. It can only be described using measures. We will come back to this function when we develop the integration theory, but for now let us say that if we define the associated set function

$$\delta_{\{a\}}(A) = \begin{cases} 1 & \text{if } a \in A \\ 0 & \text{otherwise} \end{cases}$$

this is in fact a probability measure with the property

$$\int_{-\infty}^{\infty} f(x) d\delta_{\{a\}}(x) = f(a), \quad \text{for all continuous functions } f.$$

This will be written later as $\mathbf{E}^{\delta_{\{a\}}}[f] = f(a)$. (In other sciences: $\delta_{\{a\}}(f) = f(a)$).

Also, note that $\delta_{\{a\}}(A)$ is a set function (a is fixed) and has the same value as the indicator $\mathbf{1}_A(a)$ which is a regular function (A is fixed).

2.3 Existence of Random Variables with Prescribed Distribution. Skorohod Representation of a Random Variable

In the previous section, we have seen that any random variable has a distribution function F, which is called in other classes the cdf. Recall the essential properties of this function from Proposition 2.6 on page 47: right continuity, increasing, taking values between 0 and 1. An obvious question is: given a function F with these properties, can we construct a random variable with the desired distribution?

In fact, yes, we can, and this is the first step in a very important theorem we shall see later in this course: the Skorohod representation theorem. However, recall that a random variable has to have as domain some probability space. It actually is true that we can construct random variables with the prescribed distribution on any space, but recall that the purpose of creating random variables was to have a uniform way of treating probability. It is actually enough to give the Skorohod's construction on the probability space $([0, 1], \mathscr{B}([0, 1]), \lambda)$, where λ is the Lebesgue measure.

On this space, for any ω, define the following random variables:

$$X^+(\omega) = \inf\{z \in \mathbb{R} : F(z) > \omega\}$$

$$X^-(\omega) = \inf\{z \in \mathbb{R} : F(z) \geq \omega\}$$

Note that, in statistics, X^- would be called the 100ω-percentile of the distribution F.

For most of the outcomes ω, the two random variables are identical. Indeed, if at z with $\omega = F(z)$, the function F is nonconstant, and then the two variables take the same values $X^+(\omega) = X^-(\omega) = z$. The two important cases when the variables take different values are depicted in Figure 2.11.

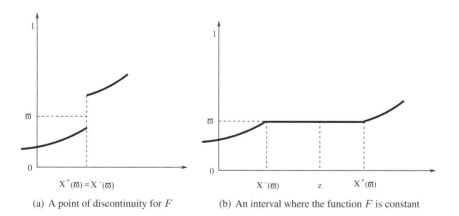

(a) A point of discontinuity for F (b) An interval where the function F is constant

Figure 2.11   Points where the two variables $X^{\pm}$ may have different outcomes

We need to show that the two variables have the desired distribution. To this end, let $x \in \mathbb{R}$. Then we have

$$\{\omega \in [0, 1] : X^-(\omega) \leq x\} = [0, F(x)]$$

Indeed, if ω is in the left set, then $X^-(\omega) \leq x$. By the definition of X^-, then $\omega \leq F(x)$ and we have the inclusion $\subseteq$. If, on the other hand, $\omega \in [0, F(x)]$, then $\omega \leq F(x)$ and, again by definition and right continuity of F, $X^-(\omega) \leq x$, thus we obtain $\supseteq$. Therefore, the distribution is

$$\lambda(\{\omega \in [0, 1] : X^-(\omega) \leq x\}) = \lambda([0, F(x)]) = F(x) - 0 = F(x).$$

Finally, X^+ also has distribution function F, and furthermore

$$\lambda(X^+ \neq X^-) = 0.$$

By definition of X^+

$$\{\omega \in [0, 1] : X^-(\omega) \leq x\} \supseteq [0, F(x)),$$

and so $\lambda(X^+ \leq x) \geq F(x)$. Furthermore, since $X^- \leq X^+$ we have

$$\{\omega \in \mathbb{R} : X^-(\omega) \neq X^+(\omega)\} = \bigcup_{x \in \mathbb{Q}} \{\omega \in \mathbb{R} : X^-(\omega) \leq x < X^+(\omega)\}.$$

But for every such $x \in \mathbb{Q}$

$$\lambda(\{\omega \in \mathbb{R} : X^-(\omega) \leq x < X^+(\omega)\}) = \lambda(\{X^- \leq x\} \backslash \{X^+ \leq x\}) \leq F(x) - F(x) = 0.$$

Since $\mathbb{Q}$ is countable, and since any countable union of null sets is a null set, the result follows.

Percentiles, quantiles of a distribution

As you can imagine, the results in the previous section are very useful for practical applications as well. In particular, generating numbers with a specified distribution is a very useful tool used in simulations. Who does not know today what the function *rand* does in any computer language[4]? In the next chapter we shall see practical applications and algorithms for generating random variables. But for now, let us formalize the quantile definition.

Definition 2.18 *Suppose F is a distribution function. Let $\alpha \in (0, 1)$ be a constant. Define*

$$x^+(\alpha) = \inf\{z \in \mathbb{R} : F(z) > \alpha\}$$
$$x^-(\alpha) = \inf\{z \in \mathbb{R} : F(z) \geq \alpha\}$$

---

[4]It generates a U(0,1) random variable.

Note that since α is fixed, there is nothing random in the definition above; they are just numbers.

*If $x^+(\alpha) = x^-(\alpha) := x_\alpha$, then the distribution admits a unique α-percentile (equal to x_α). If $x^-(\alpha) < x^+(\alpha)$, which happens in one of the two cases described in Figure 2.11 on page 65, then we need to define a **generalized** α-percentile. In fact, looking at Figure 2.11(a) is no problem (previous case) and the issue is with the α level corresponding to the horizontal line in Figure 2.11(b). In this case, in fact any point between $x^-(\alpha)$ and $x^+(\alpha)$ is an α-percentile. To have a single number corresponding to the statistic calculated from data, usually one uses $x^-(\alpha) = \inf\{z \in \mathbb{R} : F(z) \ge \alpha\}$ as the choice for the percentile.*

Note that we could just define the α-percentile as the number

$$x_\alpha = \inf\{x \in \mathbb{R} \mid F(x) \ge u\},$$

or in fact $x^-(\alpha)$ from the previous definition and be done with it. But this actually obscures the problems with the percentile definition.

A percentile as defined above is a special case of a quartile. Both notions are born in statistics by practical applications where, by observing repeated outcomes of a random variable, one tries to infer its distribution. Quantiles are outcomes corresponding to dividing the probabilities into equal parts. To be more specific, if we divide the probability into two equal parts, we obtain three points of which only the middle one is interesting (the others are the minimum and the maximum). This point is the median.

Similarly, if we divide the total probability into three equal parts, the two points that correspond to the 0.33 and 0.67 probability are called terciles. Quartiles are the better known quantiles with 0.25 and 0.75 probability corresponding to first and third quartile, respectively. The second quartile is of course the median. To sum up, here are the most popular types of quantiles:

- The 2-quantile is called the median denoted M.

- The 3-quantiles (T_1, T_2) are called terciles.

- The 4-quantiles $(Q_1, M\ Q_3)$ are called quartiles.

- The 5-quantiles $(QU_1, \ldots, QU_4)$ are called quintiles.

- The 6-quantiles $(S_1, \ldots, S_5)$ are called sextiles.

- The 10-quantiles $(D_1, \ldots, D_9)$ are called deciles.

- The 12-quantiles $(Dd_1, \ldots, Dd_{11})$ are called duo-deciles.

- The 20-quantiles $(V_1, \ldots, V_{19})$ are called vigintiles.

- The 100-quantiles are called percentiles ($\%$).

- The 1000-quantiles are called permilles ($\%_0$).

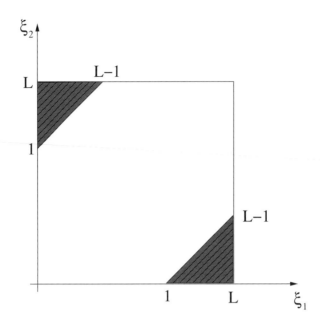

Figure 2.12   The area we need to calculate. The shaded parts need to be deleted.

2.4 Independence

In this section we extend the idea of independence originally defined for events to random variables. In order to do this, we have to explain the joint distribution of several variables.

■ **EXAMPLE 2.5   The idea of joint distribution**

Suppose two points ξ_1, ξ_2 are tossed at random and independently onto a line segment of length L (ξ_1, ξ_2 are independent identically distributed (i.i.d.). What is the probability that the distance between the two points does not exceed 1?

Proof (Solution): [Solution] If $L \leq 1$, then the probability is trivially equal to 1.

Assume that $L > 1$ (the following also works if 1 is substituted by a $l \leq L$). What is the distribution of ξ_1 and ξ_2? They are both $Unif[0, L]$. We want to calculate $\mathbf{P}(|\xi_1 - \xi_2| \leq 1)$.

We plot the surface we need to calculate in Figure 2.12. The area within the rectangle and not shaded is exactly the area we need. If we pick any point from within this area, it will have the property that $|\xi_1 - \xi_2| \leq 1$. Since the points are chosen uniformly from within the rectangle, the chance of a point being chosen is the ratio between the "good" area and the total area.

The unshaded area from within the rectangle is $L^2 - \frac{(L-1)^2}{2} - \frac{(L-1)^2}{2} = 2L - 1$. Therefore, the desired probability is

$$\mathbf{P}(|\xi_1 - \xi_2| \leq 1) = \frac{2L - 1}{L^2}.$$

■

This geometrical proof works because the distribution is uniform and furthermore the points are chosen independently of each other. However, if the distribution is anything else, we need to go through the whole calculation. We shall see this entire calculation after we define joint probability. We also need this definition for the concept of independence.

Joint distribution

We talked about σ-algebras in Chapter 1. Let us come back to them. If there is any hope of rigorous introduction into probability and stochastic processes, they are *unavoidable*. Later, when we will talk about stochastic processes we will find out the *crucial* role they play in quantifying the information available up to a certain time. For now, let us play a bit with them.

Definition 2.19 (σ-algebra generated by a random variable) *For a random variable X, we define the σ-algebra generated by X, denoted $\sigma(X)$ or sometimes $\mathscr{F}_X$, the smallest σ-field $\mathscr{G}$ such that X is measurable on $(\Omega, \mathscr{G})$. It is the σ-algebra generated by the pre-images of Borel sets through X (recall that we have already presented this concept earlier in Definition 1.4 on page 12). Because of this, we can easily show[5] that*

$$\sigma(X) = \sigma(\{\omega | X(\omega) \leq x\}, \text{ as } x \text{ varies in } \mathbb{R}).$$

Similarly, given $X_1, X_2, \ldots, X_n$ random variables, we define the sigma algebra generated by them as the smallest sigma algebra such that all are measurable with respect to it. It turns out we can show easily that it is the sigma algebra generated by the union of the individual sigma algebras or, put more specifically, $\sigma(X_i, i \leq n)$ is the smallest sigma algebra containing all $\sigma(X_i)$, for $i = 1, 2, \ldots, n$, or $\sigma(X_1) \vee \sigma(X_2) \vee \cdots \vee \sigma(X_n)$. Again, recall Proposition 1.6 on page 14.

In Chapter 1, we defined Borel sigma algebras corresponding to any space Ω. We consider the special case when $\Omega = \mathbb{R}^n$. This allows us to define a random vector on $(\mathbb{R}^n, \mathscr{B}(\mathbb{R}^n), \mathbf{P})$ as $(X_1, X_2, \ldots, X_n)$, where each X_i is a random variable. The probability $\mathbf{P}$ is defined on $\mathscr{B}(\mathbb{R}^n)$.

---

[5]Remember that the Borel sets are generated by intervals of the type $(-\infty, x]$.

We can talk about its distribution (the *"joint distribution"* of the variables $(X_1, X_2, \ldots, X_n)$) as the function

$$F(x_1, x_2, \ldots, x_n) = \mathbf{P} \circ (X_1, X_2, \ldots, X_n)^{-1} ((-\infty, x_1] \times \cdots \times (-\infty, x_n])$$
$$= \mathbf{P}(X_1 \leq x_1, X_2 \leq x_2, \ldots, X_n \leq x_n),$$

which is well defined for any $x = (x_1, x_2, \ldots, x_n) \in \mathbb{R}^n$.

In the special case when F can be written as

$$F(x_1, x_2, \ldots, x_n) = \int_{-\infty}^{x_1} \int_{-\infty}^{x_2} \cdots \int_{-\infty}^{x_n} f_X(t_1, \cdots, t_n) dt_1 \cdots dt_n,$$

we say that the vector X has a *joint density* and f_X is the joint pdf of the random vector X.

Definition 2.20 (Marginal distribution) *Given the joint distribution of a random vector $X = (X_1, X_2, \ldots, X_n)$, we define the marginal distribution of X_1 as*

$$F_{X_1}(x_1) = \lim_{\substack{x_2 \to \infty \\ \cdots \\ x_n \to \infty}} F_X(x_1 \cdots x_n)$$

and similarly for all the other variables.[6]

Independence of random variables

We can now introduce the notions of independence and joint independence using the definition in Section 1.3, the probability measure $= \mathbf{P} \circ (X_1, X_2, \ldots, X_n)^{-1}$, and any Borel sets. Writing more specifically, that definition is transformed here:

Definition 2.21 *The variables $(X_1, X_2, \ldots, X_n, \ldots)$ are independent if for every subset $J = \{j_1, j_2, \ldots, j_k\}$ of $\{1, 2, 3, \ldots\}$ we have*

$$\mathbf{P}\left(X_{j_1} \leq x_{j_1}, X_{j_2} \leq x_{j_2}, \ldots, X_{j_k} \leq x_{j_k}\right) = \prod_{j \in J} \mathbf{P}(X_j \leq x_j).$$

Remark 2.22 *The formula in the Definition 2.20 allows us to obtain the marginal distributions from the joint distribution. The converse is generally false, meaning that if we know the marginal distributions we cannot regain the joint.*

However, there is one case when this is possible: when X_i's are independent. In this case, $F_X(x) = \prod_{i=1}^{n} F_{X_i}(x_i)$. That is why the i.i.d. case is the most important in probability (we can regain the joint from the marginals without any other special knowledge).

---

[6]We can also define it simply as $\int_{-\infty}^{x_1} \int_{-\infty}^{\infty} \cdots \int_{-\infty}^{\infty} f_X(t_1, \cdots, t_n) dt_1 \cdots dt_n$ if the joint pdf exists.

Independence (specialized cases)

(i) If X and Y are discrete r.v.s with joint probability mass function $p_{X,Y}(\cdot,\cdot)$, then they are independent if and only if

$$p_{X,Y}(x,y) = p_X(x)p_Y(y), \quad \forall x, y.$$

(ii) If X and Y are continuous r.v.s with joint probability density function f, then they are independent if and only if

$$f_{X,Y}(x,y) = f_X(x)f_Y(y), \quad \forall x, y$$

where we used the obvious notations for marginal distributions. The above definition can be extended to n-dimensional vectors in an obvious way.

I.I.D. r.v.'s: (Independent Identically Distributed Random Variables).

Many of the central ideas in probability involve sequences of random variables which are i.i.d., that is, a sequence of random variables $\{X_n\}$ such that X_n are independent and all have the same distribution function, say $F(x)$.

Finally, we answer the question we asked in the earlier example: What to do if the variables ξ_1, ξ_2 are not uniformly distributed?

Suppose that ξ_1 had distribution F_{ξ_1}, and ξ_2 had distribution F_{ξ_2}. Assuming that the two variables are independent, we obtain the joint distribution

$$F_{\xi_1\xi_2}(x_1, x_2) = F_{\xi_1}(x_1)F_{\xi_2}(x_2).$$

(If they are not independent, we have to be given or infer the joint distribution).

The probability we are looking for is the area of the surface

$$\{(\xi_1, \xi_2)|\xi_1 \in [0, L], \xi_2 \in [0, L], \xi_1 - 1 \le \xi_2 \le \xi_1 + 1\}.$$

We shall find out later how to calculate this probability using the general distribution functions F_{ξ_1} and F_{ξ_2}. For now, let us assume that the two variables have densities f_1 and f_2. Then, the desired probability is

$$\int_0^L \int_0^L \mathbf{1}_{\{x_1 - 1 \le x_2 \le x_1 + 1\}}(x_1, x_2) f_{\xi_1}(x_1) f_{\xi_2}(x_2) dx_1 dx_2$$

which can be further calculated as

- When $L - 1 < 1$ or $1 < L < 2$:

$$\int_0^{L-1} \int_0^{x_1+1} f_{\xi_1}(x_1) f_{\xi_2}(x_2) dx_2 dx_1 + (2 - L)L$$
$$+ \int_1^L \int_{x_1-1}^L f_{\xi_1}(x_1) f_{\xi_2}(x_2) dx_2 dx_1;$$

- When $L - 1 > 1$ or $L > 2$:

$$\int_0^1 \int_0^{x_1+1} f_{\xi_1}(x_1) f_{\xi_2}(x_2) dx_2 dx_1 + \int_1^{L-1} \int_{x_1-1}^{x_1+1} f_{\xi_1}(x_1) f_{\xi_2}(x_2) dx_2 dx_1$$

$$+ \int_{L-1}^{L} \int_{x_1-1}^{L} f_{\xi_1}(x_1) f_{\xi_2}(x_2) dx_2 dx_1.$$

The above is given to remind ourselves about the calculation of a two-dimensional integral.

2.5   Functions of Random Variables. Calculating Distributions

Measurable functions allow us to construct new random variables. These new random variables possess their own distribution. This section is dedicated to calculating this new distribution. At this time, it is not possible to work with abstract spaces (for that we will give a general theorem – the "transport formula" in the next chapter), so all our calculations will be done in $\mathbb{R}^n$.

One-dimensional functions

Let X be a random variable defined on some probability space $(\Omega, \mathcal{F}, \mathbf{P})$. Let $g : \mathbb{R} \longrightarrow \mathbb{R}$ be a Borel measurable function. Let $Y = g(X)$, which is a new random variable. Its distribution is deduced as

$$\mathbf{P}(Y \le y) = \mathbf{P}(g(X) \le y) = \mathbf{P}(g(X) \in (-\infty, y]) = \mathbf{P}\left(X \in g^{-1}((-\infty, y])\right)$$
$$= \mathbf{P}\left(\{\omega : X(\omega) \in g^{-1}((-\infty, y])\}\right),$$

where $g^{-1}((-\infty, y])$ is the pre-image of $(-\infty, y]$ through the function g, i.e.,:

$$\{x \in \mathbb{R} : g(x) \le y\}.$$

If the random variable X has pdf f, then the probability has a simpler formula

$$\mathbf{P}(Y \le y) = \int_{g^{-1}(-\infty, y]} f(x) dx$$

■ EXAMPLE 2.6

Let X be a random variable distributed as a normal (Gaussian) with mean zero and variance 1, $X \sim N(0, 1)$. Let $g(x) = x^2$, and take $Y = g(X) = X^2$. Then

$$\mathbf{P}(Y \le y) = \mathbf{P}(X^2 \le y) = \begin{cases} 0 & \text{if } y < 0 \\ \mathbf{P}(-\sqrt{y} \le X \le \sqrt{y}) & \text{if } y \ge 0 \end{cases}$$

Note that the pre-image of $(-\infty, y]$ through the function $g(x) = x^2$ is either $\emptyset$ if $y < 0$ or $[-\sqrt{y}, \sqrt{y}]$ if $y \ge 0$. This is how we obtain the above. In the nontrivial case $y \geq 0$, we get

$$\mathbf{P}(Y \leq y) = \Phi(\sqrt{y}) - \Phi(-\sqrt{y}) = \Phi(\sqrt{y}) - [1 - \Phi(\sqrt{y})] = 2\Phi(\sqrt{y}) - 1,$$

where Φ is the cdf of X, which is a $N(0, 1)$ random variable. In this case, $\Phi(x) = \int_{-\infty}^{x} \frac{1}{\sqrt{2\pi}} e^{-t^2/2} dt$.

Since the function Φ is derivable, Y has a pdf which can be obtained as

$$
\begin{aligned}
f_Y(y) &= \frac{d}{dy}[2\Phi(\sqrt{y})] = 2\Phi'(\sqrt{y}) \frac{1}{2\sqrt{y}} \\
&= \frac{1}{\sqrt{y}} \Phi'(\sqrt{y}) = \frac{1}{\sqrt{y}} \frac{1}{\sqrt{2\pi}} e^{-y/2} \\
&= \frac{1}{\sqrt{2\pi y}} e^{-y/2}.
\end{aligned}
$$

■

We note that a random variable Y with the pdf described above is said to have a chi-squared distribution with one degree of freedom (the notation is χ_1^2).

Two and more dimensional functions

If the variable X does not have a pmf or a pdf, there is not much we can do. The same relationship holds as in the one-dimensional case. Specifically, if X is a n-dimensional random vector and $g : \mathbb{R}^n \longrightarrow \mathbb{R}^n$ is a measurable function which defines a new random vector $Y = g(X)$, then its distribution is determined using

$$\mathbf{P}(Y \leq y) = \mathbf{P}(g(X) \leq y) = \mathbf{P}\left(\{\omega : X(\omega) \in g^{-1}((-\infty, y])\}\right),$$

which is the same relationship as before.

In the case when the vector X has a density, then things become more specific. We will exemplify using R^2, but the same calculation works in n dimensions with no modification (other than the dimension of course). Suppose that a two-dimensional random vector (X_1, X_2) has joint density f. Let $g : \mathbb{R}^2 \longrightarrow \mathbb{R}^2$ be a measurable function

$$g(x_1, x_2) = (g_1(x_1, x_2), g_2(x_1, x_2)).$$

Suppose first that the function g is one to one[7].

Define a random vector $Y = (Y_1, Y_2) = g(X_1, X_2)$. First we find the support set of Y (i.e., the points where Y has nonzero probability). To this end, let

$\mathcal{A} = \{(x_1, x_2) : f(x_1, x_2) > 0\}$

$\mathcal{B} = \{(y_1, y_2) : y_1 = g_1(x_1, x_2) \text{ and } y_2 = g_2(x_1, x_2), \text{ for some } (x_1, x_2) \in \mathcal{A}\}.$

---

[7]This is why we use the same dimension n for both X and Y vectors.

This $\mathcal{B}$ is the image of $\mathcal{A}$ through g; it is also the support set of Y. Since g is one to one, when restricted to $g : \mathcal{A} \to \mathcal{B}$, it is also surjective and therefore forms a bijection between $\mathcal{A}$ and $\mathcal{B}$. Thus, the inverse function $g^{-1}(y_1, y_2) = (g_1^{-1}(y_1, y_2), g_2^{-1}(y_1, y_2))$ is a unique, well-defined function.

To calculate the density of Y, we need the derivative of this g^{-1}, and that role is played by the Jacobian of the transformation (the determinant of the matrix of partial derivatives):

$$J = J_{g^{-1}}(y_1, y_2) = \begin{vmatrix} \dfrac{\partial g_1^{-1}}{\partial y_1}(y_1, y_2) & \dfrac{\partial g_2^{-1}}{\partial y_1}(y_1, y_2) \\ \dfrac{\partial g_1^{-1}}{\partial y_2}(y_1, y_2) & \dfrac{\partial g_2^{-1}}{\partial y_2}(y_1, y_2). \end{vmatrix}$$

Then, the joint pdf of the vector Y is given by

$$f_Y(y_1, y_2) = f\left(g_1^{-1}(y_1, y_2), g_2^{-1}(y_1, y_2)\right) |J| \, \mathbf{1}_{\mathcal{B}}(y_1, y_2)$$

where we used the indicator notation and $|J|$ is the absolute value of the Jacobian.

Suppose that the function g is not one to one

In this case we recover the previous one-to-one case by restricting the function. Specifically, define the sets $\mathcal{A}$ and $\mathcal{B}$ as before. Now, the restricted function $g : \mathcal{A} \to \mathcal{B}$ is surjective. We partition $\mathcal{A}$ into $\mathcal{A}_0, \mathcal{A}_1, \mathcal{A}_2, \ldots, \mathcal{A}_k$. The set $\mathcal{A}_0$ may contain several points which are difficult to deal with; the only condition is that $\mathbf{P}((X_1, X_2) \in \mathcal{A}_0) = 0$ (it is a null set). Furthermore, for all $i \neq 0$, each restriction $g : \mathcal{A}_i \to \mathcal{B}$ is one to one. Thus, for each such $i \geq 1$, an inverse can be found $g_i^{-1}(y_1, y_2) = (g_{i1}^{-1}(y_1, y_2), g_{i2}^{-1}(y_1, y_2))$. This ith inverse gives for any $(y_1, y_2) \in \mathcal{B}$ a unique $(x_1, x_2) \in \mathcal{A}_i$ such that $(y_1, y_2) = g(x_1, x_2)$. Let J_i be the Jacobian associated with the ith inverse transformation. Then the joint pdf of Y is

$$f_Y(y_1, y_2) = \sum_{i=1}^{k} f\left(g_{i1}^{-1}(y_1, y_2), g_{i2}^{-1}(y_1, y_2)\right) |J_i| \, \mathbf{1}_{\mathcal{B}}(y_1, y_2).$$

◼ EXAMPLE 2.7

Let (X_1, X_2) have some joint pdf $f(\cdot, \cdot)$. Calculate the density of $X_1 X_2$.

Let us take $Y_1 = X_1 X_2$ and $Y_2 = X_1$, that is, $g(x_1, x_2) = (x_1 x_2, x_1) = (y_1, y_2)$. The function thus constructed $g : \mathbb{R}^2 \to \mathbb{R}^2$ is bijective, so $\mathcal{B} = \mathbb{R}^2$. To calculate its inverse, we have

$$x_1 = y_2$$
$$x_2 = \frac{y_1}{x_1} = \frac{y_1}{y_2},$$

which gives

$$g^{-1}(y_1, y_2) = \left(y_2, \frac{y_1}{y_2} \right).$$

We then get the Jacobian

$$J_{g^{-1}}(y_1, y_2) = \begin{vmatrix} 0 & \frac{1}{y_2} \\ 1 & -\frac{y_1}{y_2^2} \end{vmatrix} = 0 - \frac{1}{y_2} = -\frac{1}{y_2}.$$

Thus, the joint pdf of $Y = (Y_1, Y_2)$ is

$$f_Y(y_1, y_2) = f\left(y_2, \frac{y_1}{y_2} \right) \left| \frac{1}{y_2} \right|,$$

where f is the given pdf of X. To obtain the distribution of $X_1 X_2 = Y_1$, we simply need the marginal pdf obtained immediately by integrating out Y_2:

$$f_{Y_1}(y_1) = \int_{-\infty}^{\infty} f\left(y_2, \frac{1}{y_2} \right) \cdot \frac{1}{|y_2|} dy_2.$$

∎

◩ EXAMPLE 2.8   A more specific example

Let X_1, X_2 be independent Exp(λ). Find the joint density of $Y_1 = X_1 + X_2$ and $Y_2 = \frac{X_1}{X_2}$. Also show that the variables Y_1 and Y_2 are independent.

Let $g(x_1, x_2) = \left(x_1 + x_2, \frac{x_1}{x_2} \right) = (y_1, y_2)$. Let us calculate the domain of the transformation.

Remember that the pdf of the exponential distribution is

$$f(x) = \lambda e^{-\lambda x} \mathbf{1}_{(0,\infty)}(x),$$

thus $A = (0, \infty) \times (0, \infty)$. Since $x_1, x_2 > 0$, we get that $x_1 + x_2 > 0$ and $\frac{x_1}{x_2} > 0$, and so $B = (0, \infty)^2$ as well. The function g restricted to this sets is bijective as we can easily show by solving the equations $y_1 = x_1 + x_2$ and $y_2 = \frac{x_1}{x_2}$. We obtain

$$x_1 = x_2 y_2 \Rightarrow y_1 = x_2 y_2 + x_2$$
$$\Rightarrow x_2 = \frac{y_1}{1 + y_2}$$
$$\Rightarrow x_1 = \frac{y_1 y_2}{1 + y_2}.$$

Since the solution is unique, the function g is one to one. Since the solution exists for all $(y_1, y_2) \in (0.\infty)^2$, the function is surjective. Its inverse is precisely

$$g^{-1}(y_1, y_2) = \left(\frac{y_1 y_2}{1 + y_2}, \frac{y_1}{1 + y_2} \right).$$

Furthermore, the Jacobian is

$$J_{g^{-1}}(y_1, y_2) = \begin{vmatrix} \frac{y_2}{1+y_2} & \frac{1}{1+y_2} \\ \frac{y_1}{(1+y_2)^2} & -\frac{y_1}{(1+y_2)^2} \end{vmatrix} = -\frac{y_1 y_2}{(1+y_2)^3} - \frac{y_1}{(1+y_2)^3} = -\frac{y_1}{(1+y_2)^2}.$$

Thus the desired pdf is

$$\begin{aligned} f_Y(y_1, y_2) &= f\left(\frac{y_1 y_2}{1+y_2}, \frac{y_1}{1+y_2}\right) \left|-\frac{y_1}{(1+y_2)^2}\right| \mathbf{1}_{(y_1,y_2)\in(0,\infty)^2} \\ &= \lambda e^{-\lambda \frac{y_1 y_2}{1+y_2}} \lambda e^{-\lambda \frac{y_1}{1+y_2}} \frac{y_1}{(1+y_2)^2} \mathbf{1}_{\{y_1,y_2>0\}} \\ &= \lambda^2 e^{-\lambda y_1} \frac{y_1}{(1+y_2)^2} \mathbf{1}_{\{y_1,y_2>0\}}. \end{aligned}$$

Finally, to end the example it is enough to recognize that the pdf of Y can be decomposed into a product of two functions, one of them only of the variable y_1, and the other only a function of the variable y_2. Thus, if we apply the next lemma, the example is solved. ∎

Lemma 2.23 *If the joint distribution f of a random vector (X, Y) factors as a product of functions of only x and y, that is, there exist $g, h : \mathbb{R} \to \mathbb{R}$ such that $f(x,y) = g(x)h(y)$, then the variables X, Y are independent.*

Proof: Problem 2.16. ∎

◪ EXAMPLE 2.9

X and Y are independent random variables. X is uniform on $(-1, 1)$, and Y is exponential with mean 1. This means that their densities are

$$g_X(x) = \begin{cases} \frac{1}{2} & -1 \le x \le 1 \\ 0 & \text{for other } x\text{'s} \end{cases}$$

$$g_Y(y) = \begin{cases} e^{-y} & y \ge 0 \\ 0 & \text{for other } y\text{'s} \end{cases}.$$

Find the density of $W = X/Y$.

From the example 2.7 suitably modified to X/Y, we obtain the pdf of X/Y if $\mathbf{P}(Y = 0) = 0$ as

$$g_{X/Y}(z) = \int_{-\infty}^{\infty} f(zx, x)|x|dx = \int_{-\infty}^{\infty} \frac{1}{2} e^{-x} \mathbf{1}_{[-1,1]\times(0,\infty)}(zx, x)|x|dx,$$

where we used that the two variables are independent and the joint pdf is the product of marginal pdfs. This integral needs to be calculated.

Case $z > 0$.

$$g_{X/Y}(z) \overset{x \in [-1/z, 1/z]; x > 0}{=} \int_0^{1/z} \frac{1}{2} e^{-x} x \, dx$$

$$= \dots \text{ integration by parts } \dots = \frac{1}{2}\left(1 - \left(1 + \frac{1}{z}\right) e^{-\frac{1}{z}}\right).$$

Case $z < 0$.

$$g_{X/Y}(z) = \int_{-1/z}^0 \frac{1}{2} e^{-x} x \, dx$$

$$= \dots \text{ integration by parts } \dots = \frac{1}{2}\left(1 - \left(1 - \frac{1}{z}\right) e^{\frac{1}{z}}\right).$$

The case $z = 0$ may be treated separately or just as a limiting case of both the above results. The density is then

$$g_W(z) = \frac{1}{2}\left(1 - \left(1 + \frac{1}{z}\right) e^{-\frac{1}{z}}\right) \mathbf{1}_{\{z \geq 0\}} + \frac{1}{2}\left(1 - \left(1 - \frac{1}{z}\right) e^{\frac{1}{z}}\right) \mathbf{1}_{\{z < 0\}}.$$

∎

◫ EXAMPLE 2.10

Let X, Y be two random variables with joint pdf $f(\cdot, \cdot)$. Calculate the density of $X + Y$.

Let $(U, V) = (X + Y, Y)$. We can easily calculate the domain and the inverse $g^{-1}(u, v) = (u - v, v)$. The Jacobian is

$$J_{g^{-1}}(u, v) = \begin{vmatrix} 1 & -1 \\ 0 & 1 \end{vmatrix} = 1.$$

As a result, the desired pdf is

$$f_U(u) = \int_{-\infty}^{\infty} f(u - v, v) \, dv.$$

∎

We will refer to this particular example later when we talk about convolutions.

◫ EXAMPLE 2.11

Let X_1 and X_2 be i.i.d. $N(0, 1)$ random variables. Consider the function $g(x_1, x_2) = \left(\frac{x_1}{x_2}, |x_2|\right)$. Calculate the joint distribution of $Y = g(X)$ and the distribution of the ratio of the two normal, X_1/X_2.

First, $\mathcal{A} = R^2$ and $\mathcal{B} = \mathbb{R} \times (0, \infty)$. Second, note that the transformation is not one to one. Also note that we have a problem when $x_2 = 0^8$. Fortunately, we know how to deal with this situation. Take a partition of $\mathcal{A}$ as follows:

$$\mathcal{A}_0 = \{(x_1, 0) : x_1 \in \mathbb{R}\}, \quad \mathcal{A}_1 = \{(x_1, x_2) : x_2 < 0\}, \quad \mathcal{A}_1 = \{(x_1, x_2) : x_2 > 0\}.$$

$\mathcal{A}_0$ has the desired property since $\mathbf{P}((X_1, X_2) \in \mathcal{A}_0) = \mathbf{P}(X_2 = 0) = 0$ (X_2 is a continuous random variable). Restricted to each $\mathcal{A}_i$, the function g is bijective and we can calculate its inverse in both cases:

$$g_1^{-1}(y_1, y_2) = (-y_1 y_2, -y_2)$$

$$g_2^{-1}(y_1, y_2) = (y_1 y_2, y_2).$$

In either case, the Jacobian is identical, $J_1 = J_2 = y_2$. Using the pdf of a normal with mean zero and variance 1 ($f(x) = \frac{1}{\sqrt{2\pi}} e^{-x^2/2}$), and that X_1 and X_2 being independent, the joint pdf is the product of marginals and we obtain

$$f_Y(y_1, y_2) = \left(\frac{1}{2\pi} e^{-(-y_1 y_2)^2/2} e^{-(-y_2)^2/2} |y_2| + \frac{1}{2\pi} e^{-(y_1 y_2)^2/2} e^{-(y_2)^2/2} |y_2|\right) \mathbf{1}_{\{y_2 > 0\}}$$

$$= \frac{y_2}{\pi} e^{-\frac{(y_1^2 + 1) y_2^2}{2}} \mathbf{1}_{\{y_2 > 0\}}, \quad y_1 \in \mathbb{R},$$

and this is the desired joint distribution. To calculate the distribution of X_1/X_2, we calculate the marginal of Y_1 by integrating out y_2:

$$f_{Y_1}(y_1) = \int_0^\infty \frac{y_2}{\pi} e^{-\frac{(y_1^2 + 1) y_2^2}{2}} dy_2 \quad (\text{ Change of variables } y_2^2 = t)$$

$$= \int_0^\infty \frac{1}{2\pi} e^{-\frac{(y_1^2 + 1)}{2} t} dt = \frac{1}{2\pi} \frac{2}{y_1^2 + 1}$$

$$= \frac{1}{\pi(y_1^2 + 1)}, \quad y_1 \in \mathbb{R}.$$

But this is the distribution of a Cauchy random variable. Thus we have just proven that the ratio of two independent $N(0, 1)$ r.v.s has a Cauchy distribution. ∎

We conclude this chapter with a nontrivial application of the Borel–Cantelli lemmas. We have postponed this example until this point since we needed to learn about independent random variables first.

---

^{8}0 is in $\mathcal{A}$, since $f_{X_2}(0) > 0$.

■ **EXAMPLE 2.12**

Let $\{X_n\}$ a sequence of i.i.d. random variables, each exponentially distributed with rate 1, that is,

$$\mathbf{P}(X_n > x) = e^{-x}, \quad x > 0.$$

We wish to study how large these variables are when $n \to \infty$. To this end, take $x = \alpha \log n$, for some $\alpha > 0$ and for any $n \geq 1$. Substitute into the probability above to obtain

$$\mathbf{P}(X_n > \alpha \log n) = e^{-\alpha \log n} = n^{-\alpha} = \frac{1}{n^\alpha}.$$

But we know that the sum $\sum_n \frac{1}{n^\alpha}$ is divergent for the exponent $\alpha \leq 1$ and convergent for $\alpha > 1$. So we can apply the Borel–Cantelli lemmas since the events in question are independent. Thus,

If $\alpha \leq 1$, the sum is divergent and so $\sum_n \mathbf{P}(X_n > \alpha \log n) = \infty$, thus

$$\mathbf{P}\left(\frac{X_n}{\log n} > \alpha \text{ i.o.}\right) = 1.$$

If $\alpha > 1$, the sum is convergent, and $\sum_n \mathbf{P}(X_n > \alpha \log n) < \infty$, thus

$$\mathbf{P}\left(\frac{X_n}{\log n} > \alpha \text{ i.o.}\right) = 0.$$

We can express the same thing in terms of $\limsup$, so

$$\mathbf{P}\left(\limsup_n \frac{X_n}{\log n} > \alpha\right) = \begin{cases} 0 & , \text{ if } \alpha > 1 \\ 1 & , \text{ if } \alpha \leq 1. \end{cases}$$

Since for all $\alpha \leq 1$ we have that $\mathbf{P}\left(\limsup_n \frac{X_n}{\log n} > \alpha\right) = 1$, then we necessarily have

$$\mathbf{P}\left(\limsup_n \frac{X_n}{\log n} \geq 1\right) = 1.$$

Take $\alpha = 1 + \frac{1}{k}$ and look at the other implication: $\mathbf{P}\left(\limsup_n \frac{X_n}{\log n} > 1 + \frac{1}{k}\right) = 0$, and this happens for all $k \in \mathbb{N}$. But we can write

$$\left\{\limsup_n \frac{X_n}{\log n} > 1\right\} = \bigcup_{k \in \mathbb{N}} \left\{\limsup_n \frac{X_n}{\log n} > 1 + \frac{1}{k}\right\},$$

and since any countable union of null sets is itself a null set, the probability of the event on the left must be zero. Therefore, $\limsup_n \frac{X_n}{\log n} \leq 1$ a.s., and combining with the finding above we have

$$\limsup_n \frac{X_n}{\log n} = 1, \quad a.s.$$

This is very interesting since, as we will see in the chapter dedicated to the Poisson process, these X_n are the inter-arrival times of this process. The example above tells us that if we look at the realizations of such a process, then they form a sequence of numbers that has the upper limiting point equal to 1, or, put differently, there is no subsequence of inter-arrival times that in the limit is greater than the $\log n$. ■

■ EXAMPLE 2.13

Let X and Y be independent random variables with the same distribution function $F(x)$. U is the value of X or Y, which is closest to zero. Find the distribution function of U.

First note that since X and Y are independent, $\mathbf{P}(X \leq x, Y \leq y) = F(x)F(y)$. Let $u \in \mathbb{R}$. Then

$$\mathbf{P}(U \leq u) = \mathbf{P}(U \leq u, X \leq u, Y \leq u) + \mathbf{P}(U \leq u, X > u, Y \leq u)$$
$$+ \mathbf{P}(U \leq u, X \leq u, Y > u) + \mathbf{P}(U \leq u, X > u, Y > u).$$

Denote the four terms above by (1) to (4). Let us analyze them separately. For any $u \in \mathbb{R}$ we can calculate (1) and (4) in the following way:

$$(1) = \mathbf{P}(U \leq u, X \leq u, Y \leq u) = \mathbf{P}(U \leq u, X \leq u, Y \leq u, \{X \text{ closer to } 0 \text{ than } Y\})$$
$$+ \mathbf{P}(U \leq u, X \leq u, Y \leq u, \{Y \text{ closer to } 0 \text{ than } X\})$$
$$= \mathbf{P}(X \leq u, Y \leq u, \{X \text{ closer to } 0 \text{ than } Y\})$$
$$+ \mathbf{P}(X \leq u, Y \leq u, \{Y \text{ closer to } 0 \text{ than } X\})$$
$$= \mathbf{P}(X \leq u, Y \leq u) = F^2(u).$$

Using a similar argument, we will eventually obtain

$$(4) = \mathbf{P}(U \leq u, X > u, Y > u) = \cdots = 0.$$

For the other two terms, we need to consider separately the cases when $u > 0$ and when $u \leq 0$ to determine which of the variables X or Y is closer to 0. First, let $u \leq 0$.

Note that $-u \geq 0$ in this case. We have

$$(2) = \mathbf{P}(U \leq u, X > u, Y \leq u)$$
$$= \underbrace{\mathbf{P}(U \leq u, X \in (u, -u), Y \leq u)}_{=0 \ (U = X \text{ since is closer to } 0)} + \mathbf{P}(U \leq u, X \geq -u, Y \leq u)$$
$$= \underbrace{\mathbf{P}(U \leq u, X \geq -u, Y \leq u, X \leq -Y)}_{=0 \ (U = X \text{ since is closer to } 0)} + \mathbf{P}(U \leq u, X \geq -u, Y \leq u, X > -Y)$$
$$= \mathbf{P}(Y \leq u, X \geq -u, X > -Y).$$

In an absolutely similar way, we obtain

$$(3) = \cdots = \mathbf{P}(X \leq u, Y \geq -u, -X < Y).$$

Note that because X and Y are independent and have the same distribution, the expressions (2) and (3) are the same and both are equal to

$$= \int_{-\infty}^{u} \int_{\max\{-x, -u\}}^{\infty} dF(y)dF(x) \overset{x < u \Rightarrow -x > -u}{=} \int_{-\infty}^{u} \int_{-x}^{\infty} dF(y)dF(x)$$
$$= \int_{-\infty}^{u} (1 - F(-x)) \, dF(x) = F(u) - \int_{-\infty}^{u} F(-x)dF(x).$$

In the case when $u > 0$, we just repeat the reasoning:

$$(2) = \mathbf{P}(U \leq u, X > u, Y \leq u)$$
$$= \mathbf{P}(U \leq u, X > u, Y \in (-u, u)) + \mathbf{P}(U \leq u, X > u, Y \leq -u)$$
$$= \mathbf{P}(Y \leq u, X > u, Y \in (-u, u)) + \underbrace{\mathbf{P}(U \leq u, X > u, Y \leq -u, X \leq -Y)}_{=0 \ (U = X \text{ since is closer to } 0)}$$
$$+ \mathbf{P}(U \leq u, X > u, Y \leq -u, X > -Y)$$
$$= \mathbf{P}(Y \in (-u, u), X > u) + \mathbf{P}(Y \leq -u, X > u, X > -Y)$$
$$= (F(u) - F(-u))(1 - F(u)) + \int_{-\infty}^{-u} \int_{\max\{u, -y\}}^{\infty} dF(x)dF(y)$$
$$= (F(u) - F(-u))(1 - F(u)) + F(u) - \int_{-\infty}^{-u} F(-y)dF(y).$$

Thus putting it all together, we obtain the very complicated distribution function for U:

$$F_U(u) = \begin{cases} F^2(u) + 2F(u) - 2\int_{-\infty}^{u} F(-x)dF(x), & \text{if } u \leq 0 \\ 4F(u) - 2F(-u) - F^2(u) - F(u)F(-u) \\ \quad -2\int_{-\infty}^{-u} F(-x)dF(x), & \text{if } u > 0. \end{cases}$$

∎

■ **EXAMPLE 2.14**

Let W be a unit exponential random variable. Let U be uniformly distributed on $[0, 1]$. Note that the point $(U, 1 - U)$ is uniformly distributed on the line segment between $(0, 1)$ and $(1, 0)$ in the plane $\mathbb{R}^2$. Now construct the line through the origin and the point $(U, 1-U)$. Let (X, Y) be the point where this line intersects the circle centered at the origin and with radius W. Find the joint density of X and Y.

First note that the point of coordinates (X, Y) is uniquely determined from W and U. How?

$$\begin{cases} X^2 + Y^2 = W^2 & \text{on the circle of radius } W \\ Y = \frac{1-U}{U} X & \text{on the line connecting } (0,0) \text{ and } (U, 1-U) \end{cases} \qquad (2.7)$$

Solving the equations above and using the fact that the point is in the first quadrant $(x > 0, y > 0)$, we find the transformation

$$(U, W) \longrightarrow \left(X = \frac{WU}{\sqrt{U^2 + (1-U)^2}}, Y = \frac{W(1-U)}{\sqrt{U^2 + (1-U)^2}} \right)$$

The inverse transformation is $g^{-1}(x, y) = (\frac{x}{x+y}, \sqrt{x^2 + y^2})$.
The Jacobian is then

$$J_{g^{-1}}(x, y) = \begin{vmatrix} \frac{y}{(x+y)^2} & -\frac{x}{(x+y)^2} \\ \frac{x}{\sqrt{x^2+y^2}} & \frac{y}{\sqrt{x^2+y^2}} \end{vmatrix} = \frac{x^2 + y^2}{(x+y)^2 \sqrt{x^2 + y^2}} = \frac{\sqrt{x^2 + y^2}}{(x+y)^2}.$$

The joint density function of (U, W) is

$$f_{U,W}(u, w) = f_U(u) f_W(u) = e^{-w} \mathbf{1}_{(0,1)}(u) \mathbf{1}_{(0,\infty)}(w).$$

Therefore,

$$f_{X,Y}(x, y) = f_{U,W}\left(\frac{x}{x+y}, \sqrt{x^2 + y^2}\right) |J_{g^{-1}}| = e^{-\sqrt{x^2+y^2}} \frac{\sqrt{x^2 + y^2}}{(x+y)^2}$$

for $x > 0, y > 0$.

Problems

2.1   Prove Proposition 2.6. That is, prove that the function F in Definition 2.5 is increasing, right continuous, and taking values in the interval $[0, 1]$, using only Proposition 1.20 on page 21.

2.2   Show that any piecewise constant function is Borel measurable. (See the description of piecewise constant functions in Definition 2.9.)

2.3   Give an example of two distinct random variables with the same distribution function.

2.4   Buffon's needle problem.
Suppose that a needle is tossed at random onto a plane ruled with parallel lines a distance L apart, where by a "needle" we mean a line segment of length $l \leq L$.
What is the probability of the needle intersecting one of the parallel lines?

Hint: Consider the angle that is made by the needle with the parallel lines as a random variable α uniformly distributed in the interval $[0, 2\pi]$ and the position of the midpoint of the needle as another random variable ξ also uniform on the interval $[0, L]$. Then express the condition "needle intersects the parallel lines" in terms of the position of the midpoint of the needle and the angle α. Do a calculation similar to Example 2.5.

2.5   A random variable X has distribution function

$$F(x) = a + b \arctan \frac{x}{2} \quad, \quad -\infty < x < \infty$$

Find
   a) The constants a and b,
   b) The probability density function of X.

2.6   What is the probability that two randomly chosen numbers between 0 and 1 will have a sum no greater than 1 and a product no greater than $\frac{15}{64}$?

2.7   We know that the random variables X and Y have joint density $f(x, y)$. Assume that $\mathbf{P}(Y = 0) = 0$. Find the densities of the following variables:

   a) $X + Y$
   b) $X - Y$
   c) XY
   d) $\frac{X}{Y}$.

2.8   A density function is defined as

$$f(x, y) = \begin{cases} K(x + 2y) & \text{if } 0 < y < 1 \text{and} 0 < x < 2 \\ 0 & \text{otherwise.} \end{cases}$$

   a) Find the value of K which makes the function a probability density function.
   b) Find the marginal distributions of X and Y.
   c) Find the joint cdf of (X, Y).
   d) Find the pdf of the random variable

$$Z = \frac{5}{(X + 1)^2}.$$

2.9   The random variables X and Y have the joint distribution

		X		
		1	2	3
	12	$\frac{1}{12}$	$\frac{1}{6}$	$\frac{1}{12}$
Y	13	$\frac{1}{3}$	0	0
	44	$\frac{1}{9}$	$\frac{1}{9}$	$\frac{1}{9}$

a) Calculate the marginal distributions of X and Y.
b) Show that the random variables X and Y are dependent.
c) Find two random variables U and V which have the same marginal distributions as X and Y but are independent.

2.10   Suppose we toss a fair coin repeatedly. Let X denote the number of trials to get the first head and let Y the number needed to get two heads in repeated tosses. Are the two variables independent?

2.11   Every morning, John leaves for work between 7:00 and 7:30. The trip always takes between 40 and 50 min. Let X denote the time of departure and let Y denote the travel time. Assume that both variables are independent and both are uniformly distributed on the respective intervals. Find the probability that John arrives at work before 8:00.

2.12   Choose a point A at random in the interval $[0, 1]$. Let L_1 (respectively, L_2) be the length of the bigger (respectively, smaller) segment determined by A on $[0, 1]$. Calculate

a) $\mathbf{P}\left(L_1 \leq x\right)$ for $x \in \mathbb{R}$.
b) $\mathbf{P}\left(L_2 \leq x\right)$ for $x \in \mathbb{R}$.

2.13   Two friends decide to meet at the Castle Gate of Stevens Institute. They each arrive at that spot at some random time between a and $a + T$. They each wait for 15 min, and then leave if the other does not appear. What is the probability that they meet?

2.14   Let $X_1, X_2, \ldots, X_n$ be independent $U(0, 1)$ random variables. Let $M = \max_{1 \leq i \leq n} X_i$. Calculate the distribution function of M.

2.15   The random variable whose probability density function is given by

$$f(x) = \begin{cases} \frac{1}{2}\lambda e^{\lambda x} & , \quad \text{if } x \leq 0 \\ \frac{1}{2}\lambda e^{-\lambda x} & , \quad \text{if } x > 0, \end{cases}$$

is said to have a Laplace, sometimes called a *double exponential*, distribution.

a) Verify that the density above defines a proper probability distribution.
b) Find the distribution function $F(x)$ for a Laplace random variable.

Now, let X and Y be independent exponential random variables with parameter λ. Let I be independent of X and Y and equally likely to be 1 or -1.

c) Show that $X - Y$ is a Laplace random variable.

d) Show that IX is a Laplace random variable.

e) Show that W is a Laplace random variable, where

$$W = \begin{cases} X & , & \text{if } I = 1 \\ -Y & , & \text{if } I = -1. \end{cases}$$

2.16 Give a proof of Lemma 2.23 on page 76.

2.17 Let X and Y be independent, $N(0, 1)$ random variables.

a) Verify that X^2 is distributed as a χ_1^2 random variable.

b) Find $\mathbf{P}(X^2 < 1)$.

c) Find the distribution of $X^2 + Y^2$.

d) Find $\mathbf{P}(X^2 + Y^2 < 1)$.

2.18 Let (X, Y) have the joint density $f(x, y)$. Let $U = aX + b$ and $V = cY + d$, where the constants a, b, c, d are fixed and $a > 0$, $c > 0$. Show that the joint density of U and V is

$$f_{U,V}(u, v) = \frac{1}{ac} f\left(\frac{u - b}{a}, \frac{v - d}{c}\right).$$

2.19 A bacterial solution contains two types of antiviral cell: type A and type B. Let X denote the lifetime of cell type A and Y denote the lifetime of cell type B. Whenever they are grown in the same culture, the death of one type will cause the death of the other type. Therefore, it is not possible to observe both variables. Instead, we may only observe the variables

$$Z = \min X, Y$$

and

$$W = \begin{cases} 1 & \text{if } Z = X \\ 0 & \text{if } Z = Y. \end{cases}$$

In other words, we can observe the lifetime of the organism which dies first and we know which is that organism.

a) Find the joint distribution of Z and W.

b) Show that Z and W are independent random variables.

Hint: Show that $\mathbf{P}(Z \leq z \mid W = i) = \mathbf{P}(Z \leq z)$ for both $i = 0$ and $i = 1$.

2.20 All children in Bulgaria are given IQ tests at ages 8 and 16. Let X be the IQ score at age 8 and let Y be the IQ score for a randomly chosen Bulgarian 16-year-old. The joint distribution of X and Y can be described as follows. X is normal with mean 100 and standard deviation 15. Given that $X = x$, the conditional distribution of Y is normal with mean $0.8x + 30$ and standard deviation 9.

Among Bulgarian 16-year-olds with $Y = 120$, what fraction have $X \geq 120$?

2.21   Find a density function $f(x, y)$ such that if (X, Y) has density f then $X^2 + Y^2$ is uniformly distributed on (0,10).

2.22   Let X be a unit exponential random variable (with density $f(x) = e^{-x}, x > 0$) and let Y be an independent $U[0, 1]$ random variable. Find the density of $T = Y/X$.

2.23   Let X, Y be independent $N(0, 1)$.
a) Calculate the distribution of $\frac{X}{X+Y}$. This is called the Cauchy distribution.
b) Find the distribution of $\frac{X}{|Y|}$.

2.24   We generate a point in the plane according to the following algorithm:
Step 1: Generate R^2 which is ξ_2^2 (chi squared with 2 degrees of freedom)
Step 2: Generate θ according to the uniform distribution on the interval $(0, 2\pi)$
Step 3: Let the coordinates of the point be

$$(X, Y) = (R \cos \theta, R \sin \theta).$$

Find the joint distribution of (X, Y).

2.25   You have two opponents A and B with whom you **alternately** play games. Whenever you play A, you win with probability p_A; whenever you play B, you win with probability p_B, where $p_B > p_A$. If your objective is to *minimize the number of games you need to play to win two in a row*, should you start playing with A or with B?

2.26   Let X_1 and X_2 be independent, unit exponential random variables (so the common density is $f(x) = e^{-x}, x > 0$). Define $Y_1 = X_1 - X_2$ and $Y_2 = X_1/(X_1 - X_2)$. Find the joint density of Y_1 and Y_2.

2.27   Let Y be a LogN(0,1) random variable, that is,

$$f_Y(y) = \frac{1}{y\sqrt{2\pi}e^{-\frac{(\log y)^2}{2}}}, \quad y > 0.$$

Show that $X = log(Y)$ is normally distributed.

2.28   Let X be a random variable with a normal distribution with parameters μ and σ. Show that $Y = e^X$ has a log-normal distribution.

2.29   Depending on the weather conditions, the probability that an egg hatches is a random variable P distributed according to a Beta distribution with parameters a and b. A hen deposits 20 eggs and it is reasonable to assume that the total number of eggs that hatch is a random variable X, which has a binomial distribution with parameters 20 and P.
a) Calculate the joint distribution of X and P.
b) Calculate the marginal distribution of X (this distribution is called the *beta-binomial* distribution).

CHAPTER 3

APPLIED CHAPTER: GENERATING RANDOM VARIABLES

In this chapter we talk about methods used for simulating random variables. In to-day's world, where computers are part of any scientific activity, it is very important to know how to simulate a random experiment to find out what expectations one may have about the results of the phenomenon. We shall see later that the Central Limit Theorem (Theorem 7.43) and the Monte Carlo method allow us to draw conclusions about the expectations even if the distributions involved are very complex.

We assume as given a Uniform(0,1) random number generator. Any software can produce such a uniform random variable and the typical name for a uniform random number is RAND. For more current references and a very efficient way to gener-ate exponential and normal random variables without going to uniform, we refer the reader to Rubin and Johnson (2006). The ziggurat method developed by Marsaglia and Tsang (2000) remains one of the most efficient ways to produce normals and it is used by MATLAB®. The Mersene twister is an efficient way to create uniform random numbers. In this chapter, we will present pseudo-code and code written in R for some of these random variable generation methods. R is an open source program which, in my opinion, is one of the strongest statistical programs ever developed.

Probability and Stochastic Processes, First Edition. Ionuţ Florescu

3.1   Generating One-Dimensional Random Variables by Inverting the cdf

Let X be a one-dimensional random variable defined on any probability space $(\Omega, \mathscr{F}, \mathbf{P})$ with distribution function $F(x) = \mathbf{P}(X \leq x)$. All the distribution generation methods in this section are based on the following lemma:

Lemma 3.1 *The random variable* $U = F(X)$ *is distributed as a* $U(0,1)$ *random variable. If we let* $F^{-1}(u)$ *denote the inverse function, that is,*

$$F^{-1}(u) = \{x \in \mathbb{R} \mid F(x) = u\},$$

then the variable $F^{-1}(U)$ *has the same distribution as* X.

Note $F^{-1}(u)$ as defined in the lemma is a set. If we wish to have a single number representing the set, we obtain the definition of the quantile. Recall the quantile discussion in Section 2.3 on page 66 and the issue mentioned there in the case when the distribution function is not continuous.

Proof: The proof is simple if F is a bijective function. Note that in this case we have

$$\mathbf{P}(U \leq u) = \mathbf{P}(F(X) \leq u)$$

But recall that F is a probability itself, so the result above is zero if $u < 0$ and 1 if $u \geq 1$. If $0 < u < 1$, since F is an increasing function, we can write

$$\mathbf{P}(U \leq u) = \mathbf{P}(X \leq F^{-1}(u)) = F(F^{-1}(u)) = u$$

and this is the distribution of a $U(0,1)$ random variable.

   If F is not bijective, the proof still holds but we have to work with sets. The relevant case is, once again, $0 < u < 1$. Recall that F is increasing. Because of this and using the definition of F^{-1} above, we have

$$F^{-1}\left((-\infty, u]\right) = \{x \in \mathbb{R} \mid F(x) \leq u\}$$

If the set $F^{-1}(u)$ has only one element (F is bijective at u), there is no problem and the set above is just $(-\infty, F^{-1}(u)]$. Therefore the same derivation above works. If $F^{-1}(u) = \{x \in \mathbb{R} \mid F(x) = u\}$ has more than one element, then let $x_{\max}$ be the maximum element in the set. This element exists and it is in the set since $u < 1$. We may write

$$F^{-1}\left((-\infty, u]\right) = (-\infty, x_{\max}].$$

Thus we have

$$\begin{aligned}
\mathbf{P}(U \leq u) = \mathbf{P}(F(X) \leq u) &= \mathbf{P}\left(X \in F^{-1}\left((-\infty, u]\right)\right) \\
&= \mathbf{P}\left(X \in (-\infty, x_{\max}]\right) = \mathbf{P} \circ X^{-1}((-\infty, x_{\max}]) \\
&= F(x_{\max})
\end{aligned}$$

by the definition of distribution function. Now, recall that $x_{\max} \in F^{-1}(u)$, thus $F(x_{\max}) = u$. Once again, we reach the distribution of a uniform random variable. ∎

This lemma is very useful for generating random variables with prescribed distribution. We only need to figure out the inverse cumulative distribution function (cdf) function F^{-1} to generate random variables with any distribution starting with the uniform density. This approach works best when the distribution function F has an analytical formula and, furthermore, when F is a bijective function. One of the best examples of this situation is the exponential distribution.

◨ EXAMPLE 3.1 Generating an exponential random variable

Suppose we want to generate an Exponential(λ) random variable, that is, a variable with density:

$$f(x) = \lambda e^{-\lambda x} \mathbf{1}_{\{x > 0\}}$$

Note that the expectation of this random variable is $1/\lambda$. This distribution can also be parameterized using $\lambda = 1/\theta$, in which case the expectation will be θ. The two formulations are equivalent.

We may calculate the distribution function in this case as

$$F(x) = (1 - e^{-\lambda x}) \mathbf{1}_{\{x > 0\}}$$

We need to restrict this function to $F : (0, \infty) \to (0, 1)$ to have a bijection. In this case, for any $y \in (0, 1)$, the inverse is calculated as

$$F(x) = 1 - e^{-\lambda x} = y \Rightarrow x = -\frac{1}{\lambda} \log(1 - y)$$

$$\Rightarrow F^{-1}(y) = -\frac{1}{\lambda} \log(1 - y)$$

So, to generate an Exponential(λ) random variable, first generate U a Uniform(0,1) random variable and simply calculate

$$-\frac{1}{\lambda} \log(1 - U);$$

this will have the desired distribution.

As a note, a further simplification may be made since $1 - U$ has the same distribution as U, and we obtain the same exponential distribution by taking

$$-\frac{1}{\lambda} \log U.$$

Note that we should use one of the forms $-\frac{1}{\lambda} \log U$ or $-\frac{1}{\lambda} \log(1 - U)$, but not both, since the two variables are related and therefore not independent.

For all discrete random variables, the distribution function is a step function. In this case, the F is not bijective, so we need to restrict it somehow to obtain the desired distribution. The main issue is that the function is not surjective, so we need to know what to do when a uniform is generated.

■ **EXAMPLE 3.2   Rolling the die**

Suppose we want to generate the rolls of a fair six-sided die. We know that the probability distribution function is (see example 2.4)

$$F(x) = \begin{cases} 0 & \text{if } x < 1 \\ i/6 & \text{if } x \in [i, i+1) \text{ with } i = 1, \cdots, 5 \\ 1 & \text{if } x \geq 6 \end{cases}$$

The inverse function is then

$$F^{-1}(0) = (-\infty, 1)$$

$$F^{-1}\left(\frac{1}{6}\right) = [1, 2)$$

$$\cdots\cdots$$

$$F^{-1}\left(\frac{5}{6}\right) = [5, 6)$$

$$F^{-1}(1) = [6, \infty)$$

We can pick a point in the codomain but that would not help since the inverse function will only be defined on the discrete set $\{0, 1/6, \ldots, 5/6, 1\}$. Instead, we extend the inverse function to $(0, 1)$ in the following way:

$$F^{-1}(y) = \begin{cases} 1 & \text{if } y \in \left(0, \frac{1}{6}\right) \\ i+1 & \text{if } y \in \left[\frac{i}{6}, \frac{i+1}{6}\right) \text{ with } i = 1, \cdots, 5 \end{cases}$$

Thus, we first generate U a Uniform(0,1) random variable. Depending on its value, the roll of the die is simulated as

$$Y = i + 1, \quad \text{if } U \in \left[\frac{i}{6}, \frac{i+1}{6}\right) \text{ with } i = 0, \cdots, 5$$

■ **EXAMPLE 3.3   Generating any discrete random variable with finite number of outcomes**

The example above can be easily generalized to any discrete probability distribution. Suppose we need to generate the outcomes of the discrete random variable Y which takes n values $a_1, \ldots, a_n$ each with probability $p_1, \ldots, p_n$, respectively, so that $\sum_{j=1}^{n} p_j = 1$.

To generate such outcomes, we first generate a random variable U as a Uniform(0,1) random variable. Then we find the index j such that

$$p_1 + \cdots + p_{j-1} \leq U < p_1 + \cdots + p_{j-1} + p_j.$$

The generated value of the Y variable is the outcome a_j.

Note that, theoretically, the case when the generated uniform values of U are exactly equal to p_j does not matter since the distribution is continuous and the probability of this event happening is zero. However, in practice such events do matter since the cycle used to generate the random variable is finite and thus the probability of the event is not zero – it is extremely small but not zero. This is dealt with by throwing away 1 if it is generated and keeping the rest of the algorithm as above.

Remark 3.2 *The previous example also covers the most commonly encountered need for generating random variables – the tossing of a coin or generating a Bernoulli(p) random variable. Specifically, generate U a Uniform(0,1). If $U < p$, output 1, else output 0.*

3.2   Generating One-Dimensional Normal Random Variables

Generating normal (Gaussian) random variables is important because this distribution is the most widely encountered distribution in practice. In the Monte Carlo methods, one needs to generate millions of normally distributed random numbers. That means that the precision with which these numbers are generated is quite important. Imagine that one in 1000 numbers is not generated properly. This translates on average into about 1000 numbers being bad in 1 million numbers simulated. Depending on the simulation complexity, a single Monte Carlo path may need 1 million generated numbers, and therefore the simulated path has 1000 places where the trajectory is not what it ought to be.

Let us first remark that, if we know how to generate X a standard normal variable with mean 0 and variance 1, that is, with density

$$f(x) = \frac{1}{\sqrt{2\pi}} e^{-\frac{x^2}{2}},$$

then we know how to generate any normal variable Y with mean μ and standard deviation σ. This is accomplished by simply taking

$$Y = \mu + \sigma X.$$

Thus, to generate any normal it is enough to learn how to generate $N(0, 1)$ random variables. The inversion methodology presented in the previous section cannot be applied directly since the normal cdf does not have an explicit functional form and therefore inverting it directly is impossible.

As we are writing this, we need to mention that in fact one of the fastest and better methods (the default in the R programming language) uses a variant of the inversion method. Specifically, the algorithm developed by Wichura (1988) calculates quantiles corresponding to the generated probability values. The algorithm has two subroutines to deal with the hard-to-estimate quantiles from the tails of the Gaussian distribution.

First, the algorithm generates p using a Uniform(0,1) distribution. Then it calculates the corresponding normally distributed value z_p by inverting the distribution function

$$p = \int_{-\infty}^{z_p} \frac{1}{\sqrt{2\pi}} e^{-x^2/2} dx = \Phi(z_p),$$

$z_p = \Phi^{-1}(p)$. The respective subroutines PPND7 or PPND16 are chosen depending on the generated uniform value p, more specifically if $|p - 0.5| \leq 0.425$, respectively, greater than 0.425. These routines are polynomial approximations of the inverse function $\Phi^{-1}(\cdot)$. The algorithm has excellent precision (of the order 10^{-16}) and it runs relatively fast (only requires a logarithmic operation besides the polynomial operations).

More traditional methods of generating normally distributed numbers are presented below. All of them take advantage of transformations of random variables. Some are particular cases of more general methodology presented later in this chapter.

Taking advantage of the central limit theorem to generate a number n of Uniform(0,1) random numbers. Then calculate their sum $Y = \sum_{i=1}^{n} U_i$. The exact distribution of Y is the so-called *Irwin–Hall* distribution, named after Joseph Oscar Irwin and Philip Hall, which has the probability density

$$f_Y(y) = \frac{1}{2(n-1)!} \sum_{k=0}^{n} (-1)^k \binom{n}{k} (y-k)^{n-1} \mathrm{sign}(y-k),$$

where $\mathrm{sign}(x)$ is the sign function. The sign function is defined on $(-\infty, 0) \cup (0, \infty)$ as

$$\mathrm{sign}(x) = \frac{|x|}{x} = \begin{cases} 1 & \text{if } x > 0 \\ -1 & \text{if } x < 0 \end{cases}$$

This Irwin–Hall distribution has mean $n/2$ and variance $n/12$, as it is very easy to verify using the individual values for the uniforms. However, the Central Limit Theorem (Theorem 7.43) guarantees that, as $n \to \infty$, the distribution of Y approaches the normal distribution. So a simple algorithm will generate, say, 12 uniforms, and then it would calculate the standardized random variable

$$Y = \frac{\sum_{i=1}^{12} U_i - 12/2}{\sqrt{12/12}} = \sum_{i=1}^{12} U_i - 6$$

This new random variable is therefore approximately distributed as a N(0,1) random variable. Since we use the 12 uniforms, the range of the generated values is $[-6, 6]$, which is different from the real normal random numbers which are distributed on $\mathbb{R}$. Of course, taking a larger n will produce normals with better precision. However, recall that we need n uniform variables to create *one* normally distributed number, so the algorithm slows down considerably as n gets larger.

The Box–Muller method   This method of generating normally distributed random numbers is named after George Edward Pelham Box and Mervin Edgar Muller, who developed the algorithm in 1958. The algorithm uses two independent Uniform(0,1) random numbers U and V. Then two random variables X and Y are calculated using

$$X = \sqrt{-2 \ln U} \cos 2\pi V$$

$$Y = \sqrt{-2 \ln U} \sin 2\pi V$$

Then the two random numbers X and Y have the standard normal distribution, and are independent.

This result is easy to derive since, for a bivariate normal random vector (XY), the variable $X^2 + Y^2$ is distributed as a chi-squared random variable with two degrees of freedom. A chi-squared with two degrees of freedom is in fact the same as the exponential distribution with parameter $1/2$; note that, in fact, the generated quantity $-2 \ln U$ has this distribution. Furthermore, the projection of this quantity on the two axes is determined by the angle that is made by the point and the origin, and this angle is chosen by the random variable V. The angle is uniform between $[0, \pi]$.

From the computational perspective, we have good and bad news. The good news is that unlike the previous method the Box–Muller method uses two independent uniforms and produces two independent normal random numbers which may both be used in the algorithms. The bad news is that the method requires a logarithm operation, a square root operation, and two trigonometric function calculations, and these may be quite slow when repeated many times (recall that we need a ton of random numbers in any simulations).

The polar rejection method   This method is due to Marsaglia, and is, in fact, a simple modification of the Box–Muller algorithm. Recall that one of the things slowing down the Box–Muller method was the calculation of the trigonometric functions sin and cos. The polar rejection method avoids the trigonometric functions calculations, replacing them with the rejection sampling method presented in the next section.

In this method, two random numbers U and V are drawn from the Uniform$(-1, 1)$ distribution. Note the difference from the previous uniform numbers. Then, the quantity

$$S = U^2 + V^2$$

is calculated. If S is greater than or equal to 1, then the method starts over by regenerating two uniforms until it generates two uniforms with the desired property. Once those are obtained, the numbers

$$X = U\sqrt{-2\frac{\ln S}{S}}$$

$$Y = V\sqrt{-2\frac{\ln S}{S}}$$

are calculated. These X and Y are independent, standard, normal random numbers.. Intuitively, we need to generate coordinates of the points inside the unit circle, which is why we perform the accept/reject step.

Marsaglia's polar rejection method is both faster and more accurate (because it does not require the approximation of three functions) than the Box–Muller method. The drawback of the method is that, unlike the Muller method, it may require multiple sets of uniform numbers until it reaches values that are not rejected. In fact, for the one-dimensional case presented, we can easily calculate the probability that the generated pair is accepted. This is the probability that a two-dimensional uniform vector falls inside the unit circle and it is the ratio of the area of this circle over the square with center at origin and sides 2 (from -1 to 1). Mathematically,

$$\mathbf{P}(S < 1) = \mathbf{P}(U^2 + V^2 < 1) = \frac{\pi}{2^2} = \pi/4 \approx 0.79.$$

Therefore, about 79% of the generated pairs fall inside the circle and are used in the algorithm. Note that the number of times one needs to generate these pairs to obtain a usable one is easily seen as having a $Geometric(0.79)$ distribution.

One of the most popular methods of generating normals today is the ziggurat algorithm (due to Marsaglia and Tsang (2000)). This is the default algorithm in MATLAB® for generating normals. We shall talk about this algorithm after we present the rejection sampling method in the next section. The ziggurat algorithm is based on this method.

3.3   Generating Random Variables. Rejection Sampling Method

The polar method is a particular case of the rejection sampling method. In rejection sampling (also named the *accept-reject method*), the objective is to generate a random variable X having the *known* density function $f(x)$. The idea of this method is to use a different but easy-to-generate-from distribution $g(x)$. The method is very simple and was originally presented by John von Neumann. You can see the idea of this algorithm in the Buffon needle problem (throwing the needle and accepting or rejecting depending whether the needle touches the lines or not).

First determine a constant M such that

$$\frac{f(x)}{g(x)} < M, \quad \forall x$$

Once such M is determined, the algorithm is as follows:

Step 1. Generate a random variable Y from the distribution $g(x)$.

Step 2. Accept $X = Y$ with probability $f(Y)/Mg(Y)$. If reject, go back to step 1.

The accept-reject step can be easily accomplished by a Bernoulli random variable. Specifically, step 2 is: generate $U \sim \text{Uniform}(0, 1)$ and accept if

$$U < \frac{f(Y)}{Mg(Y)},$$

go back to step 1 if reject.

Proposition 3.3 *The random variable X created by the rejection sampling algorithm above has the desired density $f(x)$.*

Proof: Let N be the number of necessary iterations to obtain the final number X. Let us calculate the distribution of X. Since each trial is independent, we have

$$\mathbf{P}\{X \le x\} = \mathbf{P}\left\{Y \le x \,\middle|\, U \le \frac{f(Y)}{Mg(Y)}\right\} = \frac{\mathbf{P}\left(\{Y \le x\} \cap \left\{U \le \frac{f(Y)}{Mg(Y)}\right\}\right)}{\mathbf{P}\left\{U \le \frac{f(Y)}{Mg(Y)}\right\}}$$

Now, the numerator is

$$\mathbf{P}\left(\left\{U \le \frac{f(Y)}{Mg(Y)}\right\} \mid \{Y \le x\}\right) \mathbf{P}\left(\{Y \le x\}\right)$$

$$= \int_{-\infty}^{x} \mathbf{P}\left(\left\{U \le \frac{f(y)}{Mg(y)}\right\} \mid \{Y = y\}\right) g(y)\,dy$$

$$= \int_{-\infty}^{x} \frac{f(y)}{Mg(y)} g(y)\,dy = \frac{1}{M} \int_{-\infty}^{x} f(y)\,dy$$

Similarly, the denominator is

$$\mathbf{P}\left\{U \le \frac{f(Y)}{Mg(Y)}\right\} = \int_{-\infty}^{\infty} \mathbf{P}\left\{U \le \frac{f(y)}{Mg(y)} \mid \{Y = y\}\right\} g(y)\,dy$$

$$= \int_{-\infty}^{\infty} \frac{f(y)}{Mg(y)} g(y)\,dy = \frac{1}{M} \int_{-\infty}^{\infty} f(y)\,dy = \frac{1}{M}$$

Now taking the ratio shows that X has the desired distribution.   ∎

Note that calculating the denominator in the proof above shows that the probability of accepting the generated number is always $1/M$. So, if the constant M is close to 1, then the method works very efficiently. However, this is dependent on the shape of the densities f and g. If the density g is close in shape to f, then the method works very well. Otherwise, a large number of generated variates are needed to obtain one random number with density f.

Corollary 3.4 (Slight generalization) *Suppose that we need to generate from density $f(x) = Cf_1(x)$, where we know the functional form f_1 and C is a normalizing constant, potentially unknown. Then suppose we can find a density $g(x)$ easy to generate from and a constant M such that*

$$\frac{f(x)}{g(x)} < M, \quad \forall x.$$

Then the rejection sampling procedure described above will create random numbers with density f.

The corollary can be proved in exactly the same way as the main proposition. Sometimes, the constant is hard to calculate and this is why the corollary is useful in practice.

Marsaglia's Ziggurat method

The rejection sampling is a pretty basic method. One needs to find a distribution $g(x)$ such that

1. it is easy to generate random numbers, and

2. it dominates the target density $f(x)$ – eventually using a constant M.

So, the first basic idea when trying to apply it for the normal random variables' (when $f(x)$ is the $N(0, 1)$) pdf is to try taking $g(x)$ a Uniform density. This density is very easy to generate, and indeed Figure 3.1 illustrates an example where we use a scaled uniform (dashed lines) which dominates a normal density.

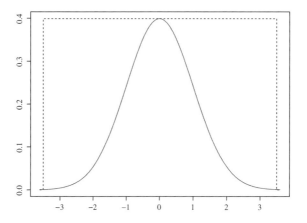

Figure 3.1   Rejection sampling using a basic Uniform.

However, there are two problems with this approach:

1. Recall that the probability with which we accept the generated number is equal to the ratio between the two pdfs at the respective generated number. Looking at the Figure 3.1, it is easy to see that this ratio gets drastically close to 0 if the generated number is in the normal tails. Thus, the algorithm would get stuck trying to generate such numbers.

2. Another problem is the tails of the normal. The Uniform generates on a fixed interval while the normal takes values all the way to $\pm\infty$.

The ziggurat method deals with both these problems separately. It also uses several tricks to simplify as well as speed up the generation of numbers. From my personal experience, I am not impressed with the implementation in MATLAB®. Nevertheless, let us investigate the algorithm.

The idea   The ziggurat method is actually general and can be applied to any density. The name ziggurat comes from the name of the shape of the temples constructed by Sumerians, Babylonians, and Assyrians. This is a pyramidal structure where the sides of the pyramid are made of steps. The Mayans of Central America are famous for building this type of pyramids. As we shall see in the next figures, the construction resembles this type of pyramid.

In the case of the normal distribution, since the distribution is symmetric, a first simplification is made by restricting to the positive axis. If we generate a positive normal number, then we just generate its sign by generating a separate $+1$ or -1 each with probability $1/2$.

The method first proceeds by constructing a dominating distribution $g(x)$ in the way plotted in Figure 3.2. The plot contains a g constructed using $n = 8$ areas. We shall describe the algorithm using this g to understand better the idea. A more realistic situation would use 256 bars. The resulting g distribution is plotted in Figure 3.3.

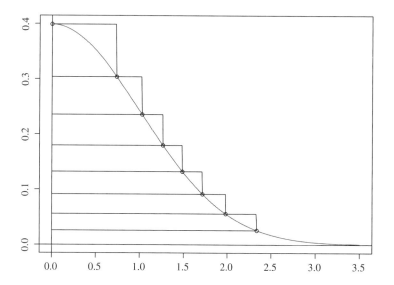

Figure 3.2   The ziggurat distribution for $n = 8$.

First, the points determining the rectangles plotted are chosen according to the uniform distribution on the g area. The rectangles themselves are constructed in such a way that it is easy to choose the point and to use the reject method. We illustrate the method for $n = 8$ (refer to Figure 3.2).

First, we choose $x_1, x_2, \ldots, x_7$, such that the areas of the rectangles in the figure are all equal and equal to the bottom area containing the infinite tail as well. We shall come back to this grid construction in a minute. For now, let us assume that the points are known, namely $0 = x_0 < x_1 < x_2 < \ldots < x_7$, and denote the rectangles in Figure 3.2 with $R_1, R_2, \ldots R_7$ and with R_8 for the bottom one which is a rectangle

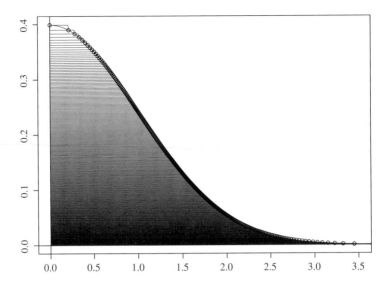

Figure 3.3 The ziggurat distribution for $n = 256$.

plus the tail. This last one needs to be dealt with differently. So the rectangle R_i is determined by points x_{i-1} and x_i.

The point with distribution g is generated in the following way:

1. We first choose one rectangle from $R_1, R_2, \ldots R_7$ and R_8 at random (we can do this since all have the same area).

2. Suppose we generated $i \le 7$. Next we need to generate a point in the rectangle R_i. Note that the x coordinate of such a point is between 0 and x_i and uniformly distributed. To generate the x coordinate of this point, just generate

$$X = Ux_i,$$

where U is $Uniform(0, 1)$. We need to see if the generated point is rejected or not. We have two situations possible:

   (a) If the generated $X < x_{i-1}$, then clearly its corresponding y coordinate (thus the g density) is less than f and therefore the point is accepted;

   (b) If the generated X is between x_{i-1} and x_i, then we need to calculate $f(x)$ and decide if we reject or not. If we reject, we go back to the beginning (choosing an i). In practice, if n is large, then the interval x_{i-1} to x_i is very small, so this step happens very rarely.

3. If the chosen i is 8, thus corresponding to the bottom area R_8, the previous step is modified a bit. Denote the last point x_7 with r, and denote with v the area of

all rectangles (which is the same for all). We generate the x coordinate of the point by taking

$$X = \frac{v}{f(v)} U,$$

where f is the normal pdf and once again U is uniform on 0 to 1. Again, we have two situations:

(a) If $X < r$, then similarly we accept the generated X.

(b) If $X > r$, we return an X from the tail. To do so, we may use Marsaglia's procedure (polar rejection again – but only in the tail). Since we are in the tail, the curve there is less than an exponential, so we can use that distribution to generate from the tail. Specifically, keep generating U_1, U_2 uniform between $(0, 1)$, and calculate

$$X = -\frac{\log U_1}{r}$$

$$Y = -\log U_2$$

until $2Y > X^2$; then return $r + X$ as the generated number.

Then, given the sequence $0 = x_0 < x_1 < \ldots < x_{n-1}$, the pseudo-code for the ziggurat algorithm is as follows:

1. Generate i a discrete uniform from $\{1, \ldots, n\}$. Generate $U \sim Unif(0, 1)$. Set $X = Ux_i$.

2. For all i: If $X < x_{i-1}$, deliver X; jump to step 4.

3. (a) If $i \in \{1, \ldots, n-1\}$, generate $V \sim Unif(0, 1)$.

   i. If

$$\frac{f(X) - f(x_i)}{f(x_{i-1}) - f(x_i)} > V$$

   deliver the generated X; jump to step 4.

   ii. Else we reject; go back to step 1.

   (b) If $i = n$ (bottom rectangle), return an X from the tail (see above).

4. Generate $W \sim Unif(0, 1)$. If $W < 0.5$, set $X = -X$.

The last step is to ensure that we generate positive and negative values. The $f(x_i)$ values can be precalculated, so really the cost of the algorithm is just the generation of uniforms and the calculation of $f(X)$.

Note that this is not the default method in R, which uses a faster method as described above.

The rectangle construction   Choosing the points x_i so that the rectangles all have the same area is not trivial. In fact, on surveying books and articles on the subject (including the original Mersene paper), one can see that the algorithm they have has a mistake. This mistake baffled me for about two hours until I decided to write everything down starting with basic principles. In any case, here is how the determination of these points is made.

Given n, find points $0 = x_0 < x_1 < \cdots < x_{n-1}$ such that all rectangles have same area. Specifically, denote v this common area (which we do not know at the moment). For this v, the points must satisfy

$$
\begin{aligned}
v &= x_1(f(0) - f(x_1)) \\
&= x_2(f(x_1) - f(x_2)) \\
&\;\;\vdots \\
&= x_{n-1}(f(x_{n-2}) - f(x_{n-1})) \\
&= x_{n-1}f(x_{n-1}) + \int_{x_{n-1}}^{\infty} f(x)dx
\end{aligned}
\tag{3.1}
$$

where in the last one R_n we added the tail to the rectangle.

Here we have n equations with n unknowns ($x_1, \ldots, x_{n-1}$ and v) so the system should be solvable. The problem is that the equations are highly nonlinear. An approximation method needs to be used.

Marsaglia proposes the following method. Denote the last point x_{n-1} with r. Simplify everything in terms of r and make the system as close to exact as possible.

For any r, define a function $z(r)$ by doing the following. First calculate

$$
v = rf(r) + \int_{r}^{\infty} f(x)dx
$$

the common area size as a function of r. Then by setting each of the expressions in (3.1) equal to this v, we obtain the rest of the points $x_1, \ldots, x_{n-2}$ as

$$
x_{n-2} = f^{-1}\left(\frac{v}{r} + f(r)\right)
$$

$$
\vdots,
$$

$$
x_1 = f^{-1}\left(\frac{v}{x_2} + f(x_2)\right)
$$

Note that we did not use the first equation. If the r is the right one, then this first equation would be verified as well, therefore we need to find the r which makes this equation as close to 0 as possible. Specifically, output the value of the function $z(r)$ as

$$
z(r) = v - x_1(f(0) - f(x_1)),
$$

where v and x_1 have just been calculated in terms of r.[1]

---

[1] It is here that most of algorithms published are wrong – they use 1 instead of $f(0)$, that is, the normal density calculated at 0 and that is not 1.

To calculate the grid points, one needs to search for the r which makes the last expression as close to 0 as possible. In practice, this is very difficult. The function f^{-1} is the inverse of the normal pdf and it is

$$f^{-1}(y) = \sqrt{-(\log 2\pi + 2 \log y)}$$

When using this function to compute the points and the final probability, everything is fine if r is actually larger than the optimal r (in this case the $z(r)$ is negative). However, if r is less than the optimal, the expression under the square root becomes negative and the value of $z(r)$ becomes a complex number. This is why finding the grid for any n is very difficult in practice. For reference, when implementing the grid in R, with 16 bit precision I obtained for $n = 8$ the optimal $r = 2.3383716982$ and the corresponding $z(r) = 6.35934 \times 10^{-11}$. Using $n = 256$ I obtained $r = 3.654152885361009$ and $z(r) = 5.50774 \times 10^{-16}$.

We will finish this section with R code to calculate the function $z(r)$, and given the points to make the plots in the section. Any line which starts with the # character is a comment line.

```
#The function below calculates for given $r$ and $n$ the value
#of the function z(r)

ziggurat.z=function(r,n)
{v=r*dnorm(r)+pnorm(-r); points=r;
for(i in 1:(n-2))
    {points=c(points,invdnorm(v/points[i]+dnorm(points[i])))};
return(c(points,v-points[n-1]\*(dnorm(0)-dnorm(points[n-1]))))
}


#Since there is a problem if r is less than the optimal one,
#the procedure is the following:
# set a fix n
#Find the r such that the last point in the output is as
#close to 0 as possible (bisection method by hand);


#Finally, here is the function to create the ziggurat plot;
#Last lines drawn are for the bottom rectangle and the axes

plot.ziggurat=function(passedgrid)
{x=seq(0,3.5,by=.001);
plot(x,dnorm(x),type=''l'',col=''blue'',lty=1,xlab='''',ylab='''');
points(passedgrid,dnorm(passedgrid))
forplot=c(0,passedgrid)

for(i in 1:(length(forplot)-1))
    {lines(c(seq(0,forplot[i+1],by=0.01),forplot[i+1]),
    c(rep(dnorm(forplot[i]),length(seq(0,forplot[i+1],
    by=0.01))),
    dnorm(forplot[i+1])))}
```

```
lines ( seq ( 0 , forplot [ length ( forplot )] , by =0.01 ),
    rep (dnorm ( forplot [ length ( forplot )]) ,
    length ( seq ( 0 , forplot [ length ( forplot )] , by =0.01 ))))

abline ( 0 ,0 ); abline ( v =0)
}


#We need to pass the grid points ( x _1 ,... , x _{n−1})
#in increasing order so here is for example how to call:

$n=8$
pict .8=( ziggurat . z (2.3383716982 , n ))[ n :1]
plot . ziggurat ( pict .8)

$n=256$
pict .256=( ziggurat . z (3.6541528852885361009 , n ))[ n :1]
plot . ziggurat ( pict .256)
```

Examples for rejection sampling

▉ EXAMPLE 3.4

Let us exemplify the corollary. Suppose I want to generate from the density

$$f(x) = Cx^2(\sin x)^{\cos x}|\log x|, \quad x \in \left(\frac{\pi}{6}, \frac{\pi}{2}\right)$$

The constant C is chosen to make the density f integrate to 1. Note that actually calculating C is impossible. A plot of this density may be observed in Figure 3.4.

We wish to apply the rejection sampling to generate from the distribution f. To this end, we will use the uniform distribution in the interval $\left(\frac{\pi}{6}, \frac{\pi}{2}\right)$, and we shall calculate the constant M so that the resulting function is majoring the distribution. To do so, we calculate the maximum of the function

$$m = \max_{x \in \left(\frac{\pi}{6}, \frac{\pi}{2}\right)} x^2(\sin x)^{\cos x}|\log x| = 1.113645$$

and we take

$$M = Cm\left(\frac{\pi}{2} - \frac{\pi}{6}\right).$$

With this constant, M we are guaranteed that $f(x) < Mg(x)$ for every $x \in \left(\frac{\pi}{6}, \frac{\pi}{2}\right)$. To see this, recall that the density of the uniform on the desired interval $x \in \left(\frac{\pi}{6}, \frac{\pi}{2}\right)$ is constant $g(x) = \left(\frac{\pi}{2} - \frac{\pi}{6}\right)^{-1}$. Furthermore, the ratio that needs to be calculated is

$$\frac{f(x)}{Mg(x)} = \frac{x^2(\sin x)^{\cos x}|\log x|}{m}.$$

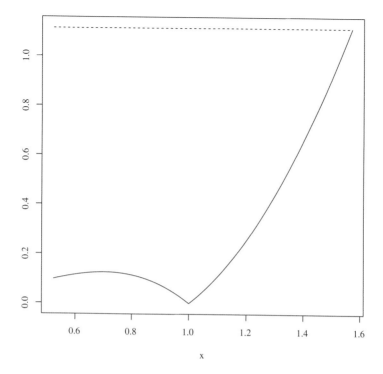

Figure 3.4 The function defining the density $f(\cdot)$ (continuous line) and the uniform distribution $M * g(\cdot)$ (dashed line) without the scaling constant C.

Obviously, this ratio is very good (approaches 1) when x is close to $\pi/2$, and it is close to 0 (as it should) when x is close to 1.

The following code is written in R and implements the rejection sampling for the example. Any line which starts with the # character is a comment line.

```
##Example 6.4: Rejection Sampling R code
#We calculate the constant m used later

m=max(x^2*(sin(x)^cos(x))*abs(log(x)))

# Next defines the function used to calculate the ratio
# f(x)/M*g(x)

ratio.calc=function(x)
{return(x^2*(sin(x)^cos(x))*abs(log(x))/m)}

#The next function returns n generated values from the
#distribution f(x)

random.f=function(n)
{GeneratedRN=NULL;
```

```
for (i in 1:n)
    {OK=0;
     while (OK!=1){
            Candidate=runif(1,pi/6,pi/2);
            U=runif(1);
     if (U<ratio.calc(Candidate))
            {OK=1;GeneratedRN=c(GeneratedRN,Candidate)}
        }
      }
return(GeneratedRN)
}
```

#Now we call the function we just created to generate
#10,000 numbers

estimated.dist=random.f(10000)

#Finally, to check here is the histogram of these numbers

hist(estimated.dist,nclass=75)

Next, we present a seemingly more complex example which has a simpler solution than the rejection sampling method.

■ **EXAMPLE 3.5**

As another example, let us generate from the mixture of beta distributions

$$f(x) = 0.5\,\beta(x; 10, 3) + 0.25\,\beta(x; 3, 15) + 0.25\,\beta(x; 7, 10), \quad x \in (0, 1),$$

where the $\beta(x, a, b)$ denotes the Beta distribution pdf with shape parameters a and b:

$$\beta(x, a, b) = \frac{\Gamma(a+b)}{\Gamma(a)\Gamma(b)} x^{a-1}(1-x)^{b-1},$$

and $\Gamma(\cdot)$ is the Gamma function:

$$\Gamma(x) = \int_0^\infty t^{x-1} e^{-t}\, dt.$$

A plot of the resulting distribution may be observed in Figure 3.6.

The mixture of distributions is always a distribution since the individual pdfs integrate to 1. Note that the Beta distribution is always distributed on $(0, 1)$, and since we can use the uniform distribution on (0.1) to generate candidate values, the constant M can be chosen as the maximum value of the mixture gamma density function f. The code presented next uses rejection sampling to generate random numbers from the desired distribution.

Histogram of estimated.dist

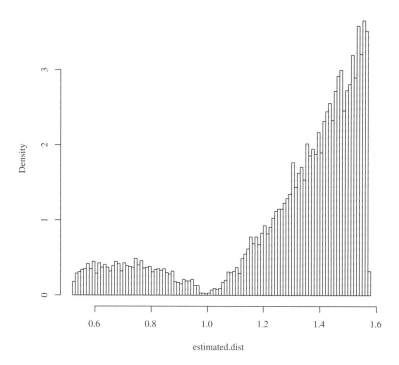

Figure 3.5   The resulting histogram of the generated values. This should be close in shape to the real function if the simulation is working properly. Note that this is a proper distribution and contains the scaling constant C.

```
#We implement the mixture density
f=function(x)
{return(0.5*dbeta(x,10,3)+0.25*dbeta(x,3,15)+0.25
    *dbeta(x,7,10))}

#We calculate the constant M
M=max(f(x))

#The next routine generates n numbers with desired distribution
random.mixturebeta=function(n)
{GeneratedRN=NULL;
for(i in 1:n)
{ OK=0;
while(OK!=1){
Candidate=runif(1);
U=runif(1);
if(U<f(Candidate)/M){OK=1;GeneratedRN=c(GeneratedRN,Candidate)}}
}
```

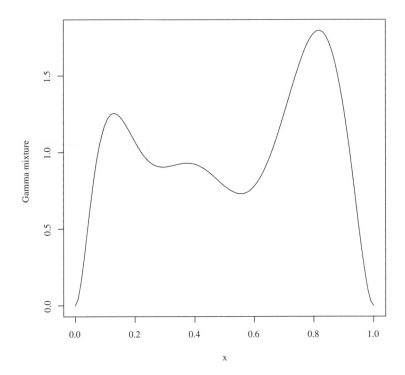

Figure 3.6   The mixture gamma density function (continuous line).

return (GeneratedRN)}

#Finally, we verify by plotting the histogram of generated values
estimated . dist=random . mixturebeta (100000)
hist (estimated . dist , nclass =100, freq=F)

The resulting histogram may be observed in Figure 3.7. Once again, the simulated values look good; however, there are ways in which the simulation of this type of distribution may be made faster.

Note that the distribution in Example 3.5 is a type of special distribution obtained by mixing three classical distributions. Such random variables are much easier to generate (and much faster) as the next section details.

Generating from a mixture of distributions

Suppose that the density we need to generate from is a mixture of elemental and easy-to-generate-from distributions. Specifically,

$$f(x) = \sum_{i=1}^{n} w_i f_i(x|\theta_i)$$

Histogram of estimated.dist

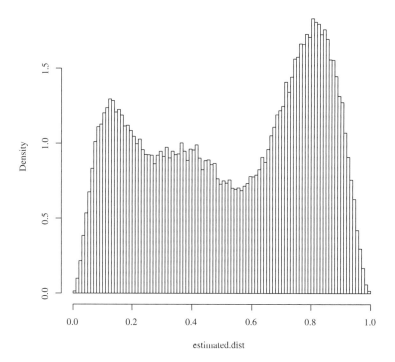

estimated.dist

Figure 3.7   The resulting histogram of the generated values from the Gamma mixture density. We had to use 100,000 generated values to see the middle hump.

where the weights w_i sum to 1, and the densities f_i may all be different and dependent on the vectors of parameters θ_i. It is much easier to generate from such distributions provided that we have implemented already generators for each of the distributions f_i. The idea is that the weights determine which distribution generates the respective random number.

Specifically, we first generate a random variable U as a Uniform(0,1) random variable. Then we find the weight index j such that

$$w_1 + \cdots + w_{j-1} \leq U < w_1 + \cdots + w_{j-1} + w_j.$$

Next, the desired random number is generated from the distribution f_j. Let us exemplify this generating strategy by continuing the example 3.5.

■ EXAMPLE 3.6   (Continuation of example 3.5)

The objective is once again to generate random numbers from the mixture Beta distribution

$$f(x) = 0.5 \, \beta(x; 10, 3) + 0.25 \, \beta(x; 3, 15) + 0.25 \, \beta(x; 7, 10), \quad x \in (0, 1)$$

We will implement the code in R once again and we will take this opportunity to display some advanced R programming features.

The following code implements the generation of random numbers one by one, just as one would do it in C or some other low level language.

```
## Method of generation for mixture distributions.

random.mixturebeta.v2=function(n)
{GeneratedRN=NULL;
for(i in 1:n)
{U=runif(1);
RandomVal=ifelse(U<0.5,rbeta(1,10,3),ifelse(U<0.75,
                                    rbeta(1,3,15),rbeta(1,7,10)))
GeneratedRN=c(GeneratedRN,RandomVal)
}
return(GeneratedRN)}

#And calling the function
random.mixturebeta.v2(100000)
```

The function $ifelse(CONDITION, VALUEIFYES, VALUEIFNO)$ is R-specific but the code above does not take advantage of the amazing strength of R, which is working with vectors and large objects. The next function accomplishes the same thing but it is much faster as we shall see.

```
## Method of generation for mixture distributions (optimized
#  code).
random.mixturebeta.v2.optimal=function(n)
{U=runif(n);   GeneratedRN=rep(0,n)
beta1=(U<0.5);beta2=(U>=0.5)&(U<0.75);beta3=(U>=0.75);
n1=GeneratedRN[beta1];  n2=GeneratedRN[beta2];
   3=GeneratedRN[beta3];
GeneratedRN[beta1]=rbeta(n1,10,3);
GeneratedRN[beta2]=rbeta(n2,3,15)
GeneratedRN[beta3]=rbeta(n3,7,10)
return(GeneratedRN)}
```

In the code above, the $beta1, beta2, beta3$, are vectors containing values TRUE and FALSE depending on whether the respective condition is satisfied. When such vectors (containing TRUE and FALSE values) are applied as indices as in $GeneratedRN[beta1]$, they select from the vector $GeneratedRN$ only those values which correspond to the TRUE indices. This allows us to operate inside vectors very fast without going through the vectors one by one. Furthermore, the code takes advantage of the internal R method of generating beta distributions which is one of the best and fastest available in any statistical software.

We did not plot the resulting histograms of the generated values for the latter two functions since they are very similar to those in Figure 3.7. However, Table 3.1

provides the running times for the two methods as well as the optimized algorithm above.

Table 3.1   Average running time in seconds for 30 runs of the three methods. Each run generates a vector of 100,000 random numbers.

	Rejection sampling	Mixture gen	Mixture gen (optimal)
Average time (s)	20.122	20.282	0.034
Standard deviation	0.409	0.457	0.008

Table 3.1 presents some interesting conclusions when looking at the two method and at the R implementation. Each method was run 30 times, and a garbage collection was performed before each run so that interferences from the other processes run by the operating system were minimized. The first two columns directly compare the two methods.

For each number having the desired distribution, the rejection sampling procedure generates a minimum of two uniform random numbers and since it may reject the numbers produced the number of uniforms generated may actually be larger. The mixture generating algorithm, on the other hand, always generates one uniform and one beta distributed number. Both algorithms produce numbers one after another until the entire set of 100,000 values are produced. The fact that the times are so close to each other tells us that generating uniforms only (even more than two) may be comparable in speed with generating from a more complex distribution.

When comparing the numbers in the second column with the numbers in the third column, and recalling that it actually is the same algorithm with the only difference being that the optimized version works with vectors, we can see the true power of R on display. The 100,000 numbers are generated in about one-third of a second, which basically means that simulating a path takes nothing at all if done in this way and thus long simulations may be made significantly quicker by rethinking the code.

3.4   Generating Random Variables. Importance Sampling

Rejection sampling works even if one only knows the approximate shape of the target density. As we have seen from the examples, any candidate distribution $g(\cdot)$ may be used, but in all examples we used the uniform distribution for it. This happens to always be the case in practice. There are two reasons for this. One is the fact that generating random uniforms is the most common random number generator as well as the fastest. Two, the constant M must be chosen so that $f(x) < Mg(x)$ for all x in support of f. This is generally hard to assess unless one uses the uniform distribution, and the constant M becomes related to the maximum value of $f(\cdot)$ over its support as we have already seen. However, this translates into rejecting a number of generated values, which is *proportional* to the difference in the area under the constant function

M and the area under $f(\cdot)$ (look at the ratio between the difference in areas and the area under the dashed line in Figure 3.4).

As we have seen, this is not so bad for one-dimensional random variables. However, it gets really bad quickly as the dimension increases. Importance sampling tries to deal with this by sampling more from certain parts of the $f(\cdot)$ density.

It is important to realize that, unlike the methods presented thus far, the importance sampling method does not generate random numbers with specific density $f(\cdot)$. Instead, the purpose of the importance sampling method is to estimate expectations. Specifically, suppose that X is a random variable (or vector) with a known density function $f(x)$ and h is some other known function on the domain of the random variable X; then the importance sampling method will help us to estimate

$$\mathbf{E}[h(X)].$$

If we recall that the probability of some set A may be expressed as an expectation

$$\mathbf{P}(X \in A) = \mathbf{E}[\mathbf{1}_A(X)],$$

then we see that the importance sampling method may be used to calculate any probabilities related to the random variable X as well. For example, the probability of the tails of the random variable X is

$$\mathbf{P}(|X| > M) = \mathbf{E}[\mathbf{1}_{(-\infty, -M)}(X)] + \mathbf{E}[\mathbf{1}_{(M, \infty)}(X)],$$

for some suitable M, and both may be estimated using importance sampling.

The idea and estimating expectations using samples

The laws of large numbers (either weak or strong) say that if $X_1, \ldots, X_n$ are i.i.d. random variables drawn from the distribution $f(\cdot)$ with a finite mean $\mathbf{E}[X_i]$, then the sample mean converges to the theoretical mean $\mathbf{E}[X]$ in probability or a.s. Either way, the theory says that, if we have a way to draw samples from the distribution $f(\cdot)$, be these $x_1, \ldots, x_n$, then for each function $h(\cdot)$ defined on the codomain of X we must have

$$\frac{1}{n} \sum_{i=1}^{n} h(x_i) \to \int h(x) f(x)\, dx = \mathbf{E}[h(X)].$$

Therefore, the idea of estimating expectations is to use generated numbers from the distribution $f(\cdot)$. However, in many cases we may not draw from the density f and, instead, use some easy-to-sample-from density g. The method is modified using the following observation:

$$\mathbf{E}_f[h(X)] = \int h(x) f(x)\, dx = \int h(x) \frac{f(x)}{g(x)} g(x)\, dx = \mathbf{E}_g\left[h(X) \frac{f(X)}{g(X)} \right],$$

where we used the notations $\mathbf{E}_f$ and $\mathbf{E}_g$ to denote expectations with respect to density f and g, respectively. The expression above is correct only if the support of g includes the support of f; otherwise, we can have points where $f(x) \neq 0$ and $g(x) = 0$, and thus the ratio $f(x)/g(x)$ becomes undefined.

The algorithm description   Combining the approximating idea with the expression above, it is now easy to describe the importance sampling algorithm to estimate $\mathbf{E}_f[h(X)]$.

1. Find a distribution g which is easy to sample from and its support includes the support of f (i.e., if $f(x) = 0$ for some x will necessarily imply $g(x) = 0$)

2. Draw n sampled numbers from the distribution g: $x_1, \ldots, x_n$.

3. Calculate and output the estimate:

$$\frac{1}{n} \sum_{i=1}^{n} h(x_i) \frac{f(x_i)}{g(x_i)} = \sum_{i=1}^{n} h(x_i) \frac{f(x_i)}{ng(x_i)}.$$

The reason why this method is called importance sampling is due to the so-called importance weight $\frac{f(x_i)}{ng(x_i)}$ given to x_i. The ratio $\frac{f(x_i)}{g(x_i)}$ may be interpreted as the number modifying the original weight $1/n$ given to each observation x_i. Specifically, if the two densities are close to each other at x_i, then the ratio $\frac{f(x_i)}{g(x_i)}$ is close to 1 and the overall weight given to x_i is close to the weight $1/n$ (the weight of x_i if we would be able to draw directly from f). Suppose that x_i is in a region of f which is very unlikely (small values of f). Then the ratio $\frac{f(x_i)}{g(x_i)}$ is going to be close to 0 and thus the weight given to this observation is very low. On the other hand, if x_i is from a region where f is very likely, then the ratio $\frac{f(x_i)}{g(x_i)}$ is going to be large and thus the weight $1/n$ is much increased.

Observations   First note that the weights $\frac{f(x_i)}{ng(x_i)}$ may not sum to 1. However, their expected value is 1:

$$\mathbf{E}_g \left[\frac{f(X)}{g(X)} \right] = \int \frac{f(x)}{g(x)} g(x) \, dx = \int f(x) \, dx = 1.$$

Thus the sum $\sum_{i=1}^{n} \frac{f(x_i)}{ng(x_i)}$ tends to be close to 1.
Second, the estimator

$$\hat{\mu} = \sum_{i=1}^{n} h(X_i) \frac{f(X_i)}{ng(X_i)}.$$

is unbiased and we can calculate its variance. That is,

$$\mathbf{E}[\hat{\mu}] = \mathbf{E}_f[h(X)]$$

$$Var(\hat{\mu}) = \frac{1}{n} Var_g \left(h(X) \frac{f(X)}{g(X)} \right). \tag{3.2}$$

Third, the variance of the estimator obviously depends on the choice of the distribution g. However, we may actually determine the best choice for this distribution.

Minimizing the variance of the estimator with respect to the distribution g means minimizing

$$Var_g\left(h(X)\frac{f(X)}{g(X)}\right) = \mathbf{E}_g\left[h^2(X)\left(\frac{f(X)}{g(X)}\right)^2\right] - \mathbf{E}_f^2[h(X)].$$

The second term does not depend on g, while using the Jensen inequality in the first term provides

$$\mathbf{E}_g\left[\left(h(X)\frac{f(X)}{g(X)}\right)^2\right] \geq \left(\mathbf{E}_g\left[|h(X)|\frac{f(X)}{g(X)}\right]\right)^2 = \left(\int |h(x)|f(x)\ dx\right)^2.$$

However, the right side is not a distribution, but it does provide the *optimal importance sampling distribution*:

$$g^*(x) = \frac{|h(x)|f(x)}{\int |h(x)|f(x)\ dx}.$$

This is not really useful from a practical perspective since typically sampling from $f(x)h(x)$ is harder than sampling from $f(x)$. However, it does tell us that the best results are obtained when we sample from $f(x)$ in regions where $|h(x)|f(x)$ is relatively large. As a consequence of this, using the importance sampling is better at calculating $\mathbf{E}[h(X)]$ than using a straight Monte Carlo approximation (i.e., sampling directly from f and taking a simple average of the $h(x_i)$ values).

Practical considerations In practice, it important that the estimator has finite variance (otherwise it never improves with n). To see this, observe the formula for variance in (3.2). Here are the sufficient conditions for the finite variance of the estimator $\hat{\mu}$:

- There exists some M such that $f(x) < Mg(x)$ for all x and $Var_f(h(X)) < \infty$, or

- The support of f is compact, f is bounded above, and g is bounded below on the support of f.

Remark 3.5 *Choosing the distribution g is crucial. For example, if f has support on $\mathbb{R}$ and has heavier tails than g, the weights $w(X_i) = f(X_i)/g(X_i)$ will have infinite variance and the estimator will fail.*

Applying importance sampling

▉ EXAMPLE 3.7

For this example, we will showcase the importance of the choice of distribution g. The example is due to Nick Whiteley in his lecture notes on machine learning.

Suppose we want to estimate $\mathbf{E}[|X|]$ where X is distributed as a Student-t random variable with three degrees of freedom. In the notation used above, $h(x) = |x|$ and $f(x)$ is the t-density function is

$$\frac{\Gamma\left(\frac{\nu+1}{2}\right)}{\sqrt{\nu\pi}\Gamma\left(\frac{\nu}{2}\right)}\left(1+\frac{x^2}{\nu}\right)^{-\frac{\nu+1}{2}},$$

with degrees of freedom $\nu = 3$ and $\Gamma(x)$ is a notation for the Gamma function used earlier in this chapter.

First, note that the target density does not have compact support, thus the use of a uniform density for g is not possible. To exemplify the practical aspects of the importance sampling algorithm, we shall use two candidate densities: $g_1(x)$ the density of a t distribution with one degree of freedom, and $g_2(x)$ the standard normal density (N(0,1)). We shall also use a straight Monte Carlo where we generate directly from the distribution f. The plot of these densities may be observed in Figure 3.8.

We know that the optimal choice is $|h(x)|f(x)$; however, generating from this density is very complex. We may also observe that while the t density with one degree of freedom dominates the tails of f, the normal density does not, so we expect

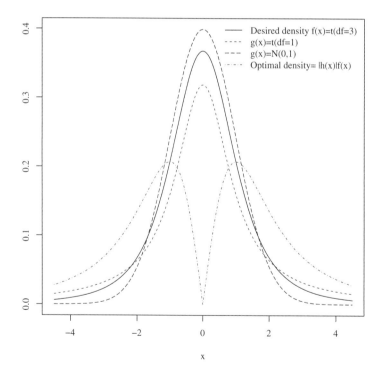

Figure 3.8 Candidate densities for the importance sampling procedure as well as the target density.

the estimator produced using the normal density to be not very good (the weights $f(X_i)/g(X_i)$ have infinite variance).

Next we present the R-code used for the importance sampling technique.

```
#Straight Monte Carlo:
n=10:1500
nsim=100

straightMC=NULL;
for(i in n)
{mu=NULL;
for(j in 1:nsim)
{a=rt(i,3);mu=c(mu,mean(abs(a)))}
straightMC=cbind(straightMC,c(i,mean(mu),sd(mu)))
}

#Importance Sampling using first candidate:

usingt1=NULL;
for(i in n)
{mu=NULL;
for(j in 1:nsim)
{a=rt(i,1);mu=c(mu,mean(abs(a)*dt(a,3)/dt(a,1)))}
usingt1=cbind(usingt1,c(i,mean(mu),sd(mu)))
}

#Importance Sampling using second candidate:

usingnorm=NULL;
for(i in n)
{mu=NULL;
for(j in 1:nsim)
{a=rnorm(i);mu=c(mu,mean(abs(a)*dt(a,3)/dnorm(a)))}
usingnorm=cbind(usingnorm,c(i,mean(mu),sd(mu)))
}
```

To have meaningful comparisons, we first generate from f distribution and simply use the average of h calculated at the generated values. To see the entire evolution of the estimator, we use n (number of generated samples) between 10 and 1500. To calculate the variance of the estimator, we also repeat the process 100 times for each value of n. We plot the evolution of the estimator with n in Figure 3.9.

Looking at the first two plots (Figs 3.9(a) and 3.9(b)), we see that in fact the importance sampling estimator is better (superoptimal) than the straight Monte Carlo estimator. The reason is easy to see by looking at the plot of densities (Fig. 3.8) and observing that the density g_1 is in fact closer to the optimal density $|h(x)|f(x)$ than f. Furthermore, Figure 3.9(d) displays the suspected poor performance of the

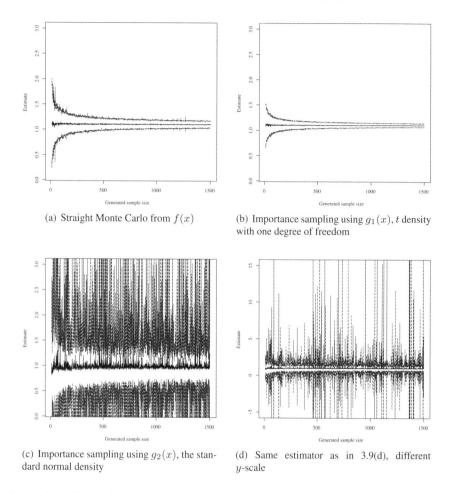

(a) Straight Monte Carlo from $f(x)$

(b) Importance sampling using $g_1(x)$, t density with one degree of freedom

(c) Importance sampling using $g_2(x)$, the standard normal density

(d) Same estimator as in 3.9(d), different y-scale

Figure 3.9   The evolution of different importance sampling estimators of $\mathbf{E}[|X|]$. The center line is the estimator, while the outside lines give an estimated 95% confidence interval. We keep the y-axis the same to have a fair comparison. The last estimator performance is horrible.

importance sampling estimator obtained using the normal density. The confidence interval never decreases in width, and the convergence of the estimator itself is very slow.

Practical consideration: normalizing distributions

Often times, in practical application, only the functional form of the distribution is known. For instance, the importance sampling methodology may be used to estimate an integral of the form

$$\int_A h(x)f(x)\,dx,$$

for some domain A. However, what if the function f does not integrate to 1 on the domain A? Or if it integrates to a finite value but is not a density (it does integrate to 1 on the whole domain). It turns out that we can still use the methodology in this case as well. However, we need to use the so-called self-normalizing weights.

Specifically, suppose that we need to estimate the expectation

$$\mathbf{E}[h(X)] = \int h(x)f(x)\,dx,$$

but the density f is only known up to a constant $f(x) = C_f \varphi(x)$, where the function $\varphi(\cdot)$ is known but the constant C_f is unknown.

The idea is to estimate the constant C_f as well. Note that, since f is a density, we must have

$$C_f = \frac{1}{\int \varphi(x)\,dx} = \frac{1}{\int \frac{\varphi(x)}{g(x)} g(x)\,dx}.$$

Thus, C_f is approximately estimated by taking samples x_i from the distribution g and constructing

$$\frac{1}{\sum_{i=1}^{n} \frac{\varphi(x_i)}{g(x_i)}}.$$

So proceeding exactly as in the straight importance sampling case, an estimator is

$$\tilde{\mu} = \frac{\sum_{i=1}^{n} h(X_i) \frac{\varphi(X_i)}{g(X_i)}}{\sum_{i=1}^{n} \frac{\varphi(X_i)}{g(X_i)}}.$$

Note that the estimator does not depend on the unknown constant C_f. Further, note that the weights associated with each variable X_i drawn from the distribution g are normalized (the weights sum to 1).

Thus, the algorithm to estimate $\mathbf{E}_f[h(X)]$ is as follows:

1. Find a distribution g which is easy to sample from and with its support including the support of f.

2. Draw n sampled numbers from the distribution g: $x_1, \ldots, x_n$.

3. Calculate and output the estimate:

$$\frac{\sum_{i=1}^{n} h(x_i) \frac{\varphi(x_i)}{g(x_i)}}{\sum_{i=1}^{n} \frac{\varphi(x_i)}{g(x_i)}}.$$

The estimator obtained is strongly consistent (i.e., it converges fast to the right estimate); however, the estimator is biased (i.e., the expected value of the estimator is under or over the target it estimates).

Lemma 3.6 *Suppose we are given a random variable X with density $f(x) = C_f \varphi(x)$ and a function h defined on the support of the function f. Let g be another density function such that its support includes the support of the function f, and define $w(x) = \varphi(x)/g(x)$. Suppose that $X_1, \ldots, X_n$ are i.i.d. random variables with density g. Then the estimator*

$$\tilde{\mu} = \frac{\sum_{i=1}^{n} h(X_i) w(X_i)}{\sum_{i=1}^{n} w(X_i)},$$

is strongly consistent, that is,

$$\tilde{\mu} \xrightarrow{a.s.} \mathbf{E}_f[h(X)] = \mu,$$

and furthermore

$$\mathbf{E}_g[\tilde{\mu}] = \mu + \frac{\mu Var_g(w(X_1)) - Cov_g(w(X_1), h(X_1)w(X_1))}{n} + O(n^{-2}),$$

$$Var_g(\tilde{\mu}) = \frac{Var_g\left(h(X_1)w(X_1) - \mu w(X_1)\right)}{n} + O(n^{-2}).$$

The consistency result is immediate using the strong law for each of the two integrals in the estimator's formula. However, the proof of the biasedness of the estimator is quite computational and we refer the reader to a recent work (Whiteley and Johansen, 2011) for more details and a proof.

You may also wonder how it is possible for the estimator to be biased but still converge to the right place. Just look at the expressions in Lemma 3.6 and observe that the bias goes to 0 with n.

Sampling importance resampling

As we mentioned, the importance sampling technique is primarily used to calculate expectations. However, it is possible to adapt the technique to obtain samples from the distribution f. To do this, recall that

$$\mathbf{P}(X \in A) = \mathbf{E}_f[\mathbf{1}_A(X)].$$

Thus, following the procedure already detailed, one approximates the probability with

$$\mathbf{P}(X \in A) = \frac{1}{N} \sum_{x_i} \mathbf{1}_A(x_i) \frac{f(x_i)}{g(x_i)}$$

Thus, only the generated random numbers that fall into the set A are actually counted.

One may take this argument further and obtain an approximate discrete distribution for f, by

$$\hat{p}(x) = \sum_{i=1}^{N} \frac{f(x_i)}{g(x_i)} \mathbf{1}_{\{x_i\}}(x) = \sum_{i=1}^{N} w(x_i) \mathbf{1}_{\{x_i\}}(x)$$

where the x_i values are generated from the density g and obviously the corresponding weights $w(x_i) = \frac{f(x_i)}{g(x_i)}$ may not sum to 1. To generate M independent random variables from the distribution f, one normalizes this discrete distribution $\hat{p}$ and just generates from it M values where M is much larger than N. Once the M new values are obtained, denoted by say $\{y_1, \ldots, y_M\}$, the new estimated f distribution is

$$\tilde{p}(x) = \frac{1}{M} \sum_{i=1}^{M} \mathbf{1}_{\{y_i\}}(x).$$

Note that, since the values y_i are generated from the x_i values, the new sample $\{y_1, y_2, \ldots y_M\}$ contains repeated observations. This method is known as *sampling importance resampling* and is due to Rubin (1998). In fact, the technique is a simple bootstrapping technique and it is not clear to us whether the technique is better that simply using the $\hat{p}$ distribution directly to generate random variables. However, the above paper is cited extensively in the computer vision and machine learning literature so we decided to include the technique here.

Adaptive importance sampling

As we have shown in Example 3.7, the proper choice of g can lead to superefficient estimation algorithms. Specifically, using a g distribution which is close to $|h(x)|f(x)$ is much more efficient than using a straight Monte Carlo method (i.e., with $g(x) = f(x)$). However, as the dimension of x increases (x becomes a random vector with many dimensions), it becomes more complicated to obtain a suitable $g(x)$ from which to draw the samples. A strategy to deal with this problem is the *adaptive importance sampling* technique, which seems to have originated in the structural safety literature (Bucher, 1988).

The method considers a parameterized distribution $g(x|\theta)$, where θ is a parameter vector which is adaptable depending on the sampling results. The idea of the method is to try and minimize the variance of the estimator $\hat{\mu}$. Specifically, consider a parameterized distribution $g(x|\theta)$. We want to minimize

$$\mathbf{E}_g[f^2(X)w^2(X)] - \mathbf{E}_f^2[h(X)],$$

with respect to g, where $w(x|\theta) = \frac{f(x)}{g(x|\theta)}$. Note that the second term does not depend on g at all, and that minimizing involves calculating the derivative of the first term with respect to θ. Since derivative and expectation commute on a probability space if the expectation exists, the problem reduces to finding the roots of the derivative

$$D(\theta) = 2\mathbf{E}_g\left[f^2(x)w(x|\theta)\frac{\partial w}{\partial \theta}(x|\theta)\right].$$

A Newton–Raphson iteration procedure would find the minimum of the original expression by using

$$\theta_{n+1} = \theta_n - (\nabla D(\theta_n))^{-1} D(\theta_n).$$

However, the expectation $D(\theta_n)$ is hard to compute exactly. Furthermore, the inverse of the gradient of D (or the Hessian) of the original expression to be minimized is even harder to calculate. Instead, the algorithm simply replaces the expression $(\nabla D(\theta_n))^{-1}$ with a learning constant and the $D(\theta_n)$ with its sample value.

To describe the algorithm, we give a pseudo-code below. The technique starts with a general distribution $g(x|\theta)$ capable of many shapes as the θ parameter varies in some parameter space Θ. Then, the method *adapts* the parameter θ to fit the problem at hand. This is done as follows:

- Start with an initial parameter value θ_0 (which produces some indifferent shape of the distribution g), and let $n = 0$.

- Do the following until the difference $|\theta_{n+1} - \theta_n| < \varepsilon$, where ε is a prespecified tolerance level.

  - Generate N values $x_1, \ldots, x_N$ from the distribution $g(x|\theta_n)$.
  - Update the θ value using

$$\theta_{n+1} = \theta_n - \alpha \frac{1}{N} \sum_{i=1}^{N} f^2(x_i) w(x|\theta_n) \frac{\partial w}{\partial \theta_n}(x_i|\theta_n).$$

- Check the condition $|\theta_{n+1} - \theta_n| < \varepsilon$. If not satisfied, let $n = n + 1$ and repeat the loop.

This technique is extremely powerful and we shall come back to it in the stochastic processes part when we will talk about Markov chain Monte Carlo (MCMC) algorithms.

Problems

3.1 Look at the Box–Muller and the two resulting variables X and Y. Calculate the joint and marginal distributions of these variables and show that they are independent.

3.2 Look at the polar rejection method. Show that the two variables given by this algorithm are independent.

3.3 Consider the following normal mixture density:

$$f(x) = 0.7 \frac{1}{\sqrt{2\pi 9}} e^{-\frac{(x-2)^2}{18}} + 0.3 \frac{1}{\sqrt{2\pi 4}} e^{-\frac{(x+1)^2}{8}}$$

a) Calculate the expected value of a random variable with this distribution.
b) Write code and implement it to generate random variables with this distribution. Use your mind or whatever methods you learned in this chapter.
c) Use the previous part to generate 1000 independent random numbers with this density. Calculate the sample mean (average of generated numbers). Compare with answer in part a).
d) Repeat the previous part, but this time generate 10, 000 numbers. Comment.

3.4   Consider the random variable X with the density

$$f(x) = C\cos(x)\sin(x), \quad x \in [0, \pi/2]$$

a) Calculate the constant C which makes above a probability density.
b) Sketch this density.
c) Implement the importance sampling method to generate random numbers from the density. Create a histogram by generating 1000 such numbers and compare with the previous part.
d) Generate 1000 such numbers and use them to estimate the probability

$$\mathbf{P}(X > \pi/12)$$

e) Calculate the same probability by integrating the density. Are the two numbers close?
f) Generate 10, 000 numbers and use them to estimate

$$\mathbf{E}\left[e^{\sin(x)}\right]$$

3.5   The pdf of a logistic random variable with parameter $\lambda > 0$ is

$$f(x) = \frac{\lambda e^{-\lambda x}}{(1 + e^{-\lambda x})^2}, \quad \text{for } x > 0$$

a) Calculate the cdf of the distribution.
b) Calculate the inverse cdf F^{-1}.
c) Write a code to generate random numbers with the logistic distribution.
d) Calculate the expectation

$$\mathbf{E}X = \int_0^\infty x f(x)\,dx$$

and the variance

$$V(X) = \mathbf{E}(X^2) - (\mathbf{E}X)^2 = \int_0^\infty x^2 f(x)\,dx - \left(\int_0^\infty x f(x)\,dx\right)^2$$

using either the simulated random numbers or integration.

3.6 Let X be a random variable with a Beta distribution with parameters a, b.

$$f(x) = \frac{\Gamma(a+b)}{\Gamma(a)\Gamma(b)} x^{a-1}(1-x)^{b-1}, \text{ where } 0 < x < 1$$

Calculate the mode of the distribution (the number x for which the pdf $f(x)$ is maximum). Using this mode, calculate a number M such that $f(x) \leq M$ for all x. Design a rejection sampling algorithm that will use this M and a uniform distribution to create random samples from the Beta distribution with parameters a and b.

To test, use parameters $a = 2$, $b = 3$, and approximate the expected value of the Beta distribution. Compare with the theoretical value, which is $\frac{a}{a+b} = \frac{2}{5}$.

3.7 Design a scheme to generalize the Box–Muller scheme to generate four-dimensional random normals with mean vector 0 and covariance matrix the identity matrix.

3.8 You want to design an experiment where you simulate bacteria living in a certain medium. To this end, you know that the lifetime of one bacterium is a random variable X (in hours) distributed with exponential density $\frac{1}{2}e^{-x/2}$. However, you also know that all of these peculiar bacteria live at least 1 h and die after 10 h. Thus you need to restrict the generated numbers to the interval $(1, 10)$ by using a conditional density.

(a) Give the exact distribution (or density function) for a random variable which you may use to generate such numbers.

(b) Use any method and write a code in any programming language that allows you to generate random numbers with this particular conditional density.

(c) Now, suppose in addition that each of the individual bacterium when it dies (and only then) either divides and creates two new individuals with probability $1/2$, or it just dies without any descendants with probability $1/2$. Create a program using the lifetime in the previous part that will keep track of the individuals living at any moment in time.

(d) With the previous program, start with one individual at time $t = 0$ h and simulate all its offspring until time $t = 240$ h (10 days). Calculate an approximate value for the expectation of the number of living bacteria 10 days after seeding the culture with 1 bacterium.

(e) Now start with 100 bacteria at time $t = 0$. Simulate each and report the number. Could you use the previous part? Explain.

CHAPTER 4

INTEGRATION THEORY

In the previous chapter, we learned about random variables and their distributions. This distribution completely characterizes a random variable. But, in general, distributions are very complex functions. The human brain cannot comprehend such things easily. So the human brain wants to talk about one typical value. For example, one can give a distribution for the random variable representing players' salaries in the NBA. Here, the variability (probability space) is represented by the specific player chosen. However, probably one is not interested in such a distribution. One simply wants to know what the typical salary is in the NBA. The person probably contemplates a career in sports and wants to find out if as an athlete he should go for basketball or baseball. Thus, he is much better served by comparing only one number corresponding to each of these distributions. Calculating such a number is hard (which number?). In this chapter, we construct a theory which allows us the calculation of any number we want from a given distribution. Paradoxically, to calculate a simple number we need to understand a very complex theory.

Probability and Stochastic Processes, First Edition. Ionuţ Florescu

4.1   Integral of Measurable Functions

Recall that the random variables are nothing more than measurable functions. Let $(\Omega, \mathscr{F}, P)$ be a probability space. We wish to define for any measurable function f an integral of f with respect to the measure P.

Notation. We shall use the following notations for this integral:

$$\int_{\Omega} f(\omega)\mathbf{P}(d\omega) = \int f d\mathbf{P}$$

If a set $A \in \mathscr{F}$ we use the notation

$$\int_A f(\omega)\mathbf{P}(d\omega) = \int_A f d\mathbf{P} = \int f\mathbf{1}_A d\mathbf{P},$$

where $\mathbf{1}_A$ denotes the indicator function of the set A in Ω.

As a note, the Dirac delta we have defined previously allows us to treat summation as another kind of integral. Specifically, let $\{a_n\}$ be a sequence of real numbers. Let $\Omega = \mathbb{R}, \mathscr{F} = \mathscr{B}(\mathbb{R})$, and define a measure on this set as

$$\delta(A) = \sum_{i=1}^{\infty} \delta_i(A),$$

where $\delta_i(A)$ is 1 if i is in the set $A \subseteq \mathbb{R}$ and 0 if it is not.

Then, the function $i \mapsto a_i$ is integrable if and only if $\sum a_i < \infty$. This is evident because we can write

$$\int_{-\infty}^{\infty} a_x d\delta(x) = \int_{-\infty}^{\infty} a_x \sum_{n=1}^{\infty} d\delta_n(x)$$

$$= \sum_{n=1}^{\infty} \int_{-\infty}^{\infty} a_x d\delta_n(x) = \sum_{n=1}^{\infty} a_n$$

What is the point of this? The simple argument above shows that any "discrete" random variable (in the undergraduate text definition) may be treated as a "continuous" random variable.

Integral of simple (elementary) functions   If $A \in \mathscr{F}$, we know that we can define a measurable function by its indicator $\mathbf{1}_A$. We define the integral of this measurable function,

$$\int \mathbf{1}_A d\mathbf{P} = \mathbf{P}(A).$$

We note that this variable has the same distribution as that of the Bernoulli random variable. The variable takes the values 0 and 1 and we can easily calculate the probability that the variable is 1 as

$$\mathbf{P} \circ \mathbf{1}_A^{-1}(\{1\}) = \mathbf{P}\{\omega : \mathbf{1}_A(\omega) = 1\} = \mathbf{P}(A).$$

Therefore, the variable is distributed as a Bernoulli random variable with parameter $p = \mathbf{P}(A)$.

Definition 4.1 (Simple function) *f is called a* simple *(elementary) function if and only if f can be written as a finite linear combination of indicators or, more specifically, there exist sets $A_1, A_2, \ldots, A_n$ all in $\mathscr{F}$ and constants $a_1, a_2, \ldots, a_n$ in $\mathbb{R}$ such that*

$$f(\omega) = \sum_{k=1}^{n} a_k \mathbf{1}_{A_k}(\omega)$$

If the constants a_k are all positive, then f is a positive simple function.

Note that the sets A_i do not have to be disjoint, but an easy exercise (Problem 4.1) shows that f could be written in terms of disjoint sets. The representation is, however, not unique; a simple function is such if it may be represented as a linear combination of indicator functions.

Suppose now that f is any simple function. We define its integral

$$\int f d\mathbf{P} = \sum_{k=1}^{n} a_k \mathbf{P}(A_k) < \infty$$

We adopted the conventions $0 * \infty = 0$ and $\infty * 0 = 0$ in the above summation.
   We need to check that the above definition is proper. Recall that there exist many representations of a simple function and we need to make sure that any such representation produces the same integral value. Furthermore, the linearity and monotonicity properties of the integral may be proven. We skip these results since they are simple to prove and do not bring any additional insight.

Integral of positive measurable functions   For every f positive measurable function $f : \Omega \longrightarrow [0, \infty)$, we define

$$\int f d\mathbf{P} = \sup \left\{ \int h d\mathbf{P} : h \text{ is a simple function, } h \leq f \right\}$$

For a given positive measurable function, can we find a sequence of simple functions that converge to it? The answer is yes, and is provided by the next simple exercise.

Exercise 2 *Let $f : \Omega \to [0, \infty]$ be a positive, measurable function. For all $n \geq 1$, we define*

$$f_n(\omega) := \sum_{k=0}^{n2^n - 1} \frac{k}{2^n} \mathbf{1}_{\left\{ \frac{k}{2^n} \leq f(\omega) < \frac{k+1}{2^n} \right\}}(\omega) + n\mathbf{1}_{\{f(\omega) \geq n\}} \tag{4.1}$$

1. Show that f_n is a simple function on $(\Omega, \mathscr{F})$, for all $n \geq 1$.

2. *Show that the sets present in the indicators in Equation (4.1) form a partition of Ω, for all $n \geq 1$.*

3. *Show that the sequence of simple functions is increasing $g_n \leq g_{n+1} \leq f$, for all $n \geq 1$.*

4. *Show that $g_n \uparrow f$ as $n \to \infty$. Note that this is not an a.s. statement and is true for all $\omega \in \Omega$.*

The solution to this exercise is not complicated, and in fact it is an assigned problem (Problem 4.3).

The following lemma is a very easy to understand and useful tool.

Lemma 4.2 *Suppose f is a positive measurable function and that its integral is zero, that is,*

$$\int f d\mathbf{P} = 0.$$

Then $\mathbf{P}\{f > 0\} = 0$ (or put otherwise $f = 0$ a.s.).

Proof: We have $\{f > 0\} = \bigcup_{n \geq 0}\{f > \frac{1}{n}\}$. Since the events are increasing by the monotone convergence property of measure, we must have

$$\mathbf{P}\{f > 0\} = \lim_{n \to \infty} \mathbf{P}\{f > \frac{1}{n}\}.$$

If we assume by absurd that $\mathbf{P}\{f > 0\} > 0$, then there must exist an n such that $\mathbf{P}\{f > \frac{1}{n}\} > 0$. However, in this case, by the definition of the integral of positive measurable functions,

$$\int f d\mathbf{P} \geq \int \frac{1}{n} \mathbf{1}_{\{f > \frac{1}{n}\}} d\mathbf{P} > 0,$$

which is a contradiction. ∎

The next theorem is one of the most useful in probability theory. In our immediate context, it tells us that the integral of positive measurable functions is well defined.

Theorem 4.3 (Monotone convergence theorem) *If $\{f_n\}_n$ is a sequence of measurable positive functions such that $f_n \uparrow f$, then f is measurable and*

$$\int_\Omega f_n(\omega) \mathbf{P}(d\omega) \uparrow \int_\Omega f(\omega) \mathbf{P}(d\omega)$$

Proof: We start by showing that the limit f is a measurable function. Let $I = (-\infty, a]$ be a semiclosed interval. We want to show that the pre-image through f of this interval is in the σ-algebra $\mathscr{F}$.

Remember that f_n is an increasing sequence, thus $f_n(\omega) \leq f(\omega)$ for all n and ω. Therefore, for any ω such that $f(\omega) \in I = (-\infty, a)$, we also have $f_n(\omega) \in I$ for all n. The reciprocal argument reads from end to the beginning, and so we must have $f^{-1}(I) = \cap_n f_n^{-1}(I)$. Since each of the f_n's are measurable, each set $f_n^{-1}(I) \in \mathscr{F}$. Since $\mathscr{F}$ is a sigma algebra, their countable intersection is also in $\mathscr{F}$, which shows that $f^{-1}(I) \in \mathscr{F}$.

So, for any generator $I = (-\infty, a]$ of the Borel sigma algebra, we have $f^{-1}(I) \in \mathscr{F}$, and a standard argument used previously shows that the result holds for any I set in the Borel sigma algebra, and therefore f is measurable. Since all f_n's are positive and $f(x) = \lim_{n \to \infty} f_n(x)$, we clearly have that f must also be nonnegative at every point. f is measurable and positive, therefore the integral is well defined.

Now let us prove the main result. By definition,

$$\int f d\mathbf{P} = \sup \left\{ \int h d\mathbf{P} : h \text{ is a simple function}, \ h \le f \right\}$$

Since $f_n(x) \le f(x)$, at every point x we have

$$\left\{ \int h d\mathbf{P} : h \text{ is simple f.}, \ h \le f_n \right\} \subseteq \left\{ \int h d\mathbf{P} : h \text{ is simple f.}, \ h \le f \right\}.$$

Since the supremum on a subset is not greater than the supremum of the whole set and going to the limit, we get

$$\lim_n \int f_n d\mathbf{P} \le \int f d\mathbf{P},$$

where the limit exists since the sequence is monotonically increasing and bounded by the supremum.

Finally, we want to show the inequality in the other direction, namely that $\lim_n \int f_n d\mathbf{P} \ge \int f d\mathbf{P}$. By the definition of the integral of f, there exists a non-decreasing sequence h_k of positive simple functions such that $h_n \le f$ for all n and

$$\lim_n \int h_n d\mathbf{P} = \int f d\mathbf{P}$$

We want to prove that, if $h_n(x) \le \lim_k f_k(x)$ for all x, then

$$\int h_n d\mathbf{P} \le \lim_k \int f_k d\mathbf{P}.$$

This will imply the reverse inequality.

Recall that for a simple function $h_n = \sum a_i \mathbf{1}_{A_i}$ the integral is defined as

$$\int h_n d\mathbf{P} = \sum_i a_i \mathbf{P}(A_i)$$

Let us pick the support sets A_i in such a way that they form a partition of Ω. Recall that the sets must be disjoint and their union is Ω. Let us look at one such set denoted

$A \in \mathcal{F}$. Suppose that the corresponding constant term in h_n is a. For a fixed $\varepsilon > 0$, define the sets

$$A_k = \{x \in A \mid f_k(x) \geq a - \varepsilon\}.$$

Since the integral is monotone and everything is positive, we have

$$\int_A f_k d\mathbf{P} \geq \int_A (a - \varepsilon) \mathbf{1}_{A_k} d\mathbf{P} = (a - \varepsilon)\mathbf{P}(A_k),$$

or

$$\mathbf{P}(A_k) \leq \frac{1}{a - \varepsilon} \int_A f_k d\mathbf{P}$$

Since for all x we must have $h_n(x) \leq \lim_k f_k(x)$, then for any x in A we will find a sufficiently large k such that $x \in A_k$. But this is just saying that

$$A \subseteq \bigcup_k A_k,$$

and since each A_k was made with elements from A, then the two sets must be equal. Since f_k is an increasing sequence, we must have that A_k is an increasing sequence of sets (i.e., $A_k \subseteq A_{k+1}$ – just follow the definition). By the monotone convergence properties of probability, we have

$$\mathbf{P}(A) = \lim_k \mathbf{P}(A_k).$$

So we can write

$$a\mathbf{P}(A) = a \lim_k \mathbf{P}(A_k) \leq \frac{a}{a - \varepsilon} \int_A f_k d\mathbf{P}.$$

This is true for all $\varepsilon > 0$ and the only way this is happening is if

$$a\mathbf{P}(A) \leq \int_A f_k d\mathbf{P}.$$

Repeating this argument with every set in the partition, we finally obtain that

$$\int h_n d\mathbf{P} = \sum_i a_i \mathbf{P}(A_i) \leq \sum_i \int_{A_i} f_k d\mathbf{P} = \int f_k d\mathbf{P},$$

since A_i form a partition of Ω. Taking $k \to \infty$, we obtain that for every n we must have

$$\int h_n d\mathbf{P} \leq \lim_k \int f_k d\mathbf{P},$$

which finally concludes the proof. ∎

Integral of measurable functions   Let f be any measurable function. Then we write $f = f^+ - f^-$, where

$$f^+(s) = \max\{f(s), 0\}$$
$$f^-(s) = \max\{-f(s), 0\}$$

Then f^+ and f^- are positive measurable functions and $|f| = f^+ + f^-$. Since they are positive measurable, their integrals are well defined by the previous part.

Definition 4.4   *We define* $L^1(\Omega, \mathscr{F}, P)$ *as being the space of all functions f such that*

$$\int |f| d\mathbf{P} = \int f^+ d\mathbf{P} + \int f^- d\mathbf{P} < \infty$$

For any f in this space, which we will shorten to $L^1(\Omega)$ *or even simpler to* L^1, *we define*

$$\int f d\mathbf{P} = \int f^+ d\mathbf{P} - \int f^- d\mathbf{P}$$

Note: With the above, it is trivial to show that $|\int f d\mathbf{P}| \leq \int |f| d\mathbf{P}$.

Linearity   If $f, g \in L^1(\Omega)$ with $a, b \in \mathbb{R}$, then

$$af + bg \in L^1(\Omega)$$
$$\int (af + bg) d\mathbf{P} = a \int f d\mathbf{P} + b \int g d\mathbf{P}$$

This follows immediately from the way the integral is constructed.

Lemma 4.5 (Fatou's Lemma for measurable functions)   *If one of the following is true,*

a) $\{f_n\}_n$ *is a sequence of positive measurable functions, or*

b) $\{f_n\} \subset L^1(\Omega)$,

then

$$\int \liminf_n f_n d\mathbf{P} \leq \liminf_n \int f_n d\mathbf{P}$$

Proof: Note that $\liminf_n f_n = \lim_{m \to \infty} \inf_{n \geq m} f_n$, where $\lim_{m \to \infty} \inf_{n \geq m} f_n$ is an increasing sequence.
Let $g_m = \inf_{n \geq m} f_n$, and $n \geq m$:

$$f_n \geq \inf_{n \geq m} f_m = g_m \Rightarrow \int f_n d\mathbf{P} \geq \int g d\mathbf{P} \Rightarrow \int g_m d\mathbf{P} \leq \inf_{n \geq m} \int f_n d\mathbf{P}$$

Now g_m increases, so we may use the Monotone Convergence Theorem and get

$$\int \lim_{m\to\infty} g_m d\mathbf{P} = \lim_{m\to\infty} \int g_m d\mathbf{P} \leq \lim_{m\to\infty} \inf_{n\geq m} \int f_n d\mathbf{P} = \liminf_n \int f_n d\mathbf{P}$$

∎

Theorem 4.6 (Dominated convergence theorem) *If f_n, f are measurable, $f_n(\omega) \to f(\omega)$ for all $\omega \subset \Omega$ and the sequence f_n is dominated by $g \in L^1(\Omega)$:*

$$|f_n(\omega)| \leq g(\omega), \qquad \forall \omega \in \Omega, \forall n \in \mathbb{N}$$

then

$$f_n \to f \text{ in } L^1(\Omega) \qquad \left(\text{i.e. } \int |f_n - f| d\mathbf{P} \to 0 \right)$$

Thus, $\int f_n d\mathbf{P} \to \int f d\mathbf{P}$ and $f \in L^1(\Omega)$.

The standard argument   This argument is the most important argument in probability theory. Suppose we want to prove that some property holds for all functions h in some space such as $L^1(\Omega)$ or the space of measurable functions.

1. Show that the result is true for all indicator functions.

2. Use linearity to show that the result holds true for all f simple functions.

3. Use the Monotone Convergence Theorem to obtain the result for measurable positive functions.

4. Finally, from the previous step and writing $f = f^+ - f^-$, we show that the result is true for all measurable functions.

4.2 Expectations

Since a random variable is just a measurable function, we just need to particularize the results of the previous section. An integral with respect to a probability measure is called an *expectation*. Let $(\Omega, \mathscr{F}, P)$ be a probability space.

Definition 4.7 *For X a r.v. in $L^1(\Omega)$, define*

$$\mathbf{E}(X) = \int_\Omega X d\mathbf{P} = \int_\Omega X(\omega) d\mathbf{P}(\omega) = \int_\Omega X(\omega) \mathbf{P}(d\omega)$$

This expectation has the same properties of the integral defined before and some extra ones since the space has finite measure. We restate the theorems we proved earlier in the expectation context since we shall use them later in this book.

Convergence theorems

Theorem 4.8 (Monotone convergence theorem) *If X_n is any sequence of random variables which are positive (take values in $[0, \infty)$, increasing ($X_n(\omega) \uparrow X(\omega)$ for any ω), and the expectation exists ($X_n \in L^1$ for all n), then*

$$\mathbf{E}(X_n) \rightarrow \mathbf{E}(X) \leq \infty.$$

Theorem 4.9 (Dominated convergence theorem) *Let X_n be a convergent sequence of random variables $X_n(\omega) \rightarrow X(\omega)$ for all $\omega \in \Omega$ with the property*

$$|X_n(\omega)| \leq Y(\omega), \ \forall \omega \in \Omega$$

where Y is some integrable random variable ($Y \in L^1(\Omega)$). Then it follows that

$$\mathbf{E}(|X_n - X|) \rightarrow 0 \ (\text{and thus } \mathbf{E}[X_n] \rightarrow \mathbf{E}[X])$$

The theorem remains true if all sure statements ($\forall \omega \in \Omega$) are replaced with almost sure statements.

The next theorem is just a consequence of the dominated convergence theorem. We are stating it here because we shall refer to it later in this book.

Theorem 4.10 (Bounded convergence theorem) *If X_n is a convergent sequence of random variables $X_n(\omega) \rightarrow X(\omega)$ for all $\omega \in \Omega$ and there exists a constant M such that $|X_n(\omega)| \leq M, \forall \omega \in \Omega$, then*

$$\mathbf{E}(|X_n - X|) \rightarrow 0 \ (\text{and thus } \mathbf{E}[X_n] \rightarrow \mathbf{E}[X])$$

Once again, the theorem remains true if all sure statements ($\forall \omega \in \Omega$) are replaced with almost sure statements.

Lemma 4.11 (Fatou's lemma) *If either $\{X_n\}_n$ is a sequence of positive random variables or $X_n \in L^1(\Omega)$ for any n, then*

$$\mathbf{E}(\liminf_{n \rightarrow \infty} X_n) \leq \liminf_{n \rightarrow \infty} \mathbf{E}(X_n)$$

▮ EXAMPLE 4.1

Let $f(x, y)$ be a continuous function defined on the 2D unit square

$$D = \{(x, y) : 0 \leq x \leq 1, \ 0 \leq y \leq 1\}.$$

Suppose that $\mathbf{E}[f(U_1, U_2)] = 0$ when U_1 and U_2 are independent $U[0, 1]$ random variables. Give a detailed, rigorous proof that there must exist a point $(x_0, y_0) \in D$ for which $f(x_0, y_0) = 0$.

This is a mathematical example that shows that the basic analysis reasoning is working. Since the two variables U_1, U_2 are independent, their joint distribution is just the product of the marginal, and since they are both uniform, we may express the expectation according to the definition as

$$\mathbf{E}[f(U_1, U_2)] = \int_0^1 \int_0^1 f(x, y)dxdy = 0$$

If there is no point in D for which the value of f is zero since the function is continuous, then the function must be either strictly positive or strictly negative on the entire square. In either case, the integral cannot be zero.

Therefore there must exist a zero of the function in the closed square D. The point is in fact in the interior of D using the intermediate value theorem or just applying a reasoning as above.   ∎

Inequalities involving expectations

Now let us present specific properties of the expectation. This is to be expected since the space has finite measure and therefore we can obtain more specific properties.

Proposition 4.12 (Markov inequality:) *Let Z be a r.v. and let $g : \mathbb{R} \longrightarrow [0, \infty]$ be an* increasing *measurable function. Then*

$$\mathbf{E}\left[g(Z)\right] \geq \mathbf{E}\left[g(Z)\mathbf{1}_{\{Z \geq c\}}\right] \geq g(c)\mathbf{P}(Z \geq c)$$

Thus,

$$\mathbf{P}(Z \geq c) \leq \frac{\mathbf{E}[g(Z)]}{g(c)}$$

for all g increasing functions and $c > 0$.

▣ EXAMPLE 4.2   Special cases of the Markov inequality

If we take $g(x) = x$ an increasing function and X a positive random variable, then we obtain

$$\mathbf{P}(Z \geq c) \leq \frac{\mathbf{E}(Z)}{c}.$$

To get rid of the necessity that $X \geq 0$, take $Z = |X|$. Then we obtain the classical form of the Markov inequality

$$\mathbf{P}(|X| \geq c) \leq \frac{\mathbf{E}(|X|)}{c}.$$

If we take $g(x) = x^2$, $Z = |X - \mathbf{E}(X)|$, and we use the variance definition (which we will see in a minute), we obtain the Chebyshev inequality

$$\mathbf{P}(|X - \mathbf{E}(X)| \geq c) \leq \frac{Var(X)}{c^2}.$$

If we denote $\mathbf{E}(X) = \mu$ and $Var(X) = \sigma^2$, and we take $c = k\sigma$ in the previous inequality, we will obtain the classical Chebyshev inequality presented in undergraduate courses:

$$\mathbf{P}(|X - \mu| \geq k\sigma) \leq \frac{1}{k^2}.$$

If $g(x) = e^{\theta x}$, with $\theta > 0$, then

$$\mathbf{P}(Z \geq c) \leq e^{-\theta c}\mathbf{E}(e^{\theta z}).$$

This inequality states that the tail of the distribution decays exponentially in c if Z has finite exponential moments. With simple manipulations, one can obtain Chernoff's inequality using it.

The following definition is just a reminder:

Definition 4.13 *A function $g : I \longrightarrow \mathbb{R}$ is called a convex function on I (where I is any open interval in $\mathbb{R}$), if its graph lies below any of its chords. Mathematically, for any $x, y \in I$ and for any $\alpha \in (0, 1)$, we have*

$$g(\alpha x + (1 - \alpha)y) \leq \alpha g(x) + (1 - \alpha)g(y).$$

Some examples of convex functions on the whole $\mathbb{R}$ are $|x|$, x^2, and $e^{\theta x}$ with $\theta > 0$.

Lemma 4.14 (Jensen's inequality) *Let f be a convex function and let X be a r.v. in $L^1(\Omega)$. Assume that $\mathbf{E}(f(X)) \leq \infty$: then*

$$f(\mathbf{E}(X)) \leq \mathbf{E}(f(X))$$

Proof: Skipped. The classical approach indicators $\rightarrow$ simple functions $\rightarrow$ positive measurable $\rightarrow$ measurable is a standard way to prove Jensen's inequality. ∎

■ **EXAMPLE 4.3**

Let X be a random variable with $\mathbf{P}(-1 < X < 1) = 1$ and with $E(X)$ finite. Let U be independent of X with a uniform distribution on $(-1, 1)$. Show that $E(|X|) < E(|X - U|)$.

At first glance, this example may seem like a straight application of a general theorem. The clue that this is not the case is provided by the specific use of the uniform distribution and the special property of the random variable (takes values only between

−1 and 1). So let us calculate the expectation instead.

$$\mathbf{E}[|X - U|] = \int_{-1}^{1} \int_{-1}^{1} |x - u| \frac{1}{2} \, du \, dF(x)$$

$$= \frac{1}{2} \int_{-1}^{1} \left(\int_{-1}^{x} (x - u) \, du + \int_{x}^{1} (u - x) \, du \right) dF(x)$$

$$= \frac{1}{2} \int_{-1}^{1} (x^2 + 1) \, dF(x)$$

$$= \frac{1}{2} \left(\mathbf{E}[X^2] + 1 \right),$$

the last equality being due to the fact that $\mathbf{P}(-1 < X < 1) = 1$.

Now looking back at what we need to prove, we realize that it is sufficient to show that, for any variable with $\mathbf{P}(-1 < X < 1) = 1$, we have $\mathbf{E}|X| < \frac{1}{2}\left(\mathbf{E}[X^2] + 1\right)$. This inequality may be rewritten as

$$\mathbf{E}\left(X^2 - 2|X| + 1\right) > 0.$$

Take a random variable X with the desired property. We write

$$\mathbf{E}\left(X^2 - 2|X| + 1\right) = \mathbf{E}\left(\left(X^2 - 2|X| + 1\right)\mathbf{1}_{\{X>0\}} + \left(X^2 - 2|X| + 1\right)\mathbf{1}_{\{X\leq 0\}}\right)$$

$$= \mathbf{E}\left(\left(X - 1\right)^2 \mathbf{1}_{\{X>0\}}\right) + \mathbf{E}\left(\left(X + 1\right)^2 \mathbf{1}_{\{X\leq 0\}}\right) \geq 0$$

as a sum of expectations of positive terms. The only case when we have equality is when both terms are zero and that can only happen when the random variable X takes values in the set $\{-1, 1\}$. But this contradicts $\mathbf{P}(-1 < X < 1) = 1$, so done. ∎

L^p spaces. Inequalities

We generalize the L^1 notion presented earlier in the following way: For $1 \leq p \leq \infty$, we define the space

$$L^p(\Omega, \mathscr{F}, P) = L^p(\Omega) = \left\{ X : \Omega \longrightarrow \mathbb{R} : \mathbf{E}\left[|X|^p\right] = \int |X|^p d\mathbf{P} < \infty \right\}.$$

On this space, we define a norm called the p-norm as

$$||X||_p = \mathbf{E}\left[|X|^p\right]^{1/p}$$

Lemma 4.15 (Properties of L^p spaces) *We have*

(i) L^p is a vector space. (i.e., if $X, Y \in L^p$ and $a, b \in \mathbb{R}$ then $aX + bY \in L^p$).

(ii) L^p is complete (every Cauchy sequence in L^p is convergent).

Lemma 4.16 (Cauchy–Bunyakovsky–Schwarz inequality) *If $X, Y \in L^2(\Omega)$, then $X, Y \in L^1(\Omega)$ and*

$$|\mathbf{E}[XY]| \leq \mathbf{E}[|XY|] \leq ||X||_2 ||Y||_2.$$

A historical remark. This inequality, which is one of the most famous and useful in any area of analysis (not only probability), is usually credited to Cauchy for sums and Schwartz for integrals and is usually known as the Cauchy–Schwartz inequality. However, the Russian mathematician Victor Yakovlevich Bunyakovsky (1804–1889) discovered and first published the inequality for integrals in 1859 (when Schwartz was 16). Unfortunately, he was born in Eastern Europe... However, all who are born in Eastern Europe (including myself) learn the inequality by its proper name.

Proof: The first inequality is clear by the Jensen inequality. We need to show that

$$\mathbf{E}[|XY|] \leq (\mathbf{E}[X^2])^{1/2} (\mathbf{E}[Y^2])^{1/2}$$

Let $W = |X|$ and $Z = |Y|$; then $W, Z \geq 0$.
Truncation:
   Let $W_n = W \wedge n$ and $Z_n = Z \wedge n$; that is

$$W_n(\omega) = \begin{cases} W(\omega), & \text{if } W(\omega) < n \\ \\ n, & \text{if } W(\omega) \geq n. \end{cases}$$

Clearly, defined in this way, W_n, Z_n are bounded. Let $a, b \in \mathbb{R}$ two constants. Then

$$0 \leq \mathbf{E}[(aW_n + bZ_n)^2] = a^2 \mathbf{E}(W_n^2) + 2ab\mathbf{E}(W_n Z_n) + b^2 \mathbf{E}(Z_n^2)$$

If we let $a/b = c$, we get

$$c^2 \mathbf{E}(W_n^2) + 2c\mathbf{E}(W_n Z_n) + \mathbf{E}(Z_n^2) \geq 0 \quad \forall c \in \mathbb{R}$$

This means that the quadratic function in c has to be positive. But this is possible only if the discriminant of the quadratic function is negative and the leading coefficient $\mathbf{E}(W_n^2)$ is strictly positive, the latter condition being obviously true. Thus we must have

$$4(\mathbf{E}(W_n Z_n))^2 - 4\mathbf{E}(W_n^2)\mathbf{E}(Z_n^2) \leq 0$$

$$\Rightarrow (\mathbf{E}(W_n Z_n))^2 \leq \mathbf{E}(W_n^2)\mathbf{E}(Z_n^2) \leq \mathbf{E}(W^2)\mathbf{E}(Z^2) \quad \forall n$$

If we let $n \uparrow \infty$ and use the monotone convergence theorem, we get

$$(\mathbf{E}(WZ))^2 \leq \mathbf{E}(W^2)\mathbf{E}(Z^2).$$

∎

A more general inequality is

Lemma 4.17 (Hölder inequality) *If* $1/p + 1/q = 1$, $X \in L^p(\Omega)$ *and* $Y \in L^q(\Omega)$ *then* $XY \in L^1(\Omega)$, *and*

$$\mathbf{E}|XY| \leq \|X\|_p \|Y\|_q = (\mathbf{E}|X|^p)^{\frac{1}{p}} (\mathbf{E}|Y|^q)^{\frac{1}{q}}$$

Proof: The proof is simple and uses the following inequality (Young inequality). If a and b are positive real numbers and p, q are as in the theorem, then

$$ab \leq \frac{a^p}{p} + \frac{b^q}{q},$$

with equality if and only if $a^p = b^q$.

Taking this inequality as given (not hard to prove), define

$$f = \frac{|X|}{\|X\|_p}, \quad g = \frac{|Y|}{\|Y\|_p}.$$

Note that the Hölder inequality is equivalent to $\mathbf{E}[fg] \leq 1$ ($\|X\|_p$ and $\|Y\|_q$ are just numbers that can be taken in and out of integral by the linearity property). To prove this, apply the Young inequality to $f \geq 0$ and $g \geq 0$ and then integrate to obtain

$$\mathbf{E}[fg] \leq \frac{1}{p}\mathbf{E}[f^p] + \frac{1}{q}\mathbf{E}[g^q] = \frac{1}{p} + \frac{1}{q} = 1$$

$\mathbf{E}[f^p] = 1$, and similarly for g may be easily checked. Finally, the extreme cases ($p = 1$, $q = \infty$, etc.) may be treated separately. ∎

This inequality and the Riesz representation theorem create the notion of conjugate space. This notion is only provided to create links with real analysis. For further details, we recommend Royden (1988).

Definition 4.18 (Conjugate space of L^p) *For* $p > 0$, *let* $L^p(\Omega)$ *define the space on* $(\Omega, \mathcal{F}, \mathbf{P})$. *The number* $q > 0$ *with the property* $1/p + 1/q = 1$ *is called the* conjugate index *of* p. *The corresponding space* $L^q(\Omega)$ *is called the* conjugate space *of* $L^p(\Omega)$.

Any of these spaces is a metric space (with the distance induced by the norm, i.e., $d(f, g) = \|f - g\|_p$). The fact that this is a properly defined linear space is implied by the triangle inequality in L^p, which is the next theorem.

Lemma 4.19 (Minkowski inequality) *If* $X, Y \in L^p$, *then* $X + Y \in L^p$, *and furthermore*

$$\|X + Y\|_p \leq \|X\|_p + \|Y\|_p$$

Proof: We clearly have

$$|X + Y|^p \le 2^{p-1}(|X|^p + |Y|^p).$$

For example, use the definition of convexity for the function x^p with $x = |X|, y = |Y|$, and $\alpha = 1/2$. Now integrating implies that $X + Y \in L^p$. Now we can write

$$\|X + Y\|_p^p = \mathbf{E}[|X + Y|^p] \le \mathbf{E}\left[(|X| + |Y|)|X + Y|^{p-1}\right]$$

$$= \mathbf{E}\left[|X||X + Y|^{p-1}\right] + \mathbf{E}\left[|Y||X + Y|^{p-1}\right]$$

$$\overset{\text{Hölder}}{\le} (\mathbf{E}[|X|^p])^{1/p}\left(\mathbf{E}\left[|X + Y|^{(p-1)q}\right]\right)^{1/q}$$

$$+ (\mathbf{E}[|Y|^p])^{1/p}\left(\mathbf{E}\left[|X + Y|^{(p-1)q}\right]\right)^{1/q}$$

$$\overset{\left(q = \frac{p}{p-1}\right)}{=} (\|X\|_p + \|Y\|_p)\,(\mathbf{E}[|X + Y|^p])^{1 - \frac{1}{p}}$$

$$= (\|X\|_p + \|Y\|_p)\frac{\mathbf{E}[|X + Y|^p]}{\|X + Y\|_p}$$

Now, identifying the left and right hand after simplifications, we obtain the result. ■

The case of L^2 The case when $p = 2$ is quite special. This is because 2 is its own conjugate index ($1/2 + 1/2 = 1$). Because of this, the space is quite similar to the Euclidian space. If $f, g \in L^2$, we may define the inner product

$$< f, g >= \int fg\,d\mathbf{P}$$

which is well defined by the Cauchy–Bunyakovsky–Schwartz inequality. The existence of the inner product and the completeness of the norm makes L^2 a Hilbert space with all the benefits that follow. In particular, the notion of orthogonality is well defined (f and g in L^2 are orthogonal if and only if $< f, g >= 0$) and this allows the Fourier representation and in general representations in terms of an orthonormal basis of functions in L^2. We do not wish to present more details than necessary; consult (Billingsley, 1995, Section 19) for further references.

Examples

📖 **EXAMPLE 4.4 Applying Jensen's Inequality**

Take X a discrete random variable with N outcomes $x_1, \ldots, x_N$ and associated probabilities $p_1, \ldots, p_N$. Apply Jensen inequality for this X and $f(x) = x^2$ to obtain

$$\left(\sum_{i=1}^{N} x_i p_i\right) \le \sum_{i=1}^{N} x_i^2 p_i$$

or, written in simpler terms, if all p_i's are equal

$$\frac{1}{N}\sum_{i=1}^{N} x_i \leq \sqrt{\frac{1}{N}\sum_{i=1}^{N} x_i^2}$$

This inequality may also be obtained directly from Cauchy-Buniakovsky–Schwartz inequality by taking $Y = 1$.

If we take $f(x) = 1/x$, we obtain the classical inequality between the regular mean and the harmonic mean *for positive numbers*. The harmonic mean of numbers $x_1, \ldots, x_N$ is defined as

$$\left(\frac{1}{N}\sum_{i=1}^{N} x_i^{-1}\right)^{-1} = \frac{N}{\frac{1}{x_1} + \frac{1}{x_2} + \cdots + \frac{1}{x_N}}$$

We need positive numbers because the function $1/x$ is convex only on the positive axis (it is concave on $(-\infty, 0)$). Applying Jensen inequality to the random variable X which takes only positive values $x_1, \ldots, x_N$ each with probability $1/N$ and using the function $1/x$, we obtain

$$\frac{1}{\frac{1}{N}\sum_{i=1}^{N} x_i} \leq \sum_{i=1}^{N} \frac{1}{x_i}\frac{1}{N} = \frac{\sum_{i=1}^{N} \frac{1}{x_i}}{N}.$$

Rewriting this gives

$$\frac{N}{\frac{1}{x_1} + \frac{1}{x_2} + \cdots + \frac{1}{x_N}} \leq \frac{1}{N}\sum_{i=1}^{N} x_i$$

or simply put, the harmonic mean of positive numbers is always less than the regular mean of the same numbers. Clearly, the same result holds if one replaces equal weights with unequal probabilities p_i.

We may also obtain a similar inequality for the geometric mean. The geometric average of N positive numbers is defined as

$$\sqrt[N]{x_1 x_2 \ldots x_N}$$

To see this, we use the function $\ln(x)$ which is concave when defined for positive numbers with Jensen's inequality, and we obtain

$$\mathbf{E}[\ln X] \leq \ln(\mathbf{E}[X]).$$

Again, if we assume that the random variable X is discrete with positive values x_i, this inequality is written as

$$\sum_{i=1}^{N} (\ln x_i)\frac{1}{N} \leq \ln\left(\sum_{i=1}^{N} x_i\frac{1}{N}\right).$$

Now exponentiating both sides and using $e^{\ln x} = x$, we get

$$\sqrt[N]{x_1 x_2 \ldots x_N} \leq \frac{1}{N} \sum_{i=1}^{N} x_i$$

or the well-known relation that geometric average is less than arithmetic average.

◼ EXAMPLE 4.5   Examples from Finance using Jensen

We recall that the sample variance of n i.i.d. random variables $X_1, X_2, \ldots, X_n$ is defined as

$$S_n^2 = \frac{1}{n-1} \sum_{i=1}^{n} (X_i - \bar{X})^2,$$

where $\bar{X} = \frac{1}{n} \sum X_i$ is the sample mean or average of the n numbers. The sample standard deviation is defined as the square root of the sample variance, or

$$S_n = \sqrt{\frac{1}{n-1} \sum_{i=1}^{n} (X_i - \bar{X})^2}.$$

We say that an estimator is *unbiased* if its expectation is equal to the parameter it estimates. In the context above, it may be checked that the sample variance is an unbiased estimator for the variance of the distribution of X_i's denoted with σ^2, or mathematically

$$\mathbf{E}[S_n^2] = \sigma^2.$$

A further useful property of estimators is *consistency*. We say that an estimator is consistent if it converges in some way to the parameter it estimates as the sample size n goes to infinity. The sample estimator is in fact consistent as well.

In finance, the observations used are typically $X_1, \ldots, X_n$ returns on a stock for some intervals $t, t + \Delta t$. To be specific, if the data is equity sampled at intervals Δt, $E_0, E_{\Delta t}, E_{2\Delta t}, \ldots, E_{n\Delta t}$, then the continuously compounded returns X_i are defined as

$$X_i = \log \left(\frac{E_{i\Delta t}}{E_{(i-1)\Delta t}} \right)$$

If we assume that these returns are i.i.d., then the standard deviation is an important parameter called the *volatility* of the equity. Naturally, one would use the sample standard deviation to estimate the volatility. Despite the fact that the sample variance is an unbiased estimate of the true variance, the sample standard deviation is biased low as an estimate of the true volatility. This can be shown easily using the Jensen's Inequality, since the square root is a concave function:

$$\mathbf{E}[S_n] = \mathbf{E}[\sqrt{S_n^2}] < \sqrt{\mathbf{E}[S_n^2]} = \sqrt{\sigma^2} = \sigma$$

As a parenthesis, the S_n estimator is consistent (i.e., it converges to the true parameter as $n \to \infty$.

◼ EXAMPLE 4.6   Portfolio Maximization

Let us continue the series of finance examples with a classical problem in finance, namely that of maximizing a portfolio return. Often, agents or portfolio managers are modeled using a risk-averse utility function to do this. More specifically, each agent is endowed with a particular deterministic utility function U which is an increasing and concave function. The function defines the strategy the agent has for choosing portfolio weights. The agent chooses the weights that maximize the expected utility of the portfolio return. Suppose $W(\lambda_1, \ldots, \lambda_k)$ is the random variable denoting future wealth if the invested weight in equity i in the portfolio is λ_i. The agents find the optimal weights λ_i by solving the following problem:

$$\max_{\lambda_1, \ldots, \lambda_k} \mathbf{E}[U(W(\lambda_1, \ldots, \lambda_k))].$$

The optimal weights are the maximizing values of this function.

The properties of the utility functions ensure that the agents have two basic characteristics that are reasonable. First, since the utility function is increasing, the agents always prefer more to less. Second, since the utility function is concave, the agents are risk-averse. Specifically, applying the Jensen inequality we always have

$$\mathbf{E}[U(W)] \leq U(\mathbf{E}[W]).$$

Basically, a risk-averse agent likes more to get a certain expected wealth ($\mathbf{E}[W]$ which is a fixed amount) rather than the same amount on average (i.e., with variability around that quantity).

◼ EXAMPLE 4.7   Siegel's paradox

In this example, we talk about exchange rates. To understand what these are, suppose $r = 1.31$ is the euro–dollar rate or the exchange rate between Euros and US dollars NOW. That means for each 1 Euro exchanged, one receives 1.31 US dollars. This number is the spot rate, and typically we have that the exchange rate for the reverse operation (exchanging dollars for euro) is the inverse of this rate, that is,

$$\text{Rate US dollars to euro} = \frac{1}{1.31} = 0.76$$

However, suppose that my mom living in Romania wishes to travel to visit me in the summer. She could exchange euros for dollars right now using the spot rate, though she will lock up the amount in dollars which she cannot spend for 3 months. So she decides to buy a financial instrument which is called a *forward rate*.

In finance, one can buy such a contract called a *future* on an exchange rate. This contract will allow to exchange a fixed sum (called the *principal amount*) in the future at some predetermined time T (in this case 3 months) for this forward rate r agreed today. However, note that, unlike the spot rate which is known now, the forward rate is in fact a random variable. In fact, if we denote by r_t the value of this random variable at time t, this is a stochastic process and we will talk about these processes in the second part of the book. For now, let us note that r_T is a random variable, which is the exchange rate at time T. The actual forward rate today is the expected value of this random variable given the information available today. In mathematical terms, we have

$$r = \mathbf{E}[r_T \mid \mathscr{F}_0],$$

where the sigma algebra $\mathscr{F}_0$ contains the information available today. If we make the assumption that the information $\mathscr{F}_0$ is already incorporated in the rate today, then, in fact, the forward rate

$$r = \mathbf{E}[r_T].$$

The Siegel paradox remarks an interesting fact about this forward rate. One would think that the forward exchange rate USD to EU will have the same relationship with the inverse rate EU to USD. However, that is not so, and the culprit is the Jensen inequality, once again. If we use the obvious notations r_{USDEU} and r_{EUUSD} and recall the relationship that exists between the spot rates,

$$r_{USDEU} = \mathbf{E}\left[r_{USDEU}(T)\right]$$
$$= \mathbf{E}\left[\frac{1}{r_{EUUSD}(T)}\right] > 1/E[r].$$

Therefore, the future rates do not follow the obvious relationship that exists between the spot rates. Jeremy Siegel showed this relationship for the first time and as a consequence the currency forward rate cannot be an unbiased estimate of the future spot rate because an expected increase in one exchange rate implies an expected decrease of smaller magnitude in its reciprocal.

Therefore, even if expected changes in the spot rate are distributed symmetrically around the forward rate from the perspective of one investor, "Siegel's paradox" guarantees that the forward rate will be biased from the perspective of the investor on the other side of the exchange rate.

◼ EXAMPLE 4.8 due to Erdós

Suppose there are 17 fence posts around the perimeter of a field and exactly 5 of them are rotten. Show that irrespective of which of these five are rotten, there should exist a row of seven consecutive posts of which at least three are rotten.

Proof (Solution): First we label the posts $1, 2 \cdots 17$. Now, define

$$I_k = \begin{cases} 1 & \text{if post } k \text{ is rotten} \\ 0 & \text{otherwise} \end{cases}$$

For any fixed k, let R_k denote the number of rotten posts among $k+1, \cdots, k+7$ (starting with the next one). Note that when any of $k+1, \cdots, k+7$ is larger than 17, we start again from 1 (i.e., modulo $17+1$).

Now pick a post at random; this obviously can be done in 17 ways with equal probability. Then after we pick this post, we calculate the number of rotten boards. We have

$$\mathbf{E}(R_k) = \sum_{k=1}^{17} (I_{k+1} + \cdots + I_{k+7}) \frac{1}{17}$$

$$= \frac{1}{17} \sum_{k=1}^{17} \sum_{j=1}^{7} I_{k+j} == \frac{1}{17} \sum_{j=1}^{7} \sum_{k=1}^{17} I_{j+k}$$

$$= \frac{1}{17} \sum_{j=1}^{7} 5 \qquad \text{(the sum is 5 since we count all the rotten posts in the fence)}$$

$$= \frac{35}{17}$$

Now, $35/17 > 2$, which implies $\mathbf{E}(R_k) > 2$. Therefore, $\mathbf{P}(R_k > 2) > 0$ (otherwise the expectation is necessarily bounded by 2) and since R_k is integer valued, $\mathbf{P}(R_k \geq 3) > 0$. So there exists some k such that $R_k \geq 3$.

Of course, now that we see the proof, we can play around with numbers and see that there exists a row of 4 consecutive posts in which at least 2 are rotten, or that there must exist a row of 11 consecutive posts in which at least 4 are rotten, and so on (row of 14 containing all 5 are rotten ones). ∎

▉ EXAMPLE 4.9

Let us give an example that will use the Borel–Cantelli lemmas. Suppose we look at a sequence of i.i.d. random variables $X_1, X_2, \ldots$. These can have any distribution; the interesting case is when they are not discrete (in that case the result is obvious).

We are concerned with the running maximum, in particular, when do we have a new larger value. To study this, define

$$M_n = \begin{cases} 1, & \text{if } X_n > X_i \text{ for all } i \in \{1, 2, \ldots, n-1\} \\ 0, & \text{otherwise.} \end{cases}$$

In other words, M_n indicates whether or not at time n we have a new running maximum.

Now let $A_n = \{M_n = 1\}$ be the event that indicates a new maximum. Then, $\mathbf{P}(A_n) = 1/n, \ \forall n$. This is due to the fact that the maximum must occur somewhere, and since the distribution is the same, each has the same chance of being the maximum. The assertion will be proven directly using the concept of order statistics. Then we have $\sum_n \mathbf{P}(A_n) = \sum_n 1/n = \infty$. Also note that the sets

A_n are independent, and thus by the Borel–Cantelli Lemma 1.29 we obtain the perhaps evident result that the new maximum happens infinitely often.

However, let us look at two records in a row. In this case, we can write

$$\mathbf{E}[M_n M_{n-1}] = 1 \; \mathbf{P}(A_n)\mathbf{P}(A_{n-1}) = \frac{1}{n(n-1)},$$

and

$$\mathbf{E}\left(\sum_{n=1}^{\infty} M_n M_{n-1}\right) = \sum_{n=1}^{\infty} \frac{1}{n(n-1)} < \infty.$$

Therefore, there are only a finitely many numbers of maximums occurring consecutively. This result is not trivial.

4.3 Moments of a Random Variable. Variance and the Correlation Coefficient

Definition 4.20 *The variance or the dispersion of a random variable $X \in L^2(\Omega)$ is*

$$V(X) = \mathbf{E}[(X - \mu)^2] = \mathbf{E}(X^2) - \mu^2$$

Where $\mu = \mathbf{E}(X)$.

More generally, we have the following:

Definition 4.21 (Moments of a random variable) *Let X be a random variable in $L^p(\Omega)$ for some $p \geq 1$, an integer. We define the p-moment of X as the constant $\mathbf{E}[X^p]$. We define the p central moment of the random variable X as the constant: $\mathbf{E}[(X - \mu)^p]$.*

With this more general definition, it is evident the variance is the second central moment of a random variable ($p = 2$).

Definition 4.22 *Given two random variables X, Y, we call the covariance between X and Y the quantity*

$$Cov(X, Y) = \mathbf{E}[(X - \mu_X)(Y - \mu_Y)]$$

where $\mu_X = \mathbf{E}(X)$ and $\mu_Y = \mathbf{E}(Y)$.

Definition 4.23 *Given random variables X, Y, we call the correlation coefficient*

$$\rho = Corr(X, Y) = \frac{Cov(X, Y)}{\sqrt{V(X)V(Y)}} = \frac{\mathbf{E}[(X - \mu_X)(Y - \mu_Y)]}{\sqrt{\mathbf{E}[(X - \mu_X)^2]\mathbf{E}[(Y - \mu_Y)^2]}}$$

From the Cauchy–Bunyakovsky–Schwartz inequality applied to $X - \mu_X$ and $Y - \mu_Y$, we get $|\rho| < 1$ or $\rho \in [-1, 1]$.

The variable X and Y are called **uncorrelated** if the covariance (or equivalently the correlation) between them is zero.

Proposition 4.24 (Properties of expectation) *The following are true:*

(i) *If X and Y are integrable r.v.s then for any constants α and β, the r.v. $\alpha X + \beta Y$ is integrable and $\mathbf{E}[\alpha X + \beta Y] = \alpha \mathbf{E} X + \beta \mathbf{E} Y$.*

(ii) $V(aX + bY) = a^2 V(X) + b^2 V(Y) + 2ab Cov(X, Y)$.

(iii) *If X, Y are independent, then $\mathbf{E}(XY) = \mathbf{E}(X)\mathbf{E}(Y)$ and $Cov(X, Y) = 0$.*

(iv) *If $X(\omega) = c$ with probability 1 and $c \in \mathbb{R}$ a constant, then $\mathbf{E} X = c$.*

(v) *If $X \geq Y$ a.s., then $\mathbf{E} X \geq \mathbf{E} Y$. Furthermore, if $X \geq Y$ a.s. and $\mathbf{E} X = \mathbf{E} Y$, then $X = Y$ a.s.*

Proof: Exercise. Note that the reverse of the part (iii) above is not true; that is, if the two variables are uncorrelated, this does not mean that they are independent. In fact, in Problem 4.4 you are required to provide a counterexample.   ∎

◼ EXAMPLE 4.10

A random variable X has finite variance σ^2. Show that for any number c,

$$P(X \geq t) \leq \frac{E[(X+c)^2]}{(t+c)^2}, \text{ if } t > -c.$$

Show that if $E(X) = 0$, then

$$P(X \geq t) \leq \frac{\sigma^2}{\sigma^2 + t^2}, \forall t > 0.$$

Proof (Solution): Let us use a technique similar to the Markov inequality to prove the first inequality. Let $F(x)$ be the distribution function of X. For any $c \in \mathbb{R}$, we may write

$$\mathbf{E}\left[(X+c)^2\right] = \int_{-\infty}^{t} (x+c)^2 dF(x) + \int_{t}^{\infty} (x+c)^2 dF(x).$$

The first integral is always positive, and if $t > -c$, then $t + c > 0$ and on the interval $x \in (t, \infty)$ the function $(x+c)^2$ is increasing. Therefore we may continue

$$\mathbf{E}\left[(X+c)^2\right] \geq \int_{t}^{\infty} (t+c)^2 dF(x) = (t+c)^2 \mathbf{P}(X > t).$$

Rewriting the final expression gives the first assertion. To show the second, note that, if $\mathbf{E}[x] = 0$, then $V(X) = \mathbf{E}[X^2]$ and thus $\mathbf{E}\left[(X+c)^2\right] = \sigma^2 + c^2$. Thus the inequality we just proved reads in this case

$$P(X \geq t) \leq \frac{\sigma^2 + c^2}{(t+c)^2}, \text{ if } t > -c.$$

Now take $c = \frac{\sigma^2}{t}$. This is a negative value for any positive t, so the condition is satisfied for any positive t. Substituting after simplifications we obtain exactly what we need. You may wonder (and should wonder) how we came up with the value $\frac{\sigma^2}{t}$. The explanation is simple, that is, the value of c which minimizes the expression $\frac{\sigma^2+c^2}{(t+c)^2}$: in other words, the value of c which produces the best bound. ∎

EXAMPLE 4.11

X is a bounded random variable; that is, $\mathbf{P}(|X| < c) = 1$. The variance of X is 1. Show that

$$|E(x)| \leq \sqrt{c^2 - 1}.$$

Can this bound be attained?

We may follow the same ideas as in the Markov inequality. We have

$$\mathbf{E}[X^2] = \mathbf{E}\left(X^2 \mathbf{1}_{\{X \leq c\}}\right) + \mathbf{E}\left(X^2 \mathbf{1}_{\{X > c\}}\right)$$

$$= \mathbf{E}\left(X^2 \mathbf{1}_{\{X \leq c\}}\right) + 0 \text{ (integral over a set of probability 0)}$$

$$\leq c^2 \mathbf{E}\left(\mathbf{1}_{\{X \leq c\}}\right) = c^2. \tag{4.2}$$

Now, using the definition of variance, we have

$$\left(\mathbf{E}[X]\right)^2 = \mathbf{E}[X^2] - Var(X) = \mathbf{E}[X^2] - 1 \leq c^2 - 1$$

Taking the squared root, we obtain the inequality.

The question of whether we may attain the bound may be answered by analyzing how we obtained the result. Note that the only place we have the inequality is in (4.2). In order that the inequality be an equality, we need to have $\mathbf{E}[X^2] = c^2$ but this is impossible since X is bounded by c (since $\mathbf{P}(|X| < c) = 1$). So, the bound is not attainable. ∎

4.4 Functions of Random Variables. The Transport Formula

In Section 2.5 on page 72, we showed how to calculate distributions and in particular pdfs for continuous random variables. We have also promised a more general result. Well, here it is. This general result allows us to construct random variables and in particular distributions in any space. This is the result that allows us to claim that studying random variables on $([0, 1], \mathscr{B}([0, 1]), \lambda)$ is enough. We had to postpone presenting the result until this point since we had to learn first how to integrate.

Theorem 4.25 (General transport formula) *Let $(\Omega, \mathbb{R}, P)$ be a probability space. Let f be a measurable function such that*

$$(\Omega, \mathscr{F}) \xrightarrow{f} (S, \mathscr{G}) \xrightarrow{\varphi} (\mathbb{R}, \mathscr{B}(\mathbb{R})),$$

where $(S, \mathscr{G})$ is a measurable space. Assuming that at least one of the integrals exists, we then have

$$\int_\Omega \varphi \circ f d\mathbf{P} = \int_S \varphi d\mathbf{P} \circ f^{-1},$$

for all φ measurable functions.

Proof: We will use the standard argument technique discussed above.

1. Let φ be the indicator function. $\varphi = \mathbf{1}_A$ for $A \in \mathscr{G}$:

$$\mathbf{1}_A(\omega) = \begin{cases} 1 & \text{if } \omega \in A \\ 0 & \text{otherwise} \end{cases}$$

   Then, we get

$$\int_\Omega \mathbf{1}_\mathbf{A} \circ f d\mathbf{P} = \int_\Omega \mathbf{1}_A(f(\omega)) d\mathbf{P}(\omega) = \int_\Omega \mathbf{1}_{f^{-1}(A)}(\omega) d\mathbf{P}(\omega)$$

$$= \mathbf{P}(f^{-1}(A)) = \mathbf{P} \circ f^{-1}(A) = \int_S \mathbf{1}_A d(\mathbf{P} \circ f^{-1})$$

   recalling the definition of the integral of an indicator.

2. Let φ be a simple function $\varphi = \sum_{i=1}^n a_i \mathbf{1}_{A_i}$, where a_i's are constant and $A_i \in \mathscr{G}$.

$$\int_\Omega \varphi \circ f d\mathbf{P} = \int_\Omega \left(\sum_{i=1}^n a_i \mathbf{1}_{A_i} \right) \circ f d\mathbf{P}$$

$$= \int_\Omega \sum_{i=1}^n a_i (\mathbf{1}_{A_i} \circ f) d\mathbf{P} = \sum_{i=1}^n a_i \int_\Omega \mathbf{1}_{A_i} \circ f d\mathbf{P}$$

$$\overset{\text{(part 1)}}{=} \sum_{i=1}^n a_i \int_S \mathbf{1}_{A_i} d\mathbf{P} \circ f^{-1}$$

$$= \int_S \sum_{i=1}^n a_i \mathbf{1}_{A_i} d\mathbf{P} \circ f^{-1} = \int_S \varphi d\mathbf{P} \circ f^{-1}$$

3. Let φ be a positive measurable function and let φ_n be a sequence of simple functions such that $\varphi_n \nearrow \varphi$; then

$$\begin{aligned}
\int_\Omega \varphi \circ f d\mathbf{P} &= \int_\Omega (\lim_{n\to\infty} \varphi_n) \circ f d\mathbf{P} \\
&= \int_\Omega \lim_{n\to\infty} (\varphi_n \circ f) d\mathbf{P} \overset{\text{monotone convergence}}{=} \lim_{n\to\infty} \int \varphi_n \circ f d\mathbf{P} \\
&\overset{\text{(part 2)}}{=} \lim_{n\to\infty} \int \varphi_n d\mathbf{P} \circ f^{-1} \overset{\text{monotone convergence}}{=} \int \lim_{n\to\infty} \varphi_n d\mathbf{P} \circ f^{-1} \\
&= \int_S \varphi d(\mathbf{P} \circ f^{-1})
\end{aligned}$$

4. Let φ be a measurable function; then $\varphi^+ = \max(\varphi, 0)$, $\varphi^- = \max(-\varphi, 0)$, which then gives us $\varphi = \varphi^+ - \varphi^-$. Since at least one integral is assumed to exist, we get that $\int \varphi^+$ and $\int \varphi^-$ exist. Also, note that

$$\begin{aligned}
\varphi^+ \circ f(\omega) = \varphi^+(f^{-1}(\omega)) &= \max(\varphi(f(\omega)), 0) \\
\max(\varphi \circ f(\omega), 0) &= (\varphi \circ f)^+(\omega)
\end{aligned}$$

Then

$$\begin{aligned}
\int \varphi^+ d\mathbf{P} \circ f^{-1} &= \int \varphi^+ \circ f d\mathbf{P} = \int (\varphi \circ f)^+ d\mathbf{P} \\
\int \varphi^- d\mathbf{P} \circ f^{-1} &= \int \varphi^- \circ f d\mathbf{P} = \int (\varphi \circ f)^- d\mathbf{P}
\end{aligned}$$

These equalities follow from part 3 of the proof. After subtracting both, we have

$$\int \varphi d\mathbf{P} \circ f^{-1} = \int \varphi \circ f d\mathbf{P}$$

■

Exercise 3 *If X and Y are independent random variables defined on $(\Omega, \mathbb{R}, P)$ with $X, Y \in L^1(\Omega)$, then $XY \in L^1(\Omega)$:*

$$\int_\Omega XY d\mathbf{P} = \int_\Omega X d\mathbf{P} \int_\Omega Y d\mathbf{P} \qquad (\mathbf{E}(XY) = \mathbf{E}(X)\mathbf{E}(Y))$$

Proof (Solution): This is an exercise that you have seen before. Here is presented to exercise the standard approach. ■

EXAMPLE 4.12

Let us solve the previous exercise using the transport formula. Let us take $f : \Omega \to \mathbb{R}^2$, $f(\omega) = (X(\omega), Y(\omega))$, and $\varphi : \mathbb{R}^2 \to \mathbb{R}$, $\varphi(x, y) = xy$. Then we have from the transport formula

$$\int_\Omega X(\omega)Y(\omega) dP(\omega) \overset{(T)}{=} \int_{\mathbb{R}^2} xy dP \circ (X, Y)^{-1}$$

The integral on the left is $\mathbf{E}(XY)$, while the integral on the right can be calculated as

$$\int_{\mathbb{R}^2} xy d(P \circ X^{-1}, P \circ Y^{-1}) = \int_{\mathbb{R}} x dP \circ X^{-1} \int_{\mathbb{R}} y dP \circ Y^{-1}$$
$$\overset{(T)}{=} \int_{\Omega} X(\omega) dP(\omega) \int_{\Omega} Y(\omega) dP(\omega) = \mathbf{E}(X)\mathbf{E}(Y)$$

■ EXAMPLE 4.13

Finally, we conclude with an application of the transport formula which will produce one of the most useful formulas. Let X be a r.v. defined on the probability space $(\Omega, \mathscr{F}, \mathbf{P})$ with distribution function $F(x)$. Show that

$$\mathbf{E}(X) = \int_{\mathbb{R}} x dF(x),$$

where the integral is understood in Riemann–Stieltjes sense.

Proving the formula is immediate. Take $f : \Omega \to \mathbb{R}$, $f(\omega) = X(\omega)$, and $\varphi : \mathbb{R} \to \mathbb{R}$, $\varphi(x) = x$. Then from the transport formula

$$\mathbf{E}(X) = \int_{\Omega} X(\omega) d\mathbf{P}(\omega) = \int_{\Omega} x \circ X(\omega) d\mathbf{P}(\omega) \overset{(T)}{=} \int_{\mathbb{R}} x d\mathbf{P} \circ X^{-1}(x)$$
$$= \int_{\mathbb{R}} x dF(x)$$

Clearly, if the distribution function $F(x)$ is derivable with $\frac{dF}{dx}(x) = f(x)$ or $dF(x) = f(x)dx$, we obtain the lower level classes formula for calculating expectation of a "continuous" random variable:

$$\mathbf{E}(X) = \int_{\mathbb{R}} x f(x) dx$$

4.5 Applications. Exercises in Probability Reasoning

The next two theorems are presented to observe the proofs. They are both early exercises in probability. We will present later much stronger versions of these theorems (and we will also see that these convergence types have very precise definitions), but for now we lack the tools to give general proofs to these stronger versions.

Theorem 4.26 (Law of large numbers) *Let $(\Omega, \mathscr{F}, P)$ be a probability space, and let $\{X_n\}_n$ be a sequence of i.i.d. random variables with $\mathbf{E}(X_i) = \int_{\Omega} X_i d\mathbf{P} = \mu$. Assume that the fourth moment of these variables is finite and $\mathbf{E}(X_i^4) = K_4$ for all i. Then*

$$\bar{X} = \frac{\sum_{i=1}^{n} X_i}{n} = \frac{X_1 + \cdots + X_n}{n} \overset{a.s}{\longrightarrow} \mu$$

Proof: Recall what it means for a statement to hold almost surely. In our specific context, if we denote $S_n = X_1 + \cdots + X_n$, then we need to show that $\mathbf{P}(S_n/n \to \mu) = 1$.

First step. Let us show that we can reduce to the case of $\mathbf{E}(X_i) = \mu = 0$. Take $Y_i = X_i - \mu$. If we prove that $\frac{Y_1 + \cdots + Y_n}{n} \to 0$, then substituting back we shall obtain $\frac{S_n - n\mu}{n} \to 0$, or $\frac{S_n}{n} \to \mu$, which gives our result. Thus we assume that $\mathbf{E}(X_i) = \mu = 0$.

Second step. We want to show that $\frac{S_n}{n} \xrightarrow{\text{a.s}} 0$. We have

$$\mathbf{E}\left(S_n^4\right) = \mathbf{E}\left((X_1 + \cdots + X_n)^4\right) = \mathbf{E}\left(\sum_{i,j,k,l} X_i X_j X_k X_l\right)$$

If any factor in the sum above appears with power 1, from independence we will have $\mathbf{E}(X_i X_j X_k X_l) = \mathbf{E}(X_i)\mathbf{E}(X_j X_k X_l) = 0$. Thus, the only terms remaining in the sum above are those with power larger than 1.

$$\mathbf{E}\left(\sum_{i,j,k,l} X_i X_j X_k X_l\right) = \mathbf{E}\left(\sum_i X_i^4 + \sum_{i<j}\binom{4}{2}X_i^2 X_j^2\right)$$

$$= \sum_i \mathbf{E}(X_i)^4 + 6\sum_{i<j}\mathbf{E}(X_i^2 X_j^2)$$

Using the Cauchy–Bunyakovsky–Schwartz inequality, we get

$$\mathbf{E}(X_i^2 X_j^2) \leq \mathbf{E}(X_i^4)^{1/2}\mathbf{E}(X_j^4)^{1/2} = K_4 < \infty$$

Then

$$\mathbf{E}(S_n^4) = \sum_{i=1}^n \mathbf{E}(X_i)^4 + 6\sum_{i<j}\mathbf{E}(X_i^2 X_j^2) \leq nK_4 + 6\binom{n}{2}\cdot K_4$$

$$= (n + 3n(n-1))K_4 - (3n^2 - 2n)K_4 \leq 3n^2 K_4$$

Therefore

$$\mathbf{E}\left(\sum_{n=1}^\infty \left(\frac{S_n}{n}\right)^4\right) = \sum_{n=1}^\infty \frac{\mathbf{E}(S_n^4)}{n^4} \leq \sum_{n=1}^\infty \frac{3n^2 K}{n^4} = \sum_{n=1}^\infty \frac{3K}{n^2} < \infty$$

Since the expectation of the random variable is finite, we must have the random variable finite with the exception of a set of measure 0 (otherwise the expectation will be infinite). This implies

$$\sum_n \left(\frac{S_n}{n}\right)^4 < \infty \quad \text{a.s.}$$

But a sum can be convergent only if the term under the sum converges to zero. Therefore

$$\lim_{n \to \infty} \left(\frac{S_n}{n} \right)^4 = 0 \quad \text{a.s.}$$

and consequently

$$\frac{S_n}{n} \xrightarrow{\text{a.s}} 0$$

■

■ EXAMPLE 4.14

I cannot resist giving a simple application of this theorem. Let A be an event that appears with probability $\mathbf{P}(A) = p \in (0, 1]$. For example, roll a fair six-sided die and let A be the event roll a 1 or a 6 ($\mathbf{P}(\{1, 6\}) = 1/3$). Let γ_n denote the number of times A appears in n *independent* repetitions of the experiment. Then

$$\lim_{n \to \infty} \frac{\gamma_n}{n} = p$$

This is an important example in statistics. Suppose, for instance, that we do not know that the die is fair but we have our suspicions. How do we test? All we have to do is roll the die many times ($n \to \infty$) and look at the average number of times 1 or 6 appears. If this number stabilizes around a different value than $1/3$, then the die is tricked. The next theorem will also tell us how many times to roll the dies to be confident in our assessment.

To prove the result, we simply apply the previous theorem. Define X_i as

$$X_i = \begin{cases} 1 & \text{if event } A \text{ appears in repetition } i \\ 0 & \text{otherwise} \end{cases}$$

Then $\mathbf{P}(X_i = 1) = p$ and $\mathbf{P}(X_i = 0) = 1 - p$, so that $\mathbf{E}(X_i) = 1 \cdot p + 0 \cdot (1 - p) = p$. Clearly, the fourth moment is finite as well, and applying the theorem $\gamma_n = \sum_{i=1}^{n} X_i$ will converge to the stated value.

4.6 A Basic Central Limit Theorem: *The DeMoivre–Laplace Theorem:*

In order to prove the theorem, we need the following lemma:

Lemma 4.27 (Stirling's formula) *For large n, it can be shown that*

$$n! \sim \sqrt{2\pi n} \cdot n^n e^{-n}$$

The proof of this theorem is only of marginal interest to us.

Theorem 4.28 (DeMoivre–Laplace) *Let $\xi_1 \cdots \xi_n$ be n independent r.v.s each taking value 1 with probability p and 0 with probability $1 - p$ (Binomial(p) random variables). Let*

$$S_n = \sum_{i=1}^{n} \xi_i$$

and

$$S_n^* = \frac{S_n - \mathbf{E}(S_n)}{\sqrt{V(S_n)}} = \frac{S_n - np}{\sqrt{np(1 - p)}}$$

Then for any $x_1, x_2 \in \mathbb{R}$, $x_1 < x_2$

$$\lim_{n \to \infty} \mathbf{P}(x_1 \leq S_n^* \leq x_2) = \Phi(x_2) - \Phi(x_1)$$

$$= \int_{x_1}^{x_2} \frac{1}{\sqrt{2\pi}} e^{-x^2/2} dx$$

Note that Φ is the distribution function of an $N(0, 1)$ random variable. This is exactly the statement of the regular Central Limit Theorem applied to Bernoulli random variables.

Proof: Notice that $S_n \sim$ Binomia(n, p) and $S_n^* = (S_n - np)/\sqrt{np(1 - p)}$ is distributed equidistantly in the total interval $[\frac{-np}{\sqrt{np(1-p)}}, \frac{n-np}{\sqrt{np(1-p)}}]$. The length between two such consecutive values is $\Delta x = 1/\sqrt{np(1 - p)}$.

For k large and $n - k$ large, we have

$$\mathbf{P}(S_n = k) = \binom{n}{k} p^k (1 - p)^{n-k} = \frac{n!}{k!(n - k)!} p^k (1 - p)^k$$

$$= \frac{\sqrt{2\pi n} \cdot n^n e^{-n}}{\sqrt{2\pi k} \cdot k^k e^{-k} \sqrt{2\pi(n - k)} \cdot (n - k)^{n-k} e^{-(n-k)}} p^k (1 - p)^{n-k}$$

$$\tag{4.3}$$

$$= \underbrace{\frac{1}{\sqrt{2\pi}} \sqrt{\frac{n}{k(n - k)}}}_{\text{Term I}} \underbrace{\left(\frac{np}{k}\right)^k \left(\frac{n(1 - p)}{n - k}\right)^{n-k}}_{\text{Term II}}$$

Equation (4.3) follows from Stirling's formula. Remember that, for $S_n = k$, the x value of $S_n^* = (S_n - np)/\sqrt{np(1 - p)}$ is

$$x = \frac{k - np}{\sqrt{np(1 - p)}} \Rightarrow k = np + x\sqrt{np(1 - p)}$$

$$\Rightarrow \frac{k}{np} = 1 + x\sqrt{\frac{1 - p}{np}}$$

Likewise, we may express

$$n - k = n - np - x\sqrt{np(1 - p)} \Rightarrow n - k = n(1 - p) - x\sqrt{np(1 - p)}$$

$$\Rightarrow \frac{n - k}{n(1 - p)} = 1 - x\sqrt{\frac{p}{n(1 - p)}}$$

Using these two expressions in Term II of equation (4.3), we get

$$\log\left(\left(\frac{np}{k}\right)^k\left(\frac{n(1-p)}{n-k}\right)^{n-k}\right) = -k\log\frac{k}{np} - (n-k)\log\frac{n-k}{n(1-p)}$$

$$= -k\log\left(1 + x\sqrt{\frac{1-p}{np}}\right) - (n-k)\log\left(1 - x\sqrt{\frac{p}{n(1-p)}}\right)$$

If we approximate $\log(1+\alpha) \simeq \alpha - \frac{\alpha^2}{2}$, we continue

$$\simeq -k\left(x\sqrt{\frac{1-p}{np}} - \frac{x^2}{2}\frac{1-p}{np}\right) - (n-k)\left(-x\sqrt{\frac{p}{n(1-p)}} - \frac{x^2}{2}\frac{p}{n(1-p)}\right) \quad (4.4)$$

Finally, we substitute k and $n-k$, and after calculations (skipped) we obtain

$$\lim_{n\to\infty}\log\left(\frac{np}{k}\right)^k\left(\frac{n(1-p)}{n-k}\right)^{n-k} = -\frac{x^2}{2}$$

Also note that

$$\sqrt{\frac{n}{k(n-k)}} \simeq \sqrt{\frac{n}{np\cdot n(1-p)}} = \frac{1}{\sqrt{np(1-p)}}$$

Putting both terms together, we obtain

$$\lim_{n\to\infty}\mathbf{P}(S_n^* = x) = \frac{1}{\sqrt{2\pi}}e^{-x^2/2}\Delta x$$

where $\Delta x = \frac{1}{\sqrt{np(1-p)}}$.

Thus

$$\lim_{n\to\infty}\mathbf{P}(x_1 \le S_n^* \le x_2) = \lim_{n\to\infty}\sum_{x_1\le x\le x_2}\mathbf{P}(S_n^* = x) = \lim_{n\to\infty}\sum\frac{1}{\sqrt{2\pi}}e^{-x^2/2}\Delta x$$

$$= \frac{1}{\sqrt{2\pi}}\int_{x_1}^{x_2}e^{-x^2/2}dx$$

∎

Problems

4.1   It is well known that 23 "random" people have a probability of about 1/2 of having at least one shared birthday. There are $365 \times 24 \times 60 = 525{,}600$ min in a year. (We'll ignore leap days.) Suppose each person is labeled by the minute in which he or she was born, so that there are 525,600 possible labels. Assume that a "random" person is equally likely to have any of the 525,600 labels, and that different "random" people have independent labels.

a) About how many random people are needed to have a probability greater than 1/2 of at least one shared birth-minute? (A numerical value is required.)

b) About how many random people are needed to have a probability greater than 1/2 of at least one birth-minute shared by three or more people? (Again, a numerical value is required. You can use heuristic reasoning, but explain your thinking.)

4.2   Show that any simple function f can be written as $\sum_i b_i \mathbf{1}_{B_i}$ with B_i disjoint sets (i.e., $B_i \cap B_j = \emptyset$, if $i \neq j$).

4.3   Prove the four assertions in Exercise 2 on page 125.

4.4   Give an example of two variables X and Y which are uncorrelated but not independent.

4.5   Prove the properties (i)–(v) of the expectation in Proposition 4.24 on page 144.

4.6   Suppose an event A has a probability 0.3. How many independent trials must be performed to assert with probability 0.9 that the relative frequency of A differs from 0.3 by no more than 0.1.

4.7   A box contains three balls, numbered 1, 2, 3. Ann randomly chooses a ball and writes down its number (without returning the ball to the box). Call the number X_1. Then Bob randomly chooses a ball and writes down its number, which we'll call Y_1. The drawn balls are returned to the box. This procedure is done over and over, for a total of 1000 times, generating random variables $X_1, X_2, \ldots, X_{1,000}$ and $Y_1, Y_2, \ldots, Y_{1,000}$.

Let

$$S = \sum_1^{1,000} X_i \quad \text{and} \quad T = \sum_1^{1,000} Y_i$$

be Ann's and Bob's respective totals. Find the correlation between S and T.

4.8   Let ν be the total number of spots which are obtained in 1000 independent rolls of an unbiased die.

1. Find $\mathbf{E}[\nu]$.

2. Estimate the probability $\mathbf{P}\left(3450 < \nu < 3550\right).$

4.9   Urn A and Urn B initially contain four marbles, of which two are white and two are black.

Urn A   Urn B

A machine simultaneously chooses one marble from each urn ("at random," and independently for the two urns) and exchanges them. What is the expected number of exchanges until all the white marbles are in one urn?

4.10   Show, using the Cantelli lemma, that when you roll a die the outcome $\{1\}$ will appear infinitely often. Also show, using the DeMoivre–Laplace theorem, that eventually the average of all rolls up to roll n will be within ε of 3.5, where $\varepsilon > 0$ is any arbitrary real number.

4.11   Ann and Bob each attempt 100 basketball free throws. Ann has probability 0.60 of success on each attempt and Bob has probability 0.50 of success on each attempt. The 200 attempts are independent.

What is the approximate numerical probability that Ann and Bob make exactly the same number of free throws?

4.12   A man goes to Atlantic City, and as part of the trip he receives a special 20$ coupon. This coupon is special in the sense that only the gains may be withdrawn from the slot machine. So the man decides to play a game of guessing red/black. For simplicity, let us assume that every time he plays one dollar. The probability of winning is $1/2$. Each time he plays the 1$ game, he has the same strategy. If he loses, 1$ is subtracted from the current coupon amount. If he wins, the coupon value remains the same and he immediately withdraws 1$ to count toward the total gains. He plays the game until there are no more funds on the coupon. Find the expected amount of his gains and the expected number of times he plays the 1$ game.

4.13   The king of Probabilonia has sentenced a criminal to the following punishment. A box initially contains $999,999$ black balls and one white ball. On the day of sentencing, the criminal draws a ball at random. If the ball is white, the punishment is over and the criminal goes free. If the ball is black then two things happen:

(i) the criminal is forced to eat a live toad, and

(ii) the black ball drawn is painted white and returned to the box.

This process is repeated on successive days until the criminal finally draws a white ball. Let X be the number of toads eaten before the punishment ends.

(a) Write down a formula which gives $P\{X = k\}$ exactly.

(b) Estimate the median of X to within three significant digits.

4.14   A robot arm solders a component on a motherboard. The arm has small tiny errors when locating the correct place on the board. This exercise tries to determine the magnitude of the error, so that we know the physical limitations for the size of the component connections. Let us say that the right place to be soldered is the origin $(0,0)$, and the actual location the arm goes to is (X, Y). We assume that the errors X and Y are independent and have the normal distribution with mean 0 and a certain standard deviation σ.

(a) What is the density function of the distance

$$D = \sqrt{X^2 + Y^2} \ ?$$

(b) Calculate its expected value and variance:

$$\mathbf{E}D \text{ and } Var D.$$

(c) Calculate

$$\mathbf{E}\left[|X^2 - Y^2|\right].$$

CHAPTER 5

PRODUCT SPACES. CONDITIONAL DISTRIBUTION AND CONDITIONAL EXPECTATION

In this chapter, we look at the following type of problems: If we know something extra about the experiment, how does that change our probability calculations. An important part of statistics (Bayesian statistics) is to build using conditional distributions. However, what about the more complex and abstract notion of conditional expectation? In principle, if we have the conditional distribution, then calculating the conditional expectation is done using the same methodology as in the previous chapter. However, in many cases we are not capable of coming up with a formula for the conditional expectation, yet we may still calculate and work with conditional expectation. This is due to its properties, which we will learn in this chapter.

Why do we need conditional expectation?

Conditional expectation is a fundamental concept in the theory of stochastic processes. The simple idea is the following: suppose we have no information about a certain variable; then our best guess about it would be some sort of regular expectation. However, in real life it often happens that we have some partial information about the random variable (or in time we come to know more about it). Then what we should do is, every time there is new information, the sample space Ω or the σ-algebra $\mathscr{F}$ is changing so they need to be recalculated. That will in turn change the probability $\mathbf{P}$ which will change the expectation of the variable. The conditional

Probability and Stochastic Processes, First Edition. Ionuţ Florescu
© 2015 John Wiley & Sons, Inc. Published 2015 by John Wiley & Sons, Inc.

expectation provides a way to recalculate the expectation of the random variable given any new "consistent" information without going through the trouble of recalculating $(\Omega, \mathcal{F}, \mathbf{P})$ every time.

It is also easy to reason that, since we calculate with respect to more precise information, it will be depending on this more precise information and thus it is going to be a random variable itself, "adapted" to this information.

5.1 Product Spaces

Before we learn about conditional expectation, we need to understand the concept of joint probability and joint spaces.

Let $(\Omega_1, \mathcal{F}_1, \mu_1)$ and $(\Omega_2, \mathcal{F}_2, \mu_2)$ be two σ-finite measure spaces. Define

$$\Omega = \Omega_1 \times \Omega_2 \text{ the cartesian product}$$
$$\mathcal{F} = \sigma(\{B_1 \times B_2 : B_1 \in \mathcal{F}_1, B_2 \in \mathcal{F}_2\})$$

Let $f : \Omega \to \mathbb{R}$ be $\mathcal{F}$ measurable such that

$$\forall \omega_1 \in \Omega_1 \ f(\omega_1, \cdot) \text{ is } \mathcal{F}_2 \text{ measurable on } \Omega_2$$
$$\forall \omega_2 \in \Omega_2 \ f(\cdot, \omega_2) \text{ is } \mathcal{F}_1 \text{ measurable on } \Omega_1$$

Then we define

$$I_1^f(\omega_1) = \int_{\Omega_2} f(\omega_1, \omega_2) \mu_2(d\omega_2)$$
$$I_2^f(\omega_2) = \int_{\Omega_1} f(\omega_1, \omega_2) \mu_1(d\omega_1)$$

which are a kind of partial integrals, well defined by the measurability of the restrictions.

Theorem 5.1 (Fubini's theorem) *Define a measure*

$$\mu(F) = \int_{\Omega_1} \int_{\Omega_2} 1_F(\omega_1, \omega_2) \mu_1(d\omega_1) \mu_2(d\omega_2).$$

Then μ is the unique measure defined on $(\Omega, \mathcal{F})$ called the product measure with the property
$$\mu(A_1 \times A_2) = \mu_1(A_1) \mu_2(A_2) \qquad \forall A_i \in \mathcal{F}_i,$$
and as a consequence
$$\int_\Omega f d\mu = \int_{\Omega_1} I_1^f(\omega_1) \mu(d\omega_1) = \int_{\Omega_2} I_2^f(\omega_2) \mu(d\omega_2)$$

Proof: Skipped. Apply the standard argument. Note that the first step is already given.   ∎

◼ **EXAMPLE 5.1   Application of Fubini's theorem**

Let X be a positive r.v. on $(\Omega, \mathscr{F}, P)$. Consider $P \times \lambda$ on $(\Omega, \mathscr{F}) \times ([0, \infty),$ $\mathscr{B}((0, \infty]))$, where λ is the Lebesgue measure. Let $A := \{(\omega, x) : 0 \leq x < X(\omega)\}$. Note that A is the region under the graph of the random variable X. Let the indicator of this set be denoted with $h = 1_A$. Then

$$I_1^h(\omega) = \int_{[0,\infty)} 1_A(\omega, x) d\lambda(x) = \int_0^\infty 1_{\{0 \leq x < X(\omega)\}}(x) d\lambda(x)$$

$$= \int_0^{X(\omega)} d\lambda(x) = X(\omega)$$

$$I_2^h(x) = \int_\Omega 1_A(\omega, x) d\mathbf{P}(\omega) = \int_\Omega 1_{\{0 \leq x < X(\omega)\}}(\omega) d\mathbf{P}(\omega)$$

$$= \mathbf{P}\{\omega : X(\omega) > x\},$$

since X is a positive r.v.

We now apply Fubini's Theorem, and we get

$$\mu(A) = \int_\Omega \int_{[0,\infty)} 1_A(x, \omega) d\mu(x) d\mathbf{P}(\omega)$$

$$= \int_\Omega X(\omega) d\mathbf{P}(\omega) = \int_0^\infty \mathbf{P}(X > x) dx$$

thus reading the last line above:

$$\mathbf{E}(X) = \int_0^\infty \mathbf{P}(X > x) dx$$

This result is actually so useful that we will state it separately.

Corollary 5.2 *If X is a **positive** random variable with distribution function $F(x)$ and we denote $\overline{F}(x) = 1 - F(x)$, then we have*

$$\mathbf{E}(X) = \int_0^\infty \overline{F}(x) dx$$

◼ **EXAMPLE 5.2**

Let Z_1 and Z_2 be independent, standard, normal random variables. Let $Y = \max(Z_1, Z_2)$. Show that the mean of Y is $\sqrt{1/\pi}$.

This is actually a simple example. When the two variables are independent, the joint space is very simple to construct. So let us just calculate.

$$\mathbf{E}[Y] = \mathbf{E}[\max(Z_1, Z_2)] = \int_{-\infty}^\infty \int_{-\infty}^\infty \max(z_1, z_2) f(z_1) f(z_2) dz_1 dz_2$$

$$= \int_{-\infty}^\infty \int_{-\infty}^{z_1} z_1 \frac{1}{2\pi} e^{-\frac{z_1^2}{2}} e^{-\frac{z_2^2}{2}} dz_2 dz_1 + \int_{-\infty}^\infty \int_{z_1}^\infty z_2 \frac{1}{2\pi} e^{-\frac{z_1^2}{2}} e^{-\frac{z_2^2}{2}} dz_2 dz_1$$

The Fubini theorem above shows that the two integrals are identical. To calculate one of them, we have

$$
\int_{-\infty}^{\infty} \int_{z_1}^{\infty} z_2 \frac{1}{2\pi} e^{-\frac{z_1^2}{2}} e^{-\frac{z_2^2}{2}} dz_2 dz_1 = \frac{1}{2\pi} \int_{-\infty}^{\infty} e^{-\frac{z_1^2}{2}} e^{-\frac{z_1^2}{2}} dz_1
$$
$$
= \frac{1}{2\pi} \int_{-\infty}^{\infty} e^{-z_1^2} dz_1
$$
$$
= \frac{1}{2\sqrt{\pi}},
$$

which uses the fact that the density of a normal with mean 0 and variance $1/2$ integrates to 1. Now, using that $\mathbf{E}[Y]$ is twice the integral calculated, we obtain the desired result.

◼ EXAMPLE 5.3

Let $Y_1, Y_2, \ldots$ be an i.i.d. sequence of nonnegative random variables. Suppose that there exist positive constants M, C, and a, such that, for all $x > M$

$$
P(Y_i > x) \le C e^{-ax}.
$$

Let N be an integer valued, nonnegative random variable independent of the Y_i's with the distribution

$$
P(N = k) = (1 - p)p^{k-1}, \; k = 1, 2, 3, \ldots
$$

($Geometric(p)$ "number of trials" random variable).

Prove that $\sum_{i=1}^{N} Y_i$ has all moments.

Proof (Solution): This is not strictly a Fubini Theorem application but rather of the Corollary 5.2. As we shall see in the stochastic processes part, the Y_i's may be thought as some sort of reward and the N represents the number of times to play the game. We stop when we reach N. The sum represents the total reward. There is another solution (perhaps simpler) using the Wald lemma from Example 5.6 in this chapter or the more general Corollary 11.9 on page 339. However, we present here a different solution using the specifics of the problem.

Use Y to denote any of the variables Y_i, and let $\overline{F}(x) = \mathbf{P}(Y > x)$. From Corollary 5.2, we may write, since Y_i's are positive,

$$
\mathbf{E}[Y^n] = \int_0^{\infty} \mathbf{P}(Y^n > x) \, dx = \int_0^{\infty} \mathbf{P}(Y > x^{\frac{1}{n}}) \, dx = \int_0^{\infty} \overline{F}(x^{\frac{1}{n}}) \, dx
$$
$$
= \int_0^{\infty} \overline{F}(y) n y^{n-1} \, dy
$$

after a change of variable $x = y^n$. The idea is to use the inequality given. Note that the inequality is true only if $x > M$, so we need to split the integral.

$$\mathbf{E}[Y^n] = n \int_0^M \underbrace{y^{n-1}}_{\leq M^{n-1}} \underbrace{\overline{F}(y)}_{\leq 1} \, dy + n \int_M^\infty y^{n-1} \underbrace{\overline{F}(y)}_{\leq Ce^{-ay}} \, dy$$

$$\leq n \int_0^M M^{n-1} \, dy + n \int_M^\infty y^{n-1} Ce^{-ay} \, dy$$

$$= nM^n + n \left(-y^{n-1} \left(\frac{C}{a} e^{-ay} \right) \Big|_M^\infty + \frac{(n-1)C}{a} \int_M^\infty y^{n-2} e^{-ay} \right)$$

$$= nM^n + \frac{nC}{a} M^{n-1} e^{-aM} + \frac{n(n-1)C}{a} \int_M^\infty y^{n-2} e^{-ay}$$

$$= \cdots$$

$$= Constant(n, M, a, C)$$

$$\leq C_1 n \, n! \, M^n$$

where the last is just an upper bound that dominates the entire sum. C_1 is a constant, and for simplicity of notation we will drop it from further calculation (it does not influence further limits). Now let $n \in \mathbb{N}$ be an integer. We may write

$$\mathbf{E}\left[\left(\sum_{i=1}^k Y_i \right)^n \right] \leq k^{n-1} \sum_{i=1}^k \mathbf{E}[Y_i^n]$$

for all k fixed integers using the simple algebraic inequality

$$(y_1 + \ldots + y_k) \leq k^{n-1}(y_1^n + \ldots + y_k^n)$$

But N is a random variable, so it is not fixed. Therefore we need to deal with that.

$$\mathbf{E}\left[\left(\sum_{i=1}^N Y_i \right)^n \right] = \sum_{k=1}^\infty \mathbf{E}\left[\left(\sum_{i=1}^N Y_i \right)^n \Big| N = k \right] \mathbf{P}(N = k)$$

$$= \sum_{k=1}^\infty \mathbf{E}\left[\left(\sum_{i=1}^k Y_i \right)^n \right] \mathbf{P}(N = k)$$

$$\leq \sum_{k=1}^\infty k^{n-1} \sum_{i=1}^k \underbrace{\mathbf{E}[Y_i^n]}_{\leq n! \, M^n} \mathbf{P}(N = k)$$

$$\leq \sum_{k=1}^{\infty} k^{n-1} k \, n \, n! \, M^n (1-p)^{k-1} p$$

$$= n \, n! \, M^n p \sum_{k=1}^{\infty} k^n (1-p)^{k-1} < \infty$$

The last sum is easily shown to be convergent using, for example, a ratio criterion. Therefore, the sum has finite moments. ∎

5.2 Conditional Distribution and Expectation. Calculation in Simple Cases

We shall give a general formulation of conditional expectation that will be most useful in the second part of this textbook. But, until then, we will present the cases that actually allow the explicit calculation of conditional distribution and expectation.

Let X and Y be two discrete variables on $(\Omega, \mathscr{F}, P)$.

Definition 5.3 (Discrete conditional distribution) *The conditional distribution of Y, given $X = x$: $F_{Y|X}(\cdot|x)$, is*

$$F_{Y|X}(y|x) = \mathbf{P}(Y \leq y | X = x)$$

The conditional probability mass function of $Y|X$ is

$$f_{Y|X}(y|x) = \mathbf{P}(Y = y | X = x) = \frac{f_{X,Y}(x,y)}{f_X(x)}$$

Note: In the case when $\mathbf{P}(X = x) = 0$, we cannot define the conditional probability.

Definition 5.4 (Discrete conditional expectation) *Let $\psi(x) = \mathbf{E}(Y|X = x)$; then $\psi(X) = \mathbf{E}[Y|X]$ is called the* conditional expectation.

Remark 5.5 *The conditional expectation is a random variable.*

Definition 5.6 (Continuous conditional distribution) *Let X, Y be two continuous random variables. The conditional distribution is defined as*

$$F_{Y|X}(y|x) = \int_{-\infty}^{y} \frac{f_{X,Y}(x,v)}{f_X(x)} dv$$

The function $f_{Y|X}(y|x) = \frac{f(x,y)}{f_X(x)}$ is the conditional probability density function.

Definition 5.7 (Continuous conditional expectation) *The conditional expectation for two continuous random variables is $\psi(X) = \mathbf{E}[Y|X]$, where the function ψ is calculated as*

$$\psi(x) = \mathbf{E}(Y|X = x) = \int_{-\infty}^{\infty} y f_{Y|X}(y|x) dy$$

In these simple cases, we can give the following theorem as well, which helps with the calculations.

Theorem 5.8 *Suppose X and Y are two random variables, and we can calculate easily the distribution of $Y|X$. Then we have*

$$\mathbf{E}[Y] = \mathbf{E}[\mathbf{E}[Y|X]],$$

$$Var(Y) = \mathbf{E}[Var(Y|X)] + Var(\mathbf{E}[Y|X]).$$

☐ EXAMPLE 5.4

A point is picked uniformly from the surface of the unit sphere. Let $L =$ longitude angle θ and let $l =$ latitude angle ϕ. Let us find the distribution functions of $\theta|\phi$ and $\phi|\theta$.

Let C be a set on the sphere (or generally in $\mathbb{R}^3$). The surface area of the sphere is $4\pi r^2 = 4\pi$. The set of points from which we sample is $S(0,1) = \{(x,y,z) : x^2 + y^2 + z^2 = 1\}$. Then, since we pick the points uniformly, the position of a point chosen has the distribution

$$\mathbf{P}((x,y,z) \in C) = \int_C \frac{1}{4\pi} \mathbf{1}_{\{x^2+y^2+z^2=1\}}(x,y,z)dxdydz$$

Since we are interested in longitude and latitude, we change to spherical coordinates to obtain the distribution of these variables. We take the transformation $X = r\cos\theta\cos\phi$, $Y = r\sin\theta\cos\phi$, and $Z = r\sin\phi$, where $r \in [0,\infty)$, $\theta \in [0, 2\pi]$, and $\phi \in [-\pi/2, \pi/2]$. To obtain the distribution, we calculate the new integral. The Jacobian of the transformation is

$$J = \begin{vmatrix} -r\sin\theta\cos\phi & -r\cos\theta\sin\phi & \cos\theta\cos\phi \\ r\cos\theta\cos\phi & -r\sin\theta\sin\phi & \sin\theta\cos\phi \\ 0 & r\cos\phi & \sin\phi \end{vmatrix}$$

$$= r^2\cos^3\phi + r^2\sin^2\phi\cos\phi = r^2\cos\phi$$

Note that the indicator is 1 if and only if $r = 1$. We conclude that

$$\mathbf{P}((x,y,z) \in C) = \int_{Im\, C} \frac{1}{4\pi}|\cos\phi|d\theta d\phi,$$

where $Im C$ is the set of *polar* coordinates that make the set C. Therefore, the joint distribution function is

$$f(\theta, \phi) = \frac{1}{4\pi}|\cos\phi|, \qquad \phi \in [-\pi/2, \pi/2], \theta \in [0, 2\pi].$$

Now, we get the marginal of ϕ as

$$f_\phi(\phi) = \int_0^{2\pi} \frac{1}{4\pi} |\cos\phi| d\theta = \frac{|\cos\phi|}{2},$$

and the marginal of θ as

$$f_\theta(\theta) = \int_{-\pi/2}^{\pi/2} \frac{1}{4\pi} |\cos\phi| d\phi = \int_{-\pi/2}^{\pi/2} \frac{1}{4\pi} \cos\phi d\phi = \frac{1}{2\pi}$$

Thus, we calculate immediately the conditional distributions:

$$f_{\theta|\phi}(\theta|\phi) = \frac{1}{2\pi}, \qquad\qquad \theta \in [0, 2\pi]$$

$$f_{\phi|\theta}(\phi|\theta) = \frac{\cos\phi}{2}, \qquad\qquad \phi \in [-\pi/2, \pi/2]$$

We note that θ and ϕ are independent (the product of marginals is equal to the joint distribution) but the conditionals are different as a result of the parameterizations (this particular example is known as the *Borel paradox*). Also note that the conditional expectations are equal to the regular expectations; this is, of course, because the variables are independent. We will obtain this property in general in the following section.

EXAMPLE 5.5

Many clustering algorithms are based on random projections. For simplicity, we consider the direction of the first coordinate unit vector $\overrightarrow{e}_1$ as the best possible projection. However, the probability of finding this direction exactly is zero, so we consider a tolerance angle α_e and we say that a projection is "good enough" if it makes an angle less than α_e with $\overrightarrow{e}_1$.

We want to calculate the probability that a random direction makes an angle less than α with $\overrightarrow{e}_1$.

The example is in $\mathbb{R}^3$ but we can easily generalize it to any dimension. We assume that $0 < \alpha_e < \pi/2$, otherwise the problem becomes trivial.

Directions in $\mathbb{R}^3$ are equivalent to points on the unit sphere. Therefore, the probability to be calculated is twice the probability that a point chosen at random on the sphere belongs to the cone of angle α_e centered at the origin. Why twice? Because we do not care if the angle formed by the random direction is with $\overrightarrow{e}_1$ or $-\overrightarrow{e}_1$. Thus, we calculate the probability by taking the ratio of the area of the intersection of the said cone and the sphere and the total surface area of the sphere.

The area of the unit sphere in $\mathbb{R}^d$ is readily calculated as $\frac{2\pi^{d/2}}{\Gamma(d/2)}$ (e.g., Kendall (2004), $\Gamma(x) = \int_0^\infty t^{x-1} e^{-t} dt$ is the gamma function). In the particular case when $d = 3$ ($\Gamma(\frac{3}{2}) = \frac{\sqrt{\pi}}{2}$), we obtain the well-known area 4π.

To compute the support area of the cone, we switch to spherical coordinates:

$$x_1 = r \cos \theta_1$$

$$x_2 = r \sin \theta_1 \cos \theta_2$$

$$x_3 = r \sin \theta_1 \sin \theta_2$$

where $r \in [0, \infty)$, $\theta_1 \in [0, \pi]$, and $\theta_2 \in [0, 2\pi]$. Note that this transformation is slightly different from the one in the previous example. This, of course, is irrelevant, and does not matter for the final calculation.

The points of interest can be found when $r = 1$ and $\theta_1 \in [0, \alpha_e]$, and we need to double the final area found to take into account symmetric angles with respect to $\overrightarrow{e}_1$.

One can check immediately that the Jacobian of this change of variables is $r^2 \sin \theta_1$, and the probability needed is easily calculated as

$$2 \sin^2 \frac{\alpha_e}{2}$$

If we now consider K projections, then the probability that at least one is a "good enough projection" is

$$1 - \left(1 - 2 \sin^2 \frac{\alpha_e}{2}\right)^K$$

Note that the example is extendable to the more interesting $\mathbb{R}^d$ case, but in that case we do not obtain an exact formula and instead only bounds.

5.3 Conditional Expectation. General Definition

To summarize the previous section, if X and Y are two random variables, we have defined the conditional distribution and conditional expectation of one **with respect to the other**. In fact, we have defined more: the conditional expectation of one **with respect to the information contained in the other**.

More precisely, in the previous subsection we defined the expectation of X conditioned by the σ-algebra generated by Y: $\sigma(Y)$. Thus, we may write without a problem

$$\mathbf{E}[X|Y] = \mathbf{E}[X|\sigma(Y)].$$

Remark 5.9 *We note that the conditional expectation, unlike the regular expectation, is a random variable measurable with respect to the sigma algebra under which it is conditioned. In simple language, it has adapted itself to the information contained in the σ-algebra $\mathcal{K}$. In the simple cases presented in the previous section, the conditional expectation is measurable with respect to $\sigma(Y)$. But since this is a very simple sigma algebra, it has to be in fact a function of Y.*

This notion may be generalized to define conditional expectation with respect to any kind of information (σ-algebra).

Why do we need a more general version of conditional expectation?   Conditional expectation is a fundamental concept in the theory of stochastic processes as we shall see in the second part of this book. The simple idea is the following: suppose we have no information about a certain observed process. Then our best guess of the next observation will be some regular expectation. However, once we have some partial information about the dynamics of the process (be it through past observations or outside information), the best guess must change. Every time we obtain new information, the possible sample space Ω, or more precisely the σ-algebra $\mathscr{F}$, is recalculated. That will in turn change the probability $\mathbf{P}$, which will change the expectation of the variable. The conditional expectation provides a way to recalculate the expectation of the random variable given any new "consistent" information without going through the trouble of recalculating $(\Omega, \mathscr{F}, \mathbf{P})$ every time.

It is also easy to reason that, since we keep recalculating with respect to any new information, it will be depending on this more precise information, and thus it is going to be a random variable itself, "adapted" to this information.

As definition, we shall use the following theorem:

Theorem 5.10 *Let $(\Omega, \mathscr{F}, \mathbf{P})$ be a probability space, and let $\mathscr{K} \subseteq \mathscr{F}$ a sub-σ-algebra. Let X be a random variable on $(\Omega, \mathscr{F}, \mathbf{P})$ such that either X is positive or $X \in L^1(\Omega)$. Then, there exists a random variable Y, measurable with respect to $\mathscr{K}$ with the property*

$$\int_A Y dP = \int_A X dP, \quad \forall A \in \mathscr{K}$$

This Y is defined to be the conditional expectation of X with respect to $\mathscr{K}$ and is denoted $\mathbf{E}[X|\mathscr{K}]$.

The proof is not complicated but it is not of interest to us. However, there is one important result that needs to be used so that the theorem provides a well-defined object. We need to make sure that the conditional expectation defined in this way is unique in some sense. For this, we use the following lemma.

Lemma 5.11 *If X and Y are two random variables defined on the same probability space and both measurable with respect to a sigma algebra $\mathcal{K}$, and*

$$\int_A X d\mathbf{P} = \int_A Y d\mathbf{P}, \quad \forall A \in \mathcal{K},$$

then the two variables are equal almost surely (i.e., $\mathbf{P}\{\omega : X(\omega) \neq Y(\omega)\} = 0$).

Proof: To prove the lemma, we may use Lemma 4.2 which says that if the integral from a positive random variable is zero, then the random variable must be zero a.s.. Note that the statement is the same as

$$\int_A (X - Y) d\mathbf{P} = 0, \quad \forall A \in \mathcal{K},$$

and take $A_1 = \{\omega \in \mathcal{K} \mid X(\omega) - Y(\omega) > 0\}$. Because both X and Y are measurable with respect to $\mathcal{K}$, their difference is as well, and therefore the set above $A_1 \in \mathcal{K}$. Thus we must have

$$\int \mathbf{1}_{\{\omega \in \mathcal{K} \mid X(\omega) - Y(\omega) > 0\}} (X(\omega) - Y(\omega)) d\mathbf{P}(\omega) = 0.$$

But the function under the integral is strictly positive and therefore according to the Lemma 4.2 it must be zero a.s.. However, on that set $X > Y$, and thus the indicator must be zero a.s. or $\mathbf{P}(A_1) = 0$.

Now repeat the same argument for the set $A_2 = \{\omega \in \mathcal{K} \mid Y(\omega) - X(\omega) > 0\}$. It will follow that $\mathbf{P}(A_2) = 0$ and thus $X = Y$ a.s. ($\mathbf{P}\{\omega \mid X \neq Y\} \subseteq A_1 \cup A_2$.) ∎

Using the lemma, it is now clear that the conditional expectation thus defined is unique.

Proposition 5.12 (Properties of the conditional expectation) *Let $(\Omega, \mathcal{F}, \mathbf{P})$ be a probability space, and let $\mathcal{K}, \mathcal{K}_1, \mathcal{K}_2$ be sub-σ-algebras. Let X and Y be random variables of the probability space. Then we have*

(1) If $\mathcal{K} = \{\varnothing, \Omega\}$ then $\mathbf{E}[X|\mathcal{K}] = \mathbf{E}X = const.$

(2) $\mathbf{E}[\alpha X + \beta Y|\mathcal{K}] = \alpha \mathbf{E}[X|\mathcal{K}] + \beta \mathbf{E}[Y|\mathcal{K}]$ for α, β real constants.

(3) If $X \leq Y$ a.s., then $\mathbf{E}[X|\mathcal{K}] \leq \mathbf{E}[Y|\mathcal{K}]$ a.s.

(4) $\mathbf{E}[\mathbf{E}[X|\mathcal{K}]] = \mathbf{E}X.$

(5) If $\mathcal{K}_1 \subseteq \mathcal{K}_2$, then

$$\mathbf{E}\left[\mathbf{E}[X|\mathcal{K}_1]\mid \mathcal{K}_2\right] = \mathbf{E}\left[\mathbf{E}[X|\mathcal{K}_2]\mid \mathcal{K}_1\right] = \mathbf{E}[X|\mathcal{K}_1]$$

(6) If X is independent of $\mathcal{K}$, then

$$\mathbf{E}[X|\mathcal{K}] = \mathbf{E}[X]$$

(7) If Y is measurable with respect to $\mathcal{K}$, then

$$\mathbf{E}[XY|\mathcal{K}] = Y\mathbf{E}[X|\mathcal{K}]$$

After proving these properties (see Problem 5.5), they will become essential in working with conditional expectation. In fact, the definition is never used anymore.

■ **EXAMPLE 5.6**

Let us obtain a weak form of the Wald's equation (an equation that serves a fundamental role in the theory of stochastic processes) right now by a simple argument. Let $X_1, X_2, \ldots, X_n, \ldots$ be i.i.d. with finite mean μ, and let N be a random variable taking values in strictly positive integers and independent of X_i

for all i. For example, X_i's may be the results of random experiments, and N may be some stopping strategy established in advance. Let $S_N = X_1 + X_2 + \cdots + X_N$. Find $\mathbf{E}(S_N)$.

Let

$$\varphi(n) = \mathbf{E}[S_N | N = n] = \mathbf{E}[X_1 + X_2 + \cdots + X_N | N = n]$$

$$= \sum_{i=1}^{n} \mathbf{E}[X_i | N = n] = \sum_{i=1}^{n} \mathbf{E}[X_i] = n\mu$$

by independence. Therefore, $\mathbf{E}[S_N | N] = \varphi(N) = N\mu$. Finally, using the properties of conditional expectation

$$\mathbf{E}(S_N) = \mathbf{E}[\mathbf{E}[S_N | N]] = \mathbf{E}[N\mu] = \mu \mathbf{E}[N].$$

Note that we have not used any distribution form but used only the properties of the conditional expectation.

5.4   Random Vectors. Moments and Distributions

As we saw when we talked about joint distribution, a random vector $\mathbf{X}$ is any measurable function defined on the probability space $(\Omega, \mathscr{F}, \mathbf{P})$ with values in $\mathbb{R}^n$. Any random vector has a distribution function, defined similarly with the one-dimensional case. Specifically, if the random vector $\mathbf{X}$ has components $\mathbf{X} = (X_1, \ldots, X_n)$, its distribution function is

$$F_{\mathbf{X}}(\mathbf{x}) = \mathbf{P}(\mathbf{X} \le \mathbf{x}) = \mathbf{P}(X_1 \le x_1, \ldots X_n \le x_n).$$

If the n-dimensional function F is differentiable, then the random vector will have a density (the joint density) f such that

$$F(\mathbf{x}) = F(x_1, \ldots, x_n) = \int_{-\infty}^{x_1} \cdots \int_{-\infty}^{x_n} f(t_1, \ldots, t_n) dt_n \ldots dt_1.$$

Using these notions, we can of course define the moments of the distribution. In fact, suppose that $g : \mathbb{R}^n \to \mathbb{R}$ is any function, then we can calculate the expected value of random variable $g(X_1, \ldots, X_n)$ when the joint density exists explicitly as

$$\mathbf{E}[g(X_1, \ldots, X_n)] = \int_{-\infty}^{\infty} \cdots \int_{-\infty}^{\infty} g(x_1, \ldots, x_n) f(x_1, \ldots, x_n) dx_1 \ldots dx_n$$

With this, we can define further the moments of the random vector. The first moment is a vector

$$\mathbf{E}[\mathbf{X}] = \mu_{\mathbf{X}} = \begin{pmatrix} \mathbf{E}[X_1] \\ \vdots \\ \mathbf{E}[X_n] \end{pmatrix}.$$

To remember that the expectation applies to each component in the bigger object. The second moment requires calculating all the combination of the components. The result can be presented in a matrix form. The second central moment can be presented as the covariance matrix.

$$Cov(\mathbf{X}) = \mathbf{E}[(\mathbf{X} - \mu_{\mathbf{X}})(\mathbf{X} - \mu_{\mathbf{X}})^t]$$

$$= \begin{pmatrix} Var(X_1) & Cov(X_1, X_2) & \ldots & Cov(X_1, X_n) \\ Cov(X_2, X_1) & Var(X_2) & \ldots & Cov(X_2, X_n) \\ \vdots & \vdots & \ddots & \vdots \\ Cov(X_n, X_1) & Cov(X_n, X_2) & \ldots & Var(X_n) \end{pmatrix}, \quad (5.1)$$

where we used the transpose matrix notation^t, and since the $Cov(X_i, X_j) = Cov(X_j, X_i)$, the matrix is symmetric.

The matrix is also positive semidefinite (nonnegative definite) as we shall see in a minute. First, what is a positive definite matrix?

Definition 5.13 (Positive definite and semidefinite matrix) *A square $n \times n$ matrix A is called positive definite if, for any vector $u \in \mathbb{R}^n$ nonidentically zero, we have*

$$u^t A u > 0$$

A matrix A is called positive semidefinite (or nonnegative definite) if, for any vector $u \in \mathbb{R}^n$ nonidentically zero, we have

$$u^t A u \geq 0$$

So why does the covariance matrix have to be semidefinite or whatever? Take any vector $u \in \mathbb{R}^n$. Then the product $u^t\mathbf{X} = \sum u_i X_i$ is just a random variable (one dimensional). As such, its variance must be nonnegative. But let us calculate its variance. Clearly, its expectation is $\mathbf{E}[u^t\mathbf{X}] = u^t\mu_{\mathbf{X}}$. Then we can write (since for any number a: $a^2 = aa^t$)

$$Var(u^t\mathbf{X}) = \mathbf{E}\left[\left(u^t\mathbf{X} - u^t\mu_{\mathbf{X}}\right)^2\right] = \mathbf{E}\left[\left(u^t\mathbf{X} - u^t\mu_{\mathbf{X}}\right)\left(u^t\mathbf{X} - u^t\mu_{\mathbf{X}}\right)^t\right]$$

$$= \mathbf{E}\left[u^t\left(\mathbf{X} - \mu_{\mathbf{X}}\right)\left(\mathbf{X} - \mu_{\mathbf{X}}\right)^t \left(u^t\right)^t\right] = u^t Cov(\mathbf{X})\, u$$

Since variance is always nonnegative, clearly the covariance matrix must be nonnegative definite (or positive semidefinite), whatever the sign is: greater than or equal to zero, but not strictly greater. This difference is in fact important in the context of random variables since you may be able to construct a linear combination $u^t\mathbf{X}$ which is not always constant but whose variance is equal to zero.

Higher moments than the second are hard to represent. For the third moment, for instance, one would need to hold terms of the type $X_1^2 X_i$ as well as $X^1 X_i^2$. Matrix would not be enough, but a matrix of matrices would, or a three-dimensional matrix.

Examples of Multivariate Distributions

The Dirichlet distribution $Dir(\alpha)$, named after Johann Peter Gustav Lejeune Dirichlet (1805–1859), is a multivariate distribution parameterized by a vector α of positive parameters $(\alpha_1, \ldots, \alpha_n)$.

Specifically, the joint density of an n-dimensional random vector $\mathbf{X} \sim Dir(\alpha)$ is

$$f(x_1, \ldots, x_n) = \frac{1}{\mathbf{B}(\alpha)} \left(\prod_{i=1}^{n} x_i^{\alpha_i - 1} \mathbf{1}_{\{x_i > 0\}} \right) \mathbf{1}_{\{x_1 + \cdots + x_n = 1\}}$$

The components of the random vector $\mathbf{X}$ thus are always positive and have the property $X_1 + \cdots + X_n = 1$. The normalizing constant $\mathbf{B}(\alpha)$ is the multinomial beta function, that is,

$$\mathbf{B}(\alpha) = \frac{\prod_{i=1}^{n} \Gamma(\alpha_i)}{\Gamma\left(\sum_{i=1}^{n} \alpha_i\right)} = \frac{\prod_{i=1}^{n} \Gamma(\alpha_i)}{\Gamma(\alpha_0)},$$

where we used the notation $\alpha_0 = \sum_{i=1}^{n} \alpha_i$ and $\Gamma(x) = \int_0^{\infty} t^{x-1} e^{-t} dt$ for the Gamma function.

Because the Dirichlet distribution creates n positive numbers that always sum to 1, it is extremely useful to create candidates for probabilities of n possible outcomes. This distribution is very popular today in machine learning. In fact, it is very much related to the multinomial distribution which needs n numbers summing to 1 to model the probabilities in the distribution.

With the notation mentioned above and α_0 as the sum of all parameters, we can calculate the moments of the distribution. The first moment vector has coordinates

$$\mathbf{E}[X_i] = \frac{\alpha_i}{\alpha_0}$$

The covariance matrix has elements

$$Var(X_i) = \frac{\alpha_i(\alpha_0 - \alpha_i)}{\alpha_0^2(\alpha_0 + 1)},$$

and when $i \neq j$

$$Cov(X_i, X_j) = \frac{-\alpha_i \alpha_j}{\alpha_0^2(\alpha_0 + 1)}$$

The covariance matrix is singular (its determinant is zero).

Finally, the univariate marginal distributions are all beta with parameters $X_i \sim Beta(\alpha_i, \alpha_0 - \alpha_i)$.

Multinomial distribution   The multinomial distribution is a generalization of the binomial distribution. Specifically, assume that n independent distributions may result in one of the k outcomes generically labeled $S = \{1, 2, \ldots, k\}$, each with corresponding probabilities $(p_1, \ldots, p_k)$. Now define a vector $\mathbf{X} = (X_1, \ldots, X_k)$, where each of the X_i counts the number of outcomes i in the resulting sample of size n. The joint distribution of the vector $\mathbf{X}$ is

$$f(x_1, \ldots, x_k) = \frac{n!}{x_1! \ldots x_k!} p_1^{x_1} \ldots p_k^{x_k} \, \mathbf{1}_{\{x_1 + \cdots + x_k = n\}}$$

In the same way as the binomial probabilities appear as coefficients in the binomial expansion of $(p + (1 - p))^n$, the multinomial probabilities are the coefficients in the multinomial expansion $(p_1 + \cdots + p_k)^n$, so they obviously sum to 1. This expansion in fact gives the name of the distribution.

It is very easy to see that, if we label the outcome i as a success and everything else a failure, then X_i simply counts successes in n independent trials and thus $X_i \sim Binom(n, p_i)$. Thus, the first moment of the random vector and the diagonal elements in the covariance matrix are easy to calculate as np_i and $np_i(1 - p_i)$, respectively. The off-diagonal elements (covariances) are not that complicated to calculate either, and the multinomial's random vector's first two moments are

$$\mathbf{E}[\mathbf{X}] = \begin{pmatrix} np_1 \\ np_2 \\ \vdots \\ np_k \end{pmatrix} \text{ and } Cov(\mathbf{X}) = \begin{pmatrix} np_1(1 - p_1) & -np_1p_2 & \cdots & -np_1p_k \\ -np_1p_2 & np_2(1 - p_2) & \cdots & -np_2p_k \\ \vdots & \vdots & \ddots & \vdots \\ -np_kp_1 & -np_kp_2 & \cdots & np_k(1 - p_k). \end{pmatrix}$$

The one-dimensional marginal distributions are binomial as we mentioned. However, the joint distribution of $(X_1, \ldots, X_r)$, the first r components, is NOT multinomial. Instead, suppose we group the first r categories into 1 and we let $Y = X_1 + \cdots + X_r$. Because the categories are linked $(X_1 + \cdots + X_k = n)$, we also have that $Y = n - X_{r+1} - \cdots - X_k$. It is easy to see that the vector $(Y, X_{r+1}, \ldots, X_k)$, or equivalently $(n - X_{r+1} - \cdots - X_k, X_{r+1}, \ldots, X_k)$, will have a multinomial distribution with associated probabilities $(p_Y, p_{r+1}, \ldots, p_k) = (p_1 + \cdots + p_r, p_{r+1}, \ldots, p_k)$.

One last result. Consider the conditional distribution of the first r components given the last $k - r$ components. That is, the distribution of

$$(X_1, \ldots, X_r) \mid X_{r+1} = n_{r+1}, \ldots, X_k = n_k$$

This distribution is also multinomial with the number of elements $n - n_{r+1} - \cdots - n_k$ and probabilities $(p'_1, \ldots, p'_r)$, where $p'_i = \frac{p_i}{p_1 + \cdots + p_r}$.

Testing whether counts are coming from a specific multinomial distribution. Suppose we have obtained a sample of n observations $(n_1, \ldots, n_k)$ which we suspect come from the multinomial distribution with probabilities $(\pi_1, \ldots, \pi_k)$. To test the hypotheses

$$H_0(p_1, \ldots, p_k) = (\pi_1, \ldots, \pi_k)$$
$$H_a \text{ Not all probabilities are equal}$$

we may use the statistic

$$\sum_{i=1}^{k} \frac{(n_i - n\pi_i)^2}{n\pi_i},$$

which is also called the Pearson's Chi-squared statistic. This statistic is asymptotically distributed as a Chi-squared random variable with $k-1$ degrees of freedom when the sample size n goes to infinity.

Multivariate normal distribution A vector $\mathbf{X}$ is said to have a k-dimensional multivariate normal distribution with mean vector $\mu = (\mu_1, \ldots, \mu_k)$ and covariance matrix $\Sigma = (\sigma_{ij})_{ij \in \{1,\ldots,k\}}$ (denoted $MVN_k(\mu, \Sigma)$) if its density can be written as

$$f(\mathbf{x}) = \frac{1}{(2\pi)^{k/2} \det(\Sigma)^{1/2}} e^{-\frac{1}{2}(\mathbf{x}-\mu)^T \Sigma^{-1}(\mathbf{x}-\mu)},$$

where we used the usual notations for the determinant, transpose, and inverse of a matrix. The vector of means μ may have any elements in $\mathbb{R}$, but, just as in the one-dimensional case, the standard deviation has to be positive; in this multivariate case, too, the matrix Σ has to be symmetric and positive definite.

The covariance matrix Σ may not necessarily be invertible. When such is the case, the matrix is replaced by its generalized inverse and the determinant is its pseudo-determinant. We refer to Appendix A.2 for these notions.

The multivariate normal defined thus has many nice properties. The basic one is that the one-dimensional distributions are all normal $X_i \sim N(\mu_i, \sigma_{ii})$. Furthermore, just using the definition and integrating, $Cov(X_i, X_j) = \sigma_{ij}$. This is also true for any marginal. For example, if $(X_r, \ldots, X_k)$ are the last coordinates, then

$$\begin{pmatrix} X_r \\ X_{r+1} \\ \vdots \\ X_k \end{pmatrix} \sim MVN_{k-r+1} \left(\begin{pmatrix} \mu_r \\ \mu_{r+1} \\ \vdots \\ \mu_k \end{pmatrix}, \begin{pmatrix} \sigma_{r,r} & \sigma_{r,r+1} & \cdots & \sigma_{r,k} \\ \sigma_{r+1,r} & \sigma_{r+1,r+1} & \cdots & \sigma_{r+1,k} \\ \vdots & \vdots & \ddots & \vdots \\ \sigma_{k,r} & \sigma_{k,r+1} & \cdots & \sigma_{k,k} \end{pmatrix} \right)$$

So any particular vector of components is normal.

Conditional distribution of a multivariate normal is also a multivariate normal. However, the parameters are not what you may think. The following is an exercise. Suppose that $\mathbf{X}$ is a $MVN_k(\mu, \Sigma)$ using the vector notations above. Suppose we are looking at $\mathbf{X}_1 = (X_1, \ldots, X_r)$ and $\mathbf{X}_2 = (X_{r+1}, \ldots, X_k)$. Suppose we write the vector μ and matrix Σ as

$$\mu = \begin{pmatrix} \mu_1 \\ \mu2_2 \end{pmatrix} \text{ and } \Sigma = \begin{pmatrix} \Sigma_{11} & \Sigma_{12} \\ \Sigma_{21} & \Sigma_{22} \end{pmatrix},$$

where the dimensions are accordingly chosen to match the two vectors (r and $k-r$). Then the conditional distribution of $\mathbf{X}_1$ given $\mathbf{X}_2 = \mathbf{a}$, for some vector $\mathbf{a}$ is

$$\mathbf{X}_1 | \mathbf{X}_2 = \mathbf{a} \sim MVN_r \left(\mu_1 - \Sigma_{12}\Sigma_{22}^{-1}(\mu_2 - \mathbf{a}), \ \Sigma_{11} - \Sigma_{12}\Sigma_{22}^{-1}\Sigma_{21} \right).$$

Furthermore, the vectors $\mathbf{X}_2$ and $X_1 - \Sigma_{21}\Sigma_{22}^{-1}X_2$ are independent. Finally, any affine transformation $AX + b$, where A is a $k \times k$ matrix and b is a k-dimensional constant, is also a multivariate normal with mean vector $A\mu + b$ and covariance matrix $A\Sigma A^T$. This is easy to calculate by doing a joint change of variables.

We conclude this brief exposition of the multivariate normal distribution by making a simple observation. We have seen that, if the joint distribution is multivariate normal, then the marginals are one-dimensional normals. Furthermore, two components are (pairwise) independent if the corresponding element in the covariance matrix Σ is zero.

However, note that if two (or more) variables are normal, that does not mean that the vector which has them as components is multivariate normal. For example, let X be a standard $N(0, 1)$. Let W be an independent variable taking values $+1$ and -1 each with probability $1/2$. Then let $Y = WX$. It is not hard to check that Y is also $N(0, 1)$. For instance,

$$
\begin{aligned}
\mathbf{P}(Y \le y) &= \mathbf{P}(WX \le y) \\
&= \mathbf{P}(WX \le y | W = -1)\mathbf{P}(W = -1) + \mathbf{P}(WX \le y | W = 1)\mathbf{P}(W = 1) \\
&= \mathbf{P}(X \ge -y \mid W = -1)\frac{1}{2} + \mathbf{P}(X \le y \mid W = 1)\frac{1}{2} \\
&\stackrel{independence}{=} (1 - \Phi(-y))\frac{1}{2} + \Phi(y)\frac{1}{2} = \Phi(y)\frac{1}{2} + \Phi(y)\frac{1}{2} \\
&= \Phi(y),
\end{aligned}
$$

where we denoted with $\Phi(y)$ the cdf of a normal. But the joint distribution of these two variables is not bivariate normal. This is easy to see since $|Y| = |X|$ and thus the sample points lie on two lines (rather than an ellipse), as in the case of a multivariate normal.

Furthermore, this is a classic example of two variables which are uncorrelated but not independent. This is all due to the fact that even though the marginals are normal, the joint distribution is not.

Generating Multivariate (Gaussian) Distributions with Prescribed Covariance Structure

We are concluding this chapter with a technique used to generate multivariate normal vectors with prescribed mean and covariance matrix. This technique may in fact be applied to other distributions as well, but the covariance matrix needs to be calculated in terms of the parameters of the distribution.

Theory We want to generate a multivariate normal vector with a given mean vector

$$
\mu = \begin{pmatrix} \mu_1 \\ \vdots \\ \mu_n \end{pmatrix}
$$

and covariance matrix

$$\Sigma = \begin{pmatrix} \sigma_{11} & \sigma_{12} & \cdots & \sigma_{1n} \\ \sigma_{21} & \sigma_{22} & \cdots & \sigma_{2n} \\ \vdots & \vdots & \vdots & \vdots \\ \sigma_{n1} & \sigma_{n2} & \cdots & \sigma_{nn} \end{pmatrix}.$$

Suppose that we can find an n-dimensional square matrix R such that $R^T R = \Sigma$. Then the algorithm is as follows: Generate a vector

$$\mathbf{X} = \begin{pmatrix} X_1 \\ \vdots \\ X_n \end{pmatrix}$$

where the X_i's are n independent standard normal random variables $N(0, 1)$.

Then the vector

$$\mathbf{Y} = R^T \mathbf{X} + \mu$$

is an n-dimensional vector with the multivariate normal distribution required.

This is very simple to prove by just calculating the mean vector and covariance matrix of $R^T \mathbf{X} + \mu$ and using the fact that the covariance matrix of the vector $\mathbf{X}$ is the identity matrix.

Cholesky decomposition   So, to generate a multivariate normal vector with covariance matrix Σ we need to find a matrix R such that $R^T R = \Sigma$. There are in fact many ways to find the matrix R, but here we mention the most popular way, the Cholesky decomposition, which uses the eigenvalues and eigenvectors of the matrix Σ.

Lemma 5.14 (Cholesky decomposition) *Given a symmetric, positive, semidefinite matrix Σ, there exists a U upper triangular matrix and D a diagonal matrix such that*

$$\Sigma = U^T D U.$$

The decomposition is unique when Σ is positive definite.

Note that an upper triangular matrix transposed is lower triangular, and in some other books the lower triangular form is used. A matrix Σ is positive semidefinite if, for every u vector with real elements, we have

$$u^T \Sigma u \geq 0.$$

A matrix Σ is positive definite if, for every u vector nonidentically 0 with real elements, we have

$$u^T \Sigma u \geq 0.$$

This condition fits the random vectors very well. Note that for any $\mathbf{X}$ vector and any u n-dimensional vector of constants we have $u^T\mathbf{X}$ is a one-dimensional random variable. Therefore, its variance must be nonnegative. But its variance is exactly

$$Var(u^T\mathbf{X}) = \mathbf{E}[(u^T\mathbf{X} - u^T\mu)^2] = u^T\Sigma u \geq 0.$$

In fact, if u is not identically 0, then clearly the variance of the resulting random variable is positive.

Therefore, the input matrix should be symmetric (easy to check) and positive definite. Any symmetric positive definite matrix has positive eigenvalues. So, to check positive definiteness, run the function *eigen(matrixname)* in R and inspect the resulting eigenvalues. If they are all positive, then the matrix is positive definite.

There is a drawback to this check. The eigenvalues of a matrix Σ are the roots of the characteristic polynomial $f(\lambda) = \det(\Sigma - \lambda I_n)$, where I_n denotes the identity matrix. The theory says that all the roots of the characteristic polynomial are real and positive if the matrix is symmetric and positive definite. However, in practice the computer does not calculate these roots exactly; instead, it approximates them. So, there may be issues especially if one of the roots is close to 0. There is another way to check by looking at the coefficients of the polynomial $f(\lambda)$ directly. Descarte's Rule of Signs says that the number of positive roots of a polynomial equals the number of sign changes in the coefficients. Since the characteristic polynomial of a positive definite matrix has to have n positive roots, the coefficients must have exactly n changes of sign. Finally, there is another way if computing the polynomial is hard but calculating determinants of constant matrices of lower dimensions is easy. A matrix Σ is positive definite if and only if the diagonal elements are all positive and the determinants of all the upper left corners are positive.

Back to finding the R matrix. If we can calculate these U and D matrices, then we may write

$$\Sigma = U^T DU = (U^T\sqrt{D})(\sqrt{D}U) = (\sqrt{D}U)^T(\sqrt{D}U),$$

where $\sqrt{D}$ is the diagonal matrix having the elements on the diagonal equal to the square-root of the respective diagonal elements in D. Therefore, the matrix R is simply

$$R = \sqrt{D}U$$

Cholesky decomposition in practice   If we want to code the Cholesky decomposition directly, for example, using C, we have to look at the methodology or use libraries that readily compute the U and D matrices. A first (free) resource is the numerical recipes in C books (Press et al., 1992, 2007, Section 2.9).

In MATLAB® or R, the Cholesky decomposition can be accessed as a base function. In both programs, the command is *"chol()"* and the argument of the function is the Σ matrix for which the decomposition is required. Also note that the R function *"chol()"* provides the upper triangular matrix $R = \sqrt{D}U$ directly, skipping the D and $\sqrt{D}$ calculation. The matrix R is generally what is needed in practical applications.

An Example: Generating Correlated Brownian Motions

This last part is given as reference for the later chapters where we learn about Brownian motion. The increment of the Brownian motion over an interval of length Δt is a normal random variable with mean 0 and variance Δt. Successive increments are independent. So to generate a trajectory (path) we just need to generate one-dimensional random normals and we are done.

However, suppose we have multiple Brownian motions which are correlated with a known correlation structure. Such a situation is common in finance when we want to simulate multiple stocks which are evolving in a related manner. The idea is to adapt the technique in the previous section to generate correlated increments over a small interval Δt.

Specifically, Brownian motion is a process that starts from 0 such that the increments of the process over an interval Δt are i.i.d. as a normal with mean 0 and variance the size of the increment Δt. Therefore, suppose we need to generate the process value at times $0 < t_1 < \cdots < t_n = t$. To generate the path, we need to generate the increments, and then we obtain the path.

Specifically, if we generate all the increments ΔB_i over any interval of the type $[t_{i-1}, t_i]$, then to obtain the process since $B_0 = 0$ we use

$$B_{t_1} = \Delta B_1$$

$$B_{t_2} = \Delta B_1 + \Delta B_2$$

$$\vdots$$

$$B_{t_n} = \Delta B_1 + \Delta B_2 + \cdots + \Delta B_n$$

In practice, since the increments are typically obtained as a vector, the process path is obtained by using the cumulative sum function applied to these increments which in most programs is called "cumsum".

Now, suppose that we need to generate five Brownian motions but in such a way that they are correlated with a particular covariance structure. Specifically, if ΔB_t is the vector of increments for the five Brownian motions over an interval of length Δt, we need to have

$$\mathbf{E}[\Delta B_t (\Delta B_t)^T] = \Sigma \Delta t,$$

where $()^T$ denote the transpose of the matrix and Σ is a given symmetric positive definite matrix.

We can see that the problem reduces to generating correlated increments as we described above. We present a pseudo-code bellow as it applies to this problem:

- We calculate the Cholesky decomposition of the correlation matrix $\Sigma = R^T R$. Note that we can calculate very easily the Cholesky decomposition of the matrix $\Sigma \Delta t$ by simply multiplying R with $\sqrt{\Delta t}$.

- Decide on a time increment Δt.

▪ For every interval $[t, t + \Delta t]$, do the following:

1. Generate five independent Normal(0,1) random variables and construct a vector Δx with these five components.

2. Calculate the vector of increments on the interval as

$$\Delta B_t = R^T \Delta x \sqrt{\Delta t}.$$

In this expression, $R^T \Delta x$ is a matrix multiplication.

3. Use these cumsum function as described above.

Done.

Note that $R = \sqrt{D} U$ is an upper triangular matrix multiplied with the matrix R^T, which should be lower triangular. If this is not the structure, then you have made a mistake in the code.

As a remark, we could also generate independent Normal$(0, \sqrt{t})$ directly and multiply them with R^T to obtain the right increments.

Problems

5.1   An assembly line produces microprocessors each of which is fully functional with probability p. A tester is applied to each such processor. If the microprocessor is faulty, then the test detects that there is a problem with the processor with probability q. If the processor is fully functional, the test will not detect a fault. Suppose the line just produced n microprocessors. Let X denote the number of faulty microprocessors, and let Y denote the number of microprocessors that the test detects as being defective. Show that

$$\mathbf{E}[X \mid Y] = \frac{np(1-q) + (1-p)Y}{1 - pq}.$$

Note that this formula allows us to estimate the real number of processors sent for sale given the number the tester found (Y), the precision of the tester (q), and the history of performance for the assembly line (p).

5.2   Let X and Y be independent, both with mean 0. Explain the error in the following derivation:

$$\mathbf{E}[X \mid X - Y = 5] = \mathbf{E}[X \mid X = Y + 5] = \mathbf{E}[Y + 5] = 5$$

5.3   My dad has a car which breaks down all the time. In fact, he knows that the number of days until breaking down is a random variable with density

$$f(i) = \frac{1}{N}, \text{ for all } i \in \{1, 2, \dots, N\},$$

(i.e., the car will break down at some point within the next N days but when exactly is anybody's guess). Given that the car in fact did not break down in t days (where $0 < t < N$), calculate the expected number of remaining days until breaking down.

Repeat the same question for the density

$$f(i) = \frac{1}{2^i}, \text{ for all } i \in \{1, 2, \dots, \}$$

(i.e., every day there is a 50:50 chance that the car will break down).

5.4   Prove Fubini's Theorem 5.1 on page 158.

5.5   Using the heorem-Definition 5.10 on page 166, prove the seven properties of the conditional expectation in Proposition 5.12.

5.6   Let X, Y be two random variables. Show that for any measurable function φ for which the expressions below exist, we have

$$\mathbf{E}[\varphi(X)\mathbf{E}[Y \mid X]] = \mathbf{E}[Y\varphi(X)]$$

5.7   For random variables X and Y, show that

$$V(Y) = \mathbf{E}[V(Y \mid X)] + V(\mathbf{E}[Y \mid X]).$$

The variance $V(Y \mid X)$ is the variance of the random variable $Y \mid X$, while the same holds for the random variable $\mathbf{E}[Y \mid X]$.

5.8   Let X have a Beta distribution with parameters a and b. Let Y be distributed as $Binomial(n, X)$. What is $\mathbf{E}[Y \mid X]$? Describe the distribution of Y and give $\mathbf{E}[Y]$ and $V[Y]$. What is the distribution of Y in the special case when X is uniform?

5.9   For each of the following joint distributions of X and Y calculate the density of $Y|X$ and $\mathbf{E}[Y \mid X]$:

a)
$$f(x, y) = \lambda^2 e^{-\lambda(x+y)}, \quad x, y > 0$$

b)
$$f(x, y) = \lambda^2 e^{-\lambda y}, \quad y > x > 0$$

c)
$$f(x, y) = xe^{-x(y+1)}, \quad x, y > 0$$

5.10   Let X be a random variable on the probability space $(\Omega, \mathscr{F}, \mathbf{P})$. Let a set $A \in \mathscr{F}$ with $\mathbf{P}(A) \neq 0$ and the sigma algebra generated by the set denoted $\sigma(A)$. What is $\mathbf{E}[X \mid \sigma(A)]$? Let $\mathbf{1}_A$ denote the indicator of A. What is $\mathbf{E}[X \mid \mathbf{1}_A]$?

Comment: I have been given this question in an industry job interview. The problem is flawed in the sense that the answer may be given only if X is independent of $\mathbf{1}_A$.

5.11   Let X, Y, Z be three random variables with joint distribution

$$P(X = k, Y = m, Z = n) = p^3 q^{n-3}$$

for integers k, m, n satisfying $1 \leq k < m < n$, where $0 < p < 1$, $p + q = 1$. Find $E\{Z | X, Y\}$.

5.12   A circular dartboard has a radius of 1 ft. Thom throws three darts at the board until all three are sticking in the board. The locations of the three darts are independent and uniformly distributed on the surface of the board. Let T_1, T_2, and T_3 be the distances from the center to the closest dart, the next closest dart, and the farthest dart, respectively, so that $T_1 \leq T_2 \leq T_3$. Find $\mathbf{E}[T_2]$.

5.13   Suppose you pick two numbers independently at random from $[0, 1]$. Given that their sum is in the interval $[0, 1]$, find the probability that $X^2 + Y^2 < 1/4$.

5.14   Let $X_1, X_2, \ldots, X_{1000}$ be i.i.d., each taking values 0 or 1 with probability $\frac{1}{2}$. Put $S_n = X_1 + \cdots + X_n$. Find $\mathbf{E}\left[(S_{1000} - S_{300}) \mid \mathbf{1}_{\{S_{700}=400\}}\right]$ and $\mathbf{E}[(S_{1000} - S_{300})^2 \mid \mathbf{1}_{\{S_{700}=400\}}]$.

5.15   Let X be uniformly distributed on $[-1, 1]$ and let $Y = X^2$. Find $\mathbf{E}[X \mid Y]$ and $\mathbf{E}[Y \mid X]$.

5.16   Let the vector (X, Y) have the joint distribution

$$f(x, y) = \frac{1}{2\pi\sigma_1\sigma_2\sqrt{1 - \rho^2}} e^{-\frac{1}{2(1-\rho^2)}\left[\left(\frac{x-\mu_1}{\sigma_1}\right)^2 + \left(\frac{y-\mu_2}{\sigma_2}\right)^2 - 2\rho\frac{(x-\mu_1)(y-\mu_2)}{\sigma_1\sigma_2}\right]}.$$

(the general bivariate normal density). Find $\mathbf{E}[Y \mid X]$.

5.17   Let the discrete random variables X and Y have density

$$\mathbf{P}(X = i, Y = j) = \frac{1}{i(i + 1)}, \quad \text{for } i = j \in \{1, 2, \ldots\},$$

and 0 everywhere else. Show that $\mathbf{E}[Y] = \infty$ while $\mathbf{E}[Y \mid X] < \infty$.

5.18   Each child in Romania is equally likely to be a boy or a girl, independent of any other children.

   a) Suppose we know that a family has n children. Show that the expected number of boys is equal to the expected number of girls. Did you need the assumption of independence?

   b) Now suppose that, in addition of knowing that the family has n children, we also know that one of them is a boy ($n > 1$). Is it still true that the expected number of boys equals the expected number of girls? What happens if the assumption of independence is dropped?

5.19   Let X, Y and Z be i.i.d. exponentially distributed random variables with parameter λ. Calculate

$$\mathbf{P}(X < Y < Z)$$

CHAPTER 6

TOOLS TO STUDY PROBABILITY. GENERATING FUNCTION, MOMENT GENERATING FUNCTION, CHARACTERISTIC FUNCTION

In this chapter we begin studying statistics which are defined as any one dimensional combination of random variables. In particular, the most important statistic is the average of random variables and by extension (because that is how we calculate the average) the sum of random variables. As we shall see below, calculating the distribution of sums of random variables is very complicated. However, the tools we introduced in this chapter will allow us to calculate the distribution of sums quite easily provided that the sum's components are independent. These tools are also the primary instrument used in Chapter 7 which studies what happens with statistics when the number of components goes to infinity. In practice, we cannot work with infinity; nevertheless, we can determine the direction a method is going if it continues the current sampling procedure and, most importantly, we may calculate how many items we need to have a good estimate of the limit.

6.1 Sums of Random Variables. Convolutions

Any sum has at least two components. We start by analyzing the distribution of a sum of just two variables.

Probability and Stochastic Processes, First Edition. Ionuț Florescu
© 2015 John Wiley & Sons, Inc. Published 2015 by John Wiley & Sons, Inc.

Let X, Y be two random variables. Let F, G be the distribution functions of X and Y where $F(x) = \mathbf{P} \circ X^{-1}(-\infty, x]$, $G(x) = \mathbf{P} \circ Y^{-1}(-\infty, x]$. We are interested in the distribution of $X + Y$. Recall the General Transport formula (Theorem 4.25 on page 146). This theorem can help us calculate the required distribution. Specifically, consider

Definition 6.1 (Convolution) *The Convolution is the distribution function of the random variable $X + Y$. We denote this distribution function with*

$$F * G = (F, G) \circ s^{-1} = \mathbf{P} \circ (X + Y)^{-1}$$

where

$$\Omega \xrightarrow{(X(\omega), Y(\omega))} \mathbb{R}^2 \xrightarrow{s(x,y)=x+y} \mathbb{R}$$

In the special case when X and Y have joint density f, then the convolution also has a density and

$$f_{X+Y}(z) = \int f(z - y, y) dy$$

If in addition X and Y are independent and have densities f_X and f_Y then we have the density of $X + Y$:

$$f_{X+Y}(z) = f_X * f_Y(z) = \int f_X(z - y) f_Y(y) dy$$

This last expression is the better known formula of convolution for two functions.

Proposition 6.2 (Properties of convolution) *Let F, G be two distribution functions. Then,*

(i) $F * G = G * F$

(ii) $F * (G * H) = (F * G) * H$

(iii) $\delta_a * \delta_b = \delta_{a+b}$ *with $a, b \in \mathbb{R}$, and δ_a, δ_b are delta distribution functions.*

(iv) *If F, G are discrete then $F * G$ is discrete.*

This proposition is easy to prove from the definition and the proof is left as an exercise (Problem 6.3).

6.2   Generating Functions and Applications

First, we introduce a generating function for discrete probability distributions. Since the distribution of a discrete random variable is just a sequence of numbers we first introduce the generating function for any sequence of numbers.

Definition 6.3 (Regular generating function) *Let* $a = \{a_i\}_{i \in \{0,1,2,\dots\}}$ *be a sequence of real numbers and define the generating function of the sequence* a *as*

$$G_a(s) = \sum_{i=0}^{\infty} a_i s^i$$

which has the domain values $s \in \mathbb{R}$ *for which the series converges.*

Remark 6.4 *Please note that we can obtain all the terms of the sequence if we know the generating function as*

$$a_i = \frac{G_a^{(i)}(0)}{i!}$$

where $f^{(i)}(x_0)$ *is a notation for the ith derivative of f calculated at* x_0.

📖 **EXAMPLE 6.1**

Take the sequence a as

$$a_n = (\cos\alpha + i\sin\alpha)^n = \cos(n\alpha) + i\sin(n\alpha) \quad \text{(De Moivre's Theorem)}$$

Then its generating function is

$$G_a(s) = \sum_{n=0}^{\infty} s^n \cdot a_n = \sum_{n=0}^{\infty} (s(\cos\alpha + i\sin\alpha))^n$$
$$= \frac{1}{1 - s(\cos\alpha + i\sin\alpha)},$$

Provided that the power term converges to 0. This holds if $|s| < 1$ for all α and obviously the radius of convergence may be increased depending on α.

Definition 6.5 *The convolution of two sequences* $\{a_i\}$ *and* $\{b_i\}$ *is the sequence* $\{c_i\}$ *with*

$$c_n = a_0 b_n + a_1 b_{n-1} + \cdots + a_n b_0$$

We write $c = a * b$. *If* a *and* b *have generating functions* G_a *and* G_b *then*

$$G_c(s) = \sum_{n=0}^{\infty} c_n s^n = \sum_{n=0}^{\infty} s^n \sum_{k=0}^{n} a_k b_{n-k} = \sum_{n=0}^{\infty} \sum_{k=0}^{n} (a_k s^k)(b_{n-k} s^{n-k})$$
$$\overset{Fubini}{=} \sum_{k=0}^{\infty} a_k s^k \sum_{n=k}^{\infty} b_{n-k} s^{n-k} = G_a(s) G_b(s)$$

This definition shows why the generating function is useful. The generating function of a convolution of two sequences is the product of individual generating functions. Thus, while a convolution is complicate to calculate, its generating function is much simple.

Generating functions for discrete random variables   This obviously has to have a meaning in our context.

Definition 6.6 (Probability generating function(discrete)) *If X is a discrete random variable with outcomes $i \in \mathbb{Z}$ and corresponding probabilities p_i then we define*

$$G_X(s) = \mathbf{E}(s^x) = \sum_i s^i \mathbf{P}(X = i) = \sum_i s^i p_i$$

This is the probability generating function of the discrete density of X.

Proposition 6.7 (Properties of probability generating function) *Let* $G_X(s)$, *$G_Y(s)$ be the generating functions of two discrete random variables X and Y.*

(i) There exists $R \geq 0$ a radius of convergence such that the sum is convergent if $|s| < R$ and diverges if $|s| > R$. Therefore, the corresponding generating function G_X exists on the interval $(-R, R)$.

(ii) The generating function $G_X(s)$ is differentiable at any s within the domain $|s| < R$.

(iii) If two generating functions $G_X(s) = G_Y(s) = G(s)$ for all $|s| < R$ where $R = \min\{R_X, R_Y\}$ then X and Y have the same distribution. Furthermore, the common distribution may be calculated from the generating function using

$$\mathbf{P}(X = n) = \frac{1}{n!} G^{(n)}(0)$$

(iv) As particular cases of the definition we readily obtain that

$$G_X(0) = \mathbf{P}(X = 0)$$

$$G_X(1) = \sum_i p_i = 1$$

The last relationship implies that the probability generating function always exists for $s = 1$, thus the radius of convergence $R \geq 1$

■ **EXAMPLE 6.2**

Below we present the generating function for commonly encountered discrete random variables:

(i) Constant variable: $\mathbf{P}(X = c) = 1$

$$G(s) = \mathbf{E}(s^x) = s^c$$

(ii) Bernoulli (p)

$$G(s) = \mathbf{E}(s^x) = 1 - p + ps$$

(iii) Geometric (p)

$$G(s) = \sum_{k=1}^{\infty} s^k (1-p)^{k-1} p$$

$$= sp \cdot \sum_{k=1}^{\infty} (s(1-p))^{k-1}$$

$$= \frac{sp}{1 - s(1-p)}$$

(iv) Poisson (λ)

$$G(s) = \sum_{k=1}^{\infty} s^k \cdot \frac{\lambda^k}{k!} e^{-\lambda} = e^{\lambda s} \cdot e^{-\lambda} = e^{\lambda(s-1)}$$

Once again let me stress that this notion is very useful when working with discrete random variables. The following two theorems tell us why this is so.

Theorem 6.8 *If X and Y are two* independent *random variables then*

$$G_{X+Y}(s) = G_X(s) G_Y(s)$$

Theorem 6.9 *If X has generating function $G(s)$ then*

(i) $\mathbf{E}(X) = G'(1)$

(ii) $\mathbf{E}[X(X-1)\cdots(X-k+1)] = G^{(k)}(1)$ *the kth factorial moment of X.*

Proof: Exercise. ∎

▣ EXAMPLE 6.3

Let X, Y be two independent Poisson random variables with parameters λ and μ. What is the distribution of $Z = X + Y$?

Proof (Solution): We could compute the convolution $f_Z = f_X * f_Y$ using the formulas in the previous section, but using generating functions is easier.

$$G_X(s) = e^{\lambda(s-1)}$$

$$G_Y(s) = e^{\mu(s-1)}$$

By independence of X and Y and from Theorem 6.8 above

$$G_Z(s) = G_X(s) G_Y(s) = e^{\lambda(s-1)} e^{\mu(s-1)} = e^{(\lambda+\mu)(s-1)}$$

But this is the generating function of another Poisson random variable thus Z has Poisson $(\lambda + \mu)$ distribution. ∎

■ **EXAMPLE 6.4**

We know that for a Binomial (n, p) random variable X may be written as $X = X_1 + \cdots + X_n$ where X_is are independent Bernoulli(p) random variables. Therefore, the generating function of a Binomial(n, p) is

$$G_X(s) = G_{X_1}(s)G_{X_2}(s) \cdots G_{X_n}(s) = (1 - p + sp)^n$$

What if we have two independent variables X Binomial(m, p) and Y Binomial(k, p)? What is the distribution of $X + Y$ in this case?

$$G_{X+Y}(s) = (1 - p + sp)^m (1 - p + sp)^k = (1 - p + sp)^{m+k}$$

We see that this is the generating function for a Binomial random variable with parameters $m + k$ and p. This is of course as it should be if one realizes that if X can be written and a sum of n Bernoulli(p) and Y can be written and a sum of k Bernoulli(p), then of course $X + Y$ is a sum of $n + k$ Bernoulli(p) random variables.

But what if we add a random number of terms? What happens then? The next theorem helps with some of these situations.

Theorem 6.10 *If $X_1, X_2, \ldots$ is a sequence of i.i.d. random variables with generating function $G_X(s)$ and N is an integer-valued random variable independent of the X_i with generating function $G_N(s)$ then $S = X_1 + \cdots + X_N$ has generating function:*

$$G_S(s) = G_N(G_X(s))$$

■ **EXAMPLE 6.5**

A hen lays N eggs where N has the Poisson distribution with parameter λ. Each egg hatches with probability p independently of all other eggs. Let K be the number of chicks that hatch. Find $\mathbf{E}(K|N)$, $\mathbf{E}(N|K)$, and $\mathbf{E}(K)$. Furthermore, find the distribution of K.

Proof (Solution): It is given that

$$f_N(n) = \frac{\lambda^n}{n!} e^{-\lambda}$$

$$f_{K|N}(k|n) = \binom{n}{k} p^k (1 - p)^{n-k}.$$

We may then calculate the conditional expectations. Let

$$\varphi(n) = \mathbf{E}(K|N = n) = \sum_k k \cdot \binom{n}{k} p^k (1 - p)^{n-k} = np$$

since that is the formula for the expectation of a Binomial(n, p) random variable. Therefore, $\mathbf{E}(K|N) = \varphi(N) = Np$. Recall that $\mathbf{E}[\mathbf{E}[X|Y]] = \mathbf{E}[X]$. Using this we get that

$$\mathbf{E}[\mathbf{E}[K|N]] = \mathbf{E}[K] = \mathbf{E}[Np] = \lambda p$$

To find $\mathbf{E}(N|K)$ we need to find $f_{N|K}$.

$$
\begin{aligned}
f_{N|K}(n|k) = \mathbf{P}(N = n|K = k) &= \frac{\mathbf{P}(N = n, K = k)}{\mathbf{P}(K = k)} \\
&= \frac{\mathbf{P}(K = k|N = n) \cdot \mathbf{P}(N = n)}{\mathbf{P}(K = k)} \\
&= \frac{\binom{n}{k} p^k (1 - p)^{n-k} \frac{\lambda^n}{n!} e^{-\lambda}}{\sum_{m \geq k} \binom{m}{k} p^k (1 - p)^{m-k} \frac{\lambda^m}{m!} e^{-\lambda}} \\
&= \frac{(q\lambda)^{n-k}}{(n - k)!} e^{-q\lambda}
\end{aligned}
$$

if $n \geq k$ and 0 otherwise, where $q = 1 - p$ and for the denominator we have used the following:

$$
\begin{aligned}
\mathbf{P}(K = k) &= \sum_{m=0}^{\infty} \mathbf{P}(K = k, N = m) = \sum_{m=0}^{\infty} \mathbf{P}(K = k|N = m)\mathbf{P}(N = m) \\
&= \sum_{m=k}^{\infty} \mathbf{P}(K = k|N = m)\mathbf{P}(N = m)
\end{aligned}
$$

since $\mathbf{P}(K = k|N = m) = 0$ for any $k > m$ (it is not possible to hatch more chicks than eggs laid).

As a result we immediately obtain

$$
\begin{aligned}
\mathbf{E}(N|K = k) &= \sum_{n \geq k} n \cdot \frac{(q\lambda)^{n-k}}{(n - k)!} e^{-q\lambda} \stackrel{\text{change of var. } n-k=m}{=} \\
&= e^{-q\lambda} \sum_{m=0}^{\infty} (m + k) \frac{(q\lambda)^m}{m!} \\
&= e^{-q\lambda} \left(\sum_{m=0}^{\infty} m \frac{(q\lambda)^m}{m!} + k \sum_{m=0}^{\infty} \frac{(q\lambda)^m}{m!} \right) \\
&= e^{-q\lambda} \left(0 + q\lambda \sum_{m=1}^{\infty} \frac{(q\lambda)^{m-1}}{(m - 1)!} + k e^{q\lambda} \right) = q\lambda + k
\end{aligned}
$$

Which gives us $\mathbf{E}(N|K) = q\lambda + K$.

Note that above we kind of obtained the distribution of K. We still have to calculate a sum but even possible we will not pursue that approach. This chapter is about generating functions and we want to demonstrate how easy it is to calculate the distribution using them.

Let $K = X_1 + \cdots + X_N$ where the $X_i \sim$Bernoulli(p). Then using the above theorem

$$G_K(s) = G_N(G_X(s))$$

$$G_N(s) = \sum_{n=0}^{\infty} s^n \frac{\lambda^n}{n!} e^{-\lambda} = e^{\lambda(s-1)}$$

$$G_X(s) = 1 - p + ps$$

Using the above equations we conclude that

$$G_K(s) = G_N(1 - p + ps) = e^{\lambda(1-p+ps-1)} = e^{\lambda p(s-1)}$$

But since this is a generating function for a Poisson r.v. we see that $K \sim$ Poisson(λp). ∎

EXAMPLE 6.6   Quantum mechanics

In an experiment a laser beam emits photons with an intensity of λ photons/ second. Each photon travels through a deflector which polarizes (spins up) each photon with probability p. A measuring device is placed after the deflector. This device counts the number of only those photons which are spin up. Let N denote the number of photons emitted by the laser beam in a second. Let K denote the number of photons measured by the capturing device. Assume that N is distributed as a Poisson random variable with mean λ. Assume that each photon is spinning independently of any other photon. Give the distribution of K and its parameters. In particular calculate the average intensity recorded by the measuring device.

Proof (Solution): Exercise. This is a different domain but the solution is identical with Example 6.10. ∎

6.3   Moment Generating Function

Generating functions are useful when working with positive discrete random variables. This is due to the definition which uses s^n and that is useful when n is an integer thus obtaining power series. If we want to study any random variables an immediate replacement would be the function s^x. However, this is just an exponential function and it is typically convenient to work with a common base. Thus, we make the transformation $s = e^t$ in $G(s) = \mathbf{E}(s^x)$, and we reach the following definition.

Definition 6.11 (Moment generating function) *The moment generating function of a random variable X is the function $M : \mathbb{R} \to [0, \infty)$ given by*

$$M_X(t) = \mathbf{E}(e^{tx})$$

Formally:

Definition 6.12 (Laplace transform of the function f) *Suppose we are given a positive function $f(x)$. Assume that*

$$\int_0^\infty e^{-t_0 x} f(x) dx < \infty,$$

*for some value of $t_0 > 0$. Then the **Laplace transform** of f is defined as*

$$\mathcal{L}\{f\}(t) = \int_0^\infty e^{-tx} f(x) dx,$$

*for all $t > t_0$. If the random variable X has distribution function F the following defines the **Laplace–Stieltjes transform**:*

$$\mathcal{L}_{LS}\{F\}(t) = \int_0^\infty e^{-tx} dF(x)$$

again with the same caution about the condition on the existence.

With the previous definitions it is easy to see that the moment generating functions is a sum of two Laplace transforms (provided either side exists). Specifically, suppose the random variable X has density $f(x)$. Then,

$$M_X(t) = \mathbf{E}(e^{tx}) = \int_{-\infty}^\infty e^{tx} f(x) dx$$

$$= \int_{-\infty}^0 e^{tx} f(x) dx + \int_0^\infty e^{tx} f(x) dx$$

$$= \int_0^\infty e^{-tx} f(-x) dx + \int_0^\infty e^{-(-t)x} f(x) dx$$

$$= \mathcal{L}\{f_-\}(t) \mathbf{1}_{(0,\infty)}(t) + \mathcal{L}\{f_+\}(t) \mathbf{1}_{(-\infty,0]}(t)$$

where

$$f_-(x) = \begin{cases} f(-x), & \text{if } x > 0 \\ 0, & \text{else} \end{cases}$$

and,

$$f_+(x) = \begin{cases} f(x), & \text{if } x > 0 \\ 0, & \text{else.} \end{cases}$$

This result is generally needed to use the theorems related to Laplace and specially inverse Laplace transforms.

The following proposition contains the properties of the moment generating function (m.g.f.).

Proposition 6.13 (Properties) *Suppose that the moment generating function of a random variable X, $M_X(t) < \infty$ for all $t \in \mathbb{R}$.*

(i) If the moment generating function is derivable then the first moment is $\mathbf{E}(X) = M'_X(0)$*; in general if the moment generating function is k times derivable then* $\mathbf{E}(X^k) = M_X^{(k)}(0)$.

(ii) More general if the variable X is in L^p *(the first p moments exist) then we may write (Taylor expansion of* e^x *):*

$$M(t) = \sum_{k=0}^{\infty} \frac{\mathbf{E}(X^k)}{k!} t^k$$

(iii) If X, Y are two independent random variables then $M_{X+Y}(t) = M_X(t) M_Y(t)$

Remark 6.14 *It is the last property in the above proposition that makes it desirable to work with moment generating functions. However, they do possess a big disadvantage. The expectation and thus the integrals involved may not be finite and therefore the function may not exist. It is for this reason that the characteristic function is introduced.*

Nevertheless, as the examples show next if it exists it is very convenient to work with

◼ EXAMPLE 6.7   Calculating moments using the MGF

Let X be an r.v. with moment generating function

$$M_X(t) = \frac{1}{2}(t + e^{t^2}).$$

Derive the expectation and the variance of X.

Proof (Solution): Since

$$M'_X(t) = \frac{1}{2}(1 + 2te^{t^2})$$

we get for $t = 0$ and using Proposition 6.13

$$\mathbf{E}X = M'_X(0) = \frac{1}{2}.$$

By differentiating twice M_X,

$$M''_X(t) = e^{t^2} + 2t^2 e^{t^2},$$

therefore,

$$\mathbf{E}X^2 = M''_X(0) = 1.$$

Thus,

$$Var X = \mathbf{E}X^2 - (\mathbf{E}X)^2 = 1 - \frac{1}{4} = \frac{3}{4}.$$

◼

The following theorems are very useful but recall that the MGF does not always exist. We shall prove the variants of these theorems for the more general characteristic function in the next section. However, keep in mind that if the moment generating function exists it is typically easier to work with it than the characteristic function.

Theorem 6.15 (Uniqueness theorem) *If the MGF of X exists for any t in an open interval containing* 0*, then the MGF uniquely determines the C.D.F. of* X*. That is, no two different distributions can have the same values for the MGF's on an interval containing* 0*.*

The uniqueness theorem in fact comes from the uniqueness of the Laplace transform and inverse Laplace transform.

Theorem 6.16 (Continuity theorem) *Let* $X_1, X_2, \ldots$ *a sequence of random variables with c.d.f.s* $F_{X_1}(x), F_{X_2}(x), \ldots$ *and moment generating functions* $M_{X_1}(t), M_{X_2}(t), \ldots$ *(which are defined for all t). Suppose that*

$$M_{X_n}(t) \to M_X(t)$$

for all t as $n \to \infty$*. Then,* $M_X(t)$ *is the moment generating function of some random variable X. If* $F_X(x)$ *denote the c.d.f. of X then*

$$F_n(x) \to F_X(x),$$

for all x continuity points of F_X

We note that the usual hypothesis of the theorem include the requirement that $M_X(t)$ be a moment generating function. This is not needed in fact, the inversion theorem stated next makes this condition obsolete. The only requirement is that the MGF be defined at all x.

Recall that the MGF can be written as a sum of two Laplace transforms. The inversion theorem is stated in terms of the Laplace transform. This is in fact the form in which it actually is used.

Theorem 6.17 (Inversion theorem) *Let* f *be a function with its Laplace transform denoted with* f^**, i.e.,*

$$f^*(t) = \mathcal{L}\{f\}(t) = \int_0^\infty e^{-tx} f(x)dx.$$

Then we have

$$f(x) = \frac{1}{2\pi i} \lim_{T \to \infty} \int_{c-iT}^{c+iT} e^{tx} f^*(t)dt.$$

where c is a constant chosen such that it is greater than the real part of all singularities of f^**. The formula above gives* f *as the inverse Laplace transform.* $f = \mathcal{L}^{-1}\{f^*\}$*.*

Remark 6.18 *In practice, computing the complex integral is done using the Residuals Theorem (Cauchy residuals theorem). If all singularities of f^* have negative real part or there is no singularity then c can be chosen 0. In this case the integral formula of the inverse Laplace transform becomes identical with the inverse Fourier transform (given in the next section).*

6.4   Characteristic Function

The characteristic function of a random variable is a powerful tool for analyzing the distribution of sums of independent random variables. In the same way in which the MGF was related to the Laplace transform the characteristic function is related to the Fourier transform of a function.

Definition 6.19 (Characteristic function) *The characteristic function associated to a random variable X is defined as a complex-valued function $\varphi : \mathbb{R} \to \mathbb{C}$:*

$$\varphi_X(t) = \mathbf{E}(e^{itX}),$$

where $i = \sqrt{-1}$

Since $|e^{itX(\omega)}| = 1$ for every $t \in \mathbb{R}$ and $\omega \in \Omega$ it is clear that the expectation defined above exists and

$$|\varphi_X(\lambda)| \leq 1.$$

This is easy to prove using the Jensen inequality (Lemma 4.14) for the convex function $|x|$.

If X has a probability density $f_X(x)$ then the characteristic function reduces to

$$\varphi_X(\lambda) = \int_{\mathbb{R}} e^{i\lambda x} f_X(x) dx. \tag{6.1}$$

This formula (6.1) with $-\lambda$ replacing λ is also known as the Fourier transform of f_x.

The Fourier transform is in fact defined for any integrable function f, and not only for those which happen to be probability densities. Specifically, the Fourier transform of a function denoted f_X is given by

$$\hat{f}(\lambda) = \int_{\mathbb{R}} e^{-i\lambda x} f_X(x) dx \tag{6.2}$$

for every $\lambda \in \mathbb{R}$. Therefore, we can immediately see the relationship between the characteristic function and the Fourier transform:

$$\varphi_X(\lambda) = \hat{f}(-\lambda).$$

The Fourier transform is also a well-defined function for every function f integrable, even if not a probability density. Specifically, since $|e^{i\lambda x}| = 1$, we have

$$|\hat{f}(\lambda)| = \left| \int_{\mathbb{R}} e^{i\lambda x} f(x) dx \right| \leq \int_{\mathbb{R}} |f(x)| dx < \infty$$

Remark 6.20 *We note that the typical definition of the Fourier transform in mathematical analysis, partial differential equations, and mathematical physics is*

$$\hat{f}(\lambda) = \frac{1}{\sqrt{2\pi}} \int_{\mathbb{R}} e^{i\lambda x} f_X(x) dx.$$

This is the same as our definition since the constant $\sqrt{2\pi}$ is irrelevant.

Relationship with the moment generating function If the moment generating function of a random variable X exists then we obviously have

$$\varphi_X(t) = M_X(it).$$

We note that the characteristic function is a better behaved function than the moment generating function. The drawback is that it takes complex values and we need to have a modicum understanding of complex analysis to be able to work with it. For now let us remark two things. First, the complex random variable whose moment we calculate e^{itX} has modulus 1 for all t. This is easy to see since

$$e^{itX} = \cos(tX) + i\sin(tX)$$

Second, also because of the previous equality we may calculate the complex integral by computing two real integrals

$$\varphi_X(t) = \mathbf{E}(\cos(tX)) + i\mathbf{E}(\sin(tX))$$

Proposition 6.21 (Properties of the characteristic function) *Suppose that a random variable X has characteristic function φ. Then*

(i) $\varphi(0) = 1$ and $|\varphi(t)| \leq 1 \; \forall t \in \mathbb{R}$.

(ii) $\varphi(-t) = \overline{\varphi(t)}$, where $\bar{a} = x - iy$ denotes the complex conjugate of $a = x + iy \in \mathbb{C}$.

(iii) φ is uniformly continuous on $\mathbb{R}$.

(iv) φ is a nonnegative definite function, i.e., for any $t_1 \cdots t_n \in \mathbb{R}$ and for any $z_1 \cdots z_n \in \mathbb{C}$ we have $\sum_{j,k} \varphi(t_j - t_k) z_j \bar{z}_k \geq 0$

(v) If $\varphi^k(0)$ exists, then $\mathbf{E}|X^k| < \infty$ if k is even and $\mathbf{E}|X^{k-1}| < \infty$ if k is odd.

(vi) If $\mathbf{E}|X^k| < \infty$, then

$$\varphi(t) = \sum_{j=0}^{k} \frac{\mathbf{E}(X^j)}{j!} (it)^j + o(t^k)$$

As an immediate consequence $\varphi^k(0) = i^k \mathbf{E}(X^k)$.

Remark 6.22 (Notation: Big O, Little o) *We adopt the following standard notation. If f and g are two univariate functions then,*

$$f = o(g) \text{ if } \lim_{x \to 0} \frac{f(x)}{g(x)} = 0$$

$$f = O(g) \text{ if } \left| \frac{f(x)}{g(x)} \right| < C, \forall x \text{ small enough and some real constant } C$$

Finally, the reason for using characteristic functions:

Theorem 6.23 *If X and Y are two independent random variables with characteristic functions φ_X and φ_Y then the variable $X + Y$ has characteristic function:*

$$\varphi_{X+Y}(t) = \varphi_X(t)\varphi_Y(t)$$

Proof: Exercise.   ■

Theorem 6.24 *If $a, b \in \mathbb{R}$ and $Y = aX + b$, then*

$$\varphi_Y(t) = e^{itb}\varphi_X(at)$$

Definition 6.25 (Joint characteristic function) *If $\mathbf{X} = (X_1, \ldots, X_n)$ is a random vector then its characteristic function is $\varphi_{\mathbf{X}} : \mathbb{R}^n \to \mathbb{C}$,*

$$\varphi_{\mathbf{X}}(\mathbf{t}) = \mathbf{E}(e^{i<\mathbf{t},\mathbf{X}>}) = \mathbf{E}\big(e^{i(t_1 x_1 + \cdots + t_n x_n)}\big),$$

where $< \mathbf{t}, \mathbf{X} >$ denotes the scalar product between $(t_1, \ldots, t_n)$ and $(x_1, \ldots, x_n)$. In particular, for two r.v.s we get

$$\varphi_{(X,Y)}(s, t) = \mathbf{E}(e^{isX} e^{itY})$$

Theorem 6.26 *With the previous definition of joint characteristic function we may verify independence. Specifically, X and Y are two independent random variables if and only if*

$$\varphi_{X,Y}(s, t) = \varphi_X(s)\varphi_Y(t) \text{ for all } s, t.$$

▣ EXAMPLE 6.8   Calculation of characteristic function for basic variables

(i) Let $X \sim$ Bernoulli(p). Then

$$\varphi_X(t) = \mathbf{E}(e^{itx}) = e^{it0}(1 - p) + e^{it1}p = e^{it}p + (1 - p)$$

(ii) $X \sim$ Binomial(n, p). Then, we may write $X = X_1 + \cdots + X_n$ where the X_i's are independent Bernoulli(p) and

$$\varphi_X(t) = \varphi_{X_1}(t) \cdots \varphi_{X_n}(t) = ((1 - p) + pe^{it})^n$$

(iii) $X \sim$ Exponential(λ).

$$\varphi_X(t) = \int_0^\infty e^{itx} \lambda e^{-\lambda x} dx = \lambda \int_0^\infty e^{-(\lambda - it)x} dx = \frac{\lambda}{\lambda - it}.$$

(iv) If $X \sim U[0, 1]$ then

$$\varphi_X(t) = \frac{e^{it} - 1}{it}.$$

More generally, if $X \sim U[a, b]$ then

$$\varphi_X(t) = \frac{e^{itb} - e^{ita}}{it(b - a)}.$$

Since the density of the $U[0, 1]$ distribution is $g(x) = 1_{[0,1]}(x)$ we get

$$\begin{aligned} \varphi_X(t) &= \int_0^1 e^{itx} dx = \frac{1}{it} \left[e^{itx} \right]_{x=0}^{x=1} \\ &= \frac{e^{it} - 1}{it}. \end{aligned}$$

The more general result is obtained by noting that $X \sim U[a, b]$ can be written as

$$X = a + (b - a)U,$$

with $U \sim U[0, 1]$.

Note a particular case. If $a = -b$ (i.e., X has a symmetric distribution $X \sim U[-b, b]$), then

$$\begin{aligned} \varphi_X(t) &= \frac{e^{itb} - e^{-itb}}{it(b - (-b))} \\ &= \frac{\sin tb}{tb} \end{aligned}$$

and note that this is a real-valued function.

(v) $Z \sim$ Normal$(0, 1)$. Then

$$\varphi_Z(t) = e^{-\frac{t^2}{2}}$$

First, we present a physics proof (of course, complex integrals do not work this way).

$$\begin{aligned} \varphi_Z(t) &= \int e^{itx} \frac{1}{\sqrt{2\pi}} e^{-\frac{x^2}{2}} dx = \frac{1}{\sqrt{2\pi}} \int e^{-\frac{x^2}{2} + itx - \frac{(it)^2}{2} + \frac{(it)^2}{2}} dx \\ &= \frac{1}{\sqrt{2\pi}} e^{-\frac{t^2}{2}} \int e^{-\frac{(x-it)^2}{2}} dx \\ &= e^{-\frac{t^2}{2}} \end{aligned}$$

To mathematically justify this change of variables (which replace a real variable by a complex one!) we would need more complicated arguments from complex function theory.

Even though the proof needs more arguments, the result is correct. A different proof of this result, based on differential equations, is given in Problem 6.13.

(vi) $X \sim \text{Normal}(\mu, \sigma^2)$. If we know the previous characteristic formula then using Theorem 6.24 we can immediately calculate

$$\varphi_X(t) = \varphi_{\sigma Z + \mu}(t) = e^{it\mu} e^{-\frac{t^2 \sigma^2}{2}} = e^{it\mu - t^2 \frac{\sigma^2}{2}}$$

(vii) If X is a Poisson random variable with parameter $\lambda > 0$ then

$$\varphi_X(t) = e^{-\lambda(1 - e^{it})}.$$

Using the definition of the Poisson distribution, for every $\lambda \in \mathbb{R}$

$$\mathbf{E}e^{i\lambda X} = \sum_{k \geq 0} e^{i\lambda k} \frac{\lambda^k}{k!} e^{-\lambda}$$

$$= e^{-\lambda} \sum_{k \geq 0} \frac{(\lambda e^{it})^k}{k!}$$

$$= e^{-\lambda(1 - e^{it})}.$$

(viii) If X is a Gamma random variable with parameters a and λ (i.e., $X \sim \Gamma(a, \lambda)$) then

$$\varphi_X(t) = \left(\frac{1}{1 - \frac{it}{\lambda}}\right)^a.$$

Recall that the density of the Gamma distribution is

$$f(x) = \frac{\lambda^a}{\Gamma(a)} x^{a-1} e^{-\lambda x} 1_{(0,\infty)}(x).$$

We use the power series expansion of the exponential function

$$e^{itx} = \sum_{n \geq 0} \frac{(itx)^n}{n!}$$

and then (the interchange of the sum and of the integral can be rigorously argued)

$$
\begin{aligned}
\varphi_X(t) &= \frac{\lambda^a}{\Gamma(a)} \sum_{n \geq 0} \int_0^\infty dx \, x^{a-1} e^{-\lambda x} \frac{(itx)^n}{n!} \\
&= \sum_{n \geq 0} \frac{\lambda^a (it)^n}{\Gamma(a) n!} \int_0^\infty e^{-\lambda x} x^{a+n-1} dx \\
&= \sum_{n \geq 0} \frac{\lambda^a (it)^n}{\Gamma(a) n!} \lambda^{-a-n} \int_0^\infty dy \, e^{-y} y^{a+n-1} \\
&= \sum_{n \geq 0} \frac{\lambda^a (it)^n}{\Gamma(a) n!} \lambda^{-a-n} \Gamma(n+a) \\
&= \sum_{n \geq 0} C_n^a \left(\frac{it}{\lambda} \right)^n \\
&= \left(\frac{1}{1 - \frac{it}{\lambda}} \right)^a
\end{aligned}
$$

using the properties of the Gamma function and the power series expansion of the function $(1 - x)^a$.

Note that the distribution $\Gamma(1, \lambda)$ corresponds to the exponential distribution with parameter $\lambda > 0$. Therefore, if $X \sim Exp(\lambda)$ then

$$
\varphi_X(t) = \frac{1}{1 - \frac{it}{\lambda}}
$$

for every $t \in \mathbb{R}$, which is identical with the formula given above.

(ix) Suppose X has a Cauchy distribution. Then

$$
\varphi_X(t) = e^{-|t|}.
$$

We will use the integral formula

$$
\int_0^\infty \frac{\cos(tx)}{b^2 + x^2} dx = \frac{\pi}{2b} e^{-tb}, \quad t \geq 0,
$$

for every $b \in \mathbb{R}$. Note that the function under the integral is even and so

$$
\int_{\mathbb{R}} \frac{\cos(tx)}{b^2 + x^2} dx = \frac{\pi}{b} e^{-tb}, \quad t \geq 0,
$$

Recall that the density of the Cauchy distribution is

$$
f(x) = \frac{1}{\pi} \frac{1}{1 + x^2}.
$$

Since the imaginary part of φ_X vanishes (the law of X is symmetric), for $t \geq 0$

$$\varphi_X(t) = \frac{1}{\pi} \int_{\mathbb{R}} e^{itx} \frac{1}{1+x^2} dx$$
$$= \frac{1}{\pi} \int_{\mathbb{R}} \cos(tx) \frac{1}{1+x^2}$$
$$= e^{-t}$$

In a similar way for $t \leq 0$ we obtain

$$\varphi_X(t) = e^t,$$

which concludes the result.

(x) Suppose X is Laplace distributed, that is, the random variable has density function

$$f_X(x) = \frac{1}{2} e^{-|x|}.$$

Then

$$\varphi_X(t) = \frac{1}{1+t^2}.$$

The formula follows from direct computation, using the fact that the law is symmetric and thus

$$\varphi_X(t) = \frac{1}{2} \int_{\mathbb{R}} e^{itx-|x|} dx$$
$$= \frac{1}{2} \int_{\mathbb{R}} \cos(tx) e^{-|x|} dx$$
$$= \int_0^\infty \cos(tx) e^{-x} dx$$

and finally using integration by parts two times.

Using the characteristic function for the Laplace distribution we can prove a very interesting property of the Cauchy random variable.

Proposition 6.27 *Suppose $X_1, ..., X_n$ are independent Cauchy distributed r.v. Let $S_n = X_1 + ... + X_n$ and*

$$\bar{X}_n = \frac{S_n}{n}.$$

Then, the average $\bar{X}_n$ is Cauchy distributed for every n.

Proof: We obtain the characteristic function of S_n as

$$\varphi_{S_n}(t) = e^{-n|t|}$$

and from Proposition 6.24 we obtain

$$\varphi_{\bar{X}_n}(t) = \varphi_{S_n}\left(\frac{t}{n}\right) = e^{-|t|}.$$

Therefore, $\bar{X}_n$ has a Cauchy law. ∎

6.5 Inversion and Continuity Theorems

The characteristic function of a random variable uniquely characterizes the random variable (hence the name). If you recall, a random variable is uniquely characterized by its distribution. The clear consequence is that two random variables with the same characteristic function will have the same law (distribution). When the random variable has a density, this density can be recovered from the characteristic function. These results will be stated in the current section.

Theorem 6.28 (Fourier inversion theorem) *If X is a continuous random variable with density f and characteristic function φ then*

$$f(x) = \frac{1}{2\pi} \int_{-\infty}^{\infty} e^{-itx} \varphi(t) dt,$$

at every point x at which f is differentiable.

Proof: This theorem is the classical Fourier inversion theorem and it is proven in any functional analysis textbook (e.g., for a probability take see Billingsley (1995)). ∎

Remark 6.29 *Here is an example where the relationship does not hold. Suppose that X is distributed as a $Exp(1)$. Then*

$$\varphi_X(t) = \frac{1}{1 - it}$$

and φ_X is not integrable since

$$|\varphi_X(t)| \sim \frac{1}{|t|}$$

as $|t| \to \infty$. This is consistent with the above theorem because the density of X is

$$f(x) = e^{-x} 1_{(0,\infty)}(x)$$

which is not continuous (at zero).

Theorem 6.30 (Inversion theorem general case) *Let X have distribution function F and characteristic function φ. Define $\overline{F} : \mathbb{R} \to [0, 1]$ by*

$$\overline{F}(x) = \frac{1}{2}\{F(x) + \lim_{y \uparrow x} F(y)\}$$

Then

$$\overline{F}(b) - \overline{F}(a) = \lim_{N \to \infty} \int_{-N}^{N} \frac{e^{-iat} - e^{-ibt}}{2\pi it} \varphi(t) dt$$

Proof: Skipped. This is the general version of the Fourier inversion theorem applied to the case when the function f is integrable but not necessarily continuous.   ■

In the case when the distribution is continuous we obtain the more classical result:

Special cases   A particular case of the above theorem is when X is an integer-valued random variable (a discrete random variable). In this case we obtain

$$\mathbf{P}(X = k) = \frac{1}{2\pi} \int_{0}^{2\pi} \varphi_X(t) e^{-itk} dt = \frac{1}{2\pi} \int_{-\pi}^{\pi} \varphi_X(t) e^{-itk} dt$$

Corollary 6.31 (Uniqueness theorem) *Two random variables X and Y have the same characteristic function if and only if they have the same distribution function.*

The corollary essentially says that if two random variables have the same characteristic function they will have the same distribution. The next example shows how useful this is.

📖 **EXAMPLE 6.9**

Using Theorem 6.23 it is easy to check that if $X \sim \Gamma(a, \lambda)$ and $Y \sim \Gamma(b, \lambda)$ then

$$X + Y \sim \Gamma(a + b, \lambda).$$

Indeed,

$$\varphi_{X+Y}(t) = \varphi_X(t)\varphi_Y(t)$$

$$= \left(\frac{1}{1 - \frac{it}{\lambda}} \right)^a \left(\frac{1}{1 - \frac{it}{\lambda}} \right)^b$$

$$= \left(\frac{1}{1 - \frac{it}{\lambda}} \right)^{a+b}$$

which means that $X + Y \sim \Gamma(a + b, \lambda)$. This proof is much easier than a direct proof based on the calculation involving convolutions and the density of the gamma distribution.

Since two random variables may have a number of moments identical but different distributions, having the same moments is not necessarily sufficient. However, if ALL moments are the same the situation changes.

Theorem 6.32 (Uniqueness theorem for moments) *Let X, Y be two r.v. and denote by F_X, F_Y their cumulative distribution functions. If* *i. X and Y each have finite moments of all orders*

ii. their moments are the same, i.e. $\mathbf{E}X^k = \mathbf{E}Y^k = \alpha_k$ *for all* $k \geq 1$

iii. the radius R of the power series $\sum_{k=1}^{\infty} \alpha_k \frac{u^k}{k!}$ is nonzero,

then the two random variables have the same distribution.

$$F_X = F_Y.$$

Remark 6.33 *This theorem says that in principle two distributions are the same if all the moments are identical. Recall that having the same distribution is also the same as having the same characteristic function.*

However, all conditions are needed. It is possible for two random variables to have the same moments but not the same distribution.

■ **EXAMPLE 6.10   Identifying a Normal by calculating its moments**

Here is an example of the applicability of the theorem. Let X be a random variable such that its moments

$$\alpha_k := \mathbf{E}X^k$$

satisfy,

$$\alpha_k = \begin{cases} 1 \times 3 \times 5 \times (k-3) \times (k-1) = \frac{(2p)!}{2^p p!} & \text{if } k \text{ is even } = 2p \\ 0 & \text{if } k \text{ is odd} \end{cases}$$

Then, applying the theorem the random variable X is $N(0,1)$ distributed. Indeed, the distribution of a standard normal has these moments and the series in the theorem:

$$\sum_{k=1}^{\infty} \alpha_k \frac{u^k}{k!}$$

has the radius of convergence $R = \infty$.

■ **EXAMPLE 6.11**

Let X be a random variable with a finite moment generating function, $M_X(t)$, $-\infty < t < \infty$.

1. Show that
$$\mathbf{P}(X > t) \leq \frac{M_X(c)}{e^{ct}} \text{ for any } c > 0.$$

2. Suppose that X is normal with mean μ and variance σ^2. Find the best upper bound of the above form for $P(X > t)$.

If the moment generating function is finite we may write

$$M_X(c) = \mathbf{E}[e^{cX}] = \int_{-\infty}^{\infty} e^{cx} dF(x)$$

$$= \underbrace{\int_{-\infty}^{t} e^{cx} dF(x)}_{>0} + \int_{t}^{\infty} \underbrace{e^{cx}}_{\geq e^{ct} \text{ since } c \geq 0} dF(x)$$

$$\geq e^{ct} \int_{t}^{\infty} dF(x)$$

$$= e^{ct} \mathbf{P}(X > t),$$

which proves the first part.

For the second part let us get the bound in this particular case. For $X \sim N(\mu, \sigma)$, the moment generating function is $M_X(c) = e^{\mu c + \frac{\sigma^2 c^2}{2}}$, thus applying the inequality we get

$$\mathbf{P}(X > t) \leq e^{c(\mu - t) + \frac{1}{2}\sigma^2 c^2}, \quad \forall c > 0$$

Since this holds for any value of c if we minimize the expression on the right with respect to c that should give the best bound. But e^x is an increasing function therefore the minimum is obtained when the quadratic expression in the exponent is at the minimum. But this is happening when c is in the parabola vertex (if possible). Therefore, we have two cases.

If $t < \mu$, then the minimum is obtained at $c = 0$.

If $t > \mu$ then the minimum is obtained at $c = -\frac{\mu - t}{\sigma^2}$, and in this case the exponent is calculated as $-\frac{(\mu - t)^2}{2\sigma^2}$. To conclude the best upper bound for the tail probability of a gaussian random variable X with mean μ and standard deviation σ is

$$\mathbf{P}(X > t) \leq \begin{cases} 1 & \text{if } t \leq \mu \\ e^{-\frac{(\mu - t)^2}{2\sigma^2}} & \text{if } t > \mu \end{cases}$$

Note that when $t \leq \mu$ the inequality is not interesting (neither practically as well). The case when $t > \mu$ is interesting and it basically says that we may bound the entire integral – impossible to calculate explicitly in the gaussian case – with an expression of the form $e^{-constant \times t^2}$.

Exercise 4 (Stirling's formula) *Show that Stirling's formula($n! \sim n^n e^{-n} \sqrt{2\pi n}$) is equivalent with*

$$\frac{n!}{n^n e^{-n} \sqrt{2\pi n}} \to 1$$

Proposition 6.34 *Let X be an r.v. and φ_X its characteristic function. Then the law of X is symmetric if and only if φ_X is real valued.*

Remark 6.35 *Here, "symmetric" means that the distribution is symmetric about zero: the distribution of X is symmetric if X and $-X$ have the same distribution. This is the same as saying that $\mathbf{P}(X \geq a) = \mathbf{P}(X \leq -a)$ for all a. In particular, if the random variable is continuous and has a density function f, it is symmetric if and only if f is an even function (recall, this means that $f(-x) = f(x)$).*

Proof: Consider the case when X admits a continuous density f. In this case,

$$\varphi_X(t) = \int_{\mathbb{R}} \cos(tx)f(x)dx + i \int_{\mathbb{R}} \sin(tx)f(x)dx.$$

If f is an even function then

$$x \to \sin(tx)f(x)$$

is odd and consequently

$$\int_{\mathbb{R}} \sin(tx)f(x)dx = 0.$$

So φ_X is real valued.

Conversely, if φ_X is real valued, by Proposition 6.34,

$$\varphi_{-X}(t) = \varphi_X(-t) = \overline{\varphi_X}(t) = \varphi_X(t)$$

and the uniqueness theorem (Corollary 6.31) implies that X and $-X$ have the same distribution.

The general case is identical but the integrals are expressed in terms of the c.d.f $F(x)$, i.e.,

$$\int_{\mathbb{R}} \sin(tx)dF(x),$$

and

$$\int_{\mathbb{R}} \cos(tx)dF(x).$$

∎

The next theorem is very important in practical applications. Specifically, it is a very important tool to show convergence in distribution, a topic presented in more detail in the Chapter 7.

Theorem 6.36 (Continuity theorem – weak convergence) *Suppose $F_1 \cdots F_n \ldots$ is a sequence of distribution functions with corresponding characteristic functions $\varphi_1 \cdots \varphi_n \cdots$. Then*

(i) *If $F_n(x) \to F(x)$, for all x continuity points of F, where F is some distribution function with corresponding characteristic function φ then $\varphi_n(t) \to \varphi(t)$ for all t.*

(ii) *Conversely, if $\varphi(t) = \lim_{n\to\infty} \varphi_n(t)$ exists for all t and is continuous at $t = 0$ then φ is the characteristic function of some distribution function F and $F_n(x) \to F(x)$ for all x continuity points of F.*

6.6   Stable Distributions. Lévy Distribution

We conclude the chapter on characteristic functions by detailing a distribution which is defined using characteristic functions.

In finance, the continuously compounded return of an asset is one of the most studied objects. If the equity at time t is denoted with S_t then the continuously compounded return over the period $[t, t+\Delta t]$ is defined as $R_t = \log S_{t+\Delta t} - \log S_t$. There are two major advantages to studying return vs. the asset itself. First, in the regular Black Scholes model (Black and Scholes, 1973) the returns are independent, identically distributed as a gaussian random variable. Two, the return over a larger period say $[t, t+n\Delta t]$ is easily expressed as the sum of the returns calculated over the smaller periods $[t+i\Delta t, t+(i+1)\Delta t]$. If we denote X_i the return over $[t+i\Delta t, t+(i+1)\Delta t]$, then clearly

$$X(n\Delta t) = \log S_{t+n\Delta t} - \log S_t = \sum_{i=1}^{n} X_i.$$

Since the sum of i.i.d. gaussians is a Gaussian this model provides a very easy to work with distribution.

It is well known today that the Black Scholes model is not a good fit for the asset prices. This is primarily due to the nature of the gaussian distribution. When plotting histograms of the actually observed returns the probability of large observations is much greater than that of a normal density (the leptokurtic property of the distribution). However, if we use a different density we may lose the scaling property of the normal. That is the ability of rescaling the larger period returns to the same distribution. Is it possible to have a different distribution with this property?

Definition 6.37 (Stable distribution) *Consider the sum of n independent, identically distributed random variables X_i, and denote it with*

$$X(n\Delta t) = X_1 + X_2 + X_3 + \cdots + X_n.$$

Since the variables are independent, the distribution of their sum may be obtained as the n-fold convolution,

$$P[X(n\Delta t)] = P(X_1) \otimes P(X_2) \cdots \otimes P(X_n).$$

*The distribution of the X_is is called **Stable** if the functional form of the $P[X(n\Delta t)]$ is the same as the functional form of $P[X(\Delta t)]$. More specifically, if for any $n \geq 2$, there exists a positive C_n and a D_n so that*

$$P[X(n\Delta t)] = P[C_n X + D_n]$$

where X has the same distribution as X_i for $i = 1, 2, \ldots n$. If $D_n = 0$, X is said to have a strictly Stable distribution.

The writing in the definition is generic, but refer to the distribution as the c.d.f.

It can be shown (e.g., Samorodnitsky and Taqqu (1994)) that

$$C_n = n^{\frac{1}{\alpha}}$$

for some parameter α, $0 < \alpha \le 2$. This α is an important characteristic of the processes of this type (which is in fact called α-stable) but we shall not talk about it here.

Lévy (1925) and Khintchine and Lévy (1936) found the most general form of the stable distributions. The general representation is through the characteristic function $\varphi(t)$ associated to the distribution.

It is very easy to understand why this form of characteristic function determines the stable distributions. The characteristic function of a convolution of independent random variables is simply the product of the respective characteristic functions and thus owing to the special exponential form above all stable distributions are closed under convolutions for a fixed value of α. Specifically, recall that (using our previous notation)

$$\varphi_{X(n\Delta t)}(t) = (\varphi(t))^n,$$

and

$$\varphi_{C_n X + D_n}(t) = e^{iD_n t}\varphi(C_n t),$$

Thus, setting them equal in principle can provide the most general form for the stable distributions. It took the genius of Paul Lévy, to actually see the answer.

The most common parametrization of Lévy distributions is given in the next definition.

Definition 6.38 (Lévy distribution) *A random variable X is said to have a Lévy distribution if its characteristic function is given by*

$$\ln(\varphi(t)) = \begin{cases} i\mu t - \gamma^\alpha |t|^\alpha \left[1 - i\beta\frac{t}{|t|}\tan(\frac{\pi\alpha}{2})\right] & \text{if } \alpha \in (0,2] \setminus \{1\} \\ i\mu t - \gamma|t|\left[1 + i\beta\frac{t}{|t|}\frac{2}{\pi}\ln|t|\right] & \text{if } \alpha = 1 \end{cases},$$

The parameters of this distribution are

- *α is called the stability exponent or the characteristic function,*

- *γ is a positive scaling factor,*

- *μ is a location parameter,*

- *$\beta \in [-1, 1]$ is a skewness (asymmetry) parameter.*

Remark 6.39 *All such distributions are heavy tailed (leptokurtic) and have no second moments for any $\alpha < 2$. When $\alpha < 1$ the distribution does not even have the first moment. Please note that neither third nor fourth moment exists for these distributions so the usual measures of skewness and kurtosis are undefined.*

The characterization above is due to Khintchine and Lévy (1936). If we take $\beta = 0$ we obtain a (Lévy) symmetric alpha-stable distribution (Lévy, 1925). The distribution is symmetric about μ.

Special cases   In general there is no formula for the p.d.f. $f(x)$ of a stable distribution. There are, however three special cases which reduce to known characteristic functions (and therefore to known distributions).

1. When $\alpha = 2$ the distribution reduces to a Gaussian distribution with mean μ and variance $\sigma^2 = 2\gamma^2$. The skewness parameter β has no consequence.

$$f(x) = \frac{1}{\sqrt{4\pi\gamma}} e^{\frac{-(x-\mu)^2}{4\gamma^2}}.$$

2. When $\alpha = 1$ and $\beta = 0$, the distribution reduces to the Cauchy (Lorentz) distribution.

$$f(x) = \frac{1}{\pi} \frac{\gamma}{\gamma^2 + (x-\mu)^2}.$$

This is the distribution of $\gamma X + \mu$ where X is the standard Cauchy we defined before ($\mu = 0. \gamma = 1$).

3. When $\alpha = \frac{1}{2}$ and $\beta = 1$ we obtain the Lévy–Smirnov distribution with location parameter μ and scale parameter γ.

$$f(x) = \sqrt{\frac{\gamma}{2\pi}} \frac{e^{-\gamma/2(x-\mu)}}{(x-\mu)^{\frac{3}{2}}}, \text{ if } x \geq \mu$$

6.6.1   Truncated Lévy flight distribution

The leptokurtic property of the stable distributions for $\alpha < 2$ is very desirable in finance. In practice, when working with daily and more frequently sampled returns their marginal distribution has heavy tails. However, recall that the Lévy distribution in general does not have a second moment for any $\alpha < 2$, and therefore, if distributed as Lévy, the returns have infinite variance. This is an issue when working with real data. In order to avoid this problem Mantegna and Stanley (1994) consider a Lévy type distribution truncated at some parameter l. That is, for any $x > l$ the density of the distribution $f(x) = 0$. Clearly this distribution has finite variance.

This distribution was named the Truncated Lévy flight (*TLF*):

$$T(x) = cP(x)\mathbf{1}_{(-l,l)}(x)$$

where $P(x)$ is any symmetric Lévy distribution (obtained when $\beta = 0$), and $\mathbf{1}_A(x)$ is the indicator function of the set A.

The truncation parameter l is quite crucial. Clearly, as $l \to \infty$ one obtains a regular Lévy distribution. However, the *TLF* distribution itself is not a stable distribution for any finite truncation level l. However, this distribution has finite variance, thus independent variables from this distribution satisfy a regular central limit Theorem (as we shall see in later chapters). If the parameter l is large the convergence may be very slow (Mantegna and Stanley, 1994). If the parameter l is small (so that the convergence is fast) the cut that it presents in its tails is very abrupt.

In order to have continuous tails, Koponen (1995) considered a *TLF* in which the cut function is a decreasing exponential characterized by a separate parameter l. The characteristic function of this distribution when $\alpha \neq 1$ can be expressed as

$$\varphi(t) = \exp\left\{c_0 - c_1 \frac{(t^2 + 1/l^2)^{\frac{\alpha}{2}}}{\cos(\pi\alpha/2)} \cos(\alpha \arctan(l|t|))(1 + il|t|\beta \tan(t \arctan l|t|))\right\}$$

with c_1 a scale factor:

$$c_1 = \frac{2\pi \cos(\pi\alpha/2)}{\alpha \Gamma(\alpha) \sin(\pi\alpha)} At$$

and

$$c_0 = \frac{l^{-\alpha}}{\cos(\pi\alpha/2)} c_1 = \frac{2\pi}{\alpha \Gamma(\alpha) \sin(\pi\alpha)} Al^{-\alpha} t.$$

In the case of symmetric distributions $\beta = 0$. In this case, the variance of this distribution can be calculated from the characteristic function:

$$\sigma^2 = -\left.\frac{\partial^2 \varphi(t)}{\partial t^2}\right|_{t=0} = \frac{2A\pi(1 - \alpha)}{\Gamma(\alpha) \sin(\pi\alpha)} l^{2-\alpha}$$

The remaining presentation is for this symmetric case of the *TLF*.

If we use time steps Δt apart, and $T = N\Delta t$, following the discussion about returns, at the end of each interval we must calculate the sum of N stochastic variables that are independent and identically distributed. Therefore, the new characteristic function will be

$$\varphi(t, N) = \varphi(t)^N = \exp\left\{c_0 N - c_1 \frac{N(t^2 + 1/l^2)^{\alpha/2}}{\cos(\pi\alpha/2)} \cos(\alpha \arctan(l|t|))\right\}.$$

The model can be improved by standardizing it. If the variance is given by

$$\sigma^2 = -\frac{\partial^2 \varphi(t)}{\partial t^2}\Big|_{t=0}.$$

we have that

$$-\frac{\partial^2 \varphi(t/\sigma)}{\partial t^2}\Big|_{t=0} = -\frac{1}{\sigma^2}\frac{\partial^2 \varphi(t)}{\partial t^2}\Big|_{t=0} = 1.$$

Therefore, a standardized model is

$$\ln \varphi_S(t) = \ln \varphi\left(\frac{t}{\sigma}\right) = c_0 - c_1 \frac{((t/\sigma)^2 + 1/l^2)^{\alpha/2}}{\cos(\pi\alpha/2)} \cos(\alpha \arctan(l\frac{|t|}{\sigma}))$$

We leave it as an exercise to write the characteristic function in a more condensed form.

Problems

6.1   Suppose that X has a discrete uniform distribution on $\{1, 2, ..., n\}$, that is

$$\mathbf{P}(X = i) = \frac{1}{n}$$

for every $i = 1, ..., n$.
  a) Compute the moment generating function of X.
  b) Deduce that

$$\mathbf{E}X = \frac{n+1}{2} \text{ and } \mathbf{E}X^2 = \frac{(n+1)(2n+1)}{6}.$$

6.2   Suppose X has uniform distribution $U[a, b]$ with $a < b$. Calculate the MGF of X: $M_X(t)$ for all $t \in \mathbb{R}$

6.3   Prove Proposition 6.2 on page 182.

6.4   Suppose that X admits a moment generating function M_X. Prove that

$$\mathbf{P}(X \geq x) \leq e^{-tx} M_X(t) \text{ for every } t > 0$$

and

$$\mathbf{P}(X \leq x) \leq e^{-tx} M_X(t) \text{ for every } t < 0.$$

These are known as the Chernoff bounds on the probability.

6.5   Suppose X has the density

$$f(x) = \frac{a}{x^{a+1}} 1_{(x>1)}$$

where $a > 0$. This distribution is called a Pareto distribution or a power law distribution.
  a) Show that $M_X(t) = \infty$ for every $t \in \mathbb{R}$.
  b) Show that

$$\mathbf{E}X^n < \infty$$

  if and only if $a > n$.

6.6   Let $X \sim Exp(\lambda)$. Calculate $\mathbf{E}X^3$ and $\mathbf{E}X^4$. Give a general formula for $\mathbf{E}X^n$.

6.7   Prove the following facts about Laplace transform:
  a)

$$\mathcal{L}\{e^{-\lambda x}\}(t) = \frac{1}{\lambda + t}$$

  b)

$$\mathcal{L}\{t^{n-1}e^{-\lambda x}\}(t) = \frac{\Gamma(n)}{(\lambda + t)^n},$$

  where $\Gamma(n)$ denote the Gamma function calculated at n.

6.8   Suppose f^* denote the Laplace transform of the function f. Prove the following properties:

a)

$$\mathcal{L}\left\{\int_0^x f(y)dy\right\}(t) = \frac{f^*(t)}{t}$$

b)

$$\mathcal{L}\{f'(x)\}(t) = tf^*(t) - f(0+),$$

where $f(0+)$ denote the right limit of f at 0.

c)

$$\mathcal{L}\{f^{(n)}(x)\}(t) = t^n f^*(t) - t^{n-1}f(0+) - t^{n-2}f'(0+) - \cdots - f^{(n)}(0+).$$

6.9   Suppose that $X_1 \sim Exp(\lambda_1)$ and $X_2 \sim Exp(\lambda_2)$ are independent. Given that the inverse Laplace transform of the function $\frac{1}{c+t}$ is e^{-cx} and that the inverse Laplace transform is linear calculate the pdf of $X_1 + X_2$.

6.10   Repeat the problem above with three exponentially distributed random variables, i.e., $X_i \sim Exp(\lambda_i)$, $i = 1, 2, 3$. As a note the distribution of a sum of n such independent exponentials is called the Erlang distribution and is heavily used in queuing theory.

6.11   The gamma function $\Gamma : [0, \infty) \to [0, \infty)$ is defined as

$$\Gamma(\alpha) = \int_0^\infty t^{\alpha-1}e^{-t}dt,$$

and is one of the most useful mathematical functions. In statistics it is the basis of the gamma and beta distributions (see the following problems). The gamma function has several interesting properties which you will prove next.

1. Show that $\Gamma(\alpha + 1) = \alpha\Gamma(\alpha)$, $\alpha > 0$.

2. Show that $\Gamma(n) = (n - 1)!$ for any integer $n > 0$.

3. Show that $\Gamma(\frac{1}{2}) = \sqrt{\pi}$ and accordingly calculate the general formula for $\Gamma(\frac{n}{2})$ for any integer $n > 0$.

4. Show that

$$f(t) = \frac{t^{\alpha-1}e^{-t}}{\Gamma(\alpha)}1_{\{t\in(0,\infty)\}},$$

is a proper density function for all $\alpha > 0$.

6.12   The $Gamma(\alpha, \beta)$ distribution may be obtained from the gamma function in exercise 6.11 but it has two parameters α also called the shape parameter and β also called the scale parameter due to their influence on the peakedness, respectively, spread of the distribution. Specifically, its density is

$$f(x) = \frac{1}{\beta^\alpha\Gamma(\alpha)}x^{\alpha-1}e^{-x/\beta}.$$

Note that when $\alpha = 1$ we obtain the Exponential($\frac{1}{\beta}$) distribution. When $\alpha = p/2$ and $\beta = 2$ where p is an integer we obtain the chi-squared distribution with p degrees of freedom χ_p^2, which is another important distribution.

1. Show that the convolution of two gamma distributions with the same scale parameter is a gamma. Specifically, if $X_1 \sim Gamma(\alpha_1, \beta)$ and $X_2 \sim Gamma(\alpha_2, \beta)$ then $X_1 + X_2$ has the distribution $Gamma(\alpha_1 + \alpha_2, \beta)$.

2. Generalize the previous part to n $Gamma$ random variables with the same scale parameter.

3. Calculate the expectation and variance of a $Gamma$ random variable

4. Show that the moment generating function of a $Gamma(\alpha, \beta)$ is

$$M_X(t) = \left(\frac{1}{1 - \beta t} \right)^\alpha, \quad t < \frac{1}{\beta}$$

5. Calculate the characteristic function of a $Gamma$ random variable.

6. Show that if $X \sim Gamma(\alpha, \beta)$ then if $Y = kX$ for a constant k then $Y \sim Gamma(\alpha, k\beta)$

6.13   In this problem we shall derive the characteristic function of a normal using a differential equation approach.

a) Let $f(x) = e^{-\frac{x^2}{2}}$. Show that the function f satisfies

$$\frac{d}{dx} f + xf = 0.$$

b) Show that if $Z \sim N(0, 1)$ then its characteristic function φ_Z satisfies the same equation.

c) Deduce that

$$\varphi_Z(t) = Ce^{-\frac{t^2}{2}}.$$

d) Compute the constant C from the relation

$$C = \varphi_Z(0) = 1.$$

Hint: Part a) is trivial since

$$\frac{d}{dx} \left(e^{-\frac{x^2}{2}} \right) = -xe^{-\frac{x^2}{2}}.$$

6.14   Let Y be an uniform random variable on $\{-1, 1\}$. Show that its characteristic function is

$$\varphi_Y(t) = \cos(t)$$

for every $t \in \mathbb{R}$.

6.15   Compute the first four moments of a standard normally distributed random variable using the characteristic function.

6.16   Let X, Y be two independent random variables with uniform law $U[-1, 1]$.
   a) Calculate the characteristic function of $X + Y$.
   b) Identify the density of $X + Y$.
   c) Show that the function

$$f(x) = \frac{1}{\pi} \left(\frac{\sin x}{x} \right)^2$$

   is a density.

6.17   Let (X, Y) be a random vector with joint density

$$f(x, y) = \frac{1}{4} \left[1 + xy(x^2 - y^2) \right] \mathbf{1}_{(|x|<1,|y|<1)}.$$

   a) Compute the marginal distributions of X and Y.
   b) Are X and Y independent?
   c) Compute the characteristic functions of X and Y.
   d) Compute the characteristic function of $Z = X + Y$.
   e) What can you conclude?

6.18   Let X, Y be two independent random variables. Denote by φ_X, φ_Y their characteristic functions, respectively. Show that

$$\varphi_{XY}(t) = \mathbf{E}\varphi_X(tY) = \mathbf{E}\varphi_Y(tX) \text{ for every } t \in \mathbb{R}.$$

6.19   Let X be an r.v. with characteristic function φ. Then for every $x \in \mathbb{R}$ we have

$$\mathbf{P}(X = x) = \lim_{T \to \infty} \frac{1}{2T} \int_{-T}^{T} e^{-itx} \varphi(t) dt.$$

Deduce that if $\varphi(t) \to 0$ as $|t| \to \infty$ then

$$\mathbf{P}(X = x) = 0 \text{ for all } x \in \mathbb{R}.$$

6.20   Use Problem 6.19 to calculate $\mathbf{P}(X = k)$ in the case when X has a binomial distribution $B(n, p)$.

6.21   Let Y be a random variable with density

$$f(y) = Ce^{-|y|}.$$

Calculate C and show that the characteristic function of Y is

$$\varphi_Y(t) = \frac{1}{1 + t^2}.$$

State the name of the distribution with this characteristic function and density.

6.22   Let $X_i, 1 \le i \le 4$ be independent identically distributed $N(0, 1)$ random variables. Denote with

$$D = \begin{vmatrix} X_1 & X_2 \\ X_3 & X_4 \end{vmatrix},$$

the determinant of the matrix.

    a) Show that the characteristic function of the random variable $X_1 X_2$ is

$$\varphi(t) = \frac{1}{\sqrt{1 + t^2}}$$

    b) Calculate the characteristic function of D and state the distribution of D.

6.23   Determine the formula for the distribution with characteristic function

$$\varphi(t) = \cos t$$

for every $t \in \mathbb{R}$.

6.24   Determine the distribution with characteristic function

$$\varphi(t) = \frac{t + \sin t}{2t}$$

for every $t \in \mathbb{R}$.

CHAPTER 7

LIMIT THEOREMS

In the previous chapter, we talked about a sequence of distribution functions converging to a target distribution, which is, as we shall see, the concept of convergence in distribution. It turns out that this type of convergence is one of the most useful concepts in applications. Often, we hypothesize a model which will have parameters estimated using statistics obtained from real data. It is essential that the approximations (statistics) converge to the desired parameters, and it is crucial that they do so fast. The way in which they converge and the rate of convergence are covered in this chapter. The chapter is dedicated to the review and discovery of the most important types of convergence in probability theory.

7.1 Types of Convergence

Let $(\Omega, \mathscr{F}, \mathbf{P})$ be a probability space, and let $X_n : \Omega \to \mathbb{R}$ be a sequence of random variables. Furthermore, let $X : \Omega \to \mathbb{R}$ be a target random variable. Throughout this chapter, we shall take $n \in \mathbb{N}$ to denote the index of the sequence of random variables. In general, the sequence may be indexed by a different index set, the only condition being that the concept of order is well defined. That is, for every

Probability and Stochastic Processes, First Edition. Ionuţ Florescu
© 2015 John Wiley & Sons, Inc. Published 2015 by John Wiley & Sons, Inc.

two different elements of the index set, i_i and i_2, we either have $i_i < i_2$ or $i_2 < i_1$. With this observation, the results are entirely similar for any such index set.

We mention this not as a curiosity but as something we shall need in the stochastic processes part. In the second part of the book, the sequences may be indexed in time $t \in (0, \infty)$, as in the stochastic process X_t; therefore, it is important to realize that the same notions we are talking about in this chapter are applicable to that case as well. Furthermore, this is why we are talking about a sequence of vectors $(X_t^1, X_t^2)_{t \in \mathbb{R}}$ with the index in $\mathbb{R}$ on which order exists and not about X_{t_1, t_2} with $(t_1, t_2) \in \mathbb{R}^2$, because order in $\mathbb{R}^2$ is not well defined (for example, which is larger $(1, 4)$ or $(2, 3)$).

When we are talking about the convergence in probability theory, we need to realize that for any particular $\omega \in \Omega$ we are talking about a sequence of numbers $X_n(\omega)$ and whether they converge to $X(\omega)$. However, to have a convergence concept we need to discuss what happens for all the $\omega \in \Omega$.

7.1.1 Traditional deterministic convergence types

Recall that essentially any random variable is a function from $\Omega \to \mathbb{R}$. If we disregard the probability information, the traditional types of convergence from real analysis apply. The following two definitions serve as a reminder:

Definition 7.1 (Uniform Convergence) *The sequence of random variables $\{X_n\}_{n \in \mathbb{N}}$ is said to converge to X uniformly if and only if for all $\varepsilon > 0$ there exists an $n_\varepsilon \in \mathbb{N}$ such that*

$$|X_n(\omega) - X(\omega)| < \varepsilon, \quad \forall n \geq n_\varepsilon, \forall \omega \in \Omega.$$

Note that in the previous definition the number n_ε is the same for all $\omega \in \Omega$, thus the convergence takes place similarly throughout the whole space Ω. There is no point $\omega \in \Omega$ where the sequence $X_n(\omega)$ is away from the target $X(\omega)$, and in fact at all points the distance from the target is similar for any n. This is why the convergence is called uniform – uniform on the whole Ω. This is the most powerful convergence type.

Definition 7.2 (Pointwise Convergence) *The sequence $\{X_n\}_n$ is said to converge to X pointwise if for any ω fixed, for all $\varepsilon > 0$ there exists $n_\varepsilon(\omega)$ such that*

$$|X_n(\omega) - X(\omega)| < \varepsilon, \quad \forall n \geq n_\varepsilon(\omega).$$

As in the previous definition, we have convergence at all points in Ω. Unlike the uniform convergence at some points ω, we may need a large n in order to be as close to the target as for other points.

These two limiting concepts are related to *all the points*, and they do not take into consideration any information about the probability measure. This is why they are called *deterministic notions*. If the convergence does not hold for one point, then the whole concept fails; so one point *determines* the fate of the whole concept.

7.1.2 Convergence of moments of a random variable. Convergence in L^p

As we already know, in probability a statement does not have to hold at all points in order to be valid. If the points are irrelevant (have small negligible probability), we do not need to care about them. Real analysis provides a notion of convergence which is applicable to probability theory. The integral of a function may have a limit even though the function itself may not have a limit at some points. The integral of a random variable is called *expectation* (or *moment*) in probability theory. If we only need the convergence of moments, we reach the notion of convergence in L^p.

Definition 7.3 (Convergence in L^p) *The sequence of random variables $\{X_n\}_n$ is said to converge to X in L^p (denoted $X_n \xrightarrow{L^p} X$ or $X_n \to X$ in L^p) if and only if*

$$\lim_{n \to \infty} \mathbf{E}|X_n - X|^p = 0$$

As mentioned in Subsection 4.3 on page 134, the particular case when $p = 2$ is special. It is extremely common and therefore it has many other names such as convergence in L^2, mean square convergence, convergence in quadratic mean, and so on.

Any L^p space is a complete, normed vector space. For an $X \in L^p(\Omega)$, the norm is defined as

$$\|X\|_p = (\mathbf{E}[X^p])^{\frac{1}{p}} = \left(\int X^p d\mathbf{P} \right)^{\frac{1}{p}}.$$

This is interesting from the real analysis perspective. For our goals, the following result is important:

Proposition 7.4 *Let X denote a random variable on $(\Omega, \mathscr{F}, \mathbf{P})$. The sequence of norms $\|X\|_q$ is nondecreasing (increasing) in q. Therefore, if a variable is in L^q for some q fixed, then it also is in any L^r for any $r \leq q$. More specifically, if $X \in L^q$, then for all $r \leq q$ we have*

$$\|X\|_r \leq \|X\|_q$$

As a consequence, we may write as spaces of random variables

$$L^1(\Omega) \supseteq L^2(\Omega) \supseteq L^3(\Omega) \dots.$$

Proof: Let $q \geq r$. Then the function $f(x) = |x|^{q/r}$ is convex (check this) and we can apply Jensen's inequality (Lemma 4.14) to the nonnegative r.v. $Y = |X|^r$. Writing the lemma out immediately yields the desired result. ∎

The following is a corollary of this result applied to sequences of random variables.

Corollary 7.5 *If $X_n \xrightarrow{L^p} X$ and $p \geq q$, then $X_n \xrightarrow{L^q} X$*

Proof: Exercise. ∎

7.1.3   Almost sure (a.s.) convergence

Starting with this subsection, we present convergence types which are specific to probability theory. They mimic the definitions of uniform and pointwise convergence but taking into account the probability measure defined on the space. As we shall see, one of the next definitions is superfluous.

Definition 7.6 (Almost Uniform Convergence) *The sequence of random variables* $\{X_n\}_n$ *is said to converge to* X *almost uniformly* ($X_n \xrightarrow{a.u.} X$ *or* $X_n \to X$ *a.u.*) *if for all* $\varepsilon > 0$ *there exists a set* $N \in \mathcal{F}$ *with* $\mathbf{P}(N) < \varepsilon$ *and* $\{X_n\}$ *converges to* X *uniformly on* N^c.

Definition 7.7 (Almost Sure Convergence) *The sequence of random variables* $\{X_n\}_n$ *is said to converge to* X *almost surely* ($X_n \xrightarrow{a.s.} X$ *or* $X_n \to X$ *a.s.*) *if for all* $\varepsilon > 0$ *there exists a set* $N \in \mathcal{F}$ *with* $\mathbf{P}(N) = 0$ *and* $\{X_n\}$ *converges to* X *pointwise on* N^c; *or written mathematically,*

$$\mathbf{P}\{\omega \in \Omega : |X_n(\omega) - X(\omega)| > \varepsilon, \forall n \geq n_\varepsilon(\omega)\} = \mathbf{P}(N) = 0$$

An alternative way to write the a.s. convergence is

$$\mathbf{P}\{\omega \in \Omega : \lim_{n \to \infty} X_n(\omega) = X(\omega)\} = \mathbf{P}(N^c) = 1$$

If we look back to the deterministic notions, then we realize that these definitions are identical to the deterministic ones but restricted to a subset N^c which contains all the "probable" values.

Remark 7.8 *The reason why you will never hear about almost uniform convergence is due to the following proposition (due to Egorov). It turns out that almost uniform and almost sure convergence despite their apparent different forms are completely equivalent **on a finite measure space** and any probability space has total measure* 1.

Proposition 7.9 (Egorov Theorem) *If the space* Ω *has a finite measure (for any probability space we have* $\mathbf{P}(\Omega) = 1$*), then the sequence* $\{X_n\}_n$ *converges almost uniformly to* X *if and only if it converges almost surely to* X.

For a proof, one may consult Theorem (10.13) in Wheeden and Zygmund (1977).

Technical note   It is possible that a sequence of random variables X_n may have a limit in the a.s. sense but the limiting variable may not be a random variable itself (the limit may not be $\mathcal{B}(\mathbb{R})$-measurable anymore). To avoid this technical problem, if one assumes that the probability space is complete (as defined next), one will always obtain random variables as the limit of random sequences (if the limit exists of course).

   Throughout this book, we will always assume that the probability space we work with is complete.

Definition 7.10 (Complete probability space) *We say that the probability space* $(\Omega, \mathscr{F}, \mathbf{P})$ *is complete if any subset of a probability zero set in* $\mathscr{F}$ *is also in* $\mathscr{F}$. *Mathematically, if* $N \in \mathscr{F}$ *with* $\mathbf{P}(N) = 0$, *then* $\forall M \subset N$ *we have* $M \in \mathscr{F}$.

The issue arises from the definition of a sigma algebra which does not require the inclusion of all subsets but only of all the unions of sets already existing in the collection.

However, we can easily "complete" any probability space $(\Omega, \mathscr{F}, \mathbf{P})$ by simply adding to its sigma algebra all the sets of probability zero.

7.1.4 Convergence in probability. Convergence in distribution

In the definition of a.s. convergence, we can make it less restrictive by looking at the measure of N. If instead of requiring that this measure be zero we require to somehow converge to Zero, we obtain the next definition (convergence in probability).

Definition 7.11 (Convergence in Probability) *The sequence of random variables* $\{X_n\}$ *is said to converge to* X *in probability* ($X_n \xrightarrow{P} X$ *or* $X_n \to X$ *in probability) if and only if, for all fixed* $\varepsilon > 0$, *the sets* $N_\varepsilon(n) = \{\omega \,|\, |X_n(\omega) - X(\omega)| > \varepsilon\}$ *have the property* $P(N_\varepsilon(n)) \to 0$ *as* $n \to \infty$, *or*

$$\lim_{n \to \infty} \mathbf{P}\{\omega \in \Omega : |X_n(\omega) - X(\omega)| > \varepsilon\} = 0$$

It is remarkable that the definition of convergence in probability looks very similar to the convergence a.s. It seems that the only difference is the location of the n. However, that location makes a world of difference.

Finally, here is the weak form of convergence, aptly named thus.

Definition 7.12 (Convergence in Distribution – Weak convergence – Convergence in Law) *Consider a sequence of random variables* X_n *defined on probability spaces* $(\Omega_n, \mathscr{F}_n, \mathbf{P}_n)$ *(which might be all different) and a random variable* X, *defined on* $(\Omega, \mathscr{F}, \mathbf{P})$. *Let* $F_n(x)$ *and* $F(x)$ *be the corresponding distribution functions.* X_n *is said to converge to* X *in distribution (written* $X_n \xrightarrow{D} X$ *or* $F_n \Rightarrow F$) *if for every point* x *at which F is continuous we have*

$$\lim_{n \to \infty} F_n(x) = F(x).$$

The convergence in distribution is special. Since we only need the convergence of the distribution functions, each of the random variables X_n in the sequence may reside on a separate probability space $(\Omega_n, \mathcal{F}_n, \mathbf{P}_n)$ and yet the convergence is valid. Each distribution function $F_n : \mathbb{R} \to [0, 1]$, and thus regardless of the underlying probability spaces the distribution functions all reside in the same space.

Remark 7.13 *There are many notations for weak convergence which are used interchangeably in various books. We mention* $X_n \xrightarrow{\mathcal{L}} X$ *(convergence in law),* $X_n \Rightarrow X$, $X_n \xrightarrow{Distrib.} X$, *and* $X_n \xrightarrow{d} X$.

If all the spaces $(\Omega_n, \mathscr{F}_n, \mathbf{P}_n) = (\Omega, \mathscr{F}, \mathbf{P})$ for all n, the variables are all defined on the same probability space. In this particular case, we may talk about relations with the rest of convergence types.

Remark 7.14 *Why do we require x to be a continuity point of F? The simple answer is that, if x is a discontinuity point, the convergence may not happen at x even though we might have convergence everywhere else around it. The next example illustrates this fact.*

■ EXAMPLE 7.1

Let X_n be a constant equal to $1/n$. Then X_n has the distribution function

$$F_n(t) = \begin{cases} 0 & , \text{if } t < \frac{1}{n} \\ 1 & , \text{if } t \geq \frac{1}{n}. \end{cases}$$

Looking at this, it makes sense to say that the limit is $X = 0$ with probability 1 which has the distribution function

$$F(t) = \begin{cases} 0 & , \text{if } t < 0 \\ 1 & , \text{if } t \geq 0. \end{cases}$$

Yet, at the discontinuity point of F we have $F(0) = 1 \neq \lim_{n\to\infty} F_n(0) = \lim_{n\to\infty} \frac{1}{n} = 0$. This is why we exclude these points from the definition.

There exists a quantity where the isolated points do not matter and that is the integral. That is why we have an alternate definition for convergence in distribution given by the next theorem. We note that the definition above and the next theorem applies to random vectors as well (X_n, X taking values in $\mathbb{R}^d$).

Theorem 7.15 (The characterization theorem for weak convergence) *Let X_n be defined on probability spaces $(\Omega_n, \mathscr{F}_n, \mathbf{P}_n)$ and X, defined on $(\Omega, \mathscr{F}, \mathbf{P})$. Then $X_n \xrightarrow{D} X$ if and only if for any bounded, continuous function ϕ defined on the range of X we have*

$$\mathbf{E}[\phi(X_n)] \to \mathbf{E}[\phi(X)], \quad \text{as } n \to \infty,$$

or equivalently

$$\int \phi(t) dF_n(t) \to \int \phi(t) dF(t)$$

Remark 7.16 *A bounded, continuous function ϕ is called a* test function.

Proof: First assume that $X_n \xrightarrow{D} X$. Let D denote the set of continuity points of F. Since any distribution function has at most a countable number of discontinuities, we know that outside this set we have at most a countable number. Therefore, D is dense in $\mathbb{R}$. For any $a, b \in D$, we have

$$\mathbf{P}_n(X_N \in (a, b]) = F_n(b) - \lim_{y \downarrow a} F_n(y) \to F(b) - \lim_{y \downarrow a} F(y) = \mathbf{P}(X \in (a, b]).$$

Written differently

$$\int \mathbf{1}_{(a, b]} d\mathbf{P}_n \to \int \mathbf{1}_{(a, b]} d\mathbf{P}, \tag{7.1}$$

for any $a, b \in D$, and thus the conclusion is true for indicators of semi-open intervals with endpoints in D. Now suppose the interval does not have endpoints in D. Since D is dense in $\mathbb{R}$, there exists a sequence of intervals either increasing or decreasing such that their limit is the respective interval. By a standard argument, the limit holds for any ϕ which is a simple function over D (linear combination of indicators with endpoints in D).

Now suppose ϕ is a bounded, continuous function with support in some compact interval (i.e., $\phi(x) = 0$ for any x outside the interval). Then, given any $\varepsilon > 0$, there exists ϕ_ε, a simple function over D, such that

$$\sup_{x \in \mathbb{R}} |\phi(x) - \phi_\varepsilon(x)| \le \varepsilon$$

We need the compact support for this property to hold. This is a particular case of the Weierstrass–Stone theorem (see, e.g., Royden (1988)) but its meaning should be easily understood by looking back on the way we approximated positive measurable functions with simple functions. A bit of work is needed with points outside D, but since the set D is dense in $\mathbb{R}$, we can always find a point in D such that the difference is as small as we want it to be. Using this result, we may write for any ϕ bounded, continuous function with support in some compact interval

$$\left| \int \phi dF_n - \int \phi dF \right| \le \left| \int (\phi - \phi_\varepsilon) dF_n \right|$$
$$+ \left| \int \phi_\varepsilon dF_n - \int \phi_\varepsilon dF \right| + \left| \int (\phi_\varepsilon - \phi) dF \right|$$

Using Jensen's inequality for the first and the third integrals and the definition of ϕ_ε, both these integrals are bounded by $\varepsilon \int dF_n = \varepsilon$. The second integral converges to zero since ϕ_ε is a simple function on D. Thus, the absolute value goes to zero and the integrals converge to each other since ε is arbitrary. To go to bounded, continuous functions (functions which do not necessarily have compact support), we use the fact that any bounded, continuous function ϕ may be approximated with an increasing sequence of bounded, continuous functions with compact support.

For example, if ϕ is bounded and continuous, then

$$\phi_M(x) = \phi(x)\mathbf{1}_{[-M, M]},$$

will converge to $\phi(x)$ as $M \to \infty$.

The argument follows exactly as the one above using monotone convergence theorem for the integrals 1 and 3 and the property above for the bounded, continuous function with compact support functions ϕ_M.

For the reciprocal argument, assume that $\int \phi(t)dF_n(t) \to \int \phi(t)dF(t)$ for any bounded, continuous ϕ. Let $f = \mathbf{1}_{(a,b]}$ with $a, b \in D$. This f is not continuous, so we cannot directly apply the theorem. To apply the theorem, we bound it with two continuous functions.

Let $\varepsilon > 0$ and let f_1 be a continuous function such that it is linear between $(a, a+\varepsilon)$ and $(b - \varepsilon, b)$ and equal to f everywhere else. See Figure 7.1 for a simple image. We clearly have $f_1 \leq f \leq f_2$, and since f_1 and f_2 are bounded and continuous, by applying the hypothesis we have

$$\int f_1 dF_n \leq \int f dF_n \leq \int f_2 dF_n$$

$$\downarrow \qquad\qquad\qquad \downarrow$$

$$\int f_1 dF \leq \int f dF \leq \int f_2 dF$$

where we used the limits for the upper and lower integral and we remark that the middle one must be there since we have $f_1 \leq f \leq f_2$.

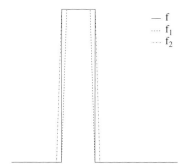

Figure 7.1   The continuous functions f_1 and f_2

The middle integral, however, must exist since it is equal to $\mathbf{P}(X \in [a, b]) = F(b) - F(a)$ by definition; recall that both a, b are in D.

We also have that the upper and the lower integrals converge to the same limit as $\varepsilon \to 0$ because

$$\int f_2 dF - \int f_1 dF \leq \int_{a-\varepsilon}^{a+\varepsilon} 1\, dF + \int_{b-\varepsilon}^{b+\varepsilon} 1\, dF$$

$$= F(a + \varepsilon) - F(a - \varepsilon) + F(b + \varepsilon) - F(b - \varepsilon)$$

and a, b are continuity points for F. Done. ■

Corollary 7.17 *As an immediate consequence of the previous theorem and the fact that composition of continuous functions is continuous, we have the following: If* $X_n \xrightarrow{D} X$ *and* g *is a continuous function, then* $g(X_n) \xrightarrow{D} g(X)$

Remark 7.18 *The continuity theorem (Theorem 6.36 on page 203 gives a criterion for weak convergence using characteristic functions. That theorem provides the main power of characteristic functions.*

We conclude this section with a characterization theorem for weak convergence of random vectors. We will not prove this theorem.

Theorem 7.19 (The characterization theorem for weak convergence: random vectors) *Let* X_n *be defined on probability spaces* $(\Omega_n, \mathscr{F}_n, \mathbf{P}_n)$ *and* X, *defined on* $(\Omega, \mathscr{F}, \mathbf{P})$. *Then* $X_n \xrightarrow{D} X$ *if and only if for any bounded, continuous function* ϕ *defined on the range of* X *we have*

$$\mathbf{E}[\phi(X_n)] \to \mathbf{E}[\phi(X)], \quad \text{as } n \to \infty,$$

or equivalently

$$\int \phi(t) dF_n(t) \to \int \phi(t) dF(t)$$

Remark 7.20 *A bounded, continuous function* ϕ *is called a* test function.

7.2 Relationships between Types of Convergence

First, the deterministic convergence types (uniform convergence and pointwise convergence) imply all the probabilistic types of convergence, with the exception of the L^p convergence. This is natural since the expectation needs first to be defined regardless of the deterministic convergence. As we saw already, almost uniform convergence and almost sure convergence are equivalent for finite measure spaces (thus in particular for probability spaces). We will not mention a.u. convergence for the remainder of this book.

The relations between various types of convergence are depicted in Figure 7.2. The solid arrows denote that one convergence implies the other one. The dashed arrows imply the existence of a subsequence that is convergent in the stronger type.

In the next sections of this chapter we present propositions that prove these relations.

7.2.1 A.S. and L^p

In general, L^p convergence and a.s. convergence are not related. The following example shows that a.s. convergence does not necessarily imply L^p convergence.

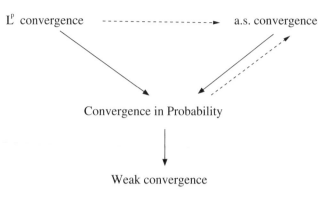

Figure 7.2 Relations between different types of convergence. Solid arrows mean immediate implication. Dashed arrows mean that there exists a subsequence which is convergent in the stronger type

■ EXAMPLE 7.2

Let $(\Omega, \mathcal{F}, \mathbf{P}) = ((0, 1), \mathcal{B}((0, 1)), \lambda)$, where λ denotes the Lebesgue measure. Define the variables X_n on this space as $X_n(\omega) = n\mathbf{1}_{(\frac{1}{n}, \frac{2}{n})}(\omega)$. Then $X_n \xrightarrow{a.s.} 0$ as $n \to \infty$ but $\mathbf{E}(X_n) = 1$ for all n so that $X_n \nrightarrow X$ in L^1, thus it does not converge in any L^p with $p \geq 1$

To show the a.s. convergence, take $\varepsilon > 0$. Then for any $\omega \in (0, 1)$ there exist a $n_\varepsilon(\omega)$ so that $\omega \notin (\frac{1}{n}, \frac{2}{n}), \forall n \geq n_\varepsilon(\omega)$. For example, take $n_\varepsilon(\omega) = [2/\omega]$, where we used the integer part notation. So for all $n \geq n_\varepsilon(\omega)$, $X_n = 0$ and thus $X_n(\omega) \longrightarrow 0$. Note that in fact the convergence is deterministic (pointwise not only a.s.).

However,

$$\mathbf{E}[X_n] = \int_0^1 n\mathbf{1}_{(\frac{1}{n}, \frac{2}{n})}(\omega)d\lambda(\omega) = n\lambda\left(\frac{1}{n}, \frac{2}{n}\right) = 1, \forall n$$

Since the expectations are always 1 regardless of n, the limit of the expectations is also 1 when $n \to \infty$ and $1 \neq \mathbf{E}[0] = 0$.

Remark 7.21 *The previous example may be read the other way as well to provide a counterexample to show L^p does not imply a.s. convergence as well. Specifically, the sequence converges in L^1 to 1, so if this would imply convergence a.s. it will result that it converges a.s. to 1 as well, which it clearly does not.*

So, are there conditions under which a sequence converging a.s. implies that it converges in L^p?

In fact, these conditions exist. Furthermore, we already know them. They are the same conditions as in the hypotheses of the dominated convergence theorem 4.6 and monotone convergence theorem 4.3 on page 126. Basically, if the hypotheses of those theorems are true, then a.s. convergence implies L^1 convergence.

Further, if the sequence is dominated by a variable in L^p, then even the convergence in probability (which, as we shall see, is weaker than convergence a.s.) implies the convergence in L^p.

Theorem 7.22 *If* $X_n \xrightarrow{P} X$ *and* $|X_n| < Y$ *a.s. for all* n *with* $Y \in L^p(\Omega)$, *then* $X_n \xrightarrow{L^p} X$.

The next example shows that a sequence converging in L^p does not necessarily converge almost surely. In fact, in this example the sequence converges in probability besides converging in L^p, but not even both types of convergence are enough for convergence a.s.

EXAMPLE 7.3

Let $(\Omega, \mathcal{F}, \mathbf{P}) = ((0,1), \mathcal{B}((0,1)), \lambda)$, where λ denotes the Lebesgue measure. Define the following sequence:

$X_1(\omega) = \omega + \mathbf{1}_{(0,1)}(\omega);$

$X_{2,1}(\omega) = \omega + \mathbf{1}_{\left(0,\frac{1}{2}\right)}(\omega); X_{2,2}(\omega) = \omega + \mathbf{1}_{\left(\frac{1}{2},1\right)}(\omega);$

$X_{3,1}(\omega) = \omega + \mathbf{1}_{\left(0,\frac{1}{3}\right)}(\omega); X_{3,2}(\omega) = \omega + \mathbf{1}_{\left(\frac{1}{3},\frac{2}{3}\right)}(\omega); X_{3,3}(\omega) = \omega + \mathbf{1}_{\left(\frac{2}{3},1\right)}(\omega);$

$\vdots$

If you want, just reindex the sequence using integers. We shall use the notation above since it is easier to follow. Furthermore, take $X(\omega) = \omega$
Then

$$\mathbf{P}\{\omega : |X_{n,k}(\omega) - X(\omega)| > \varepsilon\} = \mathbf{P}\{\omega \text{ is in some interval of length} \frac{1}{n}\} = \frac{1}{n}.$$

Therefore, $\lim_{n \to \infty} \mathbf{P}\{\omega : |X_{n,k}(\omega) - X(\omega)| > \varepsilon\} = 0$, thus $X_n \xrightarrow{P} X$. Furthermore

$$\mathbf{E}[|X_{n,k} - X|] = \mathbf{E}\left[\mathbf{1}_{\left(\frac{k}{n}, \frac{k+1}{n}\right)}\right] = 1/n,$$

and again taking $n \to \infty$ this shows that $X_n \xrightarrow{L^p} X$.
But $X_{n,k} \not\to X$ a.s.
Why? For any ω and any n, there exist an $m > n$ and a k such that $X_{m,k}(\omega) - X = 1$. The sequence at ω alternates between many ω and $\omega + 1$.

The next theorem says that, even if the convergence in probability does not imply convergence a.s., we can extract a subsequence that will converge a.s. In the previous example, if we omit the terms in the sequence with values $\omega + 1$, the resulting subsequence converges a.s.

7.2.2 Probability convergence versus a.s. and L^p convergence

Theorem 7.23 (Relation between a.s. convergence and convergence in probability) *We have the following relations:*

(1) If $X_n \xrightarrow{a.s.} X$ *then* $X_n \xrightarrow{P} X$

(2) *If $X_n \xrightarrow{P} X$, then there exist a subsequence n_k such that $X_{n_k} \xrightarrow{a.s.} X$ as $k \to \infty$*

Proof: (a) Let $N^c = \{\omega : \lim_{n \to \infty} |X_n(\omega) - X(\omega)| = 0\}$. From the definition of a.s. convergence we know that $P(N) = 0$.

Fix an $\varepsilon > 0$ and consider $N_\varepsilon(n) = \{\omega : |X_n(\omega) - X(\omega)| \geq \varepsilon\}$. Let now

$$M_k = \left(\bigcup_{n \geq k} N_\varepsilon(n) \right)^c = \bigcap_{n \geq k} N_\varepsilon(n)^c \tag{7.2}$$

- M_k's are increasing sets (because $M_k = N_\varepsilon(k)^c \cap M_{k+1}$, which implies $M_k \subseteq M_{k+1}$).

- If $\omega \in M_k$, this means that for all $n \geq k$, $\omega \in N_\varepsilon(n)^c$, or $|X_n(\omega) - X(\omega)| < \varepsilon$. By definition, this means that the sequence is convergent at ω; therefore $M_k \subseteq N^c$, $\forall k$, and thus $\bigcup_k M_k \subseteq N^c$.

I leave it as an easy exercise to take an $\omega \in N^c$ and to show that there must exist a k_0 such that $\omega \in M_{k_0}$, and therefore we will easily obtain $N^c \subseteq \bigcup_k M_k$. This will imply that $\bigcup_k M_k = N^c$ and so $P(\bigcup_k M_k) = 1$, by hypothesis.

Since the sets M_k are increasing, this implies that $P(M_k) \to 1$ when $k \to \infty$. Looking at the definition of M_k in (7.2), this clearly implies that

$$P \left(\bigcup_{n \geq k} N_\varepsilon(n) \right) \to 0, \qquad \text{as } k \to \infty,$$

therefore $P(N_\varepsilon(k)) \to 0$, as $k \to \infty$, which is the definition of the convergence in probability.

(b) For this part, we will use the Borel–Cantelli lemmas (Lemma 1.28 on page 34). We will take ε in the definition of convergence in probability of the form $\varepsilon_k > 0$ and make it to go to zero when $k \to \infty$. By the definition of convergence in probability, for every such ε_k we can find an n_k, such that $P\{\omega : |X_n(\omega) - X(\omega)| > \varepsilon_k\} < 2^{-k}$, for every $n \geq n_k$. An easy process now will construct $m_k = \min(m_{k-1}, n_k)$, so that the subsequence becomes increasing, while still maintaining the above desired property. Denote

$$N_k = \{\omega : |X_{m_k}(\omega) - X(\omega)| > \varepsilon_k\}.$$

Then, from the above construction, $P(N_k) < 2^{-k}$, which was chosen on purpose since this implies that

$$\sum_k P(N_k) < \sum_k 2^{-k} < \infty.$$

Therefore, applying the first Borel–Cantelli lemma to the N_k sets, the probability that N_k occurs infinitely often is zero.

Translated, N will stop happening infinitely often, and thus with probability 1 N_k^c will happen eventually. Or, looking at the definition of N_k, the set of ω's for which there exists a k_0 and $|X_{m_k}(\omega) - X(\omega)| < \varepsilon_k$ for all $k \geq k_0$ has probability 1. At each such ω, the subsequence is convergent and thus the set $\{\omega : X_{m_k}(\omega) \to X(\omega)\}$ has probability 1. Therefore, the subsequence X_{m_k} has the desired property and this is exactly what we needed to prove. ∎

In general, convergence in probability does not imply a.s. convergence. A counterexample was provided already in Example 7.3, but here is another variation of that example (written as an exercise).

Exercise 5 (Counterexample. $\xrightarrow{P}$ **implies** $\xrightarrow{a.s.}$ **)** *You may construct your own counterexample. For instance, take* $\Omega = (0, 1)$ *with the Borel sets on it and the Lebesgue measure (which is a probability measure for this Ω). Take now for every* $n \in \mathbb{N}$ *and* $1 \leq m \leq 2^n$

$$X_{n,m}(\omega) = \mathbf{1}_{\left[\frac{m-1}{2^n}, \frac{m}{2^n}\right]}(\omega).$$

Form a single subscript sequence by taking $Y_1 = X_{0,1}$, $Y_2 = X_{1,1}$, $Y_3 = X_{1,2}$, $Y_4 = X_{2,1}$, $Y_5 = X_{2,2}$, $Y_6 = X_{2,3}$, $Y_7 = X_{2,4}$, *and so on. Draw the graph of these variables as functions of ω on a piece of paper for a better understanding of what is going on.*

Prove that this sequence $\{Y_k\}$ has the property that $Y_k \xrightarrow{P} 0$ but $Y_k \nrightarrow Y$ a.s. In fact, it does not converge for any $\omega \in \Omega$.

Proposition 7.24 *If* $X_n \xrightarrow{L^P(\Omega)} X$, *then* $X_n \xrightarrow{P} X$.

Proof: This is an easy application of the Markov inequality (Proposition 4.12 on page 132). Take $g(x) = |x|^p$, and the random variable $X_n - X$. We obtain

$$\mathbf{P}\left(|X_n - X|^p > \varepsilon\right) \leq \varepsilon^{-p}\mathbf{E}|X_n - X|^p.$$

Therefore, if $X_n \xrightarrow{L^P(\Omega)} X$, then we necessarily have $X_n \xrightarrow{P} X$ as well. ∎

Exercise 6 *The converse of the previous result is not true in general. Consider the probability ensemble of Exercise 5.*
Let $X_n(\omega) = n\mathbf{1}_{[0, \frac{1}{n}]}(\omega)$.
Show that $X_n \xrightarrow{P} X$, *but* $X_n \nrightarrow X$ *in any L^p with $p \geq 1$.*

Finally, here is a criterion for convergence in probability using expectations.

Theorem 7.25 $X_n \xrightarrow{P} 0$ *if and only if* $\mathbf{E}\left(\frac{|X_n|}{1+|X_n|}\right) \to 0$.

Proof: The function $f(x) = \frac{|x|}{|x|+1}$ is always between 0 and 1. Using this we have

$$\mathbf{E}\left(\frac{|X_n|}{1+|X_n|}\right) = \mathbf{E}\left(\frac{|X_n|}{1+|X_n|}\mathbf{1}_{\{|X_n|>\varepsilon\}}\right) + \mathbf{E}\left(\frac{|X_n|}{1+|X_n|}\mathbf{1}_{\{|X_n|\leq\varepsilon\}}\right)$$

$$\leq \mathbf{P}\{|X_n| > \varepsilon\} + \varepsilon\,\mathbf{E}\left(\frac{1}{1+|X_n|}\mathbf{1}_{\{|X_n|\leq\varepsilon\}}\right)$$

The first term goes to 0 due to convergence in probability and the second is dominated by ε, thus the direct implication follows.

For the converse, note that, if $|x| > \varepsilon$, then $\frac{|x|}{1+|x|} > \frac{\varepsilon}{1+\varepsilon}$ since the function is strictly increasing on $(0, \infty)$ and is even thus

$$\mathbf{E}\left(\frac{|X_n|}{1+|X_n|}\right) \geq \mathbf{E}\left(\frac{|X_n|}{1+|X_n|}\mathbf{1}_{\{|X_n|>\varepsilon\}}\right) > \frac{\varepsilon}{1+\varepsilon}\mathbf{P}\{|X_n| > \varepsilon\}.$$

Take $n \to 0$ on both sides and we obtain that $\frac{\varepsilon}{1+\varepsilon}\lim_{n\to\infty}\mathbf{P}\{|X_n| > \varepsilon\} = 0$ for all $\varepsilon > 0$. But then the limit must be equal to 0 for all ε and thus $X_n \xrightarrow{P} 0$. ∎

Back to almost sure versus L^p convergence Let us return to the earlier relations for a moment. We have already shown (by providing counterexamples) that neither necessarily implies the other one. In either of these examples, the limit implied did not exist.

If both limits exist, then they necessarily must be the same.

Proposition 7.26 *If* $X_n \xrightarrow{L^p(\Omega)} X$ *and* $X_n \xrightarrow{a.s} Y$, *then* $X = Y$ *a.s.*

Proof: (Sketch) We have already proven that both types of convergence imply convergence in probability. The proof then ends by showing a.s. the uniqueness of a limit in probability. ∎

7.2.3 Uniform integrability

We have seen that convergence a.s. and convergence in L^p are generally not compatible. However, we will give an integrability condition that will make all the convergence types equivalent. It is very desirable that moments converge when $X_n \xrightarrow{a.s.} X$

Definition 7.27 (Uniform Integrability criterion) *A collection of random variables* $\{X_n\}_{n\in\mathcal{I}}$ *is called uniform integrable (U.I.) if*

$$\lim_{M\to\infty}\sup_{n\in\mathcal{I}}\mathbf{E}\left[|X_n|\mathbf{1}_{\{|X_n|>M\}}\right] = 0.$$

In other words, the tails of the expectation converge to 0 uniformly for all the family.

In general, the criterion is hard to verify. The following theorem contains two necessary and sufficient conditions for U.I.

Theorem 7.28 (Conditions for U.I.) *A family of r.v.'s* $\{X_n\}_{n\in\mathcal{I}}$ *is uniformly integrable if and only if one of the following two conditions are true:*

1. **Bounded and Absolutely Continuous family.** $\mathbf{E}[|X_n|]$ *is bounded for every* $n \in \mathcal{I}$, *and for any* $\varepsilon > 0$ *there exists a* $\delta(\varepsilon) > 0$ *such that if* $A \in \mathcal{F}$ *is any set with* $\mathbf{P}(A) < \delta(\varepsilon)$ *then we must have* $\mathbf{E}[|X_n|\mathbf{1}_A] < \varepsilon$.

2. de la Vallée–Poussin theorem. *There exists some increasing function f, $f \geq 0$, such that $\lim_{x \to \infty} \frac{f(x)}{x} = \infty$ with the property that $\mathbf{E}f(|X_n|) \leq C$ for some constant C and all n.*

We will skip the proof of this theorem. The dedicated reader may find the proof of the first result in (Chung, 2000, Theorem 4.5.3) and the second in (Grimmett and Stirzaker, 2001, Theorem 7.10.3).

Here is the main result for the uniformly integrable families.

Theorem 7.29 *Suppose that $X_n \xrightarrow{P} X$ and for a fixed $p > 0$, $X_n \in Ł^p(\Omega)$. Then the following three statements are equivalent:*

(i) *The family $\{|X_n|^p\}_{n \in \mathbb{N}}$ is U.I.*

(ii) $X_n \xrightarrow{L^p} X$.

(iii) $\mathbf{E}[X_n^p] \to \mathbf{E}[X^p] < \infty$.

Once again, we refer the reader to the proof of (Chung, 2000, Theorem 4.5.4)

We conclude the section dedicated to uniform integrability criteria by providing some examples.

■ **EXAMPLE 7.4**

Examples of U.I. families:

- Any r.v. $X \in L^1$ is U.I.
  ($\mathbf{E}|X| < \infty$ implies immediately $\mathbf{E}\left[|X|\mathbf{1}_{\{|X|>M\}}\right] \xrightarrow{M \to \infty} 0$).

- Let the family $\{X_n\}$ be bounded by an integrable random variable, $|X_n| \leq Y$ and $Y \in L^1$. Then X_n is U.I.
  Indeed, we have $\mathbf{E}\left[|X_n|\mathbf{1}_{\{|X_n|>M\}}\right] \leq \mathbf{E}\left[Y\mathbf{1}_{\{|Y|>M\}}\right]$, which does not depend on n and converges to 0 with M as in the previous example.

- Any finite collection of r.v.'s in L^1 is U.I.
  This is just an application of the previous point. If $\{X_1, X_2, \ldots, X_n\}$ is the collection of integrable r.v.'s, take, for example, $Y = |X_1| + |X_2| + \cdots + |X_n|$.

- The family $\{a_n Y\}$ is U.I. where $Y \in L^1$ and the $a_n \in [-1, 1]$ are non-random constants.

- Any bounded collection of integrable r.v.'s is U.I.

Here is an example of a family which is not U.I.

◾ EXAMPLE 7.5

Let us consider the probability space of all infinite sequences of coin tosses (we will see this space later on in reference to Bernoulli process). Assume that the coin is fair.

Let $X_n = \inf\{i : \text{such that } i > n \text{ and toss } i \text{ is a } H\}$; count the first toss after the nth toss where we obtain a head. Let M be an integer and $n \geq M$. For this n, according to the definition, we have $X_n > n \geq M$. Therefore, for this n, we have $\mathbf{E}\left[|X_n|\mathbf{1}_{\{|X_n|>M\}}\right] = \mathbf{E}[X_n] > n$, which implies that $\{X_n\}$ is not U.I.

Uniform integrability sounds like a very useful criterion. However, it is hard to use and with the exception of a few theoretical results it is rarely encountered.

7.2.4   Weak convergence and all the others

All the other three modes of convergence discussed thus far (a.s., L^p, and P) are concerned with the case when all the variables X_n as well as their limit X are defined on the same probability space. In many applications, we only care about the type of **the limiting distribution**. For this reason, in most practical applications the weak convergence is the most relevant type of convergence. Does that mean that we unnecessarily talked about the other types? Certainly not; if that were the case we would not have presented them. In many cases, one of the other types is much easier to prove. Due to the relationships that exist between convergence types, any type of convergence presented thus far will imply convergence in distribution.

I am also going to stress this: though convergence in distribution is the weakest form of convergence in the sense that it is implied by all the others, we are in fact discussing a totally different form of convergence. Recall that the variables do not have to be defined on the same probability space, something that is absolutely necessary for a.s., L^p, or convergence in probability.

The following proposition states that (if possible to express) the convergence in probability (and thus all the others) will imply convergence in distribution. That is perhaps the reason for the name "weak convergence."

Proposition 7.30 *Suppose that the sequence of random variables X_n and the random variable X are defined on the same probability space $(\Omega, \mathscr{F}, \mathbf{P})$. If $X_n \xrightarrow{P} X$, then $X_n \xrightarrow{D} X$.*

Proof: Suppose that for every $\varepsilon > 0$ we have

$$\mathbf{P}[|X_n - X| > \varepsilon] = \mathbf{P}\{\omega : |X_n(\omega) - X(\omega)| > \varepsilon\} \longrightarrow 0, \text{ as } n \to \infty.$$

Let $F_n(x)$ and $F(x)$ be the distribution functions of X_n and X. Then we may write

$$F_n(x) = \mathbf{P}[X_n \leq x] = \mathbf{P}[X_n \leq x, X \leq x + \varepsilon] + \mathbf{P}[X_n \leq x, X > x + \varepsilon]$$

$$\leq \mathbf{P}[X \leq x + \varepsilon] + \mathbf{P}[X_n - X \leq x - X, X - x > \varepsilon]$$

$$= F(x + \varepsilon) + \mathbf{P}[X_n - X \leq x - X, x - X < -\varepsilon]$$

$$\leq F(x + \varepsilon) + \mathbf{P}[X_n - X < -\varepsilon]$$

$$\leq F(x + \varepsilon) + \mathbf{P}[X_n - X < -\varepsilon] + \mathbf{P}[X_n - X > \varepsilon]$$

$$= F(x + \varepsilon) + \mathbf{P}[|X_n - X| > \varepsilon]$$

We may repeat the line of reasoning before starting with F and replacing x above with $x - \varepsilon$ (to obtain what we need):

$$F(x - \varepsilon) = \mathbf{P}[X \leq x - \varepsilon] = \mathbf{P}[X \leq x - \varepsilon, X_n \leq x] + \mathbf{P}[X \leq x - \varepsilon, X_n > x]$$

$$\leq \mathbf{P}[X_n \leq x] + \mathbf{P}[X - X_n \leq x - \varepsilon - X_n, x - \varepsilon - X_n < -\varepsilon]$$

$$\leq F_n(x) + \mathbf{P}[X - X_n < -\varepsilon] \leq F_n(x) + \mathbf{P}[|X_n - X| > \varepsilon].$$

Combining the two inequalities, we obtain

$$F(x - \varepsilon) - \mathbf{P}[|X_n - X| > \varepsilon] \leq F_n(x) \leq F(x + \varepsilon) + \mathbf{P}[|X_n - X| > \varepsilon].$$

Take $n \to \infty$ and using the hypothesis and the $\liminf$ and $\limsup$ (we don't know that the limit exists) gives

$$F(x - \varepsilon) \leq \liminf_{n \to \infty} F_n(x) \leq \limsup_{n \to \infty} F_n(x) \leq F(x + \varepsilon)$$

Recall that we only care about the x points of continuity of F. Therefore, by taking $\varepsilon \downarrow 0$ we obtain $F(x)$ in both ends of inequality, and we are done.   ∎

In general, convergence in distribution does not imply convergence in probability. We cannot even talk about it in most cases since the defining spaces for the random variables X_n may be different.

However, there is one exception. The constant random variable exists in any probability space. Suppose that $X(\omega) \equiv c$ a constant. In this case the convergence in probability reads $\lim_{n \to \infty} \mathbf{P}[|X_n - c| > \varepsilon] = 0$ and this statement makes sense even if each X_n is defined in a different probability space.

Theorem 7.31 *If $X_n \xrightarrow{D} c$ where c is a constant, then $X_n \xrightarrow{P} c$.*

Proof: If $X(\omega) \equiv c$, or in other words the constant c is regarded as a random variable, its distribution is the distribution of the delta function at c, i.e.,

$$\mathbf{P}[X \leq x] = \mathbf{1}_{(c,\infty)}(x) = \begin{cases} 0, & x \leq c \\ 1, & x > c \end{cases}$$

By hypothesis, the distribution functions of X_n converge to this function at all $x \neq c$. We may write

$$\mathbf{P}[|X_n - c| > \varepsilon] = \mathbf{P}[X_n > c + \varepsilon] + \mathbf{P}[X_n < c - \varepsilon]$$

$$= 1 - \mathbf{P}[X_n < c + \varepsilon] + \mathbf{P}[X_n < c - \varepsilon]$$

$$\leq 1 - F_n(c + \varepsilon) + F_n(c - \varepsilon)$$

Taking now $n \to \infty$ and using that the distribution of X_n converges to the indicator above, we obtain that the limit above is zero for all $\varepsilon > 0$. Done.   ∎

Finally, here is a fundamental results that explains why even though the weakest paradoxically it is also the most remarkable convergence type.

Theorem 7.32 (Skorohod's representation theorem) *Suppose $X_n \xrightarrow{D} X$. Then, there exists a probability space $(\Omega', \mathscr{F}', \mathbf{P}')$, and a sequence of random variables Y, Y_n on this new probability space, such that X_n has the same distribution as Y_n, X has the same distribution as Y, and $Y_n \to Y$ a.s. In other words, there is a representation of X_n and X on a single probability space, where the convergence occurs almost surely.*

Thus we can construct a sequence of random variables with the same distributions on a particular space, and in that space the convergence is almost sure.

7.3   Continuous Mapping Theorem. Joint Convergence. Slutsky's Theorem

Theorem 7.33 (Continuous mapping theorem) *Suppose that X_n and X are random variables. Suppose also that $\phi : \mathbb{R} \to \mathbb{R}$ is a continuous function. Then*

(i) If $X_n \xrightarrow{D} X$ then $\phi(X_n) \xrightarrow{D} \phi(X)$

(ii) If $X_n \xrightarrow{P} X$ then $\phi(X_n) \xrightarrow{P} \phi(X)$

(iii) If $X_n \xrightarrow{a.s.} X$ then $\phi(X_n) \xrightarrow{a.s.} \phi(X)$

The theorem above is also true even if g is discontinuous at certain points x_i with the condition that $\mathbf{P}(X = x_i) = 0$. We will not prove this theorem: two of the assertions we have proved earlier, and the convergence in probability implication is very simple.

In general, having $X_n \xrightarrow{D} X$ and $Y_n \xrightarrow{D} Y$ does not necessarily imply that $(X_n, Y_n) \xrightarrow{D} (X, Y)$. This is in direct contrast with the convergence in probability where, if $X_n \xrightarrow{P} X$ and $Y_n \xrightarrow{P} Y$, then we have to have $(X_n, Y_n) \xrightarrow{P} (X, Y)$. The probability assertion is easy to prove since we may write

$$\mathbf{P}\left(|(X_n, Y_n) - (X, Y)| > \varepsilon\right) = \mathbf{P}\left((X_n - X)^2 + (Y_n - Y)^2 > \varepsilon^2\right)$$
$$\leq \mathbf{P}\left(|X_n - X| > \varepsilon\right) + \mathbf{P}\left(|Y_n - Y| > \varepsilon\right)$$

However, just as in the case of convergence to a constant (above), we do have an exception, and this is quite a notable exception too. The following theorem is one of the most widely used and useful in all of probability theory.

Theorem 7.34 (Slutsky's Theorem) *Suppose that $\{X_n\}_n$ and $\{Y_n\}_n$ are two sequences of random variables such that $X_n \xrightarrow{D} X$ and $Y_n \xrightarrow{D} c$, where c is a constant and X is a random variable. Suppose that $f : \mathbb{R}^2 \to \mathbb{R}$ is a continuous functional. Then $f(X_n, Y_n) \xrightarrow{D} f(X, c)$.*

Proof: We shall use the characterization theorem for convergence in distribution (Th. 7.15 on page 218) in the proof of Slutsky's theorem. We will also use Proposition 7.35 stated next, which says that the limit in distribution is the same for any two sequences that have the same limit in probability:

We first show weak convergence of the vector (X_n, c) to (X, c). Specifically, let $\phi : \mathbb{R}^2 \to \mathbb{R}$ be a bounded, continuous function. It we take $g(x) = \phi(x, c)$, then g is itself continuous and bounded, and using Theorem 7.15 we have

$$\mathbf{E}[\phi(X_n, c)] = \mathbf{E}[g(X_n)] \to \mathbf{E}[g(X)] = \mathbf{E}[\phi(X, c)].$$

Thus $(X_n, c) \xrightarrow{D} (X, c)$. Next we have: $|(X_n, Y_n) - (X_n, c)| = |Y_n - c| \xrightarrow{D} 0$ because $Y_n \xrightarrow{D} c$. But from Theorem 7.31 we also have $|(X_n, Y_n) - (X_n, c)| \xrightarrow{P} 0$.

Finally, using the Proposition 7.35 we immediately obtain that $(X_n, Y_n) \xrightarrow{D} (X, c)$. Therefore, according to Theorem 7.15 we have

$$\mathbf{E}[\phi(X_n, Y_n)] \to \mathbf{E}[\phi(X, c)],$$

for every continuous bounded function ϕ.

Next, let $\psi : \mathbb{R} \to \mathbb{R}$ be any continuous, bounded function. Since f in the hypothesis of the theorem is a continuous function, $\psi \circ f$ is continuous. It is also bounded since ϕ is. Applying the result above to this function we obtain

$$\mathbf{E}[\psi \circ f(X_n, Y_n)] \to \mathbf{E}[\psi \circ f(X, c)],$$

which proves that $f(X_n, Y_n) \xrightarrow{D} f(X, c)$. ∎

Proposition 7.35 *Let X_n and Y_n two random sequences (they may be vector valued) such that $X_n \xrightarrow{D} X$ and $|X_n - Y_n| \xrightarrow{P} 0$. Then, we must have $Y_n \xrightarrow{D} X$*

Proof: Note that, if $|X_n - Y_n| \xrightarrow{P} 0$, then the sequence $X_n - Y_n$ also converges in distribution and therefore by the weak convergence characterization theorem 7.15, we have for any continuously bounded function g

$$\int_{\mathbb{R}^d} g(x) dF_{Y_n - X_n}(x) \to \int_{\mathbb{R}^d} g(x)\delta_{\{0\}}(dx) = g(0),$$

where $\delta_{\{0\}}$ is the delta measure at 0 and we used the definition of this particular measure to obtain $g(0)$ at the end. In fact, owing to this particular measure, the same conclusion applies if $\mathbb{R}^d$ is replaced with any interval in $\mathbb{R}^d$ containing 0.

Let x be a continuity point of $F_X(x)$. We may write $Y_n = X_n + (Y_n - X_n)$ and therefore we can write the distribution of Y_n as a convolution:

$$F_{Y_n}(x) = \mathbf{P}(Y_n \le x) = \mathbf{P}(X_n + (Y_n - X_n) \le x) = \int_{\mathbb{R}^d} F_{X_n}(x - s) dF_{Y_n - X_n}(s).$$

The distribution function $F_X(x - s)$ is bounded but not necessarily continuous at $x - s$. However, since it is continuous at x, we may find a neighborhood of 0, say U, such that $F_X(x - s)$ is continuous for any $s \in U$. If $s \notin U$, then $F_{Y_n - X_n}(s)$ may be made arbitrarily close to a constant (either 1 or 0) and thus the integral may be made negligible. Finally, we may write

$$F_{Y_n}(x) = \int_U F_{X_n}(x - s) dF_{Y_n - X_n}(s) \to \int_U F_X(x - s)\delta_{\{0\}}(ds) = F_X(x).$$

Therefore, we conclude $Y_n \xrightarrow{D} X$. ∎

☐ EXAMPLE 7.6

With the hypothesis stated in Slutsky's theorem and with $f(X_n, Y_n) = X_n + Y_n$ and $g(X_n, Y_n) = X_n Y_n$, we get as a particular case where

$$X_n + Y_n \xrightarrow{D} X + c,$$

and

$$X_n Y_n \xrightarrow{D} Xc.$$

Slutsky's theorem is most effective when coupled with the δ-theorem. We postpone the examples until later in this chapter after we learn about this delta theorem.

7.4 The Two Big Limit Theorems: LLN and CLT

7.4.1 A note on statistics

In real life, we normally do not know the underlying distribution of random variables. We usually work with observations of the random variable.

Definition 7.36 *Suppose we work on a probability space $\Omega, \mathcal{F}, \mathbf{P}$). A **simple random sample** of size n is a set of i.i.d. random variables $(X_1, X_2, \ldots, X_n)$. The outcome of the sample, or simply a sample, is a set of realizations $(x_1, x_2, \ldots, x_n)$ of these random variables. A **statistics** is any random variable Y obtained from the random sample, that is,*

$$Y = f(X_1, X_2, \ldots, X_n),$$

where f is any function $f : \mathbb{R}^n \to \mathbb{R}$.

Note that since the variables $X_1, X_2, \ldots, X_n$ are independent and have the same distribution, say F, a sample $(x_1, x_2, \ldots, x_n)$ represents a set of possible outcomes of the distribution F. Thus, it is possible to gather information about the variable by studying the sample. In fact, statistics is an entire science dedicated to the study of these empirical distributions.

Example of statistics include the following:

- Sample sum: $S_n = X_1 + X_2 + \ldots + X_n = \sum_{i=1}^{n} X_i$.

- Sample mean: $\bar{X} = \frac{S_n}{n} = \frac{1}{n} \sum_{i=1}^{n} X_i$.

- Sample variance: $S^2 = \frac{1}{n-1} \sum_{i=1}^{n} (X_i - \bar{X})^2$.

- Sample standard deviation: $S = \sqrt{S^2} = \sqrt{\frac{1}{n-1} \sum_{i=1}^{n} (X_i - \bar{X})^2}$.

- Coefficient of variance: $CV = S/\bar{X}$.

- Sample central p-moment: $S^p = \frac{1}{n} \sum_{i=1}^{n} (X_i - \bar{X})^p$.

- Sample mean absolute deviation: $MAD = \frac{1}{n} \sum_{i=1}^{n} |X_i - \bar{X}|$.

- Sample minimum and maximum: $X_{(1)} = \min\{X_1, X_2, \ldots, X_n\}$, $X_{(n)} = \max\{X_1, X_2, \ldots, X_n\}$.

- Median: the $\left[\frac{n+1}{2}\right]$ smallest observation among $X_1, X_2, \ldots, X_n$. Here $[a]$ denotes the integer part of a. When $n = 2k$ an even number, the sample median falls between two observations. In this case, it is defined as the average of the two observations: X_k and X_{k+1}.

- (100p) percentile corresponding to a certain probability p is defined as the observation such that approximately np of the observations are smaller and $n(1-p)$ observations are larger. If we denote $round(a)$ as the nearest integer of the number a[1], then the (100p) percentile is defined as the $round(np)$ smallest observation if $\frac{1}{2n} < p < \frac{1}{2}$ and the $n + 1 - round(n(1-p))$ smallest observation if $\frac{1}{2} < p < 1 - \frac{1}{2n}$. The cases when $p > 1/2$ and $p < 1/2$ are defined separately, so that the percentiles exhibit symmetry. That is, if the 5 percentile is the fifth

---

[1]If $i - \frac{1}{2} \leq a < i + \frac{1}{2}$, then $round(a) = i$.

smallest observation in a set of $n = 100$ observations, then if we use the definition above, the 95th percentile is going to be the 5th largest observation (or 95th smallest).

- Sample range: $\max\{X_1, \ldots, X_n\} - \min\{X_1, \ldots, X_n\}$.

- Sample semideviations. These are new measures and their statistical properties are not entirely understood yet. If we calculate the sample mean $\bar{X}$ and we define N_1 as the number of observations below $\bar{X}$ and N_2 as the number of observations above $\bar{X}$, then

$$\text{Lower squared semideviation} = \sqrt{\frac{1}{N_1} \sum_{\substack{0 \le i \le n \\ X_i < \bar{X}}}^{N_1} (\bar{X} - X_i)^2}$$

$$\text{Upper squared semideviation} = \sqrt{\frac{1}{N_2} \sum_{\substack{0 \le i \le n \\ X_i > \bar{X}}}^{N_2} (X_i - \bar{X})^2}.$$

Note that N_1 and N_2 are random variables and $N_1 + N_2 = n$.

There exist many other statistics; the bullet list presented above is by no means exhaustive. As you may observe, there are two predominant types in the list: statistics corresponding to some order (minimum, maximum, median percentiles) and statistics based on some sort of average (the sample mean $\bar{X}$, variance, etc.). Thus, it is very important to understand their statistical (distributional) properties. The next sections investigate them.

7.4.2 The order statistics

Let $X_1 \ldots X_n$ be n random variables. We say that $X_{(1)} \le \ldots \le X_{(n)}$ are the order statistics corresponding to $X_1 \ldots X_n$ if $X_{(k)}$ is the kth smallest value among $X_1 \ldots X_n$. Note that $X_{(1)} = \min\{X_1 \ldots X_n\}$ and $X_{(n)} = \max\{X_1 \ldots X_n\}$. Also note that the percentiles defined in the previous subsection are nothing more than these order statistics.

With this definition, we may rewrite some of the statistics above as follows:

Range $R = X_{(n)} - X_{(1)}$

(100p) Percentile $\begin{cases} X_{(round(np))} & \text{if } \frac{1}{2n} < p < \frac{1}{2} \\ X_{(n+1-round(n(1-p)))} & \text{if } \frac{1}{2} < p < 1 - \frac{1}{2n} \end{cases}$

Median When $p = 1/2$ in the above definition, we obtain the median. To obey the symmetry of the percentile, we define it in the following way:

$$\begin{cases} X_{\left(\frac{n+1}{2}\right)} & \text{if } n \text{ is odd} \\ \left(X_{\left(\frac{n}{2}\right)} + X_{\left(\frac{n}{2}+1\right)}\right)/2 & \text{if } n \text{ is even} \end{cases}$$

When $p = 0.25$ and $p = 0.75$, we obtain the so-called lower quartile and upper quartile, respectively. Sometimes, a measure of variability of a random variable is defined using the interquartile range: i.e., the distance between the upper and lower quartile.

Let us investigate the distribution of these order statistics. The first thing to notice is that we are able to calculate these distributions only when the random variables observed X_i are i.i.d. As we shall see, this primarily amounts to counting. We investigate only the case when the X_i distribution is either discrete or continuous. We give the discrete distribution case first since this case is harder than any other case. This is due to the fact that the X_i's may be equal (taking the same discrete value).

Theorem 7.37 *Let $X_1 \ldots X_n$ be n independent identically distributed random variables with discrete distribution $f(x_i) = p_i$, where $x_1 < x_2 < \ldots$ are the possible values of the X variables. We define the cumulative distribution function of the X variables as*

$$P_0 = 0$$
$$P_1 = \mathbf{P}(X \le x_1) = p_1$$
$$P_2 = \mathbf{P}(X \le x_2) = p_1 + p_2$$
$$\vdots$$
$$P_i = \mathbf{P}(X \le x_i) = p_1 + p_2 + \cdots + p_i$$
$$\vdots$$

Then the marginal distribution of the order statistics $X_{(1)} \le \ldots \le X_{(n)}$ is

$$\mathbf{P}(X_{(j)} \le x_i) = \sum_{k=j}^{n} \binom{n}{k} P_i^k (1 - P_i)^{n-k}$$

Proof: For a fixed x_i, let Y be the variable that counts how many of the $X_1 \ldots X_n$ are less than or equal to x_i. If we look at the events $\{X_1 \le x_i\}, \ldots, \{X_n \le x_i\}$ and every time one is true we call a success and every time one is false we call a failure, then Y is simply counting the number of successes in n trials. Also note that the probability of success is the same for all trials ($= P_i$) and each trial is independent of all the others. Therefore, $Y \sim Binom(n, P_i)$.

Why did we even looked at this Y? Note that $X_{(j)} \le x_i$ means that at least j of the initial variables are less than x_i. But since Y counts the number of them less than x_j, we must have

$$\mathbf{P}(X_{(j)} \le x_i) = \mathbf{P}(Y \ge j) = \sum_{k=j}^{n} \binom{n}{k} P_i^k (1 - P_i)^{n-k},$$

simply using the distribution of Y. ∎

In the continuous distribution case, we may go further since the probability of any two X_i's being equal is zero.

Theorem 7.38 *Suppose that $X_{(1)} \leq \ldots \leq X_{(n)}$ denote the order statistics of the i.i.d. variables $X_1 \ldots X_n$ which have a continuous distribution function $F(x)$ and the corresponding probability distribution function $f(x)$. Then $X_{(j)}$ is continuously distributed with pdf*

$$f_{X_{(j)}}(x) = \frac{n!}{(j-1)!(n-j)!} f(x)[F(x)]^{j-1}[1 - F(x)]^{n-j}$$

Proof: The proof works much the same way as with the previous theorem. We again fix an x, and then we let Y be the variable that counts how many of the $X_1 \ldots X_n$ are less than or equal to this x. Once again, Y counts how many of the events $\{X_1 \leq x\}, \ldots, \{X_n \leq x\}$ are successes and $Y \sim Binom(n, F(x))$, since now probability of success is $\mathbf{P}\{X_i \leq x\} = F(x)$ for any i. Once again, $X_{(j)} \leq x$ means that at least j of the X variables are less than x; so we have

$$F_{X_{(j)}}(x) = \mathbf{P}(X_{(j)} \leq x) = \mathbf{P}(Y \geq j) = \sum_{k=j}^{n} \binom{n}{k}[F(x)]^k[1 - F(x)]^{n-k}.$$

To calculate the pdf, we take the derivative with respect to x and we obtain

$$\begin{aligned} f_{X_{(j)}}(x) &= \sum_{k=j}^{n} \binom{n}{k}\left(kf(x)[F(x)]^{k-1}[1 - F(x)]^{n-k} \right. \\ &\qquad\qquad\qquad \left. - (n-k)f(x)[F(x)]^k[1 - F(x)]^{n-k-1} \right) \\ &= \binom{n}{j} jf(x)[F(x)]^{j-1}[1 - F(x)]^{n-j} \\ &\quad + \sum_{k=j+1}^{n} \binom{n}{k} kf(x)[F(x)]^{k-1}[1 - F(x)]^{n-k} \\ &\quad - \sum_{k=j}^{n-1} \binom{n}{k}(n-k)f(x)[F(x)]^k[1 - F(x)]^{n-k-1}, \end{aligned}$$

since the last term in the sum above is zero. Notice now that, if we change the summation variable $k \rightarrow l + 1$, the first summation is

$$\sum_{k=j+1}^{n} \binom{n}{k} kf(x)[F(x)]^{k-1}[1 - F(x)]^{n-k}$$

$$= \sum_{l=j}^{n-1} \binom{n}{l+1}(l+1)f(x)[F(x)]^l[1 - F(x)]^{n-l-1},$$

which looks a lot like the second sum. Noticing that

$$\binom{n}{l+1}(l+1) = \frac{n!}{k!(n-k-1)!} = \binom{n}{k}(n-k),$$

we see that the first sum in fact simplifies with the second sum and we obtain exactly the expression stated in the theorem.   ∎

We may continue providing expressions for the bivariate pdf of pairs of these order statistics, for example, of the pair (X_1, X_n). We do not present these formulas here (for these formulas one may consult Casella and Berger (2001)) but we will need the following result.

Theorem 7.39 *If X_i's are continuous random variables with a common pdf $f(x)$, then the joint density of the order statistics $X_{(1)} \dots X_{(n)}$ is given by*

$$f_{X_{(1)}, \dots, X_{(n)}}(x_1 \dots x_n) = n! f(x_1) \cdots f(x_n) 1_{\{x_1 < x_2 < \dots < x_n\}}$$

The proof of this result is quite involved. We again refer to Casella and Berger (2001) for this and other results about distribution of combinations of order statistics. This book is a classic on any result on statistical distributions.

◩ **EXAMPLE 7.7**

Let U_i be independent identically distributed random variables uniformly distributed on the interval $[0, 1]$, $i = 1, \dots, n$. Define $Y_n = U_{(n)} = \max\{U_1, \dots, U_n\}$

1. Show that the density of Y_n is

$$f_{Y_n}(y) = \begin{cases} 0, & , y \leq 0 \\ ny^{n-1} & , 0 \leq y \leq 1 \\ 0 & , y > 1 \end{cases}.$$

2. Find the constant c such that

$$Y_n \xrightarrow{P} c$$

and verify the convergence.

3. Find the asymptotic distribution of

$$Z_n = n(1 - Y_n),$$

as $n \to \infty$.

Part 1. It is just an application of the previous theorem, but since it is simple we redo the argument separately.

$$F_{Y_n}(y) = P(Y_n \leq y) = P(U_1 \leq y, \dots, U_n \leq y) = (P(U_1 \leq y))^n = y^n, \text{ if } y \in (0, 1),$$

Taking derivatives produces the stated density.

For part 2. Note that Y_n is the maximum of n numbers that are restricted to the interval $(0, 1)$. So it seems logical to investigate whether or not 1 is the limit. Let us investigate this.

$$P(|Y_n - 1| > \varepsilon) = P(1 - Y_n > \varepsilon) = P(Y_n < 1 - \varepsilon) = \int_0^{1-\varepsilon} ny^{n-1} dy = (1 - \varepsilon)^n$$

So,

$$\lim_{n \to \infty} \mathbf{P}(|Y_n - 1| > \varepsilon) = \lim_{n \to \infty} (1 - \varepsilon)^n = 0, \quad \forall \varepsilon > 0, \text{ small}$$

Therefore, by definition

$$Y_n \xrightarrow{P} 1.$$

For the third part, let Z_n be as stated. Let us actually calculate its distribution and see what happens when $n \to \infty$.

$$\mathbf{P}(Z_n \leq z) = \mathbf{P}(n(1 - Y_n) \leq z) = \mathbf{P}\left(Y_n \geq 1 - \frac{z}{n}\right),$$

and using the distribution function of Y_n which we calculated in the first part, we obtain

$$\mathbf{P}(Z_n \leq z) = \begin{cases} 0, & \text{if } z \leq 0 \\ 1 - \left(1 - \frac{z}{n}\right)^n, & \text{if } 1 - \frac{z}{n} \in (0, 1], \text{ or } z \in (0, n] \\ 1, & \text{if } z > n \end{cases}$$

If we let $n \to \infty$, then $\left(1 - \frac{z}{n}\right)^n \to e^{-z}$ and thus the limiting distribution is

$$\mathbf{P}(Z \leq z) = \begin{cases} 1 - e^{-z}, & \text{if } z > 0 \\ 0, & \text{if } z \leq 0 \end{cases}$$

that is, the Z_n sequence converges to an exponentially distributed random variable with parameter 1 (notation $Exp(1)$).

7.4.3   Limit theorems for the mean statistics

In the next two theorems, the variables are defined on the same probability space $(\Omega, \mathscr{F}, \mathbf{P})$.

Theorem 7.40 (Weak Law of Large Numbers) *If $X_1, X_2, \ldots$ are pairwise independent, identically distributed, and have finite mean, then*

$$\bar{X} = \frac{S_n}{n} \xrightarrow{P} \mathbf{E}[X_1]$$

Theorem 7.41 (Strong Law of Large Numbers, SLLN) *If $X_1, X_2, \ldots$ are i.i.d., then*

$$\text{If} \quad \mathbf{E}[|X_1|] < \infty \quad \text{then } \bar{X} = \frac{S_n}{n} \xrightarrow{a.s} \mathbf{E}[X_1] \tag{7.3}$$

$$\text{If} \quad \mathbf{E}[|X_1|] = \infty \quad \text{then } |\bar{X}| = \frac{|S_n|}{n} \xrightarrow{a.s} \infty \tag{7.4}$$

We have already proven the SLLN under an extra hypothesis (the fourth moment of the random variables is finite) in Theorem 4.26 on page 148. To remove the extra hypothesis requires a lot of work and we feel that it would detract the flow of this book. Detailed proofs of both results may be found in Florescu and Tudor (2013). However, we shall present a different proof of the SLLN under the extra hypothesis that the variance or the second moment is finite. We do this because the proof illustrates specific techniques introduced in this chapter. We also prove the converse (reciprocal) of the SLLN.

For a complete presentation, we refer the interested reader to (Chung, 2000, Chapter 5). The weak law of large numbers in the form presented above is due to Aleksandr Yakovlevich Khinchin (1894–1959). The SLLN is due to Andrey Nikolaevich Kolmogorov (1903–1987). Other researchers who contributed to the development of these and related results include William Feller (1906–1970), Boris Vladimirovich Gnedenko (1912–1995), and Paul Pierre Lévy (1886–1971).

Proof (SLLN under the assumption of second moments finite): Suppose that the random variables have $Var(X_i) = \sigma^2 < \infty$. In the same way as we did in the proof of Theorem 4.26, we can reduce it to the case when the expectation of the random variables is equal to zero. We want to show that $S_n/n = \sum_{i=1}^{n} X_i/n \xrightarrow{a.s} 0$, or there exists a set N with $\mathbf{P}(N) = 0$ such that

$$\lim_{n \to \infty} \left| \frac{S_n}{n} \right| = 0, \forall \omega \in N^c.$$

This is equivalent to proving that, $\forall \omega \in N^c$ and $\forall \varepsilon > 0$, there exists $n(\omega, \varepsilon)$ such that if $n \geq n(\omega, \varepsilon)$ we have

$$\left| \frac{S_n}{n} \right| \leq \varepsilon.$$

To prove this, it is sufficient to show that $|S_n/n| > \varepsilon$ occurs only a finitely many times (i.e., does not hold infinitely often), or

$$\lim_{N \to \infty} \mathbf{P} \left(\left| \frac{S_n}{n} \right| > \varepsilon, \text{ for some } n \geq N \right) = 0.$$

Note that this is very different from probability convergence which says that, for all $\varepsilon > 0$

$$\lim_{n \to \infty} \mathbf{P} \left(\left| \frac{S_n}{n} \right| > \varepsilon \right) = 0.$$

As a remark, the convergence in probability contains the index n outside the probability, while the convergence almost ensures all the n's are inside so it is a much stronger assertion.

Now, by Chebyshev inequality, we have

$$\mathbf{P} \left(\left| \frac{S_n}{n} \right| > \varepsilon \right) = \mathbf{P}(|S_n| > n\varepsilon) \leq \frac{Var(S_n)}{(n\varepsilon^2)} = \frac{n\sigma^2}{n^2\sigma^2} = \frac{\sigma^2}{n\varepsilon^2}$$

but the problem is that the sum $\sum 1/n$ is divergent so we cannot use the Borel–Cantelli lemmas for this sequence. However, note that the series $1/n^2$ is convergent, so we can use the lemma for a subsequence of the original X_n sequence. Specifically,

$$\mathbf{P}\left(\left|\frac{S_{n^2}}{n^2}\right| > \varepsilon\right) \le \frac{n^2\sigma^2}{n^4\varepsilon^2} = \frac{\sigma^2}{n^2\varepsilon^2}.$$

Since the events $\left\{\left|\frac{S_{n^2}}{n^2}\right| > \varepsilon\right\}$ have the property that the sum of their probabilities is convergent, by the first Borel–Cantelli (Lemma 1.28) we obtain that

$$\mathbf{P}\left(\left|\frac{S_{n^2}}{n^2}\right| > \varepsilon \text{ i.o.}\right) = 0,$$

and thus $S_{n^2}/n^2 \xrightarrow{a.s} 0$. We have proven that a subsequence converges to the desired limit. It remains to see what happens with the terms between these n^2. We define

$$D_n = \max_{n^2 \le k \le (n+1)^2} |S_k - S_{n^2}|,$$

that is, the largest possible deviation from S_{n^2} that occurs between n^2 and $(n+1)^2$. Now,

$$\mathbf{E}[D_n^2] = \mathbf{E}\left[\left(\max_{n^2 \le k \le (n+1)^2} |S_k - S_{n^2}|\right)^2\right] \le \sum_{n^2+1}^{(n+1)^2} \mathbf{E}\left[(S_k - S_{n^2})^2\right]$$

where we used $(\max(x_1, x_2, \ldots, x_n))^2 \le x_1^2 + x_2^2 + \cdots + x_n^2$.
But

$$\mathbf{E}\left[(S_k - S_{n^2})^2\right] = \mathbf{E}\left[\left(\sum_{n^2+1}^{k} X_i\right)^2\right] = (k - n^2)\sigma^2 \le (2n+1)\sigma^2,$$

with the biggest value obtained when $k = (n+1)^2$. Since there are exactly $(n+1)^2 - n^2 = 2n + 1$ terms in the sum, we obtain the bound on the expectation of the biggest squared deviation

$$\mathbf{E}[D_n^2] \le (2n+1)^2\sigma^2.$$

Using Chebyshev inequality again, we have

$$\mathbf{P}(D_n > n^2\varepsilon) \le \frac{(2n+1)^2\sigma^2}{n^4\varepsilon^2} < \frac{8\sigma^2}{n^2\varepsilon^2}.$$

Now using the same Borel–Cantelli lemma, we get that $D_n/n^2 \xrightarrow{a.s} 0$. Finally, we obtain the desired a.s. convergence since we may write for any k between n^2 and $(n+1)^2$

$$\left|\frac{S_k}{k}\right| = \frac{|S_k - S_{n^2} + S_{n^2}|}{k} \le \frac{|D_n| + |S_{n^2}|}{k} \le \frac{|D_n| + |S_{n^2}|}{n^2},$$

and each of these sequences converges to 0 a.s. ∎

Proposition 7.42 *The converse of the SLLN is true also. Suppose that $X_1, X_2, \ldots$ are i.i.d. and assume that $\frac{S_n}{n} \xrightarrow{a.s} \mu$. Then show that X_i's are in L^1 (finite expectation) and therefore $\mathbf{E}[X_i] = \mu$. Furthermore, if $\frac{S_n}{n} \xrightarrow{a.s} \infty$, then $\mathbf{E}|X_n| = \infty$.*

Proof: Clearly, if we show that $\mathbf{E}[|X_1|] < \infty$, then the X_i's have a finite mean. If this mean is anything but μ, then $\frac{S_n}{n}$ will converge to it by the regular SLLN and the sequence will have two limits, thus providing a contradiction. So let us study the behavior of $\mathbf{E}[|X_1|]$.

Without loss of generality, we may work with the case $\mu = 0$. The reasoning is identical to the one used in the proof above. Note that $|X_1|$ is a positive random variable, so by Corollary 5.2 on page 159 we may write

$$\mathbf{E}[|X_1|] = \int_0^\infty \mathbf{P}(|X_1| > x)dx = \sum_{n=0}^\infty \int_n^{n+1} \mathbf{P}(|X_1| > x)dx$$

Now, using the fact that $\mathbf{P}(|X_1| > x)$ is decreasing in x, we have

$$\mathbf{E}[|X_1|] \leq \sum_{n=0}^\infty \mathbf{P}(|X_1| > n) \leq 1 + \sum_{n=1}^\infty \mathbf{P}(|X_1| > n)$$

and

$$\mathbf{E}[|X_1|] \geq \sum_{n=0}^\infty \mathbf{P}(|X_1| > n+1) = \sum_{n=1}^\infty \mathbf{P}(|X_1| > n)$$

Therefore, $\mathbf{E}[|X_1|]$ is finite or not depends on whether or not $\sum_{n=1}^\infty \mathbf{P}(|X_1| > n)$ is finite or infinite.

Suppose that $\frac{S_n}{n} \xrightarrow{a.s} 0$. We may write

$$\frac{X_n}{n} = \frac{S_n - S_{n-1}}{n} = \frac{S_n}{n} - \frac{n-1}{n}\frac{S_{n-1}}{n-1},$$

and since $n + 1/n \to 1$, we obtain that $\frac{X_n}{n} \xrightarrow{a.s} 0$. This in turn implies that

$$\mathbf{P}\left(\left|\frac{X_n}{n}\right| > 1, \text{ i.o.}\right) = 0.$$

Since the events $\{|X_n/n| > 1\}$ are independent, the second Borel–Cantelli Lemma 1.29 implies that $\sum_{n=1}^\infty \mathbf{P}\left(\left|\frac{X_n}{n}\right| > 1\right) < \infty$, which shows that $\mathbf{E}|X_1| < \infty$.

On the other hand, if $\frac{S_n}{n} \xrightarrow{a.s} \infty$, then the tail of the series is divergent, so after a large N, $|X_n/n| > 1$ a.s. for all $n \geq N$. This proves that $\mathbf{E}[|X_1|] = \infty$ in this case. ∎

The next theorem talks about the rate of convergence to the mean in the previous theorems.

Theorem 7.43 (Central Limit Theorem) *Let* $X_1, X_2, \cdots$ *be a sequence of i.i.d. random variables with finite mean* μ *and finite nonzero variance* σ^2. *Let* $S_n = X_1 + \cdots + X_n$. *Then*

$$\frac{S_n - n\mu}{\sqrt{n\sigma^2}} \xrightarrow{D} N(0, 1)$$

Equivalently, if we denote $\bar{X} = \frac{S_n}{n}$, *we may rewrite*

$$\frac{\bar{X} - \mu}{\sigma/\sqrt{n}} \xrightarrow{D} N(0, 1)$$

Proof: Let $Y_i = \frac{X_i - \mu}{\sigma}$. Also, let $T_n = \sum_{i=1}^{n} Y_i$. This gives

$$T_n = \sum_{i=1}^{n} \frac{X_i - \mu}{\sigma} = \frac{\sum X_i - n\mu}{\sigma} = \frac{S_n - n\mu}{\sigma}$$

Note that the variables Y_i are now i.i.d. random variables with mean zero and variance 1. The above shows that proving the Central Limit Theorem is equivalent to demonstrating that $T_n/\sqrt{n} \xrightarrow{D} N(0, 1)$.

With this idea in mind, let Y_i be i.i.d. random variables such that $\mathbf{E}(Y_i) = 0$ and $Var(Y_i) = 1$. We need to show that

$$\frac{T_n}{\sqrt{n}} = \frac{1}{\sqrt{n}} \sum_{i=1}^{n} Y_i \xrightarrow{D} N(0, 1)$$

Let us calculate the characteristic function of $\frac{T_n}{\sqrt{n}}$. We denote with $\varphi(t)$ the characteristic function of any of the Y variables.

$$\varphi_{T_n/\sqrt{n}}(t) = \varphi_{Y_1 + \ldots + Y_n}\left(\frac{1}{\sqrt{n}}t\right) = \left(\varphi\left(\frac{1}{\sqrt{n}}t\right)\right)^n$$

$$= \left(\varphi(0) + \frac{t}{\sqrt{n}}\varphi'(0) + \frac{1}{2!}\left(\frac{t}{\sqrt{n}}\right)^2 \varphi''(0) + o\left(\left(\frac{t}{\sqrt{n}}\right)^2\right)\right)^n,$$

using the Taylor expansion around 0. We can do this since the characteristic function is well behaved around zero and for a large enough n the term $t/\sqrt{n}$ becomes sufficiently small.

Furthermore, we have $\varphi(0) = 1$ and

$$\varphi'(t) = \frac{d}{dt}\mathbf{E}[e^{itY_1}] = i\mathbf{E}(Y_1 e^{itY_1})$$

Thus, $\varphi'(0) = i\mathbf{E}[Y_1] = 0$, and also

$$\varphi''(t) = i^2 \mathbf{E}[Y_1^2 e^{itY_1}]$$

Therefore $\varphi''(0) = i^2\mathbf{E}[Y_1^2] = -1$. Thus we get

$$\varphi_{\frac{T_n}{\sqrt{n}}}(t) = \left(1 - \frac{t^2}{2n} + o\left(\frac{t^2}{n}\right)\right)^n$$

The above sequence converges to $e^{-\frac{t^2}{2}}$ as n goes to infinity, which is the characteristic function of $N(0, 1)$.

As a result, we obtain $\varphi_{T_n/\sqrt{n}}(t) \to \varphi_{N(0,1)}(t)$, and therefore applying the continuity theorem (Theorem 6.36), we have $T_n/\sqrt{n} \xrightarrow{D} N(0, 1)$.   ∎

There is one point in the proof of the CLT which needs clarification. In order to calculate derivatives of the characteristic function, we had to change the order of two operations: the derivation and the expectation. Is this even possible? The next theorem shows that this change of order is valid under very general conditions. In particular, the exchange of derivative and integral is permitted in any finite measure space provided that the final result may be calculated.

Lemma 7.44 (Derivation under the Lebesgue Integral) *Let $f : \mathbb{R} \times E \to \mathbb{R}$, where $(E, \mathcal{K}, P)$, is a finite measure space such that*

(i) For all $t \in \mathbb{R}$ fixed, the one-dimensional function $x \mapsto f(t, x) : E \to \mathbb{R}$ is measurable and integrable.

(ii) For all $x \in E$ fixed, the one-dimensional function $t \mapsto f(t, x) : \mathbb{R} \to \mathbb{R}$ is derivable with respect to t and the derivative is $f_{1,0}(t, x)$ a continuous function. Furthermore, there exists a g integrable on $(E, \mathcal{K}, P)$ such that

$$|f_{1,0}(t, x)| \le g(x), \quad \forall t$$

Then the function

$$\mathcal{K}(t) = \int f(t, x) dP(x) = \mathbf{E}[f(t, X)]$$

is derivable and its derivative is

$$\mathcal{K}'(t) = \int f_{1,0}(t, x) dP(x)$$

Proof: We have

$$\frac{\mathcal{K}(t+h) - \mathcal{K}(t)}{h} = \frac{1}{h} \int [f(t+h, x) - f(t, x)] dP(x)$$

By the Lagrange theorem (mean value theorem) we continue

$$= \int f_{1,0}(t + \theta h, x) dP(x)$$

for some $\theta \in (0, 1)$.

Now, if we let $h \downarrow 0$ and use the continuity of $f_{1,0}$, we obtain that $f_{1,0}(t+\theta h, x) \to f_{1,0}(t, x)$. Further use of the dominated convergence theorem provides the result we need.   ∎

Basically, this theorem is extremely useful, and it lets us take derivatives under integral signs.

■ EXAMPLE 7.8

Let $X_1, X_2, \ldots, X_{100}$ be independent random variables uniformly distributed on the interval $[0, 4]$. Find the approximate numerical probability that

$$\mathbf{P}\{\prod_{i=1}^{100} X_i \leq 1\}.$$

Here, $\prod_{i=1}^{100} X_i$ denotes the product of all 100 X_i's.

Calculating probabilities for products of random variables is a hard problem. For example, the probability above would be

$$\mathbf{P}\{\prod_{i=1}^{100} X_i \leq 1\} = \int_0^1 \cdots \int_0^1 \mathbf{1}_{\{x_1 x_2 \ldots x_{100} \leq 1\}} \frac{1}{4^{100}} dx_1 dx_2 \ldots dx_{100},$$

and even the domain of integration (limits of the n-fold integral) is hard to write. However, as we have seen in the Central Limit Theorem, approximating probabilities which involve sums of random variables is much easier. So, how do we transform products into sums? Take the logarithm.

So let us take $Y_1 = \log X_1, \ldots, Y_{100} = \log X_{100}$. Then, since logarithm is an increasing function,

$$\mathbf{P}\left\{\prod_{i=1}^{100} X_i \leq 1\right\} = \mathbf{P}\left\{\log\left(\prod_{i=1}^{100} X_i\right) \leq \log 1\right\} = \mathbf{P}\left\{\sum_{i=1}^{100} Y_i \leq 0\right\}$$

But Y_i has the distribution

$$F_{Y_i}(y) = \mathbf{P}\{Y_i \leq y\} = \mathbf{P}\{X_i \leq e^y\} = \begin{cases} \int_0^{e^y} \frac{1}{4} dx, & \text{if } e^y \leq 4 \\ 1, & \text{if } e^y > 4 \end{cases} = \begin{cases} \frac{e^y}{4}, & \text{if } y \leq \log 4 \\ 1, & \text{if } y > \log 4 \end{cases}$$

Thus the density is

$$f_{Y_i}(y) = \frac{1}{4} e^y \mathbf{1}_{(-\infty, \log 4)}$$

Therefore, each Y variable has expectation

$$\mathbf{E}[Y_i] = \int_{-\infty}^{\log 4} y e^y \frac{1}{4} dy = \cdots = \log 4 - 1,$$

and variance

$$V(Y_i) = \int_{-\infty}^{\log 4} y^2 e^y \frac{1}{4} dy - (\mathbf{E}[Y_i])^2 = \cdots = 1$$

Since X_i's are independent, so are the Y_i's, and applying the CLT we obtain the approximate distribution of their average as

$$\frac{1}{100} \sum_{i=1}^{100} Y_i \sim N\left(\log 4 - 1, \frac{1}{100}\right)$$

Therefore,

$$\mathbf{P}\left\{\sum_{i=1}^{100} Y_i \leq 0\right\} = \mathbf{P}\left\{Z \leq \frac{0 - (\log 4 - 1)}{1/10}\right\}$$
$$= \Phi(10 - 10\log 4) = 5.6 \times 10^{-5} = 0.000056,$$

where Φ denotes the $N(0, 1)$ distribution function.

7.5 Extensions of Central Limit Theorem. Limit Theorems for Other Types of Statistics

As we will see in the second part of this book concerning stochastic processes, the Central Limit Theorem is a great tool and there exist variants of CLT for renewal processes, martingales, and many other types of processes. Extensions of this theorem have also been proven for m-dependent random variables. An m-dependent sequence is a sequence where the random vector $X_1, \ldots, X_s$ is independent of the vector $X_t, X_t + 1, \ldots$ for any t, s, such that $t - s > m$. Further extensions to independent sequences but not identically distributed and many other situations have been considered, and variants of CLT have been proven, primarily due to the great importance of this theorem in practical applications.

All these theorems and extensions are looking at the distribution of the sum of random variables. However, in practical applications, more often than not, the statistics of interest is not a sum. In this situation, when the random variable may not be expressed as a sum of independent random variables, we can try to use Slutsky's theorem 7.34 on page 231. An alternative approach is the δ−method presented next.

Theorem 7.45 (δ-Method) *Suppose $Y_n \overset{D}{\longrightarrow} y$, a constant. Assume that there exists a sequence of real numbers $\{a_n\}$ such that the sequence*

$$X_n = a_n(Y_n - y)$$

converges in distribution to some random variable X (i.e., $X_n \overset{D}{\to} X$). Then, for any function f differentiable, with the derivative $f'(y) \neq 0$ and continuous (i.e., $f \in C^1$) on the range of Y_n, we have

$$a_n(f(Y_n) - f(y)) \overset{D}{\longrightarrow} f'(y)X$$

Proof: Proving this theorem is immediate using the Slutsky's theorem and the continuity theorem. Specifically, since f is derivable at y for any real number y_n, we can find a $\tilde{y}_n$ such that

$$f(y_n) - f(y) = f'(\tilde{y}_n)(y_n - y),$$

where $\tilde{y}_n$ is in between the two values y_n and y. Since this is true for any real number, we therefore can find another random sequence $\tilde{Y}_n$ such that

$$f(Y_n) - f(y) = f'(\tilde{Y}_n)(Y_n - y).$$

Since $\tilde{Y}_n$ is in between the two values Y_n and y, we can express this as

$$|\tilde{Y}_n - Y| \leq |Y_n - Y|$$

for all ω and, since $Y_n \xrightarrow{P} y$ (recall y is a constant), we immediately obtain $\tilde{Y}_n \xrightarrow{P} y$. Since f' is continuous by hypothesis, we can apply the continuity theorem and we obtain

$$f'(\tilde{Y}_n) \xrightarrow{P} f'(y).$$

On the other hand, we know that $a_n(Y_n - y) \xrightarrow{D} X$ by hypothesis. Finally, applying Slutsky's theorem to the product of $f'(\tilde{Y}_n)$ with $a_n(Y_n - y)$, we obtain the result. ∎

Remark 7.46 *Note that in the case when $f'(y) = 0$, the δ theorem gives nothing. To deal with this case, if you understand the proof, we can state a second version of the theorem where we use a second-order Taylor expansion instead of using the mean value theorem. In this form, the first derivative term will disappear and we will require that the second derivative at y be continuous. Following the same steps will produce the result.*

The δ-method may be applied to any limiting distribution of the random sequence X_n. However, in the most commonly encountered case, the sequence X_n converges to a normal (usually shown by applying the CLT). We state this special case as a corollary.

Corollary 7.47 *Suppose that the random variables $Y_n \xrightarrow{D} y$, a constant, and assume that*

$$\sqrt{n}(Y_n - y) \xrightarrow{D} N(0, \sigma^2).$$

Then for any derivable function f with continuous derivative ($f \in \mathcal{C}^1$), we have

$$\sqrt{n}(f(Y_n) - f(y)) \xrightarrow{D} N\left(0, (f'(y)\sigma)^2\right)$$

■ **EXAMPLE 7.9**

Suppose $X_1, X_2, \ldots, X_n$ are i.i.d. with finite mean μ and finite variance σ^2. Find a sequence a_n and a random variable X such that

$$a_n(\bar{X}_n^2 - \mu^2) \xrightarrow{D} X,$$

where $\bar{X}_n = \frac{1}{n} \sum_{i=1}^n X_i$ is the sample mean of the first n variables.

This is very easily handled with the δ-method. Looking at the theorem, first we need to find a sequence a_n such that $a_n(\bar{X}_n - \mu)$ has a limit in distribution. But the regular CLT gives us this result immediately. Specifically, if σ^2 is finite,

$$\frac{\bar{X}_n - \mu}{\frac{\sigma}{\sqrt{n}}} = \frac{\sqrt{n}}{\sigma}(\bar{X}_n - \mu) \xrightarrow{D} N(0, 1).$$

Next, by using the δ-theorem with the function $f(x) = x^2$ which is derivable on $\mathbb{R}$ and its derivative $f'(x) = 2x$ is continuous on $\mathbb{R}$, we obtain

$$\frac{\sqrt{n}}{\sigma}(\bar{X}_n^2 - \mu^2) \xrightarrow{D} 2\mu N(0, 1).$$

Multiplying with σ and using the scale property of the normal distribution, we obtain

$$\sqrt{n}(\bar{X}_n^2 - \mu^2) \xrightarrow{D} N(0, 4\mu^2\sigma^2).$$

■ **EXAMPLE 7.10**

In the same settings as the previous exercise, show that

$$\sqrt{n}(\log \bar{X}_n - \log \mu) \xrightarrow{D} N\left(0, \frac{\sigma^2}{\mu^2}\right)$$

and

$$\sqrt{n}(\sin \bar{X}_n - \sin \mu) \xrightarrow{D} N\left(0, (\cos \mu)^2 \sigma^2\right)$$

As you see, the idea is quite simple. In the examples presented, we still use sums. This is because the first step requires us to find the limiting distribution of the estimator and for sums this step is immediate using the CLT. We shall look at different examples in a minute, but we would like to first give a multivariate version of the theorem.

The δ-method may be extended to the case when Y_n is a random vector (i.e., one estimates multiple parameters). In this case, we require f to have partial derivatives as well as regularity conditions (similar to continuity) for these derivatives. Much recent work is dedicated to requiring weaker and weaker forms of differentiability and regularity of the derivatives.

Theorem 7.48 (Multivariate δ-method) *Suppose that $Y_1, \ldots, Y_n, \ldots$ is a sequence of d-dimensional i.i.d. random vectors. Assume that for a certain fixed y there exists a sequence a_n and a d-dimensional random variable X such that*

$$X_n = a_n(Y_n - y) \xrightarrow{D} X$$

The convergence needed by the theorem is in distribution. It is expressed in terms of the d-dimensional joint distribution function of X_n converging to the d-dimensional joint distribution function of X.

Next, suppose that $f : \mathbb{R}^d \to \mathbb{R}^k$ is a differentiable function with all derivatives continuous. Then, if we denote with J_f the Jacobian matrix of f, and if

$$\|J_f(y)\|^2 = \sum \left| \frac{\partial f_i}{\partial x_j}(y) \right|^2 \neq 0,$$

the δ-method states that

$$a_n(f(Y_n) - f(y)) \xrightarrow{D} J_f(y)X,$$

again in the k-dimensional joint distribution sense.

As a reminder, we note that the Jacobian J_f is defined as

$$J_f(x) = \begin{pmatrix} \frac{\partial f_1}{\partial x_1}(x) & \frac{\partial f_1}{\partial x_2}(x) & \cdots & \frac{\partial f_1}{\partial x_d}(x) \\ \frac{\partial f_2}{\partial x_1}(x) & \frac{\partial f_2}{\partial x_2}(x) & \cdots & \frac{\partial f_2}{\partial x_d}(x) \\ \vdots & \vdots & & \vdots \\ \frac{\partial f_k}{\partial x_1}(x) & \frac{\partial f_k}{\partial x_2}(x) & \cdots & \frac{\partial f_k}{\partial x_d}(x) \end{pmatrix}$$

which is a $k \times d$-dimensional matrix, and thus the limit has the correct dimension indeed.

The proof of the multivariate δ-method is very similar to that of the one-dimensional case. The mean value approximation used in the one-dimensional case is replaced in the multidimensional case with a Jacobian approximation as follows:

$$f(Y_n) - f(y) = J_f(y)(Y_n - y) + o(\|Y_n - y\|),$$

where $o(\|Y_n - y\|)$ is our little o notation of a function that goes to 0 when divided by the Euclidian norm

$$\|Y_n - y\| = \sqrt{\sum(Y_i^n - y_i)^2}.$$

The technical condition in the theorem ensures that the $J(y) \neq 0$ and thus the approximation works without having to go into higher order terms involving the second derivatives (Hessian) of f.

Once again, we state the more popular form of the δ-theorem as a corollary.

Corollary 7.49 *If* $Y_1, \ldots, Y_n, \ldots$ *is a sequence of d-dimensional i.i.d. random vectors, and for a certain fixed y there exists a sequence* a_n *and a d-dimensional random variable X such that*

$$a_n(Y_n - y) \xrightarrow{D} MVN_d(0, \Sigma)$$

then for any $f : \mathbb{R}^d \to \mathbb{R}^k$, *differentiable function with continuous derivatives and* $\|J_f(y)\|^2 \neq 0$, *we have*

$$a_n(f(Y_n) - f(y)) \xrightarrow{D} J_f(y)MVN_d(0, \Sigma) = MVN_k\left(0, J_f(y) \Sigma J_f(y)^t\right).$$

In the corollary, $MVN_d(0, \Sigma)$ is the notation for the multivariate normal distribution with mean vector 0 and covariance matrix Σ, i.e., its density is

$$f(y) = f(y_1, \ldots, y_d) = \frac{1}{\sqrt{(2\pi)^d \det \Sigma}} e^{-y^t \Sigma y}$$

A^t is the transpose of a matrix. Also, recall that for a random vector Y and a matrix A such that the dimensions match, we have

$$Var(AY) = \mathbf{E}[(AY - A\mu_Y)(AY - A\mu_Y)^t] = A \, Var(Y) \, A^t.$$

◼ EXAMPLE 7.11

Suppose $X_1, \ldots, X_n, \ldots$ are i.i.d. random variables with mean μ, variance σ^2, and finite fourth moment $\mathbf{E}[X_i^4] = \mu_4 < \infty$. We know that the sample variance $S_n^2 = \frac{1}{n-1} \sum_1^n (X_i - \bar{X}_n)^2$ converges to σ^2. But how does it converge in distribution? How about the sample standard deviation $S_n = \sqrt{\frac{1}{n-1} \sum_1^n (X_i - \bar{X}_n)^2}$?

To study this problem, first note that S_n actually involves two statistics: the sample mean of the first n observations $\bar{X}_n$, and the sample second moment $\frac{1}{n} \sum X_i^2$. Specifically

$$S_n^2 = \frac{1}{n-1} \sum_{i=1}^n (X_i - \bar{X}_n)^2 = \frac{n}{n-1} \left(\frac{1}{n} \sum_{i=1}^n X_i^2 - \bar{X}_n^2 \right).$$

So let us investigate the limiting behavior of the random vector

$$Y_n = \begin{pmatrix} \bar{X}_n \\ \frac{1}{n} \sum_{i=1}^n X_i^2 \end{pmatrix}$$

The first component goes to a normal clearly, and since the second is also an average of the i.i.d. variables X_i^2, it also must converge to a normal of some sorts. Fortunately, the joint distribution must be a multivariate normal then, and even more fortunately such multivariate normal only depends on its first moment and the covariance structure. All these being things we can calculate.

Specifically,

$$\sqrt{n}\left(\begin{pmatrix} \bar{X}_n \\ \frac{1}{n}\sum_{i=1}^{n} X_i^2 \end{pmatrix} - \begin{pmatrix} \mu \\ \mu_2 \end{pmatrix}\right) \xrightarrow{D} \mathrm{MVN}_2\left(0, \Sigma\right)$$

where $\mu_2 = \mathbf{E}[X_1^2] = \mu^2 + \sigma^2$ is the second moment, and the matrix Σ is

$$\Sigma = \begin{pmatrix} \sigma^2 & Cov(X_1, X_1^2) \\ Cov(X_1^2, X_1) & Var(X_1^2) \end{pmatrix}$$

To see this, it is enough to calculate

$$Cov(\bar{X}_n, \frac{1}{n}\sum_{i=1}^{n} X_i^2) = \frac{1}{n^2}\sum_{i,j=1}^{n} Cov(X_i, X_j^2) = Cov(X_1, X_1^2)$$

since X_i and X_j are independent for $i \neq j$, so it is in fact a constant for all n, and a similar calculation holds for $Var(\frac{1}{n}\sum_{i=1}^{n} X_i^2)$. Now, we take the function $f(x_1, x_2) = x_2 - x_1^2$ and apply the multivariate δ-theorem. The Jacobian is

$$J_f(x) = \begin{pmatrix} -2x_1 & 1 \end{pmatrix}$$

We may calculate $Cov(X_1, X_1^2) = \mu_3 - \mu^3 - \mu\sigma^2$ and $Var(X_1^2) = \mu_4 - \mu^4 - 2\mu^2\sigma^2 - \sigma^4$, where we denoted $\mu_3 = \mathbf{E}[X_1^3]$ and $\mu_4 = \mathbf{E}[X_1^4]$ the third and fourth moments, respectively. After much algebra, we obtain

$$J_f(\mu, \mu_2)\sigma J_f(\mu, \mu_2)^t = \begin{pmatrix} -2x_1 & 1 \end{pmatrix}\begin{pmatrix} \sigma^2 & Cov(X_1, X_1^2) \\ Cov(X_1^2, X_1) & Var(X_1^2) \end{pmatrix}\begin{pmatrix} -2x_1 \\ 1 \end{pmatrix}$$

$$= \mu_4 - 4\mu\mu_3 + 6\mu^2\sigma^2 - 3\mu^4 - \sigma^4$$

To simplify the expression, we denote with $\mu_4^c = \mathbf{E}[(X - \mu)^4]$ the fourth central moment, and calculate

$$\mu_4^c = \mathbf{E}[(X - \mu)^4] = \mu_4 - 4\mu\mu_3 + 6\mu^2\sigma^2 - 3\mu^4$$

Comparing with the expression above, we see that it is simply $\mu_4^c - \sigma^4$; therefore, applying the multivariate δ-theorem for the function f, we obtain

$$\sqrt{n}\left(\frac{1}{n}\sum_{i=1}^{n} X_i^2 - \bar{X}_n^2 - \sigma^2\right) \xrightarrow{D} N(0, \mu_4^c - \sigma^4)$$

Recall the expression for S_n^2, so we can write

$$\sqrt{n}\left(S_n^2 - \sigma^2\right) = \sqrt{n}\left(\frac{n}{n-1}\left(\frac{1}{n}\sum_{i=1}^{n} X_i^2 - \bar{X}_n^2\right) - \frac{n}{n-1}\sigma^2 + \frac{1}{n-1}\sigma^2\right)$$

$$= \frac{n}{n-1}\sqrt{n}\left(\frac{1}{n}\sum_{i=1}^{n} X_i^2 - \bar{X}_n^2 - \sigma^2\right) + \frac{\sqrt{n}}{n-1}\sigma^2,$$

and an application of the Slutsky's theorem 7.34, using that $n/(n-1) \to 1$ and $\frac{\sqrt{n}}{n-1}\sigma^2 \to 0$, gives

$$\sqrt{n}\left(S_n^2 - \sigma^2\right) \xrightarrow{D} N(0, \mu_4^c - \sigma^4).$$

From here, another application of the δ-theorem for the function $f(x) = \sqrt{x}$ will give the rate of convergence for the sample standard deviation $S_n = \sqrt{\frac{1}{n-1}\sum(X_i - \bar{X})^2}$:

$$\sqrt{n}\left(S_n - \sigma\right) \xrightarrow{D} \frac{1}{2\sigma}N(0, \mu_4^c - \sigma^4) = N\left(0, \frac{\mu_4^c - \sigma^4}{4\sigma^2}\right)$$

7.6   Exchanging the Order of Limits and Expectations

This is an important issue. In many cases, we need to take the limit under the integral sign, but are we doing it correctly? In other words, when is convergence a.s. or in probability imply convergence in L^p. The most general criterion we have seen already: it is the U.I. criterion. However, in many cases this criterion is hard to verify. Some classical results implying convergence in L^1 are presented below. We have seen these results before, but we are just reiterating them at this time to fix them in your memory.

The first two results results basically require the sequence and the limit to be integrable.

Theorem 7.50 (Dominated Convergence) *If there exists a random variable Y such that* $\mathbf{E}Y < \infty$, $X_n \le Y$ *for all n, and if we have* $X_n \xrightarrow{\mathcal{P}} X$, *then* $\mathbf{E}X_n \to \mathbf{E}X$ *as well.*

In the particular case when Y is non-random, we obtain

Corollary 7.51 (Bounded Convergence) *Suppose that* $X_n \le C$, $\forall n$ *for some finite constant C. If* $X_n \xrightarrow{\vee} X$, *then* $\mathbf{E}X_n \to \mathbf{E}X$ *as well.*

In the case of monotone (increasing) convergence of nonnegative r.v.'s, we can exchange the limit and the expectation even if X is *non-integrable*.

Theorem 7.52 (Monotone Convergence) *If* $X_n \ge 0$ *and* $X_n(\omega) \uparrow X(\omega)$ *a.s., then* $\mathbf{E}X_n \uparrow \mathbf{E}X$. *This is true even if* $X(\omega) = \infty$ *for some* $\omega \in \Omega$

Remark 7.53 *You may think that, as we have increasing convergence, we must also have decreasing convergence. We indeed have this implication as well, but the result is not that useful. It requires the extra assumption* $\mathbf{E}[X_1] < \infty$. *But if we make this assumption, the exchange of limit and integral is true already from the dominated convergence theorem. If we wish to drop the extra assumption, the result is no longer true as the next example demonstrates.*

■ EXAMPLE 7.12

Let Z be a random variable such that $\mathbf{E}Z = \infty$. Take $X_1 = Z$, and, in general, $X_n(\omega) = n^{-1}Z(\omega)$. Then we have that $\mathbf{E}X_n = \infty$ for any n but $X_n \downarrow 0$ wherever Z is finite.

Practice your understanding by solving the following exercise:

Exercise 7 *Let Y_n a sequence of nonnegative random variables. Use the monotone convergence theorem to show that*

$$\mathbf{E}\left[\sum_{n=1}^{\infty} Y_n\right] = \sum_{n=1}^{\infty} \mathbf{E}[Y_n].$$

Continue by showing that, if $X \geq 0$ a.s. and A_n are disjoint sets with $\mathbf{P}(\cup_n A_n) = 1$ (partition of Ω), then

$$\mathbf{E}[X] = \sum_{n=1}^{\infty} \mathbf{E}(X\mathbf{1}_{A_n}).$$

Furthermore, show that the result applies also when $X \in L^1$.

The last result presented below is the most useful in practice; we do not require the sequence or the limit to be integrable, nor do we require a special (monotone) form of convergence. We only require the existence of a lower bound. However, the result is restrictive; it only allows exchange of the lim inf with the expectation.

Lemma 7.54 (Fatou's Lemma) *Suppose that X_n is a sequence of random variables such that there exists a $Y \in L^1$ with $X_n > Y$ for all n. Then we have*

$$\mathbf{E}\left[\liminf_{n\to\infty} X_n\right] \leq \liminf_{n\to\infty} \mathbf{E}[X_n]$$

Here,

$$\liminf_{n\to\infty} X_n = \lim_{n\to\infty}\left\{\inf_{k\geq n} X_k\right\}.$$

Problems

7.1   Show that, if $X_n \xrightarrow{L^p} X$, then $\mathbf{E}|X_n|^p \to \mathbf{E}|X|^p$.
HINT: The $\|\cdot\|_p$ is a proper norm (recall the properties of a norm).

7.2   Show that if $f : \mathbb{R} \to \mathbb{R}$ is a continuous function and $X_n \xrightarrow{a.s.} X$, then $f(X_n) \xrightarrow{a.s.} f(X)$ as well.

7.3   Write a statement explaining why the Skorohod's theorem 7.32 on page 230 does not contradict our earlier statement that convergence in distribution does not imply convergence a.s.

7.4   Let $X_1, X_2, \ldots$ be i.i.d. with mean μ and variance σ^2 and $\mathbf{E}[X_1^4] < \infty$. Let $\mu_k^c = \mathbf{E}[(X_1 - \mu)^k]$ denote the kth central moment. Then show that

$$\sqrt{n}\left(\begin{pmatrix} \bar{X}_n \\ S_n^2 \end{pmatrix} - \begin{pmatrix} \mu \\ \sigma^2 \end{pmatrix}\right) \xrightarrow{D} \mathrm{MVN}_2\left(\begin{pmatrix} 0 \\ 0 \end{pmatrix}, \begin{pmatrix} \sigma^2 & \mu_3^c \\ \mu_3^c & \mu_4^c - \sigma^4 \end{pmatrix}\right)$$

where S_n^2 is the sample variance. Note that the exercise says that the sample mean and sample variance are asymptotically independent if the third central moment is zero.

7.5   Let $(X_n)_{n \geq 1}$ be i.i.d. random variables distributed as normal with mean 2 and variance 1.

   a) Show that the sequence

$$Y_n := \frac{1}{n} \sum_{i=1}^n X_i e^{X_i}$$

   converges almost surely and in distribution and find the limiting distribution.

   b) Answer the same question for the sequence

$$Z_n := \frac{X_1 + \cdots + X_n}{X_1^2 + \cdots + X_n^2}.$$

7.6   Let $X_1, \ldots, X_n$ be independent random variables with the same distribution given by

$$\mathbf{P}(X_i = 0) = \mathbf{P}(X_i = 2) = \frac{1}{4}$$

and

$$\mathbf{P}(X_i = 1) = \frac{1}{2}.$$

Let

$$S_n = X_1 + \ldots + X_n.$$

   a) Find $\mathbf{E}(S_n)$ and $Var(S_n)$.

   b) Give a necessary and sufficient condition on $n \in \mathbb{N}$ to have

$$\mathbf{P}\left(\frac{1}{2} \leq \frac{S_n}{n} \leq \frac{3}{2}\right) \geq 0.999.$$

7.7   Consider a sequence of i.i.d. r.v.'s Bernoulli distributed on $\{-1, 1\}$. That is,

$$P(X_1 = 1) = p \text{ and } P(X_1 = -1) = 1 - p.$$

Denote

$$S_n = X_1 + \ldots + X_n.$$

a) Show that, if $p \neq \frac{1}{2}$, then the sequence $\{|S_n|\}_{n \geq 1}$ converges almost surely.

   For the rest of the problem we take $p = \frac{1}{2}$. Let $\alpha > 0$.

b) Compute

$$\mathbf{E}e^{\alpha X_1} \text{ then } \mathbf{E}e^{\alpha S_n}.$$

   Deduce that for every $n \geq 1$,

$$\mathbf{E}\left[(\cosh \alpha)^{-n} e^{\alpha S_n}\right] = 1.$$

c) Show that

$$\lim_{n \to \infty} \left[(\cosh \alpha)^{-n} e^{\alpha S_n}\right] = 0$$

   almost surely.

7.8   Suppose $X_1, X_2, \ldots$ and $Y_1, Y_2, \ldots$ are two sequences of random variables, each sequence i.i.d., and suppose that the random variables X_1 and Y_1 are correlated with correlation coefficient ρ, while X_i and Y_j are independent if $i \neq j$. With the usual notations $\bar{X}_n, \bar{Y}_n$, denote $S_n(X) = \sqrt{\frac{1}{n} \sum_{i=1}^{n} (X_i - \bar{X}_n)^2}$ and $S_n(Y) = \sqrt{\frac{1}{n} \sum_{i=1}^{n} (Y_i - \bar{Y}_n)^2}$. The sample correlation coefficient is defined as

$$r_n = \frac{\frac{1}{n} \sum_{i=1}^{n} (X_i - \bar{X}_n)(Y_i - \bar{Y}_n)}{S_n(X) S_n(Y)} = \frac{\frac{1}{n} \sum_{i=1}^{n} X_i Y_i - \bar{X}_n \bar{Y}_n}{S_n(X) S_n(Y)}.$$

Show that there exists some ν constant such that

$$\sqrt{n}(r_n - \rho) \xrightarrow{D} N(0, \nu^2).$$

Show that in the particular case when the pairs (X_i, Y_i) are i.i.d. multivariate normal,

$$\begin{pmatrix} X_i \\ Y_i \end{pmatrix} \sim \mathrm{MVN}_2 \left(\begin{pmatrix} \mu_X \\ \mu_Y \end{pmatrix}, \begin{pmatrix} \sigma_X^2 & \rho \sigma_X \sigma_Y \\ \rho \sigma_X \sigma_Y & \sigma_Y^2 \end{pmatrix} \right)$$

The calculation of ν is explicit and

$$\sqrt{n}(r_n - \rho) \xrightarrow{D} N(0, (1 - \rho^2)^2).$$

7.9   Refer to Example 7.7. Let U_i be independent identically distributed random variables uniformly distributed on the interval $[0, 1]$, $i = 1, \ldots, n$. Define $U_{(n)} = \max\{U_1, \ldots, U_n\}$. Show that

$$n(1 - U_{(n)}) \xrightarrow{D} \mathrm{Exp}(1),$$

where $\text{Exp}(1)$ denotes an exponentially distributed random variable with mean 1. Find a sequence a_n and a random variable Z such that

$$\log U_{(n)} \xrightarrow{D} Z.$$

7.10   Consider $\{Y_n\}_{n \geq 1}$ a sequence of independent identically distributed random variables with common distribution $U([0, 1])$ (uniform on the interval). For every $n \geq 1$, we define

$$Y^n_{(1)} = \min(Y_1, \ldots, Y_n)$$

and

$$Y^n_{(n)} = \max(Y_1, \ldots, Y_n).$$

the first and last order statistics of the sequence, respectively. Let $F_n(x)$, $G_n(x)$ denote respectively the cdfs of these random variables.

  a) Show that for every x the sequence $F_n(x)$ converges, and find the limit. Use this result to conclude that the sequence of random variables $Y^n_{(1)}$ converges in distribution to 0.
  b) Show that for every x the sequence $G_n(x)$ converges, and find the limit. Use this result to conclude that the sequence of random variables $Y^n_{(n)}$ converges in distribution to 1.

7.11   Suppose X_i are i.i.d. $\text{Exp}(1)$. Find the limiting distribution of

$$\sqrt{n}(\bar{X}^2 - X_{(1)} - 1),$$

where $X_{(1)}$ is the first-order statistics.

7.12   Let $\bar{X}$ and S denote the sample mean and standard deviation of the random variables $X_1, X_2, \ldots$. Find the limiting distribution of $\frac{\bar{X}}{S}$ and $\frac{S}{\bar{X}}$ in each of the following cases:

1. The random variables are uniform on $(0, 1)$,

2. The random variables are exponential with parameter λ, i.e., their density is:

$$f(x) = \lambda e^{-\lambda x} \mathbf{1}_{\{x>0\}}$$

,

3. The random variables are chi-squared with parameter p an integer, that is,

$$f(x) = \frac{1}{2^{p/2}\Gamma(p/2)} x^{p/2} e^{-x/2} \mathbf{1}_{\{x>0\}},$$

and $\Gamma(t) = \int_0^\infty x^{t-1} e^{-x} dx$ denotes the Gamma function.

7.13   Let X_n, $n \geq 1$ be i.i.d. random variables with some distribution with finite mean μ and variance σ^2. For every $n \geq 1$, we denote

$$Z_n := \left(\prod_{i=1}^{n} e^{X_i} \right)^{\frac{1}{n}}.$$

Find the almost sure limit of the sequence $\{Z_n\}_{n \geq 1}$.

7.14   Let $\{X_n\}_{n \geq 1}$ be a sequence of i.i.d. random variables with common distribution $N(\mu, \sigma^2)$, where $\mu \in \mathbb{R}$ and $\sigma > 0$. For every $n \geq 1$, denote

$$S_n = \sum_{k=1}^{n} X_k.$$

    a) Give the distribution of S_n.

    b) Compute
$$\mathbf{E}\big((S_n - n\mu)^4 \big).$$

    c) Show that for every $\varepsilon > 0$,

$$\mathbf{P}\left(\left| \frac{S_n}{n} - \mu \right| \geq \varepsilon \right) \leq \frac{3\sigma^4}{n^2 \varepsilon^4}.$$

    d) Derive that the sequence $\left\{ \frac{S_n}{n} \right\}_{n \geq 1}$ converges almost surely and identify the limit.

    e) Do you recognize this result?

7.15   Suppose $X_1, X_2, \ldots$ are Poisson(λ). Find the limit distribution of of $e^{\bar{X}}$ and $\left(1 - \frac{1}{n} \right)^{n\bar{X}}$. Also, find the limiting distribution of $\frac{1}{\bar{X} + \bar{X}^2 + \bar{X}^3}$

7.16   Let $\{X_i\}_{i \geq 1}$ be i.i.d. random variables. Assume that the sums $S_n = \sum_{i=1}^{n} X_i$ have the property $S_n/n \to 0$ almost surely as $n \to \infty$. Show that $\mathbf{E}[|X_1|] < \infty$ and therefore $\mathbf{E}[X_1] = 0$.

7.17   Let $(X_n)_{n \geq 1}$ be i.i.d. random variables. Suppose that

$$\mathbf{P}(X_1 = 0) < 1.$$

    a) Show that there exists $\alpha > 0$ such that

$$\mathbf{P}(|X_1| \geq \alpha) > 0.$$

    b) Show that
$$P(\limsup_n \{|X_n| \geq \alpha\} = 1.$$

    Derive that
$$P(\limsup_n X_n = 0) = 0.$$

c) Show that

$$\mathbf{E}|X_1| = \int_0^\infty P(|X_1| > t)dt.$$

Deduce that X_1 is integrable if and only if

$$\sum_k P(|X_1| > k) < \infty.$$

Show that X_1 is integrable if and only if

$$P(\limsup_n \{X_n > n\}) = 0.$$

Deduce that, if X_1 is integrable, then almost surely

$$\limsup_n \frac{1}{n} X_n \le 1.$$

7.18 Let X_n be i.i.d. random variables with $\mathbf{E}X_1 = 0$ and assume that X_n are bounded. That is, there exists a $C > 0$ such that

$$|X_1| \le C \text{ a.s.}$$

Show that for every $\varepsilon > 0$,

$$P\left(\left|\frac{S_n}{n}\right| \ge \varepsilon\right) \le 2\exp\left(\left(-n\frac{\varepsilon^2}{2c^2}\right)\right).$$

Deduce that S_n/n converges in probability to zero.

7.19 Let $X, X_1, \ldots, X_n, \ldots$ be i.i.d. with

$$P(X = 2^k) = \frac{1}{2^k}$$

for $k = 1, 2, \ldots$.
   a) Show that $\mathbf{E}X = \infty$.
   b) Prove that

$$\frac{S_n}{n} \longrightarrow \frac{1}{\log 2}$$

in probability as $n \to \infty$.

7.20 Let $X \sim N(100, 15)$. Find four numbers x_1, x_2, x_3, x_4 such that

$$P(X < x_1) = 0.125 \qquad P(X < x_2) = 0.25$$

and

$$P(X < x_3) = 0.75 \qquad P(X < x_4) = 0.875.$$

7.21   A fair coin is flipped 400 times. Determine the number x such that the probability that the number of heads is between $200 - x$ and $200 + x$ is approximately .80.

7.22   In an opinion poll, it is assumed that an unknown proportion p of people are in favor of a proposed new law and a proportion $1 - p$ are against it. A sample of n people is taken to estimate p. The sample proportion $\hat{p}$ of people in favor of the law is taken as an estimate of p. Using the Central Limit Theorem, determine how large a sample will ensure that the estimate $\hat{p}$ will, with probability .95, be within .01 of the true p.

CHAPTER 8

STATISTICAL INFERENCE

We conclude the probability part of this book with a chapter dedicated to statistical applications. This chapter is not meant to replace a detailed statistical book; instead, we present some common methods that are very useful for applications and for stochastic processes. This chapter is essentially dedicated to parameter estimation methods. For a study of the relationship between two or more variables, we direct the reader to a book such as Kutner, Nachtsheim, Neter, and Li (2004).

8.1 The Classical Problems in Statistics

In general, in statistics we work with a random sample. We recall here the definition 7.36 on page 233 of a simple random sample.

Suppose that we work on a probability space $\Omega, \mathcal{F}, \mathbf{P}$. A **simple random sample** of size n is a set of independent identically distributed (i.i.d.) random variables $\mathbf{X} = (X_1, X_2, \ldots, X_n)$. The outcome of the sample, or simply a sample, is a set of realizations $(x_1, x_2, \ldots, x_n)$ of these random variables. This random sample is called *simple* because the components are independent. Obviously, in this case the joint distribution is written as the product of marginal distributions. Typically, in statistics all the random variables in the sample have a distribution which depends on a

Probability and Stochastic Processes, First Edition. Ionuț Florescu
© 2015 John Wiley & Sons, Inc. Published 2015 by John Wiley & Sons, Inc.

parameter θ. More often than not, the random variables have in fact a density which is symbolically written as $f(x|\theta)$ to express the fact that the density depends on the parameter θ. The parameter may in fact be a vector; for example, a normal density $N(\mu, \sigma^2)$ is characterized by the parameter vector $\theta = (\mu, \sigma)$ or $\theta = (\mu, \sigma^2)$.

Clearly, for a simple random sample $\mathbf{X} = (X_1, X_2, \ldots, X_n)$ the density is

$$f_{\mathbf{X}}(\mathbf{x}) = f_{\mathbf{X}}(x_1, \ldots, x_n|\theta) = f(x_1|\theta)f(x_2|\theta)\ldots f(x_n|\theta).$$

If the random sample does not have independent components, we cannot simplify in this way the joint density. Such cases arise as we shall see in the stochastic processes part when we observe a path of a process (i.e., X_1 is the value of the process at time $t = 1$, etc.). However, for the purposes of illustration, in this chapter a random sample will always have i.i.d. components, unless clearly specified otherwise.

The most important problem in statistics is the problem of estimation of the parameter θ. Specifically, we are concerned with the following questions:

- How do we calculate an estimator for θ?

- If an estimator is obtained, how good is the estimator?

- Can we tell by just observing the sample whether the parameter is equal to a specific value θ_0 or not?

In order, these questions are the problem of finding estimators, the confidence interval problem, and the testing problem. We will give details about these problems in this chapter.

8.2   Parameter Estimation Problem

We begin by giving a general definition on how we properly calculate things using only the sample.

A **statistics** is any random variable Y obtained from the random sample, i.e.,

$$Y = f(X_1, X_2, \ldots, X_n),$$

where f is any function $f : \mathbb{R}^n \to \mathbb{R}$.

Examples of commonly used statistics are given on page 233. Note that the main requirement for a statistic is to be able to calculate it by just observing the sample outcomes. For something which is not a statistics, let me give the following example:

Let $X_1, \ldots, X_n$ be i.i.d. as a t_ν random variable. The parameter here is $\theta = \nu$. Consider the expression

$$\frac{1}{\nu} \sum_{i=1}^{n} X_i^2.$$

This is not a statistics, and in fact it is useless for estimation purposes. To calculate a value we need to know what ν is, but that is the parameter we need to find in the first place.

To motivate the next concepts, consider the problem of estimating ν above and also consider the statistics $Y = 2$. Clearly, this Y is a statistics according to the definition (all constants are). However, this statistic is useless for the purposes of estimating ν since it does not use change regardless of the values in the random sample. We need some criteria which will allow us to decide whether some statistics are better than others.

Definition 8.1 (Unbiased Estimator) *On a probability space $(\Omega, \mathcal{F}, \mathbf{P})$ let $\mathbf{X} = (X_1, \ldots, X_n)$ denote a random sample. Suppose that*

$$Y = f(X_1, X_2, \ldots, X_n)$$

is some statistics, and suppose that θ is some parameter of the density of $\mathbf{X}$. We say that Y is an unbiased estimator of θ if

$$\mathbf{E}[Y] = \theta.$$

An unbiased estimator has a distribution which has mean θ. This type of estimators on average tend to be around theta due to the law of large numbers. The next lemma gives some classical examples of unbiased estimators.

Lemma 8.2 *Let $\mathbf{X} = (X_1, \ldots, X_n)$ be a random sample from some distribution with mean μ and variance σ^2. Then the following hold for the statistics: sample mean:*

$$\bar{X} = \frac{1}{n} \sum_{i=1}^{n} X_i,$$

and sample variance:

$$S^2 = \frac{1}{n-1} \sum_{i=1}^{n} (X_i - \bar{X})^2.$$

(i) $\mathbf{E}[\bar{X}] = \mu,$

(ii) $V[\bar{X}] = \frac{\sigma^2}{n},$

(iii) $\mathbf{E}[S^2] = \sigma^2.$

Proof: The proof is a simple exercise. For example ,part (iii) is

$$\mathbf{E}[S^2] = \mathbf{E}\left[\frac{1}{n-1}\sum_{i=1}^{n}(X_i - \bar{X})^2\right] = \frac{1}{n-1}\mathbf{E}\left[\sum_{i=1}^{n}X_i^2 - n\bar{X}^2\right]$$

$$= \frac{1}{n-1}\left[\sum_{i=1}^{n}\mathbf{E}[X_i^2] - n\mathbf{E}[\bar{X}^2]\right]$$

$$= \frac{1}{n-1}\left[\sum_{i=1}^{n}\left(Var(X_i) + \mathbf{E}[X_i]^2\right) - n\left(Var(\bar{X}) + \mathbf{E}[\bar{X}]^2\right)\right]$$

$$= \frac{n}{n-1}\left(\sigma^2 + \mu^2 - \frac{\sigma^2}{n} - \mu^2\right) = \sigma^2.$$

■

Clearly, Lemma 8.2 shows that the sample mean and the sample variance are unbiased estimators for the mean parameter and the variance parameter regardless of the distribution.

Definition 8.3 (Consistent Estimator) *A sequence $\hat{\theta}_n$ of estimators of some parameter θ is called* consistent *if $\hat{\theta}_n$ converges almost surely (a.s.) (strong convergence) or in probability (weak convergence) to θ.*

Consistency is a very important property. It assures us that, as the sample size increases, we obtain values closer and closer to the target we want to reach (the parameter θ). The estimators in the previous lemma are all consistent (strongly), which can be easily proven using the (strong) law of large numbers. In particular, if

$$\hat{\theta}_n = \frac{1}{n}\sum_{i=1}^{n}X_i$$

with i.i.d. observations, then clearly by the law of large numbers

$$\hat{\theta}_n \to \mu, \text{a.s.},$$

where $\mu = \mathbf{E}[X_i]$.

8.2.1 The case of the normal distribution, estimating mean when variance is unknown

As we saw from Lemma 8.2, the sample mean $\bar{X} = \frac{1}{n}\sum_{i=1}^{n}X_i$ of a random sample $\mathbf{X} = (X_1, \ldots, X_n)$ is a good candidate for estimating the mean of the distribution μ. Suppose that X_i's are all normally distributed with mean μ and variance σ^2. When X_i's are normal, we can show that for any n

$$\frac{\bar{X} - \mu}{\sigma/\sqrt{n}} \sim N(0, 1).$$

See Problem 8.2.

There is an issue with using this result. To estimate and drawing conclusions for μ using it, we need to know the σ parameter. This is of course not possible in reality and, instead, we need to replace it with the sample variance $S^2 = \frac{1}{n-1} \sum_{i=1}^{n} (X_i - \bar{X})^2$. However, this will change the distribution of the potential statistic. Since we can write

$$\frac{\bar{X} - \mu}{S/\sqrt{n}} = \frac{\bar{X} - \mu}{\sigma/\sqrt{n}} \frac{1}{\sqrt{S^2/\sigma^2}},$$

and the distribution of $\frac{S^2}{\sigma^2}$ is distributed as a $\frac{1}{n-1}\chi^2_{n-1}$ random variable (see Problem 8.2), then in the case when the variance of the sample is unknown we need to look at the distribution of the following random variable:

$$\frac{U}{\sqrt{V/p}},$$

where $U \sim N(0,1)$ and $V \sim \chi^2_p$ are random variables. William Sealy Gosset (1876–1937) studied this distribution. Upon graduating in 1899, he joined the brewery of Arthur Guinness & Son in Dublin, Ireland. Gosset worked on this test because his sample size was small and the brewery was interested in improving the quality of the beer produced based on the sample which was typically small. Another researcher at Guinness had previously published a paper containing the trade secrets of the Guinness brewery. To prevent further disclosure of confidential information, Guinness prohibited its employees from publishing any papers regardless of the contained information. However, after pleading with the brewery and explaining that his mathematical and philosophical conclusions were of no possible practical use to competing brewers, he was allowed to publish them, but under a pseudonym ("Student"), to avoid difficulties with the rest of the staff. Thus his most noteworthy achievement is now called Student's, rather than Gosset's, t-distribution.

Gosset's work was published in the 1908 paper in Biometrika. In it Gosset shows that (using the notation above) $\frac{U}{\sqrt{V/p}}$ has a t-distribution with p degrees of freedom. A random variable X distributed as a t random variable with p degrees of freedom (denoted $X \sim t_p$) has the following density:

$$f_{t_p}(x) = \frac{\Gamma(\frac{p+1}{2})}{\sqrt{p\pi}\,\Gamma(\frac{p}{2})} \left(1 + \frac{x^2}{p}\right)^{-\frac{p+1}{2}},$$

where Γ is the gamma function.

If $p = 1$, we obtain the special case of the Cauchy distribution. A random variable distributed as t_p has only $p - 1$ moments, therefore such random variable does not have a moment generating function. If $X \sim t_p$, then

$$\mathbf{E}[X] = 0, \text{ for all } p \geq 2 \text{ (does not exist for) } p = 1;$$

$$V[X] = \frac{p}{p-2} \text{ for all } p \geq 3 \text{ (does not exist for) } p = 1, 2.$$

With this result, one can easily estimate the mean of the population when each individual is normally distributed and when the variance is unknown. For example, from Gosset's result, if $X_1, \ldots, X_n$ is the sample, then

$$\frac{\bar{X} - \mu}{S/\sqrt{n}} \sim t_{n-1}.$$

So to obtain a 95% confidence interval for the mean μ, one can rewrite this expression to obtain

$$\bar{X} + t_{n-1}(0.025)\frac{S}{\sqrt{n}} \leq \mu \leq \bar{X} + t_{n-1}(0.975)\frac{S}{\sqrt{n}},$$

where the notation $t_{n-1}(0.025)$ and $t_{n-1}(0.975)$ represent the quantiles of the t_{n-1} distribution. For example, suppose that we have 10 barrels of beer with average alcohol content $\bar{X} = 7.8$ and sample variance $S^2 = 4.5$. In R, we calculate a 95% confidence interval for the average alcohol content of the entire batch of barrels that year (which could be thousands, but of course we cannot open them):

```
> 7.8+qt(0.025,df=10−1)*sqrt(4.5)/(10−1)
[1] 7.266804
> 7.8+qt(0.975,df=10−1)*sqrt(4.5)/(10−1)
[1] 8.333196
```

so we get information about the average alcohol content for that year. Note that the numbers are completely fictitious.

8.2.2   The case of the normal distribution, comparing variances

Suppose that we have two samples now, for example, $X_1, \ldots, X_n$ from a normal with mean μ_X and variance σ_X^2, and $Y_1, \ldots, Y_m$ from a normal with mean μ_Y and variance σ_Y^2.

The goal here is to compare the two variabilities and decide if population 1 (Xs) have the same variability as population 2 (Ys). To do this, we need to calculate the distribution of the following random variable:

$$F = \frac{S_X^2/\sigma_X^2}{S_Y^2/\sigma_Y^2}.$$

Note that both the numerator and denominator have χ^2 distributions with different degrees of freedom. So, essentially the distribution of this random variable F is the distribution of the ratio of two χ^2 random variables. It turns out that the random variable has an F distribution with $n-1$ and $m-1$ degrees of freedom (denoted $F(n-1, m-1)$). The probability distribution function (pdf) is

$$f_F(x) = \frac{\left(\frac{n-1}{m-1}\right)^{\frac{n-1}{2}}}{\beta\left(\frac{n-1}{2}, \frac{m-1}{2}\right)} x^{\frac{n-1}{2}-1} \left(1 + \frac{n-1}{m-1}x\right)^{-\frac{n+m-2}{2}},$$

where β denotes the Beta function.

This distribution is also known as Snedecor's F distribution or the Fisher–Snedecor distribution after R.A. Fisher and George W. Snedecor who both discovered this important distribution coming from two different directions.

Here is a theorem describing this distribution in relationship with other distributions.

Theorem 8.4 *The following relations between distributions are true:*

1. *If $X \sim F(p, q)$, then $\frac{1}{X} \sim F(q, p)$.*

2. *If $X \sim t_p$, then $X^2 \sim F(1, p)$.*

3. *If $X \sim F(p, q)$, then*

$$\frac{\frac{p}{q} X}{1 + \frac{p}{q} X} \sim \beta\left(\frac{p}{2}, \frac{q}{2}\right).$$

8.3 Maximum Likelihood Estimation Method

As we saw in the previous section, in the case when we have to estimate the expected value or the variance of the population, all we have to do is to work with the sample average $\bar{X}$ and the sample variance S^2. Not only that, but in the case when the observations are normally distributed, we can in fact obtain the distribution of these statistics exactly.

What if we want to estimate parameters which are not the mean or variance?

EXAMPLE 8.1

Suppose $X_1, \ldots, X_n$ is a sample of n i.i.d. random variables from a Cauchy distribution with parameter θ:

$$f(x|\theta) = \frac{1}{\pi(1 + (x - \theta)^2)},$$

where we wrote $f(x|\theta)$ to emphasize the pdf dependence on the parameter θ and the fact that we need it to be able to write down the actual distribution.

Suppose that we want to estimate the parameter θ. Note that the mean of the distribution does not in fact exist ($= \infty$), so it is not clear if and how we can use the methods for the normal.

The maximum likelihood estimation method is one of the most widely used method of finding parameters. Suppose $X_1, X_2, \ldots, X_n$ are i.i.d. with a density $f(x|\theta)$. Here, θ is the parameter or the vector of parameters which we need to estimate. First, we construct the so-called likelihood function

$$L(\theta) = f(x_1, \ldots, x_n|\theta) = \prod_{i=1}^{n} f(x_i|\theta).$$

Note that this is simply the joint density function but calculated using all the observed points $x_1, \ldots, x_n$. Since we substitute the observations, the function is simply a function of θ. The maximum likelihood estimator of θ is then obtained by simply maximizing $L(\theta)$, i.e., it is the value of θ that achieves the maximum of $L(\theta)$.

In many, many cases, by taking the logarithm in the likelihood function we obtain a much simpler expression to maximize. This is evident if we just look at the general formula

$$l(\theta) = \log L(\theta) = \sum_{i=1}^{n} \log f(x_i|\theta).$$

Sometimes, $\log L$ is a simpler function to deal with than L. The function l is called the **log-likelihood function**.

Since logarithm is an increasing function, maximizing $L(\theta)$ is the same as maximizing $l(\theta)$. We should always choose the expression which is simpler to deal with.

The value of θ which maximizes the log-likelihood function or the likelihood function (it is the same number) is called the *maximum likelihood estimator* or in short MLE. Note that the value must be a global maximum and not a local one. This value will be typically denoted with a hat symbol on top, for example the MLE estimator of θ is denoted using $\hat{\theta}$.

Recall that the maximum of a function is oftentimes found by calculating the derivative of the function and setting it equal to zero. Because of this principle, a third function is important:

$$U(\theta) = \frac{\partial l}{\partial \theta}(\theta),$$

which is the gradient of the log-likelihood function l. This function is called *the score function*. Recall that the gradient of a function l is the vector of derivatives of l with respect to all variables in the function.

Besides being a straightforward approach, the MLE has very nice statistical properties. If something is an MLE, it will converge to the parameter it attempts to estimate. We first look at some examples.

EXAMPLE 8.2

Suppose that in a trial we obtain the following numbers representing a random sample of 10 bricks selected from a production line:

$x_1 = 6.161520, x_2 = 4.892259, x_3 = 6.342859, x_4 = 4.673576, x_5 = 4.936554,$
$x_6 = 3.966134, x_7 = 4.351286, x_8 = 6.010079, x_9 = 4.944755, x_{10} = 5.691312.$

We assume that the distribution of each brick is a Cauchy random variable with location parameter θ:

$$f(x|\theta) = \frac{1}{\pi(1 + (x - \theta)^2)}.$$

We want to estimate the parameter based on these observations.

We will use the notation $x_1, \ldots, x_n$ since the problem may be solved for any measurements and any number of sampled bricks. We may write the likelihood function as

$$L(\theta) = \prod_{i=1}^{n} \frac{1}{\pi(1 + (x_i - \theta)^2)}.$$

To maximize this function, we will need to take the derivative with respect to θ and set it equal to 0. This is quite unpleasant because of the product, so it makes more sense to use the log-likelihood function $l(\theta)$:

$$l(\theta) = \sum_{i=1}^{n} \log \frac{1}{\pi(1 + (x_i - \theta)^2)} = -\sum_{i=1}^{n} \log(1 + (x_i - \theta)^2) - n \log \pi.$$

To maximize this expression, we take the derivative with respect to θ and set it equal to 0. We obtain the score function

$$U(\theta) = l'(\theta) = \sum_{i=1}^{n} \frac{2(x_i - \theta)}{1 + (x_i - \theta)^2}.$$

Now it remains to find θ that makes this expression 0, and then show that for that θ the function $l(\theta)$ attains its maximum. Unfortunately, this problem cannot be solved analytically; instead, one would need to solve it numerically. Here is a simple algorithm that works when the parameter is one dimensional and when we know an interval which contains the parameter.

8.3.1   The bisection method

The bisection method is a very simple method. It is based on the fact that, if the function f has a single root at x_0, then the function changes the sign from negative to positive. Suppose there exists a neighborhood of x_0 which contains only that one root x_0. Take any two points a and b in the neighborhood. Then, since the function changes sign at x_0 if x_0 is between a and b, the values $f(a)$ and $f(b)$ will have opposite signs. Therefore, the product $f(a)f(b)$ will be negative. However, if x_0 is not between a and b, then $f(a)$ and $f(b)$ will have the same sign (be it negative or positive) and the product $f(a)f(b)$ will be positive. This is the main idea of the bisection method.

The algorithm starts by finding two points a and b such that the product $f(a)f(b)$ is negative. In the case of implied volatility, the suggested starting points are 0 and 1.

Pseudo-Algorithm:

Step 1: Check if the distance between a and b (i.e., $b - a$) is less than some tolerance level ε. If Yes, STOP report either point you want (a or b). If No, step further.

Step 2: Calculate $f(\frac{a+b}{2})$. Evaluate $f(a)f(\frac{a+b}{2})$ and $f(\frac{a+b}{2})f(b)$.

Step 3: If $f(a)f(\frac{a+b}{2}) < 0$, make $b \leftarrow \frac{a+b}{2}$. Go to Step 1.

Step 4: If $f(\frac{a+b}{2})f(b) < 0$, make $a \leftarrow \frac{a+b}{2}$. Go to Step 1.

Think about what this algorithm does. Obviously, if there is more than one root in the starting interval $[a, b]$ or the root is a multiple root, the algorithm fails.

There are many improvements of this method and many other variants; for more references, consult a numerical analysis book. We will let the reader apply the bisection method to find the MLE of θ in the Cauchy distribution example (Problem 8.4 on page 286).

Remark 8.5 *There is one other problem with this density. The log-likelihood estimator is the point of global maximum of the likelihood function. However, for the Cauchy distribution in particular and for other ones in general, there are multiple roots of the derivative. Oftentimes, these are local maxima or minima, and thus caution has to be exercised especially when approximating them using numerical methods.*

■ EXAMPLE 8.3

In the previous example, we wanted to provide a realistic example because in most real applications estimators are found using some sort of optimization technique. In this example we offer a more classical case where the MLE is found explicitly.

Suppose that in a population the true percentage of people who would vote for the candidate X is p. A sample of n people is taken and asked the same question "Would you vote for Candidate X at the next elections?" Each result is recorded as a 1 if the answer is Yes, and 0 if the answer is no. What is the MLE for p?

Let each answer be denoted with x_i. A reasonable assumption about the distribution of X_i's is that they are Bernoulli(p). We can write the density of each as

$$f(x_i|p) = p^{x_i}(1-p)^{1-x_i}.$$

Verify that this indeed gives the right distribution. Therefore, the joint distribution, and thus the likelihood function, is

$$L(p) = \prod_{i=1}^{n} \left(p^{x_i}(1-p)^{1-x_i} \right) = p^{\sum_{i=1}^{n} x_i}(1-p)^{n-\sum_{i=1}^{n} x_i}.$$

Obviously, it is easier to work with the log likelihood:

$$l(p) = \sum_{i=1}^{n} x_i \log p + \left(n - \sum_{i=1}^{n} x_i \right) \log(1-p).$$

Taking derivative with respect to p gives the score function

$$U(p) = \frac{\partial l}{\partial p}(p) = \frac{1}{p}\sum_{i=1}^{n}x_i - \frac{1}{1-p}\left(n - \sum_{i=1}^{n}x_i\right)$$

$$= \frac{\sum_{i=1}^{n}x_i - np}{p(1-p)}.$$

Setting it equal to 0 and solving for p gives

$$p = \frac{1}{n}\sum_{i=1}^{n}x_i,$$

which is simply the sample proportion who voted "Yes". All that remains is to show that the likelihood function is maximized in this value which will make $\frac{1}{n}\sum_{i=1}^{n}x_i$ the MLE for p. However, looking at the derivative it is very clear that the log likelihood is increasing for $p \in [0, \frac{1}{n}\sum_{i=1}^{n}x_i)$ and decreasing on $(\frac{1}{n}\sum_{i=1}^{n}x_i, 1]$.

■ EXAMPLE 8.4

We next give a classical example: the case of a normal distribution. Suppose we have $X_1, X_2, \ldots, X_n$ i.i.d. $N(\mu, \sigma^2)$. We need to estimate the parameter vector $\theta = (\theta_1, \theta_2) = (\mu, \sigma^2)$.

To do so using the maximum likelihood method, we form the likelihood function for a particular sample $x_1, \ldots, x_n$.

$$L(\theta_1, \theta_2) = \frac{1}{\sqrt[n]{2\pi\sigma^2}}e^{-\sum_{i=1}^{n}\frac{(x_i-\mu)^2}{2\sigma^2}} = \frac{1}{\sqrt[n]{2\pi\theta_2}}e^{-\sum_{i=1}^{n}\frac{(x_i-\theta_1)^2}{2\theta_2}}$$

Clearly, this function is hard to work with, so we calculate the log-likelihood function

$$l(\theta_1, \theta_2) = -\frac{n}{2}(\log 2\pi + \log\theta_2) - \frac{1}{2\theta_2}\sum_{i=1}^{n}(x_i - \theta_1)^2.$$

The score function (gradient) is

$$U(\theta) = \frac{\partial l}{\partial\theta}(\theta) = \begin{pmatrix} \frac{1}{\theta_2}\sum_{i=1}^{n}(x_i - \theta_1) \\ \frac{1}{2\theta_2^2}\sum_{i=1}^{n}(x_i - \theta_1)^2 - \frac{n}{2\theta_2}. \end{pmatrix}$$

Setting equal to 0 and solving gives the MLEs as

$$\hat{\mu} = \bar{x} = \frac{1}{n}\sum_{i=1}^{n}x_i$$

$$\hat{\sigma}^2 = \frac{1}{n}\sum_{i=1}^{n}(x_i - \bar{x})^2. \tag{8.1}$$

To verify that indeed this is a maximum, we can calculate the Hessian or the matrix of the second derivatives and check that it is negative definite (eigenvalues are negative). That will prove that the estimates are a maximum point.

However, instead, we will plot the surface that represents the log-likelihood function. In R, we generate a sample of points with known distribution (for exemplification we take normal points with mean $\mu = 2$ and standard deviation $\sigma = 3$). We plot the log-likelihood function in Figure 8.1 with $\hat{\mu}$ on the x-axis and $\hat{\sigma}$ on the y-axis. The three different surfaces correspond to three different sample sizes.

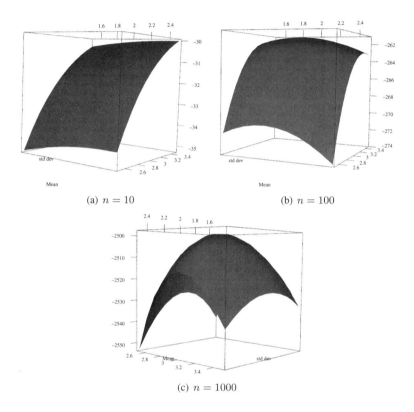

(a) $n = 10$ (b) $n = 100$

(c) $n = 1000$

Figure 8.1 The log-likelihood function for 10 observations as a function of the parameters μ and σ

We use a common domain for all when $\mu \in [2.5, 3.5]$ and $\sigma \in [3.5, 4.5]$. We clearly observe that the surfaces look different. This is easy to understand if we realize that, as n increases, the curvature of the function becomes greater and that makes it easier to calculate the MLE. Next we observe that the maximum of the functions is not attained in the parameter values $\mu = 2$ and $\sigma = 3$. Recall that, in fact, the MLE is attained at the sample mean and sample standard deviation and these can vary quite a bit from the parameter values especially if the sample size is small. In fact, for the first case plotted ($n = 10$), the sample mean is 2.396997 and the sample estimate of

MAXIMUM LIKELIHOOD ESTIMATION METHOD **271**

the standard deviation is 5.275858. This point does not belong to the domain plotted, this is why the function does not seem to have picked yet (because, in fact, it didn't).

Also note that in the function we have estimated $\hat{\theta}_2 = \hat{\sigma}^2$ the variance and not the standard deviation. So can I derive the MLE for the standard deviation σ from $\hat{\sigma}^2 = \frac{1}{n}\sum_{i=1}^{n}(x_i - \bar{x})^2$? The natural one would seem to be $\hat{\sigma} = \sqrt{\frac{1}{n}\sum_{i=1}^{n}(x_i - \bar{x})^2}$. The following theorem tells us that indeed this is the correct way to estimate σ. HOWEVER, THIS ONLY HOLDS FOR THE MLE NOT FOR THE OTHER METHODS OF ESTIMATION!!!

■ **EXAMPLE 8.5   Uniform Distribution**

Suppose $X_1, \ldots, X_n$ is a random sample from the uniform distribution $U(a, b)$. Find MLE estimates for a and b.

The likelihood function is

$$L(a, b) = \prod_{i=1}^{n} \frac{x_i - a}{b - a}.$$

The log likelihood is

$$l(a, b) = \sum_{i=1}^{n} \log \frac{x_i - a}{b - a}.$$

To maximize the function, we proceed as usual by calculating the score function

$$U(a, b) = \begin{pmatrix} \sum_{i=1}^{n} \frac{x_i - b}{(x_i - a)(b - a)} \\ -\frac{n}{b - a} \end{pmatrix}.$$

However, note that the score function is negative in both variables (recall $x_i \in (a, b)$ for all i), so we will get nothing if we set it equal to 0. However, since the derivatives in both a and b are negative, we need to look at the extremes of the domain. Note now that we have a problem. Clearly, the extremes are a and b, but those numbers are not estimators, they are the actual parameters.

This is in fact a case where we need to guess an estimator and then verify that it is the MLE. As a guess: how can we start with the sample $x_1, \ldots, x_n$ and obtain a number which is smaller (larger) than all yet still in the interval (a, b)? As far as I know, the only points guaranteed to be in the interval are the actual points of the sample. Thus, try

$$x_{(1)} = \min\{x_1, \ldots, x_n\},$$

and

$$x_{(n)} = \max\{x_1, \ldots, x_n\}.$$

Plugging back in the log-likelihood function it is easy to see that indeed these are the points of maximum for a (respectively b), thus the MLE is

$$\hat{\theta} = (\hat{a}, \hat{b}) = (x_{(1)}, x_{(n)}).$$

Theorem 8.6 (MLE Invariance property) *Suppose that $L(\theta)$ is the likelihood function for estimating a parameter vector $\theta \in \Theta$, and let $\hat{\theta}$ denote the MLE. Then, for any measurable function h, the MLE for estimating $h(\theta)$ is $h(\hat{\theta})$.*

I stated the theorem here without technical conditions. The theorem is easy to prove and left as an exercise. Applying the theorem makes it very easy to see that $\hat{\sigma}$ is the MLE for the standard deviation of the normal distribution in the example above. The paper that first summarized various results and stated the invariance property of the MLE is Zehna (1966).

To understand fully the next theorem and when it may be applied, one needs to understand the concept of family of distributions and in particular the concept of regular exponential family of distributions. We refer the reader to the best book in the area Casella and Berger (2001).

Theorem 8.7 (MLE Consistency) *Suppose $X_1, \ldots, X_n$ is a random sample from a distribution $f(x|\theta)$, which form a regular exponential family. If $\hat{\theta}_n$ is the MLE for estimating θ based on n observations, the MLE estimator is consistent, i.e.,*

$$\hat{\theta}_n \to \theta.$$

Not only the estimator is consistent but we can in fact find the rate at which it converges to the parameter. This rate is what is called the *Fisher information*.

Definition 8.8 (Fisher information) *If $X_1, \ldots, X_n$ are i.i.d. random variables with density $f(x|theta)$; then the Fisher information is defined as*

$$I(\theta) = -\mathbf{E}[U'(\theta)] = -\int \frac{\partial^2 l}{\partial \theta^2} f(x|\theta) dx = -\int \frac{\partial}{\partial \theta} \frac{\partial \log f(x|\theta)}{\partial \theta} f(x|\theta) dx.$$

We can also show two more facts which simplify the calculation of the Fisher information:

$$I(\theta) = \mathbf{E}[(U(\theta))^2] = \int \left(\frac{\partial \log f(x|\theta)}{\partial \theta} \right)^2 f(x|\theta) dx,$$

and

$$I(\theta) = n \int \left(\frac{\partial \log f(x_1|\theta)}{\partial \theta} \right)^2 f(x_1|\theta) dx,$$

where the first expression is an n-dimensional integral, while the latter is just a simple one-dimensional integral.

The Fisher information measures the curvature of the log-likelihood function around the true value of θ. The larger the quantity, the better the approximation and the fewer the observations needed. We present here the case when θ is one dimensional, the case of a parameter vector where the Fisher information becomes a matrix with elements of the type

$$I(\theta)_{ij} = \mathbf{E}\left[\left(\frac{\partial \log f(x|\theta)}{\partial \theta_i} \right) \left(\frac{\partial \log f(x|\theta)}{\partial \theta_j} \right) \right] = -\mathbf{E}\left[\frac{\partial^2 \log f(x|\theta)}{\partial \theta_i \theta_j} \right].$$

Returning to the one-dimensional case, let us justify the two assertions in the definition. For any statistics $T(X) = T(X_1, \ldots, X_n)$, we have

$$
\frac{\partial}{\partial \theta} \mathbf{E}[T(X)] = \int \frac{\partial}{\partial \theta} (T(X) f(x|\theta)) \, dx
$$

$$
= \int T(X) \frac{\partial}{\partial \theta} f(x|\theta) \frac{1}{f(x|\theta)} f(x|\theta) dx
$$

$$
= \int T(X) \frac{\partial \log f(x|\theta)}{\partial \theta} f(x|\theta) dx
$$

$$
= \mathbf{E}\left[T(X) \frac{\partial \log f(x|\theta)}{\partial \theta} \right],
$$

where we used the fact that a statistics does not depend on θ. Choosing $T(X) = 1$, we obtain

$$
\mathbf{E}\left[\frac{\partial \log f(x|\theta)}{\partial \theta} \right] = \frac{\partial}{\partial \theta} 1 = 0 = \mathbf{E}[U(\theta)]. \tag{8.2}
$$

In the last expression, we take the derivative with respect to θ to obtain

$$
0 = \frac{\partial}{\partial \theta} \mathbf{E}\left[\frac{\partial \log f(x|\theta)}{\partial \theta} \right] = \int \frac{\partial}{\partial \theta} \left(\frac{\partial \log f(x|\theta)}{\partial \theta} f(x|\theta) \right) dx
$$

$$
= \int \frac{\partial^2 \log f(x|\theta)}{\partial \theta^2} f(x|\theta) dx + \int \frac{\partial \log f(x|\theta)}{\partial \theta} \frac{\partial f(x|\theta)}{\partial \theta} \frac{1}{f(x|\theta)} f(x|\theta) dx
$$

$$
= \mathbf{E}\left[\frac{\partial^2 \log f(x|\theta)}{\partial \theta^2} \right] + \mathbf{E}\left[\left(\frac{\partial \log f(x|\theta)}{\partial \theta} \right)^2 \right],
$$

which gives us the first equality in the definition. For the second equality, note that

$$
\mathbf{E}\left[\left(\frac{\partial \log f(x|\theta)}{\partial \theta} \right)^2 \right] = \mathbf{E}\left[\left(\frac{\partial}{\partial \theta} \log \prod_{i=1}^{n} f(x_i|\theta) \right)^2 \right]
$$

$$
= \mathbf{E}\left[\left(\sum_{i=1}^{n} \frac{\partial}{\partial \theta} \log f(x_i|\theta) \right)^2 \right]
$$

$$
= \sum_{i=1}^{n} \mathbf{E}\left[\left(\frac{\partial}{\partial \theta} \log f(x_i|\theta) \right)^2 \right] + 2 \sum_{i<j} \mathbf{E}\left[\frac{\partial}{\partial \theta} \log f(x_i|\theta) \frac{\partial}{\partial \theta} \log f(x_j|\theta) \right]
$$

$$
= n \mathbf{E}\left[\left(\frac{\partial}{\partial \theta} \log f(x_1|\theta) \right)^2 \right],
$$

since the variables are i.i.d. Thus all terms in the first sum are identical and by independence the expectation of product is the product of expectations each of which is 0 by (8.2).

Why is this important? Armed with all these strange expressions, we can show that the MLE will converge to the parameter it estimates according to a normal distribution. Essentially, we can prove a central limit theorem (CLT) for the MLE. To see this, let us expand the score function U around the true parameter value denoted with θ_0. According to Taylor expansion, we have

$$U(\hat{\theta}) = U(\theta_0) + U'(\theta_0)(\hat{\theta} - \theta_0) + \frac{1}{2}U''(\tilde{\theta})(\hat{\theta} - \theta_0)^2$$

for some $\tilde{\theta}$ between $\hat{\theta}$ and θ_0. When $\hat{\theta}$ is close to θ_0, the second derivative term is multiplied with the square of a small number and should become negligible when compared with the first derivative term. Also, recall that $\hat{\theta}$ is a maximum point for the log-likelihood function and therefore it is a root of its derivative, i.e., $U(\hat{\theta}) = 0$. Rewriting, we have

$$-U'(\theta_0)(\hat{\theta} - \theta_0) \approx U(\theta_0). \tag{8.3}$$

Now let us look at $U(\theta_0)$. *This is the point where we need to use the form of the exponential distributions.* For these distributions, we can always write the log likelihood $\log f(X|\theta) = \sum_{i=1}^{n} \log f(X_i|\theta)$. Assuming that the distribution has this property, we can write

$$U(\theta_0) = \left.\frac{\partial \log f}{\partial \theta}(X|\theta)\right|_{\theta=\theta_0} = \sum_{i=1}^{n} \left.\frac{\partial \log f}{\partial \theta}(X_i|\theta)\right|_{\theta=\theta_0} = \sum_{i=1}^{n} U_i(\theta_0).$$

The point here is that we can write the score function at θ_0 as a sum of i.i.d. random variables. Since we can do this, the CLT can be used and we obtain

$$\frac{\frac{1}{n}U(\theta_0) - \mathbf{E}[U_i(\theta_0)]}{\sigma/\sqrt{n}} \to N(0, 1),$$

where $\sigma^2 = Var(U_i(\theta_0))$. However, recall we have shown the following:

$$Var(U_i(\theta_0)) = \mathbf{E}[U_i^2(\theta_0)] = -\mathbf{E}[U_i'(\theta_0)].$$

Thus we get

$$\frac{U(\theta_0)}{\sqrt{n\mathbf{E}[-U_i'(\theta_0)]}} \to N(0, 1)$$

in distribution.

Furthermore

$$-U'(\theta_0) = \sum_{i=1}^{n} -U_i'(\theta_0),$$

and since once again we wrote $-U'(\theta_0)$ as a sum or i.i.d. random variables, the law of large numbers gives

$$-\frac{1}{n}U'(\theta_0) \to \mathbf{E}[-U_i'(\theta_0)]$$

in probability, or

$$-\frac{1}{n}U'(\theta_0)\frac{1}{\mathbf{E}[-U_i'(\theta_0)]} \to 1.$$

Finally, we get back to equation (8.3) and combine all these expressions to obtain

$$\sqrt{n}(\hat{\theta} - \theta_0) \approx \frac{U(\theta_0)}{\sqrt{n}} \frac{n}{\sum -U_i'(\theta_0)} = \frac{U(\theta_0)}{\sqrt{n}\mathbf{E}[-U_i'(\theta_0)]} \frac{n\mathbf{E}[-U_i'(\theta_0)]}{\sum -U_i'(\theta_0)}.$$

Now we use Slutsky's theorem and we can finally show that

$$\sqrt{n}(\hat{\theta} - \theta_0) \to N\left(0, \frac{1}{\mathbf{E}[-U_i'(\theta_0)]}\right)$$

in distribution.

The Rao–Cramer theorem essentially uses this expression and Jensen inequality to show that $\hat{\theta}$ is consistent. Let us state this result as a theorem.

Theorem 8.9 *Under suitable regularity conditions, the likelihood equation has a unique consistent root $\hat{\theta}$ (there may be others not consistent) for the parameter θ_0. This root has the property*

$$\sqrt{n}(\hat{\theta} - \theta_0) \xrightarrow{D} N\left(0, \frac{1}{\sigma^2}\right),$$

where

$$\sigma^2 = \mathbf{E}[-U_i'(\theta_0)] = -\mathbf{E}\left[\frac{\partial \log f}{\partial \theta}(X|\theta_0)\right] = -\int \frac{\partial \log f}{\partial \theta}(x|\theta_0)f(x, \theta_0)dx.$$

▣ EXAMPLE 8.6   The Neyman–Scott Paradox

Here is an example of a distribution where the MLE is not consistent. You will be asked to prove most of the claims here in Problem 8.7. Suppose random variables $(X_1, Y_1), (X_2, Y_2), \ldots, (X_n, Y_n)$ are all independent and that for each i we have X_i, Y_i distributed as $N(\mu_i, \sigma^2)$. Note that the variables have means indexed by i but they all have the same variance. In this case, we can form the log-likelihood function and show that the MLEs for the parameters are

$$\hat{\mu}_i = \frac{X_i + Y_i}{2}, \tag{8.4}$$

$$\hat{\sigma}^2 = \frac{1}{n}\sum_{i=1}^{n} \frac{(X_i - Y_i)^2}{4}. \tag{8.5}$$

These are in fact the unique solutions of the likelihood equation.

Because of the last expression, it is easy to see that

$$\hat{\sigma}^2 \xrightarrow{P} \frac{\sigma^2}{2}$$

and therefore the MLE is not consistent (it is biased).

8.4   The Method of Moments

The method of moments is a very simple idea. Suppose that we observe $x_1, \ldots, x_n$ from a distribution $f(x|\theta)$, where θ is the vector of parameter which may be d dimensional. The idea of this method is to match the empirical moments estimated from data with theoretical moments calculated using the distribution $f(x|\theta)$.

Obviously, the theoretical moments will need to exist, and one needs a minimum of d moments to obtain d equations to be able to estimate all the components of the parameter vector θ. An example of a distribution for which this method does not work is the Cauchy distribution since it has infinite theoretical moments.

■ EXAMPLE 8.7

Suppose the random sample $x_1, \ldots, x_n$ comes from a distribution $f(x|\mu, \sigma^2)$, where μ is the expected value of the distribution and σ^2 is the variance. Estimating these parameters using the method of moments means matching the sample moments with these parameters. To estimate the mean, we use

$$\mathbf{E}[X] = \frac{1}{n} \sum_{i=1}^{n} x_i = \bar{x},$$

and for the variance

$$Var(X) = \frac{1}{n} \sum_{i=1}^{n} (x_i - \bar{x})^2,$$

where we used the law of large numbers to obtain these estimators.

The method of moments estimators are

$$\mu = \bar{x}$$

$$\sigma^2 = \frac{1}{n} \sum_{i=1}^{n} (x_i - \bar{x})^2 = \frac{n-1}{n} S^2,$$

where S^2 is the usual sample variance estimator.

■ EXAMPLE 8.8   Gamma distribution

Recall that we wrote the $Gamma(a, \lambda)$ density as

$$f(x) = \frac{\lambda^a}{\Gamma(a)} x^{a-1} e^{-\lambda x} 1_{(0,\infty)}(x).$$

We want to estimate the parameters using the method of moments.

We can calculate the first and second moments as

$$\mathbf{E}[X] = \frac{a}{\lambda},$$

$$Var(X) = \frac{a}{\lambda^2}.$$

Setting them equal to $\bar{X}$ and S^2, we obtain the moments estimators

$$\hat{a} = \frac{(\bar{X})^2}{S^2},$$

$$\hat{\lambda} = \frac{\bar{X}}{S^2}.$$

Note that these estimators have been obtained very easily. The MLE in this case is hard to calculate because it will involve the derivative

$$\frac{\partial}{\partial a} \log \Gamma(a).$$

8.5   Testing, the Likelihood Ratio Test

Suppose we observe a sample $x_1, \ldots, x_n$ from a distribution $f(x|\theta)$ where the parameter θ is known to be in some set Θ. The problem here is to test whether the parameter θ is some fixed value or, more generally, is a value in some set denoted Θ_0.

For example, suppose p is the proportion of the population that will vote for a certain candidate A at the next elections. In a simplistic model, we see all voters as i.i.d. $Bernoulli(p)$ random variables, with 1 meaning that they voted for the candidate and 0 meaning that they did not. Suppose we take a sample of voters $X_1, \ldots, X_n$. Based on this sample, we are interested in estimating p but even more importantly whether or not $p > 0.5$. Clearly, if $p > 0.5$, the candidate A will be elected. Thus in this example the set Θ_0 is going to be $(0.5, 1]$. Note: We can either choose this Θ_0 and prove that it contains theta, or we can choose Θ_0 as the set $[0, 0.5]$ and show that theta does not belong there. These two choices and conclusions seem to be the same, but in statistics this is not true. The choice of the set Θ_0 is important. We will come back to this idea in a minute.

Statistical testing involves a simple principle. The first step is to construct an estimator for theta, say $\hat{\theta}$, and to calculate its limiting distribution. After we do this, we need to calculate or estimate the following probability:

Probability of seeing the kind of observations we are obtaining if the parameter θ would be in Θ_0.

This probability is called the p-value of the test, with hypotheses that θ is in the set Θ_0 and θ is in the set $\Theta \setminus \Theta_0$. In statistics, these two hypotheses are called the *null* and *alternative* hypotheses, and they are typically written as

$$H_0 : \quad \theta \in \Theta_0,$$
$$H_a : \quad \theta \notin \Theta_0.$$

Now let us look at that probability above or the so-called p-value. If this probability is small, and since the values are actually what we are seeing, this is a direct evidence that the null hypothesis $\theta \in \Theta_0$ is false. The smaller the probability, the stronger the evidence that H_0 is false. However, if the p-value is large, the hypothesis may be true – a large p-value does not actually show that the null hypothesis is definitely true. Recall the comment we made earlier about the choice of Θ_0. Since the null hypothesis is either rejected or we do not have enough evidence to reject it, we always put the fact we want to prove in the alternative. Going back to our candidate example, if we want to show that the candidate leads in the poll, we will choose $\Theta_0 = [0, 0.5]$. Rejecting this hypothesis will mean that the true population percentage which supports candidate A is strictly greater than 50%. Let me give an example to reinforce this idea.

■ **EXAMPLE 8.9   An example of constructing a testing procedure**

Suppose in a random sample of 100 voters 60 will support candidate A. Assume the variables collected are i.i.d. $Bernoulli(p)$ and thus the sample percentage that supports A is $\hat{p} = 0.6$. In Problem 8.3, I ask you to prove that $\hat{p}$ has an asymptotic distribution:

$$\sqrt{n}(\hat{p} - p) \xrightarrow{D} N(0, p(1 - p)).$$

Testing:

$$H_0 : \quad p \leq 0.5$$
$$H_a : \quad p > 0.5$$

means that we need to calculate the probability of seeing the kind of observations we are obtaining (or more extreme) if the parameter θ would be in Θ_0. Specifically, this translates to

$$\mathbf{P}(\hat{p} > 0.6 \mid p \leq 0.5).$$

Recall that small values of this probability are evidence against H_0. Clearly, the largest probability of this type is obtained when the true parameter $p = 0.5$. So we approximate the probability under this assumption:

$$\mathbf{P}(\hat{p} > 0.6 \mid p = 0.5) = \mathbf{P}\left(\sqrt{100}(\hat{p} - p) > \sqrt{100}(0.6 - p)\Big| p = 0.5\right)$$
$$= \mathbf{P}\left(\sqrt{100}\frac{\hat{p} - 0.5}{\sqrt{0.5(1 - 0.5)}} > \sqrt{100}\frac{0.6 - 0.5}{\sqrt{0.5(1 - 0.5)}}\right)$$
$$= \mathbf{P}(Z > 2) = 0.02275013,$$

where Z is an $N(0, 1)$ random variable. Thus there is a 0.0228 probability of seeing the kind of numbers we are seeing if the null hypothesis is true. This seems to be enough evidence to conclude that in fact candidate A is leading in the opinion of the population. VERY IMPORTANT: Nothing guarantees that this sample is not in fact one of the 2.28/100 that will produce this kind of result even if $p = 0.5$.

To be more certain, one can do one of the two things:

1. Take a larger sample size, or

2. Repeat the process at a different location.

The first approach is the classical approach. Suppose we have a sample of 1000, and in fact we obtain the same $\hat{\mathbf{P}} = 0.6$; that is, 600 of respondents said they will vote for A. Repeating the calculation with this new sample size will produce a p-value of 1.269816×10^{-10}, a number that is essentially 0.

For 2) the idea is that if two tests are rejecting the null, the probability of both rejecting when in fact H_0 is true is $0.0228^2 = 0.00051984$, which is extremely small. One could also increase the sample size and combine the new observations with the old ones, but this may not be possible if the results are from two geographic locations or if some time has passed.

Finally, to understand why if the probability is large we do not accept H_0 as the true hypothesis, and instead we say "we did not find enough evidence to disprove H_0", consider the same simple example. Suppose that the true percentage of people supporting A is indeed large but only slightly: specifically, say, $p = 0.52$.

Now suppose we take a sample of 100 and obtain the expected number 52 of individuals supporting candidate A. Calculating the same probability in exactly the same way gives

$$\mathbf{P}(\hat{p} > 0.52 \mid p = 0.5) = \mathbf{P}\left(\sqrt{100}\frac{\hat{p} - 0.5}{\sqrt{0.5(1 - 0.5)}} > \sqrt{100}\frac{0.52 - 0.5}{\sqrt{0.5(1 - 0.5)}}\right)$$
$$= \mathbf{P}(Z > 0.4) = 0.3445783.$$

Interpreting it, about one in three samples (0.345) will produce the kind of results we are seeing if the true percentage is less than or equal 0.5. So it can happen, and with quite high probability. What is going on here is that the sample is not large enough to detect such small differences (0.5 versus 0.52).

Now suppose the sample size is 1000 people, and once again we obtain the expected value 520 people supporting A. In this case, the calculation is

$$\mathbf{P}(\hat{p} > 0.52 \mid p = 0.5) = \mathbf{P}\left(\sqrt{1000}\frac{\hat{p} - 0.5}{\sqrt{0.5(1 - 0.5)}} > \sqrt{1000}\frac{0.52 - 0.5}{\sqrt{0.5(1 - 0.5)}}\right)$$
$$= \mathbf{P}(Z > 1.264911) = 0.1029516,$$

or 1 in about 10 samples will produce the results we are seeing. Still not enough sample size. Say, we have 5000 people and we once again obtain $\hat{p} = 0.52$.

Calculating the p-value, we obtain 0.002338867 or about 2 in 1000 samples will produce such results if H_0 was true.

This example raises a question: How small is sufficiently small? Note that the example states 1 in 3 can happen and even 1 in 10 but not 2 in 1000. What does that even mean?

This is the key statistical problem. In the applied statistics literature, the threshold that is used the most is 0.05. Why? Nobody knows. This is the value that was deemed appropriate. In fact, I have seen many studies where the p-value obtained was 0.048 and the test rejected the null. So one needs to exercise caution in practical applications and use his or her judgment. The most important thing to remember is this: We obtained these numbers by using a sample statistics and a derived distribution. These were all obtained using a model which has very strict assumptions in most of the cases. Are those assumptions reasonable? If they aren't, fiddling over 0.048 versus 0.056 is a moot point.

8.5.1   The likelihood ratio test

In many situations, calculating the approximate distribution for the test statistics is very complicated. Furthermore, for simplicity the testing problem was presented for one-dimensional parameters. The procedure works when we have multiple parameters as well, but then one needs to look at the joint distribution. Finding the approximate joint distributions is again a very complicated task. There is, however, one approach which allows us to deal with these complicated cases. Be aware that, since the approach is more generic, it is also less precise than the testing procedure presented earlier.

Suppose $\theta = (\theta_1, \theta_2, \ldots, \theta_d)$ is a vector of parameters and we want to test

$$H_0 : \quad \theta \in \Theta_0,$$
$$H_a : \quad \theta \in \Theta \backslash \Theta_0.$$

We are given $\mathbf{X} = (X_1, \ldots, X_n)$, a sample from the distribution $f(x|\theta)$. We calculate the likelihood function as in the MLE section and form the likelihood ratio

$$\lambda(\mathbf{x}) = \frac{\sup_{\{\theta \in \Theta_0\}} L(\theta, \mathbf{x})}{\sup_{\{\theta \in \Theta\}} L(\theta, \mathbf{x})}.$$

This expression is calculated for the sample value $\mathbf{x} = (x_1, \ldots, x_n)$.

Intuitively, if the best parameter value is in Θ_0, we would obtain 1 as the value for the ratio. Having a small ratio is direct evidence that the optimal value in Θ_0 is far from the general optimal value, thus providing direct evidence that the null hypothesis is false. Again, how small is small?

The likelihood ratio test has a rejection region (a region containing observations x which will reject the null hypothesis) of the type

$$\{x \mid \lambda(x) \leq K\},$$

where K is a constant between 0 and 1. The constant K is calculated so that the probability of falsely rejecting H_0 is small. The biggest problem with likelihood ratio tests is that one needs to establish the distribution of the likelihood ratio, i.e., the distribution of

$$\lambda(\mathbf{X}) = \frac{L(\hat{\theta}_0(\mathbf{X}), \mathbf{X})}{L(\hat{\theta}(\mathbf{X}), \mathbf{X})},$$

where $\mathbf{X}$ is the vector of observed random variables, and $\theta_0(\mathbf{X})$ and $\theta(\mathbf{X})$ are, respectively, the optimal values when the maximization is done over Θ_0 and Θ.

◼ EXAMPLE 8.10   Not a likelihood ratio test

I am only giving this example because it seems to be prevalent today in online sources (Wikipedia, e.g.) as an example of likelihood ratio test. In fact, the example is wrong, and trying to obtain results based on likelihood functions is extremely hard.

Suppose we have two coins and we want to know if they have the same probability of showing heads. To verify this, we toss both coins and we record the outcomes. We summarize the results:

	Heads	Tails	Total
Coin A	n_{AH}	n_{AT}	N_A
Coin B	n_{BH}	n_{BT}	N_B

Here, n_{1H} represents the number of heads obtained when tossing Coin A. Clearly, the number of heads is a binomial random variable, and similarly for B. Let p_A be the probability of landing heads for Coin A, and similarly p_B for the other coin.

Of course, the same example can be applied to a situation where an opinion pool is collected. Then a public talk or a rally is given to the audience and a second opinion pool is collected. The test may be used to see if the public rally had an effect on the public perception. Truly, this test should be used for all the TV commercials before we are flooded by them.

The test we want to perform is

$$H_0 : \quad p_A = p_B,$$
$$H_a : \quad p_A \neq p_B.$$

Suppose we observe the values above and that the tosses for the coins are independent. Then the likelihood function is

$$L(p_A, p_B) = \binom{N_A}{n_{AH}} p_A^{n_{AH}} (1 - p_A)^{n_{AT}} \binom{N_B}{n_{BH}} p_B^{n_{BH}} (1 - p_B)^{n_{BT}}.$$

Under H_a, the two probabilities p_A and p_B are different. One would calculate the log likelihood, and then take derivatives with respect to p_A and p_B. By setting the derivatives equal to 0 and solving for the parameters p_A and p_B, it is not hard to see that the MLEs are the sample proportions

$$\hat{p}_A = \frac{n_{AH}}{N_A}, \text{ and } \hat{p}_B = \frac{n_{BH}}{N_B}.$$

On the other hand, under H_0 the two probabilities are the same, $p_A = p_B = p$, and thus the likelihood becomes

$$L(p) = \binom{N_A}{n_{AH}}\binom{N_B}{n_{BH}} p^{n_{AH}+n_{BH}} (1-p)^{n_{AT}+n_{BT}}.$$

Maximizing the corresponding log likelihood using the same method gives the MLE in this case as the natural estimate: the sample proportion of heads for both coins put together is

$$\hat{p} = \frac{n_{AH}+n_{BH}}{N_A + N_B}.$$

We know from the CLT that all these proportions are approximately normal; more specifically

$$\frac{\hat{p}_A - p_A}{\sqrt{(N_A p_A (1-p_A))}} \xrightarrow{D} N(0,1),$$

and similarly for p_B.

Based on this result, we can show that under the assumption that the hypothesis H_0 is true, i.e., $p_A = p_B = p$, the test statistics

$$T = \frac{\hat{p}_A - \hat{p}_B}{\sqrt{\frac{\hat{p}(1-\hat{p})}{N_A + N_B}}}$$

has a $t_{N_A+N_B-1}$ distribution. If $N_A + N_B$ is a large number, this distribution is approximately $N(0,1)$.

Now, to test whether or not H_0 is true, we just need to calculate the probability that the test values will take more extreme observations than the sample result. Specifically, for the two-sided hypothesis presented, since we know the distribution of T, we need

$$\mathbf{P}\left(t_n > \left|\frac{\hat{p}_A - \hat{p}_B}{\sqrt{\frac{\hat{p}(1-\hat{p})}{N_A + N_B}}}\right|\right),$$

where $n = N_A + N_B - 1$ and the quantities on the right are all calculated using the sample. We use the absolute value because it is a two-sided alternative hypothesis. If the hypotheses were

$$H_0: \quad p_A = p_B,$$
$$H_a: \quad p_A < p_B,$$

we would have calculated the p-value as

$$\mathbf{P}\left(t_n < \frac{\hat{p}_A - \hat{p}_B}{\sqrt{\frac{\hat{p}(1-\hat{p})}{N_A+N_B}}}\right).$$

In the previous example, we took advantage of the distribution of the sample proportion. However, in many cases the test statistics is hard to discover and many times it is even harder to get its distribution. This is why the likelihood ratio test is very useful. We will state the next theorem without proving it.

Theorem 8.10 (Likelihood ratio test distribution: Wilks' theorem) *Suppose that we want to test the following hypotheses:*

$$H_0 : \quad \theta \in \Theta_0,$$
$$H_a : \quad \theta \in \Theta \backslash \Theta_0,$$

where the set Θ_0 is a subset of the set Θ. We furthermore assume that the dimension of the set Θ_0 is k and the dimension of the larger set Θ is m. Let $\lambda(\mathbf{x})$ denote the likelihood ratio, and assume that the values that maximize it $\hat{\theta}_0$ and $\hat{\theta}$ are normally distributed. Then, under H_0, the distribution of

$$-2\log\lambda(\mathbf{x})$$

converges in distribution to a chi-squared random variable: χ^2_{m-k}.

Note the condition that the MLEs $\hat{\theta}_0$ and $\hat{\theta}$ are normally distributed. This is in fact not a restriction because, under the general assumptions about the distribution (exponential family), the MLE is guaranteed to be normal. I am not going to give more details here; refer to Casella and Berger (2001) for exact statements.

EXAMPLE 8.11   A likelihood ratio test

Let us look at the likelihood ratio test of the previous example and attempt to use Wilks' theorem. We already calculated the MLEs, so we only need to calculate the likelihood ratio. Note that Θ_0 is one dimensional:

$$\Theta_0 = \{p \in (0,1) \mid p_A = p_B = p\},$$

and Θ is two dimensional:

$$\Theta = \{(p_A, p_B) \mid p_A, p_B \in (0,1)\}.$$

Recall that we found the MLEs previously and showed they are normally distributed. All we need to do is to replace them in $\lambda(\mathbf{x}) = \lambda(n_{AH}, n_{BH})$, and after simplifications we obtain

$$-2 \left((n_{AH} + n_{BH}) \log \frac{n_{AH} + n_{BH}}{N_A + N_B} - n_{AH} \log \frac{n_{AH}}{N_{AH}} - n_{BH} \log \frac{n_{BH}}{N_{BH}} \right.$$

$$\left. (n_{AT} + n_{BT}) \log \frac{n_{AT} + n_{BT}}{N_A + N_B} - n_{AT} \log \frac{n_{AT}}{N_{AT}} - n_{BT} \log \frac{n_{BT}}{N_{BT}} \right) \sim \chi_1^2.$$

8.6   Confidence Sets

To calculate confidence sets, we need to find a function $g(\theta, \mathbf{X})$ where $\mathbf{X}$ is the random sample such that the distribution is the same for all values of the parameter θ. Such a function $g(\theta, \mathbf{X})$ is called a **pivot** (or pivotal quantity).

A pivotal quantity will typically depend on both the parameter vector θ and the random sample $\mathbf{X}$, but the probability

$$\mathbf{P}(g(\theta, \mathbf{X}) \in A)$$

cannot depend on θ for all sets A.

Once we determine a pivotal quantity $g(\theta, \mathbf{X})$, the $(1-\alpha)$-Confidence set (interval) for θ is determined as follows:

1. Given the sample $\mathbf{X}$, find a set A such that $\mathbf{P}(g(\theta, \mathbf{X}) \in A) = 1 - \alpha$.

2. Given that $g(\theta, \mathbf{X}) \in A$, find a set B in the Θ space such that

$$g(\theta, \mathbf{X}) \in A \quad \Leftrightarrow \quad \theta \in B.$$

Here, $(1-\alpha)$ is the desired probability value (confidence level in statistics language).

Finding a pivot is the tricky part. In general, if the pdf can be written in a particular form, one may use a typical form for the pivotal quantity. The table below summarizes some simple cases.

Table 8.1   Commonly encountered cases of pivotal quantities. $\bar{X}$ is the sample mean, and S is the sample standard deviation.

Form of the one-dimensional pdf $f(x\|\theta) =$	Pivotal quantity $g(\theta, \mathbf{X}) =$
$f(x - \mu)$	$\bar{X} - \mu$
$\frac{1}{\sigma} f(\frac{x}{\sigma})$	$\frac{\bar{X}}{\sigma}$
$\frac{1}{\sigma} f(\frac{x - \mu}{\sigma})$	$\frac{\bar{X} - \mu}{S}$

This is probably better understood by an example.

■ EXAMPLE 8.12 Confidence interval for the variance of a normal

Suppose we need to find a confidence set for the variance of a random sample normally distributed $N(\mu, \sigma^2)$.

Let $S^2 = \frac{1}{n-1} \sum_{i=1}^{n} (X_i - \bar{X})^2$ be the sample variance. In Problem 8.2, we show that the distribution of S^2/σ^2 is the same as $\frac{1}{n-1}\chi^2_{n-1}$. Put differently

$$\frac{(n-1)S^2}{\sigma^2} \sim \chi^2_{n-1},$$

therefore the function

$$g(\sigma^2, \mathbf{X}) = \frac{(n-1)S^2}{\sigma^2}$$

is a pivotal quantity since its distribution does not depend on the value of the parameter σ^2.

To find a confidence interval level $1 - \alpha$, we calculate the quantiles of the distribution of the pivotal quantity $g(\sigma^2, \mathbf{X})$: $\chi^2_{n-1}\left(\frac{\alpha}{2}\right)$ and $\chi^2_{n-1}\left(1 - \frac{\alpha}{2}\right)$, which are numbers such that the area under the χ^2_{n-1} distribution is exactly $1 - \alpha$. Then we use the pivotal quantity to obtain a confidence interval for the parameter.

$$\mathbf{P}\left(\chi^2_{n-1}\left(\frac{\alpha}{2}\right) \leq \frac{(n-1)S^2}{\sigma^2} \leq \chi^2_{n-1}\left(1 - \frac{\alpha}{2}\right)\right) = 1 - \alpha,$$

which is the same as

$$\mathbf{P}\left(\frac{(n-1)S^2}{\chi^2_{n-1}\left(1 - \frac{\alpha}{2}\right)} \leq \sigma^2 \leq \frac{(n-1)S^2}{\chi^2_{n-1}\left(\frac{\alpha}{2}\right)}\right) = 1 - \alpha$$

for the variance interval, and of course for the standard deviation

$$\mathbf{P}\left(\frac{\sqrt{(n-1)}S}{\sqrt{\chi^2_{n-1}\left(1 - \frac{\alpha}{2}\right)}} \leq \sigma \leq \frac{\sqrt{(n-1)}S}{\sqrt{\chi^2_{n-1}\left(\frac{\alpha}{2}\right)}}\right) = 1 - \alpha.$$

The left and right bounds provide the $1 - \alpha$ confidence level interval. In general, we can follow the same method for any areas to the left and right; that is, the interval from $\chi^2_{n-1}\left(\frac{\alpha_1}{2}\right)$ to $\chi^2_{n-1}\left(\frac{\alpha_2}{2}\right)$ will have area $p = \alpha_2 - \alpha_1$. We can see that we have multiple intervals that will provide the same probability. The right thing to do for a desired probability level p is to find values α_1 and α_2 that will minimize the length of the resulting confidence interval $\chi^2_{n-1}\left(\frac{\alpha_1}{2}\right)$ to $\chi^2_{n-1}\left(\frac{\alpha_2}{2}\right)$. This approach is sadly rarely seen in practice because most researchers are used to the normal distribution for the pivot quantity (this is the case of the sample mean). Recall that the normal is symmetric around the mean, and thus the best possible choice for the interval is indeed $\alpha_1 = \frac{p}{2}$ and $\alpha_2 = 1 - \frac{p}{2}$. We refer to Problem 8.23 for an example of finding this optimal interval.

Problems

8.1   Suppose $Z_1, Z_2, \ldots, Z_n$ is a random sample from a Cauchy distribution with parameters 0 and 1 (denoted $Cauchy(0,1)$):

$$f(x|(0,1)) = \frac{1}{\pi(1+x^2)}.$$

Show that $\sum_{1=1}^{n} Z_i$ has a Cauchy distribution with parameters 0 and n and also that the sample average $\bar{Z} = \frac{1}{n}\sum_{1=1}^{n} Z_i$ is $Cauchy(0,1)$. For completion, the density of a $Cauchy(\mu, \gamma)$ is

$$f(x|(\mu, \gamma)) = \frac{1}{\pi\gamma\left(1 + \left(\frac{x-\mu}{\gamma}\right)^2\right)}.$$

8.2   Let $X_1, \ldots, X_n$ a random sample from an $N(\mu, \sigma^2)$ distribution.
  a) Show that the random variable

$$\frac{\bar{X} - \mu}{\sigma/\sqrt{n}}$$

  is distributed as a $N(0,1)$.
  b) Show that the random variable

$$\frac{1}{\sigma^2}S^2 = \frac{1}{\sigma^2}\frac{1}{n-1}\sum_{i=1}^{n}(X_i - \bar{X})^2$$

  has the same distribution as $\frac{1}{n-1}\chi^2_{n-1}$, where χ^2_{n-1} is a Chi-squared random variable with $n-1$ degrees of freedom.

8.3   Suppose we are given a sequence of i.i.d. random variables $X_1, \ldots, X_n$ distributed as $Bernoulli(p)$ random variables.
  a) Calculate $\hat{p}_n$ the MLE for p based on these n random variables.
  b) Calculate the limiting distribution of $\hat{p}_n$.

8.4   Suppose that in a trial we obtain the following numbers representing a random sample of 10 bricks selected of a production line:

$x_1 = 6.161520, x_2 = 4.892259, x_3 = 6.342859, x_4 = 4.673576, x_5 = 4.936554,$
$x_6 = 3.966134, x_7 = 4.351286, x_8 = 6.010079, x_9 = 4.944755, x_{10} = 5.691312.$

We assume that the distribution of each brick is a Cauchy random variable with location parameter θ.

$$f(x|\theta) = \frac{1}{\pi(1 + (x - \theta)^2)}.$$

We want to estimate the parameter θ based on these observations. Find the MLE estimator for theta. Try and apply the method of moments to obtain an estimator for theta. Can you apply the method? Explain.

8.5   In Formulas (8.1) we gave the point of extreme for the likelihood of a normal sample. Show that the point is indeed a maximum point, by calculating the Hessian matrix and showing that it is negative definite.

8.6   Suppose that for a sample $x_1, \ldots, x_n$ coming from a distribution with density $f(x|\theta)$ we can construct the maximum likelihood function $L(\theta)$, with $\theta \in \mathbb{R}$. Suppose further that we can calculate $\hat{\theta}$ the MLE, that is, $L(\hat{\theta}) \geq L(\theta)$, for all $\theta \in \mathbb{R}$. Let $h : \mathbb{R} \to \Lambda$, where Λ denotes the codomain of the function, and let $\lambda = h(\theta)$ and $\hat{\lambda} = h(\hat{\theta})$. For each λ, let U_λ be its pre-image through h, $U_\lambda = \{\theta \in \mathbb{R} : h(\theta) = \lambda\}$, and for every λ define the function

$$M(\lambda) = \sup_{\theta \in U_\lambda} L(\theta).$$

Show that the function $M(\lambda)$ is the likelihood function for estimating $\lambda = h(\theta)$, and derive that the MLE for estimating λ is $\hat{\lambda} = h(\hat{\theta})$.

8.7   In this problem we refer to Example 8.6. Suppose the random variables $(X_1, Y_1), (X_2, Y_2), \ldots, (X_n, Y_n)$ are all independent and that for each i we have X_i, Y_i distributed as $N(\mu_i, \sigma^2)$.

    a) Set $\nu = \frac{1}{\sigma^2}$ and express the likelihood function in terms of the parameters $\mu_1, \ldots, \mu_n$, and ν.

    b) Show that the log likelihood is

$$l(\mu_1, \ldots, \mu_n, \nu) = 2n \log \frac{1}{\sqrt{2\pi}} + n \log \nu - \nu \sum_{i=1}^{n} \frac{(X_i - \mu_i)^2 + (Y_i - \mu_i)^2}{2}.$$

    c) Calculate the MLEs for the parameters $\mu_1, \ldots, \mu_n, \nu$.

    d) Derive that the MLE for σ^2 is $\hat{\sigma}^2 = \frac{1}{2n} \sum_{i=1}^{n} s_i^2$, where

$$s_i^2 = (X_i - \hat{\mu}_i)^2 + (Y_i - \hat{\mu}_i)^2.$$

    e) Show that the statistics s_i^2 are i.i.d. and calculate their expected value.

    f) Use the law of large numbers to show that

$$\hat{\sigma}^2 \xrightarrow{P} \frac{\sigma^2}{2},$$

and discuss the consistency of the MLE.

    g) Find another estimator $\tilde{\sigma}^2$ which is consistent.

8.8   Let $X_1, \ldots, X_n$ be exponential with parameter λ. Calculate the MLE of λ.

8.9   Let $X_1, \ldots, X_n$ be i.i.d. $N(\mu, 1)$, where we know that $\mu \geq 0$. Calculate $\hat{\mu}$ the MLE of the parameter μ. Show that the limiting distribution of the MLE is

$$\mathbf{P}(\sqrt{n}\hat{\mu} \leq x) \longrightarrow \begin{cases} 0 & \text{if } x < 0 \\ \int_{-\infty}^{x} \frac{1}{\sqrt{2\pi}} e^{-\frac{x^2}{2}} & \text{if } x > 0 \end{cases}.$$

What should the limiting distribution be at $x = 0$ to be a proper cdf?
Note that in this case the limiting distribution is not normal.

8.10   Let $X_1, \ldots, X_n$ be i.i.d. $N(\mu, \sigma^2)$ where we know that μ is an integer. Calculate the MLE of μ.

8.11   Let $X_1, \ldots, X_n$ be i.i.d. $U(0, \theta)$. Calculate the MLE of θ and show that

$$n(\hat{\theta} - \theta) \xrightarrow{D} Exp(\theta).$$

8.12   Let $X_1, \ldots, X_n$ be i.i.d. with density

$$f(x|\theta_1, \theta_2) = \begin{cases} Ce^{-\theta_1 x}, & \text{if } x \geq 0 \\ Ce^{\theta_2 x}, & \text{if } x < 0 \end{cases}.$$

a) Calculate the constant C so that the pdf is a proper pdf.
b) Calculate the MLEs of θ_1 and θ_2.

8.13   Let $X_1, \ldots, X_m$ distributed as $N(\mu, \sigma_1^2)$ and $Y_1, \ldots, Y_n$ distributed as $N(\mu, \sigma_2^2)$, and all random variables are independent. Calculate the MLE of $(\mu, \sigma_1^2, \sigma_2^2)$.

8.14   Two independent readers proofread this manuscript which contains $N \geq 0$ errors with N unknowns. Reader A found 50 errors, while reader B found 77 errors. Thirty-six of those errors were found by both readers. What is the MLE for N, the total number of errors in the manuscript?

8.15   Find the MLE for λ using the observations X_i independent $Poisson(\lambda \rho_i)$ where ρ_i are known and $1 \leq i \leq n$.

8.16   Suppose X_i are independent $Poisson(\lambda^i)$, for $1 \leq i \leq n$. Calculate $\hat{\lambda}$ the MLE of λ. What is the limiting distribution of $\hat{\lambda}$?

8.17   Calculate the MLE for the parameter λ if $X_1, \ldots, X_{30}$ are i.i.d. double exponentials, i.e., they have the density $f(x|\lambda) = \begin{cases} \frac{1}{2}\lambda e^{-\lambda x}, & \text{if } x \geq 0 \\ \frac{1}{2}\lambda e^{\lambda x}, & \text{if } x < 0 \end{cases}.$

8.18   Refer to the previous problem. Suppose $\lambda = 1$. Calculate the MLE in 20 repeated simulations and estimate its variance. Compare with the variance obtained by using Fisher information.

8.19 Let $X_1, \ldots, X_n$ be a random sample from a Poisson distribution with parameter denoted with θ. Obtain the likelihood ratio for testing

$$
\begin{aligned}
H_0 &: \quad \theta = \theta_0, \\
H_a &: \quad \theta \neq \theta_0.
\end{aligned}
$$

8.20 Let $X_1, \ldots, X_n$ be a random sample from a normal distribution with **known** variance σ^2. Obtain the likelihood ratio for testing

$$
\begin{aligned}
H_0 &: \quad \mu = \mu_0, \\
H_a &: \quad \mu \neq \mu_0.
\end{aligned}
$$

8.21 Let $X_1, \ldots, X_n$ be a random sample from a normal distribution with **unknown** variance σ^2. Obtain the likelihood ratio for testing

$$
\begin{aligned}
H_0 &: \quad \mu = \mu_0, \\
H_a &: \quad \mu \neq \mu_0.
\end{aligned}
$$

Hint: Note that in this case the parameter space $\Theta = \{(\mu, \sigma^2) \mid \mu \in \mathbb{R}, \sigma^2 > 0\}$ is two dimensional, while $\Theta_0 = \{(\mu, \sigma^2) \mid \mu = \mu_0, \sigma^2 > 0\}$ is one dimensional.

8.22 Let $X_1, \ldots, X_n$ be a random sample from a normal distribution $N(\mu, \sigma^2)$ where both parameters are unknown. Obtain the likelihood ratio for testing

$$
\begin{aligned}
H_0 &: \quad \sigma^2 = \sigma_0^2, \\
H_a &: \quad \sigma^2 \neq \sigma_0^2.
\end{aligned}
$$

8.23 A pulsar is a highly magnetized, rotating neutron star that emits a beam of electromagnetic radiation. Such stars are extremely useful because they can be observed by any civilization and they are distinct enough in the space to provide spatial coordinates by referring the relative distances in 3D space to them. The first radio pulsar CP 1919 (now known as PSR 1919+21), with a pulse period of 1.337 s and a pulse width of 0.04 s, was discovered in 1967. The pulses are very precise; however, suppose we observe and record the periods with a novel instrument of measure and want to determine the variability of measurements. To this end, we perform 20 observations of the pulse periods and obtain an average of 1.336 s and a sample standard deviation $\sigma = 0.00123$ s. We want to compute a 95% confidence interval for the standard deviation (an interval that will contain the true standard deviation with probability 0.95). To this end, we assume that all the measurements are i.i.d. normally distributed.

a) Give a 95% confidence interval following the method in Example 8.12.

b) Next we will give an optimized confidence interval. To this end, you need to use a computer. We will use a brute-force optimization method. Steps to follow are as follows:

1. Take a grid of α_1 values from 0 to 0.05 in steps of 0.001.

2. For each such α_1 value, calculate the corresponding α_2 such that $\alpha_2 - \alpha_1 = 0.95$.

3. For each pair α_1 and α_2 thus calculated, determine the percentiles $\chi^2_{n-1}\left(\frac{\alpha_1}{2}\right)$ and $\chi^2_{n-1}\left(\frac{\alpha_2}{2}\right)$.

4. Finally, calculate the corresponding confidence interval bounds for σ for each such pair.

> All these intervals are 95% confidence intervals. Select the interval with the smallest width. That is the approximate optimal confidence interval.

c) Compare the intervals found in the parts a) and b) of this problem. Comment.

STOCHASTIC PROCESSES

CHAPTER 9

INTRODUCTION TO STOCHASTIC PROCESSES

What is a stochastic process?

Definition 9.1 *Given a probability space* $(\Omega, \mathcal{F}, \mathbf{P})$, *a stochastic process is* **any** *collection of random variables defined on this probability space.*

More specifically, the collection of random variables $\{X(t)\}_{t \in \mathcal{I}}$ *or alternatively written* $\{X(t) : t \in \mathcal{I}\}$, *where* $\mathcal{I}$ *is the index set. We will alternate between the notations* X_t *and* $X(t)$ *to denote the value of the stochastic process at time* t.

We give here the famous *R.A. Fisher* quote, who answered the same question:

What is a stochastic process? Oh, it's just one darn thing after another.

In this chapter, we start the study of stochastic processes by presenting common properties and characteristics. Generally, possessing one or more of these properties will simplify the study of the particular stochastic process. Not all stochastic processes have these properties. In the second section of this chapter, we present the simplest stochastic process possible: the coin toss process (the Bernoulli process).

Probability and Stochastic Processes, First Edition. Ionuț Florescu

9.1 General Characteristics of Stochastic Processes

9.1.1 The index set $\mathcal{I}$

The parameter that indexes the stochastic process determines the type of stochastic process we are working with.

For example, if $\mathcal{I} = \{0, 1, 2 \ldots\}$ (or equivalent), we obtain the so-called discrete-time stochastic processes. We will often write $\{X_n\}_{n \in \mathbb{N}}$ in this case.

If $\mathcal{I} = [0, \infty)$, we obtain the continuous-time stochastic processes. We shall write $\{X_t\}_{t \geq 0}$ in this case.

If $\mathcal{I} = \mathbb{Z} \times \mathbb{Z}$, we may be describing a discrete random field. If $\mathcal{I} = [0, 1] \times [0, 1]$, we may be describing the structure of some surface.

These are the most common cases encountered in practice but the index set may be quite general.

9.1.2 The state space $\mathcal{S}$

This is the domain space of all the random variables X_t. Since we are talking about random variables and random vectors, necessarily $\mathcal{S} \subseteq \mathbb{R}$ or $\mathbb{R}^n$. Again, we have several important examples. If $\mathcal{S} \subseteq \mathbb{Z}$, we say that the process is integer-valued or a process with discrete state space. If $\mathcal{S} = \mathbb{R}$, then we say that X_t is a real-valued process or a process with a continuous state space. If $\mathcal{S} = \mathbb{R}^k$, then X_t is a k-dimensional vector process. As a side remark, the state space $\mathcal{S}$ may be more general (say an abstract Lie algebra), in which case the definitions work very similarly except that for each t we have X_t measurable functions. In these general cases, it is customary to talk about stochastic processes and specify the nature of the space $\mathcal{S}$ (as in "with values in Hadamard manifolds").

9.1.3 Adaptiveness, filtration, standard filtration

In the special case when the index set $\mathcal{I}$ possesses a total order relationship[1], we can talk about the information contained in the process X at some moment $t \in \mathcal{I}$. To quantify this information, we introduce the abstract notion of filtration.

Definition 9.2 (Filtration) *We say that a probability space* $(\Omega, \mathscr{F}, \mathbf{P})$ *is a filtered probability space if and only if there exists a sequence of sigma algebras* $\{\mathscr{F}_t\}_{t \in \mathcal{I}}$ *included in* $\mathscr{F}$ *such that it is an increasing collection, that is,*

$$\mathscr{F}_s \subseteq \mathscr{F}_t, \quad \forall s \leq t, \ s, t \in \mathcal{I}.$$

The filtration is called complete *if its first element contains all the null sets of* $\mathscr{F}$. *If, for example, 0 is the first element of the index set (the usual situation), then $\forall N \in \mathscr{F}$, with $\mathbf{P}(N) = 0 \Rightarrow N \in \mathscr{F}_0$. We shall assume that any filtration defined in this book is complete, and all filtered probability spaces are complete.*

---

[1]That is, for any two elements $x, y \in \mathcal{I}$, either $x \leq y$ or $y \leq x$.

In the particular case when the time is continuous (e.g., $I = [0, \infty)$) it makes sense to talk about what happens with the filtration when two consecutive times get close to one another. For a certain time $t \in I$, we denote

$$\mathscr{F}_{t+} = \bigcap_{u > t} \mathscr{F}_u = \lim_{u \to t; u > t} \mathscr{F}_u,$$

$$\mathscr{F}_{t-} = \sigma \left(\bigcup_{u < t} \mathscr{F}_u \right).$$

As we already know from the probability part that intersections of sigma algebras are always sigma algebras but unions are not. This is why we needed the generated sigma algebra in the definition of $\mathscr{F}_{t-}$.

Definition 9.3 (Right and Left continuous filtrations) *We say that a filtration* $\{\mathscr{F}_t\}_{t \in I}$ *is right continuous iff* $\mathscr{F}_t = \mathscr{F}_{t+}$ *for all t, and we say that the filtration is left continuous iff* $\mathscr{F}_t = \mathscr{F}_{t-}$ *for all t.*

In general, we shall assume throughout (if applicable) that any filtration is right continuous.

Definition 9.4 (Adapted stochastic process) *A stochastic process* $\{X_t\}_{t \in I}$ *defined on a filtered probability space* $(\Omega, \mathscr{F}, \mathbf{P}, \{\mathscr{F}_t\}_{t \in I})$ *is called* adapted *if and only if* X_t *is* $\mathscr{F}_t$-*measurable for any* $t \in I$.

This is an important concept. In general, $\mathscr{F}_t$ quantifies the flow of information available at any moment t. By requiring that the process be adapted, we ensure that we can calculate probabilities related to X_t based solely on the information available at time t. Furthermore, since the filtration by definition is increasing, this also says that we can calculate said probabilities at any later moment in time as well.

Remark 9.5 *On the other hand, also due to the same increasing property of filtration, it may not be possible to calculate probabilities related to* X_t *based only on the information available in* $\mathscr{F}_s$ *for a moment s earlier than t (s < t). This is why the conditional expectation is important for stochastic processes. Recall that* $\mathbf{E}[X_t | \mathscr{F}_s]$ *is* $\mathscr{F}_s$-*measurable, therefore even if we may not calculate the probabilities related to* X_t, *we can calculate probabilities related to its best guess based on the current information* $\mathbf{E}[X_t | \mathscr{F}_s]$.

Definition 9.6 (Standard filtration) *In some cases, we are only given a standard probability space (without a filtration defined on the space). This usually corresponds to the case where we assume that all the information available at time t comes from the stochastic process* X_t *itself. No external sources of information are available. In this case, we introduce the standard filtration generated by the process* $\{X_t\}_{t \in I}$ *itself. To this end, let*

$$\mathscr{F}_t = \sigma(\{X_s : s \leq t, s \in I\}),$$

where we use the notation for the sigma algebra generated by random variables given in Definition 2.19 on page 69. With this definition, the collection of sigma algebras $\{\mathscr{F}_t\}_t$ is increasing, and obviously the process $\{X_t\}_t$ is adapted with respect to it.

Notation   In the case where the filtration is not specified, we will always construct the standard filtration and denote it with $\{\mathscr{F}_t\}_t$.

In the special case where $\mathcal{I} = \mathbb{N}$, the natural numbers, and the filtration is generated by the process, we will sometimes substitute the notation $X_1, X_2, \ldots, X_n$ instead of $\mathscr{F}_n$, as in

$$\mathbf{E}[X_T^2|\mathscr{F}_n] = \mathbf{E}[X_T^2|X_n, \ldots, X_0].$$

9.1.4   Pathwise realizations

Suppose a stochastic process X_t is defined on some probability space $(\Omega, \mathscr{F}, \mathbf{P})$. Recall that, by definition, for every $t \in \mathcal{I}$ fixed, X_t is a random variable. On the other hand, for every $\omega \in \Omega$ fixed in the probability space, we shall find one realization of these variables corresponding to this ω for all t's: $\{X_t(\omega)\}_t$. Thus, for every ω we find a collection of numbers representing the realization of the stochastic process – a path. Therefore, a realization of the stochastic process is a path of the stochastic process:

$$t \mapsto X_t(\omega).$$

This means that we can identify each ω with a function from $\mathcal{I}$ into $\mathbb{R}$, and thus the space Ω may be identified as a subset of all the functions from $\mathcal{I}$ into $\mathbb{R}$.

In Figure 9.1, we plot three different paths, each corresponding to a different realization ω. Accordingly, calculating probabilities regarding the distribution of the stochastic processes is equivalent to calculating the distribution of these paths in the two-dimensional space xy. However, such a calculation is impossible when the state space is infinite or when the index set is infinite (infinite here means not countable). There is hope, however.

9.1.5   The finite distribution of stochastic processes

As we have seen, a stochastic process is just a collection of random variables. Thus, we have to ask: what quantities characterize a random variable? The answer is obviously its distribution. However, here we are working with a lot of variables. Depending on the number of elements in the index set $\mathcal{I}$, the stochastic process may have a finite or infinite number of components. In either case, we will be concerned with the joint distribution of a finite sample taken from the process. This is due to practical considerations and the fact that in general we cannot study the joint distribution of a continuum of random variables. The processes that have a continuum structure on the set $\mathcal{I}$ serve as subject for a more advanced topic in stochastic differential equations. However, even in that more advanced situation, the finite distribution of the process still constitutes a primary object of study.

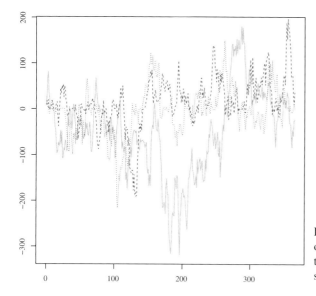

Figure 9.1 An example of three paths corresponding to three ω's for a certain stochastic process.

Let us clarify what we mean by finite-dimensional distribution. Let $\{X_t\}_{t\in\mathcal{I}}$ be a stochastic process. For any $n \geq 1$ and for any subset $\{t_1, t_2, \ldots, t_n\}$ of $\mathcal{I}$, we will write $F_{X_{t_1}, X_{t_2}, \ldots, X_{t_n}}$ for the joint distribution function of the variables $X_{t_1}, X_{t_2}, \ldots,$ X_{t_n}. The statistical properties of the process X_t are completely described by the family of distribution functions $F_{X_{t_1}, X_{t_2}, \ldots, X_{t_n}}$ indexed by the n and the t_i's. This is a famous result due to Kolmogorov in the 1930s, (the exact statement is omitted – the consistency relations are very logical, you can look them up on any stochastic processes book – for example Karlin and Taylor (1975) or Øksendal (2003)).

I will restate this result again: *If we can describe these finite-dimensional joint distributions, we completely characterize the stochastic process.* Unfortunately, in general this is a complicated task. However, there are some properties of the stochastic processes that make this calculation task much easier. Intuitively, you saw in Figure 9.1 three different paths. It is pretty clear that every time the paths are produced they are different, but note that the paths have common characteristics. For example, the paths keep coming back to the starting point and they seem to have large oscillations when the process has large values and small oscillations when the process is close to 0. These features, if they exist, help us calculate probabilities related to the stochastic processes. Next we talk about the most important types of such features.

9.1.6 Independent components

This is the most desirable property and the most useless. Let us explain. This property equivalent to saying that for any sample $\{t_1, t_2, \ldots, t_n\}$ of elements in $\mathcal{I}$, the corresponding random variables $X_{t_1}, X_{t_2}, \ldots, X_{t_n}$ are independent. Note that, in this case, the joint distribution $F_{X_{t_1}, X_{t_2}, \ldots, X_{t_n}}$ is just the product of the marginals $F_{X_{t_i}}$.

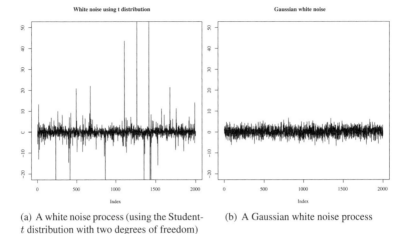

(a) A white noise process (using the Student-t distribution with two degrees of freedom)   (b) A Gaussian white noise process

Figure 9.2   We have kept the y-axes the same to illustrate the differences in these white noise processes

Thus, it is very easy to calculate. However, no reasonable real life processes have this property. In effect, every new component being random implies no structure of the process, so this is just a noise process. We give a definition of this noise for future reference.

Definition 9.7 (White noise process)   *A stochastic process* $\{X_t\}_{t \in \mathcal{I}}$ *is called a* white noise process *if it has independent components in the sense above, and for any t, the random variables* X_t *have the same distribution* $F(x)$, *with the expected value* $\mathbf{E}[X_t] = 0$.
The process is called a Gaussian white noise process *if it is a white noise process and, in addition, the common distribution of the* X_t's *is normal with mean* 0.

Note that independent components do not require the distribution to be the same for all variables. Generally speaking, in practice, this is the component in a perceived signal that one wishes to eliminate to get to the real signal process; this is why the process is defined as noise. Note also that the general white noise process may look like it contains signal. It does not. Figure 9.2 illustrates this; the large values of the process are totally random, and in fact their frequency may be calculated using the Student-t_2 distribution.

9.1.7   Stationary process

A stochastic process X_t is said to be *strictly stationary* if the joint distribution functions of

$$(X_{t_1}, X_{t_2}, \ldots, X_{t_n}) \quad \text{and} \quad (X_{t_1+h}, X_{t_2+h}, \ldots, X_{t_n+h})$$

are the same for all $h > 0$ and any arbitrary selection $\{t_1, t_2, \ldots, t_n\}$ in $\mathcal{I}$. In particular, the distribution of X_t is the same for all t. Notice that this property simplifies the calculation of the joint distribution function. The condition implies that in essence the process is in equilibrium and that the particular times at which we choose to examine the process are of no relevance.

A stochastic process X_t is said to be *wide sense stationary* or *covariance stationary* if X_t has finite second moments for any t and if the covariance function $Cov(X_t, X_{t+h})$ depends only on h for all $t \in \mathcal{I}$. This is a generalization of the notion of stationarity. A strictly stationary process with finite second moments is covariance stationary. The reverse is not true; there are plenty of examples of processes which are covariance stationary but are not strictly stationary. The notion arose from real-life processes that are covariance stationary but not stationary. In addition, we can test the stationarity of the covariance, but it is very hard to test strict stationarity.

Many phenomena can be described by stationary processes. Furthermore, many classes of processes, which will be discussed later in this book, become eventually stationary if observed for a long time.

However, some of the most common processes encountered in practice – the Poisson process and the Brownian motion – are not stationary. Instead, they have stationary (and independent) increments.

The white noise process is strictly stationary.

9.1.8 Stationary and independent increments

In order to talk about the increments of the process, we assume that the set $\mathcal{I}$ is totally ordered.

A stochastic process X_t is said to have *independent increments* if the random variables

$$X_{t_2} - X_{t_1}, \; X_{t_3} - X_{t_2}, \; \ldots, \; X_{t_n} - X_{t_{n-1}}$$

are independent for any n and any choice of the sequence $\{t_1, t_2, \ldots, t_n\}$ in $\mathcal{I}$ with $t_1 < t_2 < \cdots < t_n$.

A stochastic process X_t is said to have *stationary increments* if the distribution of the random variable $X_{t+h} - X_t$ depends only on the length h of the increment and not on the time t. Notice that this is not the same as stationarity of the process itself. In fact, except for the constant process, there exist no process with stationary *and* independent increments which is also stationary. This is illustrated in the next proposition.

Proposition 9.8 *If a process* $\{X_t, t \in [0, \infty)\}$ *has stationary and independent increments and* $X_t \in L^1$, $\forall t$, *then*

$$\begin{cases} \mathbf{E}[X_t] = m_0 + m_1 t \\ Var[X_t - X_0] = Var[X_1 - X_0]t, \end{cases}$$

where $m_0 = \mathbf{E}[X_0]$, and $m_1 = \mathbf{E}[X_1] - m_0$.

Proof: We will give the proof for the variance; the result for means is entirely similar (see Karlin and Taylor (1975)). Let $f(t) = Var[X_t - X_0]$. Then for any t, s we have

$$\begin{aligned} f(t + s) &= Var[X_{t+s} - X_0] = Var[X_{t+s} - X_s + X_s - X_0] \\ &= Var[X_{t+s} - X_s] + Var[X_s - X_0] \text{ (indep. increments)} \\ &= Var[X_t - X_0] + Var[X_s - X_0] \text{ (stationary increments)} \\ &= f(t) + f(s) \end{aligned}$$

or the function f is additive (the above equation is also called *Cauchy's functional equation*). If we assume that the function f obeys some regularity conditions[2], then the only solution is $f(t) = f(1)t$ and the result stated in the proposition holds.   ∎

9.1.9   Other properties that characterize specific classes of stochastic processes

- **Point Processes.** These are special processes that count rare events. They are very useful in practice due to their frequent occurrence. For example, look at the process that gives at any time t the number of buses passing by a particular point on a certain street, starting from an initial time $t = 0$. This is a typical rare event ("rare" here does not refer to the frequency of the event, rather to the fact that there are gaps between event occurrence). Or, look at the process that counts the number of defects in a certain area of material (say 1 cm^2). A particular case of such a process (and the most important) is the Poisson process, which will be studied in the next chapter.

- **Markov processes.** In general terms, this is a process with the property that given X_s, future values of the process (X_t with $t > s$) do not depend on any earlier X_r with $r < s$. Or, put differently, the behavior of the process at any future time when its present state is known exactly is not modified by additional knowledge about its past. The study of Markov processes constitutes a big part of this book. The finite distribution of such a process has a much simplified structure. Let us explain. Using conditional distributions, for a fixed sequence of times $t_1 < t_2 < \cdots < t_n$, we may write

---

[2]These regularity conditions are (i) f is continuous, (ii) f is monotone, or (iii) f is bounded on compact intervals. In particular, the third condition is satisfied by any process with finite second moments. The linearity of the function under condition (i) was first proven by Cauchy himself Cauchy (1821).

$$
\begin{aligned}
F_{X_{t_1},X_{t_2},\ldots,X_{t_n}} &= F_{X_{t_n}|X_{t_{n-1}},\ldots,X_{t_1}} F_{X_{t_{n-1}}|X_{t_{n-2}},\ldots,X_{t_1}} \cdots F_{X_{t_2}|X_{t_1}} F_{X_{t_1}} \\
&= F_{X_{t_n}|X_{t_{n-1}}} F_{X_{t_{n-1}}|X_{t_{n-2}}} \cdots F_{X_{t_2}|X_{t_1}} F_{X_{t_1}} \\
&= F_{X_{t_1}} \prod_{i=2}^{n} F_{X_{t_i}|X_{t_{i-1}}}
\end{aligned}
$$

which is a much simpler structure. In particular, it means that we only need to describe one-step transitions.

- **Martingales.** This is a process that has the property that the expected value of the future, given the information we have today, is going to be equal to the known value of the process today. These are some of the oldest processes studied in the history of probability due to their tight connection with gambling. In fact, in French (the origin of the name is attributed to Paul Lévy) a martingale means a winning strategy (formula).

9.2   A Simple Process – The Bernoulli Process

We will start the study of stochastic processes with a very simple process – tosses of a (not necessarily fair) coin. Historically, this is the first process ever studied.

Let $Y_1, Y_2, \ldots$ be independent identically distributed (i.i.d.) Bernoulli random variables with parameter p:

$$
Y_i = \begin{cases} 1 & \text{with probability } p \\ 0 & \text{with probability } 1 - p \end{cases}.
$$

To simplify the language, say, a head appears when $Y_i = 1$ and a tail is obtained at the ith toss if $Y_i = 0$. Let

$$
N_k = \sum_{i=1}^{k} Y_i
$$

be the number of heads up to the kth toss, which we know is distributed as a Binomial (k, p) random variable ($N_k \sim \text{Binom}(k, p)$).

A sample outcome may look like this:

Table 9.1   Sample Outcome

Y_i	0	0	1	0	0	1	0	0	0	0	1	1	1
N_i	0	0	1	1	1	2	2	2	2	2	3	4	5

Let S_n be the time at which nth head (success) occurred. Mathematically

$$S_n = \inf\{k : N_k = n\}.$$

Let $X_n = S_n - S_{n-1}$ be the number of tosses to get the nth head starting from the $(n-1)$th head. We present a sample image below:

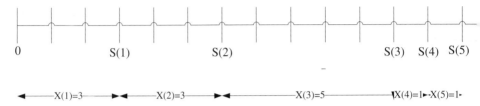

Figure 9.3 Failure and waiting time

Proposition 9.9 *We will give some simple results about these processes.*

1) "Waiting times" $X_1, X_2 \ldots$ are i.i.d. "trials" $\sim Geometric(p)$ random variables.

2) The time at which the nth head occurs is negative binomial, $S_n \sim$ negative binomial(n, p).

3) Given $N_k = n$, the distribution of $(S_1, \ldots, S_n)$ is the same as the distribution of a random sample of n numbers chosen without replacement from $\{1, 2, \ldots, k\}$.

4) Given $S_n = k$, the distribution of $(S_1, \ldots, S_{n-1})$ is the same as the distribution of a random sample of $n - 1$ numbers chosen without replacement from $\{1, 2, \ldots, k - 1\}$.

5) We have as sets

$$\{S_n > k\} = \{N_k < n\}.$$

6) Central Limit theorems:

$$\frac{N_k - \mathbf{E}[N_k]}{\sqrt{Var[N_k]}} = \frac{N_k - kp}{\sqrt{kp(1-p)}} \xrightarrow{D} N(0, 1).$$

7)

$$\frac{S_n - \mathbf{E}[S_n]}{\sqrt{Var[S_n]}} = \frac{S_n - n/p}{\sqrt{n(1-p)/p}} \xrightarrow{D} N(0, 1).$$

8) As $p \downarrow 0$,

$$\frac{X_1}{\mathbf{E}[X_1]} = \frac{X_1}{1/p} \xrightarrow{D} exponential(\lambda = 1).$$

9) As $p \downarrow 0$,

$$\mathbf{P}\left\{N_{[\frac{t}{p}]} = j\right\} \longrightarrow \frac{t^j}{j!}e^{-t}.$$

We will prove several of these properties. The rest are assigned as exercises.

For 1) and 2), the distributional assertions are easy to prove: we may just use the definition of Geometric(p) and negative binomial random variables. We need only to show that the X_i's are independent. See Problem 9.1.

Proof (Proof for 3)): We take $n = 4$ and $k = 100$ and prove this part only for this particular case. The general proof is identical (Problem 9.3). A typical outcome of a Bernoulli process looks like

$$\omega : 00100101000101110000100.$$

In the calculation of probability, we have to have $1 \le s_1 < s_2 < s_3 < s_4 \le 100$. Using the definition of the conditional probability, we can write

$$\mathbf{P}(S_1 = s_1 \dots S_4 = s_4 | N_4 = 100)$$

$$= \frac{\mathbf{P}(S_1 = s_1 \dots S_4 = s_4, \text{ and } N_{100} = 4)}{\mathbf{P}(N_{100} = 4)}$$

$$= \frac{\mathbf{P}\left(\overbrace{00\dots0}^{s_1-1} 1 \overbrace{00\dots0}^{s_2-s_1-1} 1 \overbrace{00\dots0}^{s_3-s_2-1} 1 \overbrace{00\dots0}^{s_4-s_3-1} 1 \overbrace{00\dots0}^{100-s_4}\right)}{\binom{100}{4}p^4(1-p)^{96}}$$

$$= \frac{(1-p)^{s_1-1}p(1-p)^{s_2-s_1-1}p(1-p)^{s_3-s_2-1}p(1-p)^{s_4-s_3-1}p(1-p)^{100-s_4}}{\binom{100}{4}p^4(1-p)^{96}}$$

$$= \frac{(1-p)^{96}p^4}{\binom{100}{4}p^4(1-p)^{96}} = \frac{1}{\binom{100}{4}}.$$

The result is significant since it means that if we only know that there have been four heads by the 100th toss, then any 4 tosses among these 100 are equally likely to contain the heads.   ∎

Proof (Proof for 8)):

$$\mathbf{P}\left(\frac{X_1}{1/p} > t\right) = \mathbf{P}\left(X_1 > \frac{t}{p}\right) = \mathbf{P}\left(X_1 > \left[\frac{t}{p}\right]\right)$$

$$= (1-p)^{[\frac{t}{p}]} = \left[(1-p)^{-\frac{1}{p}}\right]^{-p[\frac{t}{p}]} \longrightarrow e^{-t},$$

since

$$\lim_{p \to 0} -p\left[\frac{t}{p}\right] = \lim_{p \to 0} -p\left(\frac{t}{p} + \left[\frac{t}{p}\right] - \frac{t}{p}\right)$$

$$= -t + \lim_{p \to 0} p\underbrace{\left(\frac{t}{p} - \left[\frac{t}{p}\right]\right)}_{\in [0,1]} = -t.$$

Therefore, $\mathbf{P}\left(\frac{X_1}{1/p} \leq t\right) \to 1 - e^{-t}$, and we are done. ∎

The problems ending this chapter contain a more involved application of the Borel–Cantelli lemmas 1.28 and 1.29 to the Bernoulli process. The example is due to Dembo (2008).

Problems

9.1   Prove that the X_i's in Proposition 9.9 are in fact independent.

9.2   Using the notation in Section 9.2, give the distribution of N_3 and the joint distribution of (N_2, N_3) when $p = \frac{1}{3}$. Furthermore, calculate $\mathbf{E}(N_3)$ and $\mathbf{E}(N_n)$ for some arbitrary n.

9.3   Give a general proof for parts 3) and 4) in Proposition 9.9 for any $n, k \in \mathbb{N}$.

9.4   Show the equality of sets in part 5) of Proposition 9.9 by double inclusion.

9.5   Prove parts 6) and 7) of Proposition 9.9 by applying the Central Limit Theorem.

9.6   Prove part 9) of Proposition 9.9.

9.7   Let X_i be distributed as a normal with mean μ_i and variance σ^2, where $\mu_1 < \mu_2 < \cdots < \mu_n < \dots$. Let $\mathscr{F}_i = \sigma(X_i)$ the sigma algebra generated by the random variable X_i.
  a) Is $\mathscr{F} = \{\mathscr{F}_i\}_i$ a filtration?
  b) If it is not, how can we modify it so that it becomes a filtration for the stochastic process $X = \{X_i\}$?
  c) Is $X = \{X_i\}$ stationary? Does it have independent components?
  d) Does $X = \{X_i\}$ have stationary increments? Does it have independent increments?

9.8   Let X_i be distributed as a normal with mean μ_i and variance σ^2, where $\mu_1 < \mu_2 < \cdots < \mu_n < \dots$, as in the previous problem. Let a process $Y = \{Y_i\}_i$ be defined as the sum of first i components of X_i, that is

$$Y_i = \sum_{j=1}^{i} X_j.$$

  a) Construct a filtration $\mathscr{F} = \{\mathscr{F}_i\}_i$ for the stochastic process $Y = \{Y_i\}_i$.
  b) Is $Y = \{Y_i\}_i$ stationary? Does it have independent components?
  c) Does $Y = \{Y_i\}_i$ have stationary increments? Does it have independent increments?

9.9   Let $S_n = \sum_{i=1}^{n} X_i$, where X_i's are i.i.d. $Uniform(0, 1)$ random variables. Define the stochastic process

$$N_t = \sup\{n \in \mathbb{N} \mid S_n \leq n\}.$$

N_t is an example of a point process.

    a) Does S_n have continuous paths? Continuous index set?

    b) Does N_t have continuous paths? Continuous index set?

    c) Show that

$$\{\omega \mid N_t(\omega) \leq n\} = \{\omega \mid S_n(\omega) \geq t\}$$

    as equality of sets (use double inclusion).

    d) Show that

$$\mathbf{E}[N_t] = \sum_{n=1}^{\infty} n\mathbf{P}(N_t = n) = \sum_{n=1}^{\infty} \mathbf{P}(N_t \leq n).$$

Exercises due to Amir Dembo about the Bernoulli process The next problems refer to the following situation. Consider an infinite Bernoulli process with $p = 0.5$, i.e., an infinite sequence of random variables $\{Y_i, i \in \mathbb{Z}\}$ with $\mathbf{P}(Y_i = 0) = \mathbf{P}(Y_i = 1) = 0.5$ for all $i \in \mathbb{Z}$. We would like to study the length of the maximum sequence of 1's. To this end, let us define some quantities.

    Let

$$l_m = \max\{i \geq 1 : X_{m-i+1} = \cdots = X_m = 1\},$$

be the length of the run of 1's up to the mth toss and including it. Obviously, l_m will be 0 if the mth toss is a tail. We are interested in the asymptotic behavior of the longest such run from 1 to n for large n.

    That is, we are interested in the behavior of L_n where

$$L_n = \max_{m \in \{1, \ldots, n\}} l_m$$
$$= \max\{i \geq 1 : X_{m-i+1} = \cdots = X_m = 1, \text{ for some } m \in \{1, \ldots, n\}\}.$$

9.10   Explain why $\mathbf{P}(l_m = i) = 2^{-(i+1)}$, for $i = 0, 1, 2, \ldots$ and any m.

9.11   Apply the first Borel–Cantelli lemma 1.28 to the events

$$A_n = \{l_n > (1 + \varepsilon) \log_2 n\}.$$

Conclude that for each $\varepsilon > 0$, with probability 1, $l_n \leq (1 + \varepsilon) \log_2 n$ for all n large enough.

Take a countable sequence $\varepsilon_k \downarrow 0$, and then conclude that

$$\limsup_{n \to \infty} \frac{L_n}{\log_2 n} \leq 1, \quad \text{a.s.}$$

9.12 Fix $\varepsilon > 0$. Let $A_n = \{L_n < k_n\}$ for $k_n = (1 - \varepsilon) \log_2 n$. Explain why

$$A_n \subseteq \bigcap_{i=1}^{m_n} B_i^c,$$

where $m_n = [n/k_n]$ (integer part) and $B_i = \{X_{(i-1)k_n+1} = \ldots = X_{ik_n} = 1\}$ are independent events.

Deduce that $\mathbf{P}(A_n) \leq \mathbf{P}(B_i^c)^{m_n} \leq \exp(-n^\varepsilon/(2 \log_2 n))$, for all n large enough.

9.13 Apply the first Borel–Cantelli lemma for the events A_n defined in problem 9.12, followed by $\varepsilon \downarrow 0$, to conclude that

$$\liminf_{n \to \infty} \frac{L_n}{\log_2 n} \geq 1 \quad \text{a.s.}$$

9.14 Combine problems 9.11 and 9.13 together to conclude that

$$\frac{L_n}{\log_2 n} \to 1 \quad \text{a.s.}$$

Therefore the length of the maximum sequence of Heads is approximately equal to $\log_2 n$, when n, the number of tosses, is large enough.

CHAPTER 10

THE POISSON PROCESS

We now start the study of a stochastic process which is absolutely vital for applications. Random number generation (Lu et al., 1996), queuing theory (Gross and Harris, 1998), quantum physics (Jona-Lasinio, 1985), electronics (Kingman, 1993), and biology (Blæsild and Granfeldt, 2002) are only some of the many areas of application for this process (Good, 1986).

The process is named in honor of Siméon-Denis Poisson (1781–1840), and, as we shall see, it is a continuous-time process (the index set $\mathcal{I} = (0, \infty)$) with discrete state space $\mathcal{S} = \mathbb{N}$.

We start with basic definitions.

10.1 Definitions

As always, we work on a probability space $(\Omega, \mathcal{F}, \mathbf{P})$ endowed with a filtration $\{\mathcal{F}_t\}_{t \geq 0}$. However, as we shall see, the Poisson process has a discrete state space. Therefore, as mentioned in the probability part of the book, the sigma algebra and the filtration are not essential, and we shall take $\mathcal{F}$ formed from the pre-images of all subsets of $\mathbb{N} = \{0, 1, 2, \ldots\}$. This will allow us to have every set measurable, and we

Probability and Stochastic Processes, First Edition. Ionuţ Florescu
© 2015 John Wiley & Sons, Inc. Published 2015 by John Wiley & Sons, Inc.

will be able to calculate any probability we wish. Thus, this will be the last mention of σ-algebras in this chapter.

Definition 10.1 (Counting Process) N_t *is a counting process if and only if*

1. $N_t \in \{0, 1, 2 \ldots\}, \forall t$, *and*

2. N_t *is nondecreasing as a function of t.*

Here, N_t nondecreasing means pathwise (i.e., all the sample paths $N_t(w)$ are non-decreasing as a function of t for every $w \in \Omega$, with w fixed).

Definition 10.2 (Poisson Process) $N(t)$ *is a Poisson process if it is a counting process And, in addition,*

1. $N(0) = 0$,

2. $N(t)$ *has stationary independent increments,*

3. $P(N(h) = 1) = \lambda h + o(h)$, *and*

4. $P(N(h) \geq 2) = o(h)$.

The parameter λ that appears in the third property is called the arrival rate *(or simply the* rate*) of the Poisson process. We shall say $N(t)$ is a Poisson(λ) process.*

The fourth property means that the chance of two or more events happening simultaneously is zero. This is a defining characteristic of the Poisson process.

Explanation of the little o and big O notations in the context of the Poisson process:

$f \sim o(g)$ if and only if $\lim_{h \to 0} \frac{f(h)}{g(h)} = 0$.

$f \sim O(g)$ if and only if there exist c_1, c_2 constants, such that $c_1 \leq \frac{f(h)}{g(h)} \leq c_2$, for any h in a neighborhood of 0.

Theorem 10.3 *If $N(t)$ is a Poisson(λ) process, then*

$$P(N(t) = n) = \frac{(\lambda t)^n}{n!} e^{-\lambda t}.$$

In other words, $N(t)$ is distributed as a Poisson random variable with mean λt. We also note that $P(N(s + t) - N(s) = n) = P(N(t) = n)$ by the stationarity of the increments, so this distribution completely characterizes the entire process.

Proof: A standard proof derives and solves the Kolmogorov's forward differential equations of the Poisson(λ) process, a discrete state space Markov chain. This method will be seen later when we talk about Markov chains. See, for example, Ross (1995) for this standard proof.

Here, we will give a different proof – a more constructive one. The idea is to approximate a Poisson(λ) process with a Bernoulli process and then make the size of the intervals go to 0. To this end, we fix $t > 0$, and then partition $[0, t]$ into 2^k equally spaced intervals. Specifically, the partition is

$$\pi = \left(0 < \frac{t}{2^k} < 2\frac{t}{2^k} < \cdots < (2^k - 1)\frac{t}{2^k} < t \right).$$

Let $\widetilde{N}_k$ = the number of these 2^k intervals with at least one event in them. We want to apply Proposition 7.35 on page 231. We will show that $\widetilde{N}_k$ converges in distribution to the desired Poisson random variable and that $|\widetilde{N}_k - N(t)|$ converges in probability to 0 as $k \to \infty$. Applying the proposition the conclusion follows since $N(t)$ does not depend on k.

Step 1   Note that the probability that at least one event occurs in any one interval, by the stationarity of increments is

$$\mathbf{P}\left(N\left(\frac{t}{2^k}\right) \geq 1 \right) = \mathbf{P}\left(N\left(\frac{t}{2^k}\right) = 1 \right) + \mathbf{P}\left(N\left(\frac{t}{2^k}\right) \geq 2 \right) = \lambda\frac{t}{2^k} + 2o\left(\frac{t}{2^k}\right).$$

For each interval, we have either no event occurring or at least one event occurring. Since the increments are independent, we obtain the distribution of the approximating process: $\widetilde{N}_k \sim$ Binomial $\left(2^k, \lambda\frac{t}{2^k} + 2o\left(\frac{t}{2^k}\right)\right)$.
   To continue, we state the following result:

Exercise 8 *If W_k is a Binomial(k, p_k) random variable and $kp_k \to \lambda$ when $k \to \infty$, then*

$$W_k \xrightarrow{\mathcal{D}} Poisson(\lambda),$$

that is,

$$\lim_{k \to \infty} \mathbf{P}(W_k = n) = \frac{\lambda^n}{n!}e^{-\lambda}.$$

This result is stated as an exercise (Problem 10.1) at the end of the chapter. In our case, $\widetilde{N}_k$ has the binomial distribution and the condition $kp_k \to \lambda$ is verified since

$$2^k\left(\lambda\frac{t}{2^k} + 2o\left(\frac{t}{2^k}\right)\right) = \lambda t + 2\frac{o(t/2^k)}{t/2^k}t \xrightarrow{k \to \infty} \lambda t$$

. Therefore, the stated Exercise 8 above implies

$$\widetilde{N}_k \xrightarrow{\mathcal{D}} Poisson(\lambda t), \tag{10.1}$$

or

$$P(\widetilde{N}_k = n) \to \frac{(\lambda t)^n}{n!}e^{-\lambda t}$$

.

Step 2 We note that by the definition of $\widetilde{N}_k$ we must have

$$\widetilde{N}_k \leq N_t,$$

with equality if and only if each interval contains exactly 1 event.
Then, since both variables are integer-valued, for any small $\varepsilon > 0$ we have the event E_k, given by

$$E_k = \{|N_t - \widetilde{N}_k| > \varepsilon\} = \{|N_t - \widetilde{N}_k| > 0\} = \{N_t > \widetilde{N}_k\}$$

$$= \bigcup_{i=0}^{2^{k-1}} \underbrace{\left\{ N\left(\frac{i+1}{2^k}t\right) - N\left(\frac{i}{2^k}t\right) \geq 2 \right\}}_{\text{at least one interval with two events}}.$$

We can use double inclusion to show that the last equality holds. We therefore calculate the probability

$$\mathbf{P}(E_k) \leq \sum_{i=0}^{2^{k-1}} \mathbf{P}\left(N\left(\frac{i+1}{2^k}t\right) - N\left(\frac{i}{2^k}t\right) \geq 2 \right)$$

$$= \sum_{i=0}^{2^{k-1}} \mathbf{P}\left(N\left(\frac{1}{2^k}t\right) \geq 2 \right) \quad \text{(by stationarity)}$$

$$= 2^k \, o\left(\frac{t}{2^k}\right) = \frac{o(t/2^k)}{t/2^k} \, t \xrightarrow{k \to \infty} 0,$$

for every t fixed. So $\mathbf{P}(E_k) \to 0$ as $k \to \infty$, therefore

$$|\widetilde{N}_k - N(t)| \xrightarrow{P} 0. \tag{10.2}$$

Applying the proposition and using (10.1) and (10.2), we obtain the conclusion of the theorem. ∎

10.2 Inter-Arrival and Waiting Time for a Poisson Process

Let

$$X_1 = \text{time of the first event,}$$
$$X_2 = \text{time between the first and second event,}$$
$$\vdots$$
$$X_n = \text{time between } (n-1)\text{th and } n\text{th event.}$$

Let $S_n =$ time of the nth event. Notice that

$$S_n = \sum_{i=1}^{n} X_i,$$

$$S_n = \inf\{t|N(t) \geq n\} = \inf\{t|N(t) = n\}.$$

Proposition 10.4 *The inter-arrival times $X_1, X_2 \ldots$ of a Poisson process with rate λ are i.i.d. random variables, exponentially distributed with mean $\frac{1}{\lambda}$.*

Note that we may prove that the distribution of each inter-arrival time is exponential very easily. If $t > 0$

$$\begin{aligned}
\mathbf{P}(X_i > t) &= \mathbf{P}\{N(S_{i-1} + t) - N(S_{i-1}) = 0\} \\
&= \int_0^\infty \mathbf{P}\{N(s + t) - N(s) = 0|S_{i-1} = s\}dF_{S_{i-1}}(s) \\
&= \int_0^\infty \mathbf{P}\{N_t = 0|S_{i-1} = s\}dF_{S_{i-1}}(s) \\
&= \mathbf{P}\{N_t = 0\} = e^{-\lambda t}
\end{aligned}$$

by stationarity and independence of increments. Thus the distribution is $F(t) = 1 - e^{-\lambda t}$, which is the distribution function of an exponential with mean $1/\lambda$. However, showing that they are independent is not so simple.

10.2.1 Proving that the inter-arrival times are independent

The proof of the proposition is very interesting, and in fact is done in several steps. The steps themselves provide insight into the behavior of the process, which is why they are included as a separate subsection.

Claim–Evidence The distribution of S_n is Gamma(n, λ); more specifically, the density (pdf) of S_n is given by

$$f_{S_n}(t) = \frac{\lambda e^{-\lambda t}(\lambda t)^{n-1}}{(n-1)!} \qquad t \geq 0$$

.

Why is this statement labeled as an evidence? We note first that the exponential distribution is a special case of the Gamma distribution:

$$\text{Exponential}(\lambda) = \text{Gamma}(1, \lambda).$$

We also note that, if two variables are gamma-distributed and independent, then the distribution of their sum is

$$\text{Gamma}(\alpha_1, \beta) + \text{Gamma}(\alpha_2, \beta) \overset{\mathcal{D}}{=} \text{Gamma}(\alpha_1 + \alpha_2, \beta).$$

This last is an easy exercise (Exercise 6.12 on page 209) or check any statistics book such as Casella and Berger (2001).

Now we can see that, if Proposition 10.4 is true and the inter-arrival times are i.i.d. exponential, then we must have the distribution of the arrival times S_n as

$$S_1 = X_1 \sim \text{Gamma}(1, \lambda),$$
$$S_2 = X_1 + X_2 \sim \text{Gamma}(2, \lambda),$$
$$\vdots$$
$$S_n = X_1 + \ldots + X_n \sim \text{Gamma}(n, \lambda).$$

Therefore, proving the claim adds evidence in favor of Proposition 10.4. We shall, in fact, prove the Proposition using the claim in this subsection.

Proof (Proof of the claim-evidence): We can show very easily that $\{S_n \leq t\} = \{N(t) \geq n\}$ (problem 10.2). Thus, the (cdf)

$$F_{S_n}(t) = \mathbf{P}\{S_n \leq t\} = \mathbf{P}\{N(t) \geq n\} = \sum_{j=n}^{\infty} \frac{(\lambda t)^j}{j!} e^{-\lambda t}.$$

Take the derivative with respect to t, i.e., $\frac{d}{dt}$

$$f_{S_n}(t) = \sum_{j=n}^{\infty} \left[\left(\frac{\lambda j (\lambda t)^{j-1}}{j!} \right) e^{-\lambda t} + \frac{(\lambda t)^j}{j!} (-\lambda) e^{-\lambda t} \right]$$

$$= \lambda e^{-\lambda t} \sum_{j=n}^{\infty} \left[\frac{(\lambda t)^{j-1}}{(j-1)!} - \frac{(\lambda t)^j}{j!} \right]$$

$$= \frac{\lambda e^{-\lambda t} (\lambda t)^{n-1}}{(n-1)!}, \quad \text{and DONE.}$$

We can also illustrate the calculation using an alternative mode (this particular idea will become very useful in the next chapter).

$$f_{S_n}(t) dt \approx \mathbf{P}(t \leq S_n \leq t + dt)$$

$$= P \left(\underbrace{N(t) = n - 1}_{\text{independent}} \text{ and } \underbrace{\text{at least one event takes place in } [t, t + dt]}_{\text{independent}} \right)$$

$$= P(N(t) = n - 1) P(N(dt) \geq 1)$$

$$= \frac{(\lambda t)^{n-1} e^{-\lambda t}}{(n-1)!} [\lambda dt + o(dt) + o(dt)]$$

Dividing the last expression by dt and taking $dt \to 0$, we get

$$f_{S_n} = \frac{(\lambda t)^{n-1} e^{-\lambda t}}{(n-1)!} \left(\lambda + 2 \frac{o(dt)}{dt} \right) \xrightarrow{dt \to 0} \frac{(\lambda t)^{n-1} \lambda e^{-\lambda t}}{(n-1)!}$$

■

We shall use the claim evidence to calculate the joint distribution of the X_i's. To this end, we need to calculate first the joint density of the S_i's.

Theorem 10.5 *The joint density of* $(S_1, \ldots, S_n) = (X_1, X_1 + X_2, \ldots, \sum_{i=1}^n X_i)$ *is*

$$f_{S_1 \ldots S_n}(t_1 \ldots t_n) = \lambda^n e^{-\lambda t_n} \mathbf{1}_{\{0 \le t_1 < t_2 \ldots < t_n\}}$$

Proof: Let $0 \le t_1 < t_2 \ldots < t_n$, and let $\delta > 0$ be a constant small enough such that $0 \le t_1 < t_1 + \delta < t_2 < t_2 + \delta \ldots < t_n$[1]. Let

$$I_j = (t_j, t_j + \delta)$$

Goal   Find $\mathbf{P}(S_1 \in I_1, S_2 \in I_2, \ldots, S_n \in I_n)$ and then take $\delta \to 0$ to obtain the joint density. Note that terms can be expressed as follows:

$$\begin{cases} S_1 \in I_1 & \text{no event in } (0, t_1) \text{ and 1 event in } I_1 \\ S_2 \in I_2 & \text{no event in } (t_1 + \delta, t_2) \text{ and 1 event in } I_2 \\ \quad \vdots \\ S_n \in I_n & \text{no event in } (t_{n-1} + \delta, t_n) \text{ and at least 1 event in } I_n \end{cases}$$

Then, since the process has independent increments, we may write

$$\mathbf{P}(S_1 \in I_1, S_2 \in I_2, \ldots, S_n \in I_n)$$

$$= \underbrace{e^{-\lambda t}}_{0 \in (0, t_1)} \underbrace{\left(\frac{\lambda \delta}{1!} e^{-\lambda(t_2 - t_1 - \delta)}\right)}_{\text{one event in } I_1} \underbrace{e^{-\lambda t}}_{0 \in (t_1 + \delta, t_2)} \underbrace{\left(\frac{\lambda \delta}{1!} e^{-\lambda(t_2 - t_1 - \delta)}\right)}_{\text{1 event in } I_2} \ldots e^{-\lambda(t_n - t_{n-1} - \delta)} \underbrace{\left(1 - e^{-\lambda \delta}\right)}_{\text{at least 1 in } I_n}$$

$$= (\lambda \delta)^{n-1} e^{-\lambda \delta (n-1)} (1 - e^{-\lambda \delta}) e^{-\lambda t_n} e^{\lambda(n-1)\delta}$$

$$= (\lambda \delta)^{n-1} e^{-\lambda t_n} (1 - e^{-\lambda \delta})$$

Divide by δ^n and note that $\frac{1 - e^{-a}}{a} \to 1$ if $a \to 0$; therefore

$$\frac{\mathbf{P}(S_1 \in I_1, S_2 \in I_2 \ldots S_n \in I_n)}{\delta^n} = e^{-\lambda t_n} \lambda^{n-1} \underbrace{\frac{1 - e^{-\lambda \delta}}{\delta}}_{\to \lambda} \to \lambda^n e^{-\lambda t_n}$$

which concludes the proof of the theorem.   ■

We are finally in a position to prove the result about inter-arrival times of a Poisson process.

Proof (Proof of Proposition 10.4):   Note that $X_1 = S_1, X_2 = S_2 - S_1, \ldots, X_n = S_n - S_{n-1}$. Thus we can form an n-dimensional transformation and obtain the joint

---

[1]In other words, chose δ such that we create non-overlapping intervals.

distribution of the X variables in a very standard way. The inverse transformation is simply $g^{-1}(x_1, \ldots, x_n) = (x_1, x_1 + x_2, \ldots, x_1 + x_2 + \cdots + x_n)$; therefore, the Jacobian J of the transformation is

$$
\begin{vmatrix}
1 & 0 & 0 \ldots & 0 \\
1 & 1 & 0 \ldots & 0 \\
\vdots & & \ddots & \\
1 & 1 & 1 \ldots & 1
\end{vmatrix} = 1
$$

The joint distribution is

$$
f_{X_1 \ldots X_n}(x_1, \ldots, x_n) = f_{S_1 \ldots S_n}\left(x_1, x_1 + x_2, \ldots, \sum_{i=1}^{n} x_i\right) |J| \mathbf{1}_{\{0 \leq x_1 \leq x_1 + x_2 \ldots \leq \sum_{i=1}^{n} x_i\}}
$$

Hence,

$$
\begin{aligned}
f_{X_1 \ldots X_n}(x_1 \ldots x_n) &= f_{S_1, \ldots, S_n}\left(x_1, x_1 + x_2, \ldots, \sum_{i=1}^{n} x_i\right) \mathbf{1}_{\{0 \leq x_1 \leq x_1 + x_2 \ldots \leq \sum_{i=1}^{n} x_i\}} \\
&= \lambda^n e^{-\lambda(x_1 + x_2 \ldots x_n)} \prod_{i=1}^{n} \mathbf{1}_{\{x_i \geq 0\}} \\
&= \prod_{i=1}^{n} \lambda e^{-\lambda x_i} \mathbf{1}_{\{x_i \geq 0\}}
\end{aligned}
$$

which is the product of n separable functions. Using Lemma 2.23 on page 76, we immediately obtain that the variables are independent and calculating the marginal distributions yields the desired result. ∎

To express the significance of the results we obtained, we recall the concept of *order statistics*.

Order Statistics Let $Y_1 \ldots Y_n$ be n random variables. We say that $Y_{(1)} \ldots Y_{(n)}$ is the vector of order statistics corresponding to $Y_1 \ldots Y_n$ if each $Y_{(k)}$ is the kth smallest value among $Y_1 \ldots Y_n$.

We recall the joint distribution of the order statistics (Theorem 7.39 on page 237): If Y_i's are continuous random variables with pdf f, then the joint density of the order statistics $Y_{(1)} \ldots Y_{(n)}$ is given by

$$
f_{Y_{(1)}, \ldots, Y_{(n)}}(y_1 \ldots y_n) = n! f(y_1) \cdots f(y_n) \mathbf{1}_{\{y_1 < y_2 < \ldots < y_n\}} \tag{10.3}
$$

Corollary 10.6 *Given $S_n = t_n$, the other $n - 1$ arrival times $S_1, S_2 \ldots S_{n-1}$ have the same distribution as the order statics corresponding to $(n - 1)$ independent uniform random variables on the interval $(0, t_n)$.*

Proof: Using the formula above, we only need to verify that the conditional distribution is the desired uniform. We have

$$f_{S_1 \ldots S_{n-1} | S_n}(t_1 \ldots t_{n-1} | t_n) = \frac{f_{S_1 \ldots S_n}(t_1 \ldots t_n)}{f_{S_n}(t_n)}$$

$$= \frac{\lambda^n e^{-\lambda t_n} \mathbf{1}_{\{0 \leq t_1 < t_2 \ldots < t_n\}}}{\frac{\lambda e^{-\lambda t_n} (\lambda t_n)^{n-1}}{(n-1!)}}$$

$$= \frac{(n-1)!}{t_n^{n-1}} \mathbf{1}_{\{0 \leq t_1 < t_2 \ldots < t_n\}}$$

Using the result above, we can also calculate the joint distribution of the order statistics of $n-1$ uniform variables. The pdf is $f(t) = \frac{1}{t_n} \mathbf{1}_{\{t < t_n\}}$, and thus by substituting in (10.3) we obtain exactly the pdf calculated above. ∎

Corollary 10.7 *Given $N(t) = n$, the n arrival times $S_1, S_2 \ldots S_n$ have the same distribution as the order statics corresponding to n independent uniform random variables on the interval $(0, t)$, that is,*

$$f_{S_1 \ldots S_n | N(t)}(t_1 \ldots t_n | n) = \frac{n!}{t^n} \mathbf{1}_{\{0 \leq t_1 < t_2 \ldots < t_n < t\}}$$

Proof: Exercise. ∎

10.2.2 Memoryless property of the exponential distribution

So why was it important to show that the inter-arrival times are exponentially distributed? The exponential distribution has a peculiar property.

Definition 10.8 *A random variable X with a distribution F is said to have the memoryless property if*

$$\mathbf{P}(X > s + h | X > s) = \mathbf{P}(X > h), \quad \forall s, h \geq 0$$

Note that this is a distributional property.

Importance A variable whose distribution has this property possesses no memory. If we already know that the variable is greater than s, this has absolutely no effect on the probability that the variable is greater than h units more.

The distributions with these property There is exactly one discrete distribution and exactly one continuous distribution with the property. The two distributions are as follows:

- The $Geometric(p)$ distribution, for any p.
- The $Exponential(\lambda)$ distribution, for any λ.

Recall that the inter-arrival times for the Bernoulli process were geometric random variables and you see the close links between that process and the Poisson process.

If X_i models the time between events, this property means that the amount of time since the last event has no influence on the amount of time until the next event. If we waited for one hour already, the probability that we will wait for at least one more hour is the same as if we have been waiting for ten hours.

10.2.3   Merging two independent Poisson processes

Suppose we observe two Poisson processes in time, $N_1(t)$ and $N_2(t)$, which have rates λ_1 and λ_2, respectively. For example, we observe cars and trucks on a highway. We are interested in the number of automobiles on that highway, that is, $N(t) = N_1(t) + N_2(t)$.

Proposition 10.9 *Assume that the two Poisson processes $N_1(t)$ and $N_2(t)$ are independent. Then their sum $N(t)$ is a Poisson process with rate $\lambda_1 + \lambda_2$.*

The proof is very simple and is left as an exercise.

10.2.4   Splitting the events of the Poisson process into types

Suppose we look at the reverse problem from above. We observe two types of automobiles passing by a highway: trucks and cars. The process that counts automobiles may be approximated quite well with a Poisson process. What about the process that counts the cars? What about the process that counts the trucks? Are they also Poisson? It turns out that under quite general assumptions the answer is, yes.

Proposition 10.10 *Assume that each event of a Poisson(λ) process can be classified as either Type I with probability p or Type II with probability $1 - p$.*
If $N_i(t)$ counts the number of events of Type i, $i \in \{I,II\}$ by time t, then $N_1(t)$ and $N_2(t)$ are independent Poisson processes with rates λp and $\lambda(1 - p)$, respectively.

Proof: There is a more general version of this proposition that will be stated in the next section. To prove the current version, let us calculate the joint distribution of the two split processes.

$$\mathbf{P}(N_1(t) = m, N_2(t) = n) = \sum_{k=0}^{\infty} \mathbf{P}(N_1(t) = m, N_2(t) = n | N(t) = k)\mathbf{P}(N(t) = k)$$
$$= \mathbf{P}(N_1(t) = m, N_2(t) = n | N(t) = m + n)\mathbf{P}(N(t) = m + n)$$
$$= \binom{n + m}{m} p^m (1 - p)^n \frac{(\lambda t)^{m+n}}{(m + n)!} e^{-\lambda t}$$

The last equality is due to the fact that conditional on there being $n + m$ trials the probability of exactly m successes is a binomial probability. Calculating further, we obtain

$$\mathbf{P}(N_1(t) = m, N_2(t) = n) = \frac{(\lambda t p)^m}{m!} e^{-\lambda t p} \frac{(\lambda t (1 - p))^n}{n!} e^{-\lambda t (1 - p)}$$

To calculate the marginal distributions, we simply sum the joint probabilities and notice that each sum represents the probabilities of a Poisson random variable. Thus we obtain

$$\mathbf{P}(N_1(t) = m) = \frac{(\lambda t p)^m}{m!} e^{-\lambda t p}$$

and similarly for the type II process. Since both processes have independent stationary increments and the joint is the product of the marginal distributions, we are done. ∎

10.3   General Poisson Processes

In this section we generalize the notion of Poisson process. Note that the way it is defined in the beginning of this chapter the Poisson process can only count events in time. Suppose that we wish to model defects in a material. Or, suppose that the events occur in time but the rate with which the events arrive in time is changing (e.g., more in the morning, less in the afternoon). We need a more general definition of Poisson process that would help model these situations.

Definition 10.11 *Let $\mathcal{X}$ be a set and $\mathcal{G}$ be a σ-field on $\mathcal{X}$. A counting process on $(\mathcal{X}, \mathcal{G})$ is a stochastic process $\{N(A)\}_{A \in \mathcal{G}}$ with the following properties:*

(i) $N(A) \in \{0, 1, 2 \dots\}$

(ii) $N(\bigcup_{i=1}^{\infty} A_i, \omega) = \sum_{i=1}^{\infty} N(A_i, \omega), \forall \omega \in \Omega$ *and* $A_1, A_2 \dots$ *disjoint sets in* $\mathcal{G}$

Definition 10.12 (General Poisson Process) *Let $(\mathcal{X}, \mathcal{G}, \mu)$ be a measure space. A Poisson process on $(\mathcal{X}, \mathcal{G})$ with intensity μ is a counting process $\{N(A)\}_{A \in \mathcal{G}}$ with*

1. *$N(A)$ is a Poisson random variable with mean $\mu(A)$*

2. *The process has independent increments, in the following sense: if $A_1, A_2 \dots A_n$, are disjoint $\mathcal{G}$ sets in $\mathcal{X}$, then $N(A_1), N(A_2), \dots N(A_n)$ are independent random variables.*

Note that, unlike the simple definition of the Poisson process, this general process does not have stationary increments anymore. This is explained after we see the next theorem.

Theorem 10.13 *Let $\{N(A)\}_{A \in \mathcal{G}}$ be a counting process on $(\mathcal{X}, \mathcal{G})$. Define a set function $\mu(A) = \mathbb{E}[N(A)]$ for $A \in \mathcal{G}$. If the following three assertions are true, namely*

1. *The counting process has independent increments (as before),*

2. $\forall \epsilon > 0$, *there exists* $\delta > 0$ *such that* $\forall A$ *with* $\mathbb{E}[N(A)] < \delta$,

$$\frac{\mathbf{P}(N(A) \geq 2)}{\mu(A)} < \epsilon$$

,

3. *If* $x \in \mathcal{X}$ *with* $\mu(\{x\}) > 0$ *then* $N(\{x\}) \sim Poisson(\mu(\{x\}))$,

then $N(A)$ *is a Poisson random variable with mean* $\mu(A)$ *for all* $A \in \mathcal{G}$. *In particular, the stochastic process* $\{N(A)\}_{A \in \mathcal{G}}$ *is a generalized Poisson process with intensity* μ.

Consequence: If $\{N(A)\}_{A \in \mathcal{G}}$ satisfies the Above, then

$$P(N(A) = k) = \frac{\mu(A)^k}{k!} e^{-\mu(A)}.$$

After looking at the theorem above, we see why the process cannot be stationary. Essentially, stationarity of increments means that increments in a process over intervals of the same length should have the same distribution. In the Poisson process case presented in this chapter, that meant that $X_{t+h} - X_t$ and $X_{s+h} - X_s$ will have the same distribution Poisson(λh) because the measure of these intervals is the same:

$$\lambda h = \mathbf{E}[N((t, t+h])] = \mathbf{E}[N(t+h) - N(t)] = \mathbf{E}[N(s+h) - N(s)]$$

This is not possible, however, if the measure used is anything else than linear with respect to the Lebesgue measure. In particular, suppose that the measure of an interval $(0, t]$ is some function of the interval $\lambda(t)$ and not just λt. Then the stationarity will not work since

$$\mathbf{E}[N((t, t+h])] = \lambda(t+h) - \lambda(t) \neq \lambda(s+h) - \lambda(s) = \mathbf{E}[N(s+h) - N(s)].$$

Further, the stationarity of increments will not work with the general Poisson processes as described above since, in general, we may have two identically sized sets A_1 and A_2 and yet their associated measures $\mu(A_1)$ and $\mu(A_2)$ may be completely different.

The general definition of a Poisson process is fairly complex. Next, we give the most encountered types in applications.

10.3.1   Nonhomogenous Poisson process

Definition 10.14 *This is a straightforward generalization of the regular Poisson process. A Poisson process with the rate a function of time* $\lambda(t)$ *instead of the linear* λt *is called a* nonhomogeneous Poisson process. *In terms of the general Poisson process, we use* $\mathcal{X} = [0, \infty)$, $\mathcal{G} = \mathcal{B}([0, \infty))$, *and* $\mu(A) = \int_A \lambda(t) dt$. *Notice that this process does not have stationary increments anymore.*

The next proposition generalizes Proposition 10.10 and it is very useful in applications.

Proposition 10.15 *Assume that each event of a Poisson(λ) process can be classified as either a Type I or a Type II event. Furthermore, suppose that if an event occurs at time s, then it is classified as being Type I with probability $p(s)$ and Type II with probability $1 - p(s)$, where the probability depends on the time of arrival.*

If $N_i(t)$ is the number of events of Type i, $i \in \{I,II\}$ by time t, then $N_1(t)$ and $N_2(t)$ are independent Poisson random variables with means(rates) $\lambda t p$ and $\lambda t (1 - p)$, respectively, where

$$p = \frac{1}{t} \int_0^t p(s) ds$$

In particular, $N_1(t)$ and $N_2(t)$ are independent, nonhomogeneous Poisson processes.

Proof: The theorem may be proven using Theorem 10.13 involving general Poisson processes. We may also prove it similarly with Proposition 10.10 (see, for example, Ross (1995)). We may also prove it as a consequence of general renewal theory (next chapter). Since there are so many ways of attacking the proof, we skip it here. ∎

Corollary 10.16 *In general, if $N(t)$ is a Poisson(λ) process and events can be categorized into some category type A independently of the original process, then if $N_A(t)$ is the number of events of type A by time t, then $N_A(t)$ is a nonhomogeneous Poisson process with rate $\lambda \cdot \int_0^t p_A(s) ds$, where $p_A(s)$ is the probability that one event occurring at time s is of type A.*

Note that the original Poisson(λ) process has the mean $\mathbb{E}[N(t)] = \lambda t$.

For the process counting the events of type A, the mean may be written[2] as

$$\mathbb{E}[N_A(t)] = \underbrace{\lambda t} \cdot \underbrace{\frac{1}{t} \int_0^t p_A(s) ds} \ .$$

10.3.2 The compound Poisson process

Definition 10.17 *For each hit i of a generalized Poisson(μ_1) process N on $(\mathcal{X}, \mathcal{G})$, attach the random variables Y_i i.i.d. with cdf F_2, which give rise to the probability measure μ_2[3]. Then the process*

$$Z(A) = \sum_{i=1}^{N(A)} Y_i$$

---

[2]Note that the expectation is the product between the rate of the original Poisson(λ) process and the probability that the event is of type A.

[3]One can obtain the measure from the cdf remembering that the Borel sets are generated by intervals and using the relation $\mu((a, b]) = F(b) - F(a)$.

is a process called the compound Poisson process *on* $[0, \infty) \times \mathbb{R}$ *with intensity* $\mu = \mu_1 \times \mu_2$

This is the most general definition of the compound process.

In the particular case when N is a regular Poisson process and Y_i's are independent of the time of the event, we obtain

$$Z(t) = \sum_{i=1}^{N(t)} Y_i,$$

which is called the *(simple) compound Poisson process*.

Proposition 10.18 (For the simple Compound Poisson process) *If λ is the rate for the regular Poisson process $N(t)$ and the variables Y_i have mean μ and variance ν^2, then we can calculate the mean and standard deviation of the compound Poisson process*

$$\mathbf{E}[Z(t)] = \lambda \mu t, \quad V[Z(t)] = \lambda(\nu^2 + \mu^2)t$$

Proof: We have, using the basic properties of conditional expectation (Theorem 5.8),

$$\mathbf{E}[Z(t)] = \mathbf{E}\{\mathbf{E}[Z(t)|N(t)]\} = \mathbf{E}[N(t)]\mathbf{E}[Y_i] = \lambda t \mu$$

$$Var(Z(t)) = \mathbf{E}[Var(Z(t)|N(t))] + Var(\mathbf{E}[Z(t)|N(t)]) = \cdots = \lambda t(\nu^2 + \mu^2)$$

■

It is left as an exercise to show that the moment generating function of the $Z(t)$ variable exists if Y_i admits a moment generating function $M_Y(s)$ and that we have

$$M_{Z(t)}(s) = e^{\lambda t(M_Y(s) - 1)}$$

A more specific example: On can imagine a lot of situations where each event has associated with it a certain random quantity. For example, imagine a health insurance agency. Events may be claims, with the time of arrival modeled by the Poisson process, and with the amount of each claim modeled by the variables Y_i.

EXAMPLE 10.1

Consider a system with possible states $\{1, 2 \ldots\}$. Individuals enter the "system" according to a Poisson(λ) process. At any time after the entry, any individual is in some state $i \in \mathbb{N}^{*4}$.

Let $\alpha_i(s) = \mathbf{P}\{$An individual is in state i at time s after entry$\}$. Let $N_i(t)$ be the number of individuals in state i at time t.

Find $\mathbb{E}[N_i(t)]$.

—————

$^4\mathbb{N}^* = \mathbb{N} \setminus \{0\}.$

We can represent the state of each point in this system as the pair

$$
\left(
\underbrace{\text{entry time}}_{\text{Poisson process}},
\underbrace{\text{state at time t}}_{\text{the r.v. } Y_i}
\right)
\in [0, t] \times \mathbb{N}^*.
$$

An event is of type i if at time t it is in state i. The process N that counts the number of particles on this set $\mathcal{X} = [0, t] \times \mathbb{N}^*$ is a general Poisson process.

Using this definition, $N_i(t) = N([0, t] \times \{i\})$. Recalling Theorem 10.13, we obtain that $N_i(t)$ is a Poisson random variable with mean $\mu([0, t] \times \{i\})$. Therefore, its mean is

$$
\mu([0, t] \times \{i\}) = \lambda \int_0^t \mathbf{P}(\text{event at time } s \text{ is of type } i)ds = \lambda \int_0^t \alpha_i(t - s)ds.
$$

Let us look at this further. We have, using $r = t - s$,

$$
\begin{aligned}
\lambda \int_0^t \alpha_i(t - s) &= \lambda \int_0^t \alpha_i(r)dr \\
&= \lambda \int_0^t \mathbf{P}(\text{invidual is in state } i, r \text{ units after its entry})dr \\
&= \lambda \int_0^t \mathbf{E}\left[\mathbf{1}_{\{\text{individual is in state } i, r \text{ units after its entry}\}}\right]dr \\
\text{Fubini} \rightarrow &= \lambda \mathbf{E}\left[\underbrace{\int_0^t \mathbf{1}_{\{\ldots\}}dr}_{\text{time spent in state } i \text{ during } [0, t]}\right] \\
&= \lambda \mathbf{E}[\text{time spent in state } i \text{ during the first } t \text{ time units}]
\end{aligned}
$$

Question: What happens as $t \to \infty$?

◻ EXAMPLE 10.2

Cars enter a highway (one-way highway) according to a Poisson(λ) process in time. Each car has velocity $v(i)$ i.i.d. random variables with distribution function F.

Question: Assuming that each car travels at constant velocity, find the distribution of the number of cars on the highway between points a and b (spatial points) at time t?

Each car is described by the entry time and velocity, that is,

$$
(\text{entry time, velocity}) = (S(i), v(i)) \in [0, \infty) \times [0, \infty)
$$

A sample outcome is presented in Figure 10.1. Each event is represented by the tip of the arrow. Furthermore, the position of the car i at time t is easily

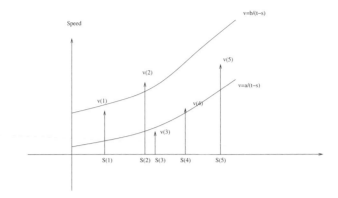

Figure 10.1   Cars enter at $S(i)$ with velocity $v(i)$. All the points within the two curves will be between spatial points a and b at time t

calculated as $v(i)(t - S(i))$. For this reason, cars that are between the spatial positions a and b at time t are those points of the generalized Poisson process in the set

$$A = \{(S(i), v(i))) : i \leq N(t), a < v(i)(t - S(i)) < b\}.$$

According to Theorem 10.13, the number desired, $N(A)$, is a Poisson random variable with mean $\mu(A)$. To calculate $\mu(A)$, we need to apply the compound Poisson result and recognize that μ is a product measure of μ_1 defined as $\mu_1((a, b]) = \lambda(b - a)$ and μ_2 with $\mu_2((a, b]) = F(b) - F(a)$. Thus

$$\mu(A) = \int_0^t \int_0^\infty \mathbf{1}_{\{a < y(t-s) < b\}} F(dy) d(\lambda s)$$

$$= \lambda \int_0^t \int_0^\infty \mathbf{1}_{\{\frac{a}{t-s} < y < \frac{b}{t-s}\}} F(dy) ds$$

$$= \lambda \int_0^t \left(F\left(\frac{b}{t-s}\right) - F\left(\frac{a}{t-s}\right) \right) ds$$

An alternative solution may be given using Corollary 10.16. We call the event i an event of Type AB if a car entering at s_i with velocity v_i is in $[a, b]$ at time t. Then, using the Corollary 10.16, we have

$N(t)$ = number of cars in $[a, b]$ = number of events of type AB

$N(t) \sim$ Poisson with the rate $= \lambda \int_0^t \mathbf{P}(\text{events enter at } s \text{ is of type } AB) ds$

Furthermore, the probability that a car that arrives at s will be in the interval $[a, b]$ at time t is

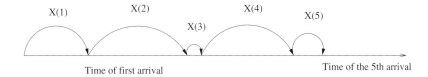

Figure 10.2 Illustration of the construction

$$\mathbf{P}\left\{v : a < v(t-s) < b\right\} = \mathbf{P}\left\{v : \frac{a}{t-s} < v < \frac{b}{t-s}\right\}$$
$$= \left[F\left(\frac{b}{t-s}\right) - F\left(\frac{a}{t-s}\right)\right] \mathbf{1}_{[0,t]}(s)$$

Therefore, $N(t)$ is a Poisson random variable with mean

$$\lambda \int_0^t \left[F\left(\frac{b}{t-s}\right) - F\left(\frac{a}{t-s}\right)\right] ds.$$

Question: What happens as $t \to \infty$?

10.4   Simulation Techniques. Constructing the Poisson Process.

We discuss methods of generating a Poisson process. It is assumed that we know how to generate variables having simple distribution (uniform, exponential, normal, etc.).

10.4.1   One-dimensional simple Poisson process

There are two ways to construct a 1D Poisson process:

Simplest way   Let $X_1, X_2 \ldots$ iid, Exponential(λ) with mean $\frac{1}{\lambda}$. Use X_i as the time between events $i - 1$ and i. (Done!)

Generate interval-wise   For each time interval $[0, 1), [1, 2) \ldots [t-1, t) \ldots$

1. Simulate[5] $N_I = N([k-1, k)) =$ number of events in $I = [k-1, k)$; this generates the number of events in each interval.

2. To get the actual times of the events, use the uniform distribution to generate times in each interval. For example, say, the generated $N([0, 1)) = 2$, then just generate two Uniform$(0, 1)$ random variables, the numbers represent the time of occurrence of the two events.

---

[5]Note that for each interval, $N([k-1, k))$ are i.i.d. Poisson$(\lambda \cdot 1)$ random variables.

As anything in life, the simple way is simple but only works with the 1D process. The interval-wise way, on the other hand, is more complicated but it can be extended to the general Poisson process. The way to do it is straightforward. Suppose we are given $(\mathcal{X}, \mathcal{G})$ a measurable space, and μ a σ-finite measure (see Definition 1.18). Partition $\mathcal{X}$ into $\{B_i\}_{i=1}^{\infty}$ such that $\mu(B_i) < \infty$. Then for each B_i, we get

1. $N(B_i) = $ the number of events in B_i which is distributed as a Poisson$(\mu(B_i))$ random variable,

2. $X_1^{(i)}, X_2^{(i)} \ldots$ i.i.d.[6] with probability distribution

$$P(X_k^{(i)} \in A) = \mu(A|B_i) = \frac{\mu(A \cap B_i)}{\mu(B_i)}$$

Then for every $A \in \mathcal{G}$, let

$$N(A) = \sum_{i=1}^{\infty} N(A \cap B_i) = \sum_{i=1}^{\infty} \left[\sum_{k=1}^{\infty} \mathbf{1}_{\{X_k^{(i)} \in A \text{ and } N(B_i) \geq k\}} \right] \tag{10.4}$$

This simply counts the number of generated points within the set A.

Theorem 10.19 *The construction above and* (10.4) *yields a Poisson process with intensity μ on $(\mathcal{X}, \mathcal{G})$.*

Proof (Sketch of the proof): We omit the detailed proof, but give below the important ideas of the proof.

The countable additivity is satisfied automatically by the definition of measure. The proof continues by demonstrating the following facts:

1. $N(A) \sim$ Poisson$(\mu(A))$ for any $A \in \mathcal{G}$

2. $N(A)$ and $N(B)$ are independent if $A \cap B = \emptyset$

First, we show these properties inside each B_i. To show 2 inside B_i, we may proceed as follows: Let $A \subset B_i$ and $C = B_i \backslash A$ (see Figure 10.3). For integers $a, c \in \mathbb{N}$, we have

---

[6]Random points positions in B_i.

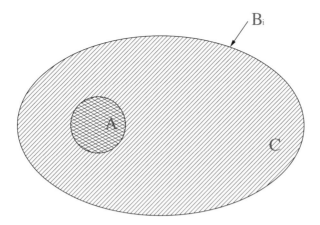

Figure 10.3   Illustration of set A and B_i

$$
\begin{aligned}
\mathbf{P}(N(A) = a, N(C) = c) &= \mathbf{P}(N(A) = a, N(C) = c, N(B_i) = a + c) \\
&= \mathbf{P}(N(A) = a, N(C) = c | N(B_i) = a + c)\mathbf{P}(N(B_i) = a + c) \\
&= \mathbf{P}(N(A) = a | N(B_i) = a + c)\mathbf{P}(N(B_i) = a + c) \quad\quad (10.5)
\end{aligned}
$$

$$
= \mathbf{P}(N(A) = a | N(B_i) = a + c)\frac{[\mu(B_i)]^{a+c}}{(a+c)!}e^{-\mu(B_i)} \quad\quad (10.6)
$$

$$
= \binom{a+c}{a}[\mu(A|B_i)]^a[\mu(C|B_i)]^c\frac{[\mu(B_i)]^{a+c}}{(a+c)!}e^{-\mu(B_i)} \quad\quad (10.7)
$$

$$
= \binom{a+c}{a}\frac{[\mu(A \cap B_i)]^a[\mu(c \cap B_i)]^c}{(a+c)!}e^{-\mu(B_i)} \quad\quad (10.8)
$$

$$
= \frac{[\mu(A \cap B_i)]^a[\mu(c \cap B_i)]^c}{a!c!}e^{-\mu(A \cap B_i) - \mu(C \cap B_i)}
$$

$$
= \underbrace{\frac{[\mu(A \cap B_i)]^a}{a!}e^{-\mu(A \cap B_i)}}_{\text{Poisson in } A} \cdot \underbrace{\frac{[\mu(C \cap B_i)]^c}{c!}e^{-\mu(C \cap B_i)}}_{\text{Poisson in } C}
$$

$$
= \frac{[\mu(A)]^a}{a!}e^{-\mu(A)}\frac{[\mu(C)]^c}{c!}e^{-\mu(C)}
$$

$$
= P(N(A) = a) \cdot P(N(C) = c)
$$

In (10.5), we removed redundant information.

In (10.6), we used the Poisson distribution to write the probability for $P(N(B_i) = a + c)$.

In (10.7), we used the binomial distribution with $n = a + c$ and $p = \mu(A|B_i)$.

In (10.8), by the definition of conditional probability, $[\mu(A|B_i)]^a = \frac{[\mu(A \cap B_i)]^a}{[\mu(B_i)]^a}$

Therefore, $N(A)$ and $N(C)$ are independent.   ∎

◼ EXAMPLE 10.3   Astronomy

Consider stars distributed in space according to a 3D Poisson process with intensity $\lambda\mu$, where μ is the Lebesgue measure on $\mathbb{R}^3$, $\lambda > 0$. Let x, y be 3D points in $\mathbb{R}^3$ (positions). Assume that light intensity exerted at x by a star located at y is

$$f(x, y, \alpha) = \frac{\alpha}{\|x - y\|^2} = \frac{\alpha}{(x_1 - y_1)^2 + (x_2 - y_2)^2 + (x_3 - y_3)^2} \quad,$$

where α is a random parameter depending on the size of the star at y.

Assume that α's associated with stars are i.i.d. random variables with the same mean μ_α and the same variance σ_α^2. Also, assume that the combined intensity at x accumulates additively.

Let $Z(x, A)$ be the total intensity at x due to stars in the region A. Then

$$Z(x, A) = \sum_{i=1}^{N(A)} \frac{\alpha_i}{\|x - Y_i\|^2} = \sum_{i=1}^{N(A)} f(x, Y_i, \alpha_i),$$

where $N(A)$ is the number of stars in the region A in space. Note that Y and α are random variables.

We have

$$\mathbf{E}[Z(x, A)] = \mathbf{E}[N(A)]\mathbf{E}[f(x, Y, \alpha)]. \tag{10.9}$$

We do not prove this result here; instead, we note that the expression is a direct consequence of the Wald's equation which will be proven in more generality later.

We have that $\mathbf{E}[N(A)] = \lambda\mu(A)$, where $\mu(A)$ is the volume of A.

$$\mathbf{E}[f(x, Y, \alpha)] = \mathbf{E}\left[\frac{\alpha}{\|x - Y\|^2}\right] = \mathbf{E}[\alpha]\mathbf{E}\left[\frac{1}{\|x - Y\|^2}\right],$$

since α and Y are independent. As a consequence of the generalized Poisson process in space, Y is going to be uniform in A and thus $\mathbf{E}[\|x - Y\|^{-2}] = \frac{1}{\mu(A)} \int_A \frac{1}{\|x-y\|^2} dy$. Finally, applying the equation (10.9), we have

$$\mathbf{E}[Z(x, A)] = \lambda\mu(A)\mu_\alpha \frac{1}{\mu(A)} \int_A \frac{1}{\|x - y\|^2} dy$$

$$= \lambda\mu_\alpha \int_A \frac{1}{\|x - y\|^2} dy$$

Problems

10.1   Prove Exercise 8 on page 309. To this end, use Stirling's approximation

$$n! \approx \sqrt{2\pi n}\, n^n e^{-n},$$

for all n large enough.

10.2   Let N_t be a Poisson process. Let S_n be the time of the nth event. Show by double inclusion that

$$\{\omega : S_n(\omega) \leq t\} = \{\omega : N_t(\omega) \geq n\}.$$

Recall that ω is a path for stochastic processes.

10.3   Let N_t be a Poisson process with rate λ. Let S_n be the time of the nth event. Find

   a) $\mathbf{E}[S_4]$,
   b) $\mathbf{E}[S_4 \mid N_1 = 2]$,
   c) $\mathbf{E}[N_4 - N_2 \mid N_1 = 3]$.

10.4   Prove Corollary 10.7 on page 315.

10.5   Suppose vehicles arrive at an intersection according to a Poisson Process with rate $\lambda = 10$ per minute.
   a) Let T_i denote the time between vehicle $i - 1$ and vehicle i. What is the distribution of T_1?
   b) Find the probability that the time between vehicles 5 and 6, T_6, is less than 30 s.
   c) Given that in 30 min there have been 27 vehicles by that intersection, what is the expected time of arrival for the 4th vehicle?
   d) Assume that 10% of all vehicles passing by that intersection are buses. What is the expected time one has to wait until the next bus shows up?
   e) What is the distribution of the number of buses arriving at that intersection in the next half hour?
   f) If 60 vehicles have passed by the intersection in an hour, what is the probability that exactly 6 of them were buses?

10.6   Suppose that buses arrive at a particular stop according to a Poisson process with rate 6 per hour. I start waiting for the buss to arrive at 10:00.
   a) What is the probability that no bus arrives in the next 20 min?
   b) How many buses are expected to arrive in the next 90 min?
   c) What is the variance of the number of buses during the next 3 h?
   d) Somebody tells me that there has been no bus for one hour. Given this information, what is the probability that the next bus will arrive by 10:15?
   e) What is the probability that no bus arrives in the next 15 min, given that I already have been waiting for 10 min for it.
   f) At 10:15 I decide to go for a coffee. I return at 10:30 and see a bus leaving. One of the people at the bus station tells me that this was the second since I left the station and there has been another bus during my leaving the station. What is the probability that the first bus I did not even see arrived at the bus station within 1 min after I left (10:15 to 10:16)?

10.7   Buses arrive to a certain stop according to a Poisson process with rate λ. If you take the bus from that stop, then it takes a time S measured from the time you enter the bus to reach home. If you walk from that bus stop, then it takes a time T to reach home. Suppose that the rule you decide to follow is: wait a deterministic amount of time s, and then if the bus has not arrived yet, you walk home.

   a) Compute the expected time to reach home from the moment you arrive at the bus stop.
   b) Show that if $T < 1/\lambda + S$, then the expectation in part a) is minimized by letting $s = 0$; and if $T > 1/\lambda + S$, then the expectation in part a) is minimized by letting $s = \infty$. Also show that, when $T = 1/\lambda + S$, all s values give the same expected time.
   c) Give an intuitive explanation why it is enough to consider 0 and ∞ as possible values for s when minimizing the expected time.
      You may assume that S and T are deterministic in this problem. For extra points, repeat the reasoning with S and T random variables. How do the conditions in part b) change?

10.8   A critical component of the next space shuttle to Europa (the Jupiter satellite) has an operating lifetime that is exponentially distributed with mean 1 year (ship time). As soon as a component fails, it is replaced by a new one having statistically identical properties. It is known that the one-way travel to Europa lasts exactly 2 years. What is the smallest number of components that the shuttle should stock if it wishes that probability of having an inoperable unit caused by failures exceeding the spare inventory throughout the time of the **voyage** to be less than 0.02?

10.9   A device recording lightning intensity is attached to a pole during a thunderstorm. Assume that during that time the bolts of lightning occur at that pole according to a Poisson process with rate λ. However, each bolt registered by the machine renders the device inoperative for a fixed length of time a and it does not register anything that may occur during that interval. Let $R(t)$ denote the number of lightning bolts that are registered by the machine at time t.

a) Find the probability that the first k bolts of lightning are all registered.
b) For $t \geq (n-1)a$, find $\mathbf{P}\{R(t) \geq n\}$
Hint: Look to Type I Type II events for the Poisson process.

10.10   Suppose that during the thunderstorm in the previous problem we monitor the device using a computer connected to a power source. Suppose the rate is $\lambda = 3$ per hour. Each time lightning hits, it would damage the computer, so we created a device which will protect the computer. However, the device needs to be taken out and serviced for exactly 10 min before it is operational again.

   a) Suppose that a single lightning event during the service period will destroy the computer. What is the probability that the computer will crash?
   b) Assume now that the computer will survive a single lightning bolt during the service period but two during that period will destroy it. What is the probability of computer crashing now?

c) Assume that the crash will not happen unless there are two bolts within 5 min of each other during the service period. Compute the probability of the system crashing.

d) Answer parts a) and b), but now assume that the service time is exponentially distributed with mean 10 min.

10.11 A gas station has only one pump. The time to fill the tank is distributed as $Exp(\lambda)$. However, after filling the tank, the customers pay by walking inside the gas station while leaving the car at the pump thus blocking all other cars from filling their tank. The service time at the counter is distributed as $Exp(\mu)$. The customer leaves immediately after paying in the gas station. When I arrive at the station, there is a customer there filling the tank of his car. What is my expected time until I leave the station?

10.12 Suppose two individuals A and B both require heart transplant. The remaining time to live if they do not receive transplants are exponentially distributed with mean μ_A and μ_B, respectively. New hearts become available through a transplant program according to a Poisson process with rate λ. It has been decided that the order in which heart transplants will take place is A then B, provided either are alive at the time of the operation. B will receive transplant if either A has already received one or if B is alive and A is not at the time of the first transplant.

a) Calculate the probability that A receives a new heart.

b) Calculate the probability that B receives a new heart.

10.13 Let X_t, $t \geq 0$, be a Poisson process of rate λ. Let $T_0 = 0$, and let T_i be the time of the ith observation (jump of X), if $i \geq 1$. Let $N = \inf\{k \geq 1 : T_k - T_{k-1} > 1\}$. Find $\mathbf{E}T_N$, $\mathbf{E}N$, and $\mathbf{E}(T_N | N = 8)$.

10.14 Using a software package, simulate a Poisson process with rate 2 events/min. Using your simulation, estimate the probabilities

$$\mathbf{P}\{N_{[2,4]} = 4\}$$

$$\mathbf{P}\{S_3 \in [3,5]\},$$

where $N_{[2,4]}$ denotes the number of events in the time interval $[2,4]$ minutes, and S_3 is the time of the third event.

Calculate what these theoretical probabilities should be and see what was the difference between your simulated probabilities and the theoretical ones for 1000, 10,000 and 100,000 repetitions, respectively. (This is called a *Monte Carlo* simulation approach.)

10.15 A two-dimensional Poisson process is a general Poisson process with $\mathcal{X} = \mathbb{R}^2$, similar to the process presented in Example 10.3. More specifically, for any region of the plane A, the number of events in A, $N(A)$, is a Poisson random variable with mean $\lambda |A|$, where $|A|$ denotes the area of the region A. Furthermore, the events in any nonoverlapping region are independent. Consider the origin $(0,0)$ of the plane, and let X denote the Euclidian distance between the nearest event and the origin. Show that

a) $\mathbf{P}\{X > r\} = e^{-\lambda \pi r^2}$,
b) $\mathbf{E}[X] = 1/(2\sqrt{\lambda})$ (you may assume a) for this part).

10.16 In the same setting as in the previous problem, let R_i denote the distance from the origin to the ith closest event to it. Prove that $Y_i = \pi R_i^2 - \pi R_{i-1}^2$, with $i \geq 1$ being independent random variables exponentially distributed with mean $1/\lambda$.

10.17 Using a software package, simulate a Poisson process on the plane suitable for the previous two problems. Use $\lambda = 2$. With the help of this simulation, answer the following questions:
a) Estimate the probability that the circle of radius 1 centered at the origin of the plane contains two events.
b) Estimate the probability in part a) of the problem 10.15. Use varying values for the distance r (say $r \in \{0.25, 0.5, 1, 2, 3, 4\}$). Use as many repetitions as you like. Do you obtain values close to what you should get?
c) Do the same thing as in part b) but for the distance $R_2 - R_1$, with the R_i's defined in the problem 10.16.

10.18 Using a software package, simulate a one-dimensional Poisson process with rate 2 events/min. Using your simulation, estimate the following probabilities:

$$\mathbf{P}\{N_{[2,4]} = 4\},$$

$$\mathbf{P}\{S_3 \in [3,5]\},$$

where $N_{[2,4]}$ denotes the number of events in the time interval $[2, 4]$ minutes, and S_3 is the time of the third event.
Calculate what these theoretical probabilities should be and see what the difference between your simulated probabilities and the theoretical ones was for 1000, $10,000$ and $100,000$ repetitions, respectively.

10.19 Simulate the following inventory problem. Assume that for a year you administer a maritime platform – oil facility off the cost of Nigeria. You have oil tankers arriving according to a Poisson process with rate $0.3/day$. Each oil tanker has a storage capacity that varies depending on the size of the tanker, according to the distribution

$$f(x) = \frac{1}{1000}\left(3 - \frac{x}{500}\right)\mathbf{1}_{\{500 < x < 1500\}},$$

and x is expressed in thousands of barrels.

The oil is loaded in the tankers at a constant rate $10,000$ barrels/hour. Furthermore, the oil wells are linked to the maritime platform and they produce (we will assume for simplicity) at a total constant rate of $10,000$ barrels/hour. Assume that the facility has infinite storage facility and it works constantly for 24 hours and 7 days a week. Each time an oil tanker finishes loading, if there is one in the queue, it will start loading automatically.
a) Assuming that the cost of oil storage is $\$0.5$ per 1000 barrels and that each 1000 barrels loaded gives you a profit of $\$2$, find the expected profit after 1 year.
b) Estimate the expected loss if the platform undergoes repairs for a full day (24 h).

CHAPTER 11

RENEWAL PROCESSES

In this chapter, we concentrate on a generalization of the Poisson processes. Recall the memoryless property of the exponential distribution. Essentially, this property makes the Poisson process restart every time we observe it. Consider a light bulb that has been in operation for two years. Suppose we assume the light bulb is replaced according to a Poisson process. This implies that the probability this light bulb lasts for one more day given that it has been operating for two years is exactly the same as the probability of a totally new light bulb operating one day. Obviously, this is not realistic, but using anything else than the exponential for the inter-arrival time distribution will lose the simplicity of the Poisson process.

If we consider some other type of inter-arrival time distribution, we obtain a renewal process. This type of process is very general. Due to their general nature, they provide some of the most powerful (and general theorems) in stochastic processes. Here, "powerful" means applicable to a wide class of processes. However, since these results are very general, in order to obtain specific results one needs to hypothesize specific distribution, and oftentimes this leads to more specialized processes. We cover renewal processes in this chapter since studying them provide strong results which will be applied later to Markov chains and processes that are widely used in practice.

Probability and Stochastic Processes, First Edition. Ionuţ Florescu
© 2015 John Wiley & Sons, Inc. Published 2015 by John Wiley & Sons, Inc.

▣ EXAMPLE 11.1   A practical example where renewal process appears

A light bulb in a room keeps burning out. Assume that a mechanism instantaneously replaces the bulb with another one as soon as it burns out. Describe the
number of light bulbs replaced by time t.

The number up to time t in the example above is a renewal process. Let us formally
define this process.

Definition 11.1 *Let $X_1, X_2 \ldots$ be i.i.d. random variables with cdf F. Suppose the
random variables X_i are positive with $\mathbb{E}[X_1] = \mu \in (0, \infty]$ and $Var(X_1) = \sigma^2$.
These variables will describe the lifetimes of the renewal process. Define*

$$S_n = \text{time to replace the } n\text{-th bulb}$$

$$S_n = \sum_{i=1}^{n} X_i \sim \underbrace{F * F * \ldots * F}_{n \text{ times}} = F_n$$

Note that F_n denotes F convoluted[1] itself n times.
In this setting, the renewal process, $N(t)$, is defined as

$$N(t) = \sup\{n : S_n \le t\}$$
$$= \text{number of renewals up to time } t$$

In the light bulb example, X_i's are the lifetimes of each light bulb replaced. It is
easy to see that the renewal process is a more general process than the a Poisson(λ)
process. The Poisson process is a special case of a renewal process obtained by taking
the X_i variables exponentially distributed. As mentioned in the introduction, a very
important drawback of the exponential distribution is its memoryless property. That
property applied to the example above would mean that the chance of a light bulb
burning out in the next half hour is the same regardless of whether the light bulb was
replaced half an hour ago or whether it has been functioning since last year. Clearly,
items of this type get old and their functionality decreases and we can easily see why
the exponential distribution would not be appropriate for the light bulb example.

For a renewal process, X_i's are i.i.d. and have distribution F. The same result that
we had for the Poisson(λ) process is still valid for renewal processes.

$$\{N(t) \ge n\} \Leftrightarrow \{S_n \le t\}.$$

Therefore, we can write

$$P(N(t) = n) = P(N(t) \ge n) - P(N(t) \ge n + 1)$$
$$= P(S_n \le t) - P(S_{n+1} \le t) = F_n(t) - F_{n+1}(t)$$

---

[1]Recall that $X, Y \sim F, G$ and with pdf f, g, then $X + Y \sim F * G(z) = G * F(z) = \int_{-\infty}^{\infty} f(x)g(z - x)dx$. Also, see Section 6.1 in this book.

which links the distributions of the renewal process with the distribution of the times of renewals. This relationship does not reduce to a simple exponential/Poisson relationship as in the case of the Poisson process.

11.0.2 The renewal function

Since the distributions involved in renewal processes are complex, the main topic of study is the renewal function. This function is defined as

$$m(t) = \mathbb{E}[N(t)] = \text{The expected number of renewals by time } t. \qquad (11.1)$$

Note that we study the expected value of $N(t)$ instead of the entire distribution.

We may express this function as

$$m(t) = \mathbb{E}[N(t)] = \mathbb{E}\left[\sum_{n=1}^{\infty} \mathbf{1}_{\{S_n \leq t\}}\right]$$

$$= \sum_{n=1}^{\infty} \mathbb{E}\left[\mathbf{1}_{\{S_n \leq t\}}\right]$$

$$= \sum_{n=1}^{\infty} P(S_n \leq t)$$

$$= \sum_{n=1}^{\infty} F_n(t)$$

Thus we just showed that

$$m(t) = \sum_{n=1}^{\infty} F_n(t) \qquad (11.2)$$

This is a function of time t that expresses the time evolution of the expected number of renewals. Clearly, this function should be increasing in time. But is it a finite number?

Proposition 11.2 *The renewal function $m(t)$ is finite for all $0 < t < \infty$ fixed.*

Proof: Since the event times X_k are not identically zero, $P(X_k = 0) < 1$. Therefore, there must exist a strictly positive value α such that $P(X_k \geq \alpha) = p > 0$. In this proof, we assume $\alpha = 1$ for simplicity of notation. Of course, the entire proof also works for an arbitrary α by replacing $2, 3, \ldots$ with multiples of α. Thus, we assume that $P(X_k \geq 1) = p > 0$.

Recall that $t \in \mathbb{R}$ is fixed. Let j be an integer such that $j - 1 \leq t \leq j$.

Claim: $N(t)$ (the number of renewals by time t) is less than or equal to a sum of j independent Geometric(p) random variables[2].

---

[2] These are Geometric number of trials until the first success random variables, and not the number of failures until first success.

Let us prove the claim. For each bulb k, consider the old lifetime X_k and construct a new process as follows:

If $X_k < 1 =$ just throw the light bulb away (it counts as a renewal)

If $X_k \geq 1 =$ use the bulb for 1 unit of time and then throw it away

Let $N^*(t)$ denote the number of bulbs replaced by time t using the protocol described above. We obviously have $N^*(t) \geq N(t)$ because by time t we counted all those bulbs we threw away. The only case when $N(t) = N^*(t)$ is when all the lifetimes are identically equal to 1. This proves the claim since $N^*(t)$ has the desired probability distribution (is a sum of geometric random variables).

Therefore, using the claim

$$
\begin{aligned}
m(t) = \mathbb{E}[N(t)] &\leq \mathbb{E}[N^*(t)] \\
&= \mathbb{E}[Y_1 + Y_2 \dots Y_j] \\
&= \underbrace{\frac{1}{p} + \frac{1}{p} \dots \frac{1}{p}}_{j \text{ times}} \\
&= \frac{j}{p} < \frac{t+1}{p} < \infty \quad (\text{because } j - 1 < t < j)
\end{aligned}
$$

We also have

$$
\begin{aligned}
\mathbb{E}[N(t)^2] &\leq \mathbb{E}[N^*(t)^2] \\
&\leq \mathbb{E}[(Y_1 + Y_2 \dots Y_j)^2] \\
&= \underbrace{\text{Var}(Y_1 + Y_2 \dots Y_j)}_{\text{Negative Binomial}(j,p)} + \underbrace{(\mathbb{E}[Y_1 + Y_2 \dots Y_j])^2}_{\text{known}} \\
&= j\frac{(1-p)}{p^2} + \left(\frac{j}{p}\right)^2 < c(t+1)^2 < \infty \qquad (11.3)
\end{aligned}
$$

Therefore, the variance of the number of renewals by time t is a finite number as well.

■

11.1   Limit Theorems for the Renewal Process

Observe Figure 11.1. It depicts a graphical representation of the current item in a renewal process at time t. We will address questions about the age of the item and the residual life of the item later. For now, let us concentrate on providing limiting results (as $t \to \infty$) for the number of items replaced by time t, $N(t)$, and correspondingly the time of the nth replacement, $S_{N(t)}$.

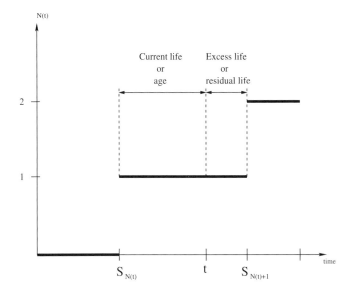

Figure 11.1   Relationship between $S_{N(t)}$, t, and $S_{N(t)+1}$

Proposition 11.3 (Limiting results) *This is a summary of the results that will be proven about the limiting behavior of renewal processes. In these results, μ is the mean and σ is the standard deviation of the lifetime distribution F.*

- $\frac{N(t)}{t} \xrightarrow{a.s.} \frac{1}{\mu}$

- $\frac{\mathbf{E}[N(t)]}{t} = \frac{m(t)}{t} \longrightarrow \frac{1}{\mu}$

- $\frac{N(t)-t/\mu}{\sigma\sqrt{t/\mu^3}} \xrightarrow{\mathcal{D}} N(0,1)$, *where $\mathcal{D}$ denotes convergence in distribution and $N(0,1)$ denotes a normal random variable with mean 0 and variance 1.*

Furthermore, in the next section we shall prove two very important results: the Blackwell renewal theorem and the key renewal theorem. The key renewal theorem is the result we shall use in most applications.

Theorem 11.4 (The Strong Law of Large Numbers for renewal processes) *Using the notation defined earlier,*

$$\frac{N(t)}{t} \longrightarrow \frac{1}{\mu} \quad a.s. \ as \ t \to \infty$$

Proof: Recall that $S_n = \sum_{i=1}^{n} X_i$ and that the X_i are all i.i.d. Thus, applying the regular strong law of large numbers for these X_i's implies that $\frac{S_n}{n} \to \mu$ a.s.

By the definition of $N(t)$, we have $S_{N(t)} \le t < S_{N(t)+1}$

Divide both sides by $N(t)$:

$$\underbrace{\frac{S_{N(t)}}{N(t)}}_{\to \mu \text{ by SLLN}} \leq \frac{t}{N(t)} < \frac{S_{N(t)+1}}{N(t)} = \underbrace{\frac{S_{N(t)+1}}{N(t)+1}}_{\to \mu \text{ by SLLN}} \cdot \underbrace{\frac{N(t)+1}{N(t)}}_{\to 1 \text{ a.s.}}$$

which implies $\frac{t}{N(t)} \to \mu$ a.s. OR $\frac{N(t)}{t} \to \frac{1}{\mu}$ a.s.

∎

Next, we want to obtain a convergence result for the renewal function $m(t) = \mathbf{E}[N(t)]$. Since we already have an a.s. convergence result for $N(t)$, can we get a result about $m(t)$? Recall from Chapter 7 that convergence a.s. does not necessarily imply convergence in L^1.

However, for our particular case the implication is true. If we apply either the dominated convergence theorem or the monotone convergence theorem, we can show that the implication is true.

Theorem 11.5 (Elementary renewal theorem) *With the earlier notations, we have*

$$\frac{m(t)}{t} \to \frac{1}{\mu}, \qquad as\ t \to \infty$$

with the convention $\frac{1}{\infty} = 0$

Proof (Unimaginative proof): Recall that we showed in (11.3) that $\mathbb{E}[N(t)^2] \leq c(t+1)^2$. Thus we have

$$\mathbb{E}\left[\left(\frac{N(t)}{t}\right)^2\right] \leq \frac{c(t+1)^2}{t^2} \leq 2c \quad \text{which is independent of } t$$

Therefore, $\frac{N(t)}{t}$ is uniformly integrable (since it is in L^2) and we may apply the dominated (or bounded) convergence theorem to obtain the desired result. ∎

We could just leave the elementary renewal theorem the way it is: after all, the proof is entirely correct. Instead, we will provide a more insightful alternate proof. To do this, we require the introduction of a new concept (stopping time) which will prove to be very useful when studying stochastic processes. Furthermore, we also need to introduce Wald theorem which is one of the most important results for applications. In Example 5.6 on page 167, we have seen an early introduction of this result. In the next section we will provide the most general version as well as the particular version that will be useful for us.

11.1.1 Auxiliary but very important results. Wald's theorem. Discrete stopping time

The first such new concept is the general Wald's theorem. This theorem has many names; this is the reason for the many alternative titles.

Theorem 11.6 (Wald's Theorem/Identity/Equation) *Let* $X_1, X_2 \ldots$, *and* W_1, $W_2 \ldots$ *be two sequences of random variables with* X_k *independent of* W_k *for any fixed* k. *If either one of the following conditions is true, namely*

1. *All X_k's and W_k's are greater than or equal to 0*

or,

2. $\sum_{k=1}^{\infty} \mathbf{E}[W_k X_k] < \infty$,

then

$$\mathbf{E}\left[\sum_{k=1}^{\infty} W_k X_k\right] = \sum_{k=1}^{\infty} \mathbf{E}[W_k]\mathbf{E}[X_k]$$

   Note that the sequences do not have to contain independent terms or identically distributed terms.

Proof (Proof of Wald's Theorem:):

If the first hypothesis is true, we have

$$\mathbf{E}\left[\sum_{k=1}^{\infty} X_k W_k\right] \underset{\text{positivity, thus monotone convergence th.}}{=} \sum_{k=1}^{\infty} \mathbf{E}[X_k W_k]$$

$$\underset{\text{independence}}{=} \sum_{k=1}^{\infty} \mathbf{E}[X_k]\mathbf{E}[W_k]$$

If the second hypothesis is true, $\sum_{k=1}^{\infty} \mathbf{E}[X_k W_k] < \infty$.

Let

$$W_k^+ = W_k \mathbf{1}_{\{W_k \geq 0\}}$$
$$W_k^- = -W_k \mathbf{1}_{\{W_k < 0\}}$$
$$X_k^+ = X_k \mathbf{1}_{\{X_k \geq 0\}}$$
$$X_k^- = -X_k \mathbf{1}_{\{X_k < 0\}}$$

Note that

$$W_k = W_k^+ - W_k^-$$
$$X_k = X_k^+ - X_k^-$$

We then have the following:

$$\sum_{k=1}^{\infty} W_k X_k = \sum_{k=1}^{\infty} W_k^+ X_k^+ - \sum_{k=1}^{\infty} W_k^+ X_k^- - \sum_{k=1}^{\infty} W_k^- X_k^+ + \sum_{k=1}^{\infty} W_k^- X_k^-$$

All X_k^+, X_k^-, W_k^+, and W_k^- are Positive; thus they are under Assumption 1 and we have

$$\mathbf{E}[\sum_k W_k^+ X_k^+] = \sum_k \mathbf{E}[W_k^+]\mathbf{E}[X_k^+]$$

$$\mathbf{E}[\sum_k W_k^+ X_k^-] = \sum_k \mathbf{E}[W_k^+]\mathbf{E}[X_k^-]$$

$$\mathbf{E}[\sum_k W_k^- X_k^+] = \sum_k \mathbf{E}[W_k^-]\mathbf{E}[X_k^+]$$

$$\mathbf{E}[\sum_k W_k^- X_k^-] = \sum_k \mathbf{E}[W_k^-]\mathbf{E}[X_k^-]$$

Recombining the terms in the expression above will finish the proof. ∎

This result has many applications. Here is an example of an immediate application.

■ EXAMPLE 11.2

Let $X_1, X_2 \ldots$ be i.i.d. with $\mathbf{E}[X_i] = \mu$. Define X_k to be the monetary gain in some game if you actually make the kth bet. Let

$$W_k = \begin{cases} 1 & \text{if you win} \\ 0 & \text{if you lose} \end{cases}$$

Then $\sum_{k=1}^{\infty} X_k W_k$ is the total gain from all bets.

Assume that $X_k > 0$ and also that W_k is determined by previous bets, that is, by $X_1 \ldots X_{k-1}$[3] and maybe by some $U \sim$ Uniform random variable[4]. Let $N = \sum_{i=1}^{\infty} W_k$ to be the number of bets you win. Then Wald's theorem says that, if $\mathbb{E}[N] < \infty$, we have

$$\mathbf{E}\left[\sum_{k=1}^{\infty} W_k X_k\right] = \sum_{k=1}^{\infty} \mathbf{E}[W_k]\underbrace{\mathbf{E}[X_k]}_{=\mu}$$

$$= \mu \sum_{k=1}^{\infty} \mathbf{E}[W_k] = \mu\mathbf{E}[N]$$

Note: Think about this and explain to yourself why this result is obvious.

Next concept is going to be even more important in the economy of stochastic processes. The notion of stopping time will help with practical calculations more than any other notion.

---

[3] In other words, it can depend on previous gains or losses but not on the current
[4] The variable U represents randomness in the current outcome.

Definition 11.7 (Discrete Stopping time) *Let $X_1, X_2 \ldots$ be a sequence of independent random variables. The discrete random variable $N \in \{0, 1, 2 \ldots\}$ is called a stopping time for this $\{X_n\}_n$ sequence if $\{N = n\}$ is independent of $X_{n+1}, X_{n+2} \ldots$. Note that this is true if $\{N = n\}$ is determined only by $X_1, X_2, \ldots, X_n$. Mathematically, this is expressed as the event $\{N = n\}$ being measurable with respect to the sigma algebra generated by $X_1, X_2, \ldots, X_n$*

Remark 11.8 *N is a stopping time for the sequence $X_1, X_2 \ldots$ if and only if the event $\{N \leq n\}$ is independent of $\{X_{n+1}, X_{n+2} \ldots\}$.*

Proof (Proof of this Remark:): is an exercise. As a hint, note that

$$\{N \leq n\} = \cup_{k=1}^{n} \{N = k\}.$$

■

After introducing the notion of stopping time, let us explain how we will use it.

Corollary 11.9 A simpler version of Wald's theorem. Useful for the renewal processes *Let $X_1, X_2 \ldots$ be a sequence of i.i.d. random variables with $\mu = \mathbf{E}[X_i]$ finite and let N be a stopping time for this sequence with $\mathbf{E}[N] < \infty$. Then*

$$\mathbf{E}\left[\sum_{k=1}^{N} X_k\right] = \mathbf{E}[X_i]\mathbf{E}[N] = \mu\mathbf{E}[N]$$

Proof: This result is a corollary of the general Wald's theorem 11.6. To apply the more general result, observe that we may write

$$\sum_{k=1}^{N} X_k = \sum_{k=1}^{\infty} X_k \mathbf{1}_{\{N \geq k\}}.$$

Thus, we need to show that $\mathbf{1}_{\{N \geq k\}}$ is independent of X_k and apply the general theorem, and be done.

Using the Remark 11.8, since N is a stopping time, the event $\{N \leq n\}$ is independent of $X_{n+1}, X_{n+2} \ldots$. Thus its complement, the set $\{N > n\}$, is also independent of $X_{n+1}, X_{n+2} \ldots$ for any n.

We may rewrite for $n - 1$ to obtain that the event $\{N > n - 1\}$ is independent of $X_n, X_{n+1} \ldots$. Since N is a discrete random variable, we have that $\{N > n-1\} = \{N \geq n\}$ and this is independent of $X_n, X_{n+1} \ldots$. In particular, $\{N \geq n\}$ is independent of X_n and therefore we can use the general Wald's theorem. We have

$$\mathbf{E}\left[\sum_{k=1}^{N} X_k\right] = \mathbf{E}\left[\sum_{k=1}^{\infty} X_k \mathbf{1}_{\{N \geq k\}}\right]$$

$$= \sum_{k=1}^{\infty} \underbrace{\mathbf{E}[X_k]}_{=\mu} \mathbf{E}[\mathbf{1}_{\{N \geq k\}}]$$

$$= \mu \sum_{k=1}^{\infty} \mathbf{E}[\mathbf{1}_{\{N \geq k\}}]$$

$$= \mu \sum_{k=1}^{\infty} P(N \geq k)$$

$$= \mu \begin{pmatrix} P(N=1) & +P(N=2) & +P(N=3) & +\dots \\ & +P(N=2) & +P(N=3) & +\dots \\ & & +P(N=3) & +\dots) \end{pmatrix}$$

$$= \mu[1 \cdot P(N=1) + 2 \cdot P(N=2) + 3 \cdot P(N=3) + \dots]$$

$$= \mu \mathbf{E}[N]$$

∎

11.1.2   An alternative proof of the elementary renewal theorem

After introducing the concept of stopping time, let us use the Wald's theorem for a proof of the elementary renewal theorem. To do so, we need to find a stopping time for the sequence of inter-arrival times $X_1, X_2, \dots$.

The next logical question is:

"Can we use $N(t) = \sup\{n : S_n \leq t\}$? Is this a stopping time for the sequence $X_1 \dots X_n \dots$?"

The short answer to this question is, No. There could be an event happening between the nth event and t.

Long answer:

$$\{N(t) = n\} = \{S_n \leq t < S_{n+1}\}$$
$$= \{X_1 + X_2 \dots + X_n \leq t < \underbrace{X_1 + X_2 \dots X_n + X_{n+1}}_{\text{not independent of } X_{n+1}}\}$$

Note that the event is determined by X_{n+1}, hence $\{N(t) = n\}$ can't be a stopping time.

Next question:

"How about $N(t) + 1$? Is this a stopping time?"

Answer: Yes! Mathematically

$$\{N(t) + 1 = n\} = \{N(t) = n - 1\} = \{S_{n-1} \le t < S_n\}$$
$$= \{X_1 \ldots + X_{n-1} \le t < X_1 \ldots X_{n-1} + X_n\}$$

Since everything inside the last $\{\cdot\}$ does not contain terms of type $n + 1$ or larger, and since the X_i's are independent, the event is independent of $X_{n+1}, \ldots$. Therefore, $\{N(t) + 1\}$ is a stopping time.

At this point, we may give the alternate proof of the elementary renewal theorem.

Proof (Alternate Proof of the Elementary Renewal Theorem):

Since $N(t) + 1$ is a stopping time, from Wald's theorem we have

$$\mathbb{E}[S_{N(t)+1}] = \mathbb{E}[N(t) + 1]\mathbb{E}[X] = \mu(m(t) + 1) \qquad (11.4)$$

Claim 1. $\liminf\limits_{t \to \infty} \frac{m(t)}{t} \ge \frac{1}{\mu}$

Proof of Claim 1 By definition, $S_{N(t)+1} > t \Rightarrow \mathbb{E}[S_{N(t)+1}] > t$. Then using (11.4), we obtain

$$\mu(m(t) + 1) > t$$
$$\frac{m(t)}{t} > \frac{1}{\mu} - \frac{1}{t}$$

Take $\liminf$ on both sides:

$$\liminf_{t \to \infty} \frac{m(t)}{t} \ge \liminf_{t \to \infty} \left(\frac{1}{\mu} - \frac{1}{t} \right) = \frac{1}{\mu}$$

Claim 2. $\limsup \frac{m(t)}{t} \le \frac{1}{\mu}$

Proof of Claim 2. Fix $M > 0$ constant. Let

$$\bar{X}_k = \begin{cases} X_k & \text{if } X_k \le M \\ M & \text{if } X_k > M \end{cases}$$

Let $\mu_M = \mathbb{E}[\bar{X}_k]$, and define $\bar{N}(t)$ to be the number of renewals up to t with these new lifetimes $\bar{X}_k$.

Note that $\bar{N}(t) \ge N(t)$ (due to shorter lifetimes[5])

$$\bar{m}(t) = \mathbb{E}[\bar{N}(t)] \ge m(t)$$

Now look at the Figure 11.2, which represents the behavior of the new process at t. Note that since we bounded the inter-arrival times by M, we have

$$\bar{S}_{\bar{N}(t)+1} > t$$
$$\bar{S}_{\bar{N}(t)+1} \le t + M$$

---

[5]Life span is limited by the upper bound M.

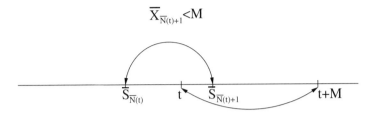

Figure 11.2 Representation of the new process

If we use (11.4) for $\bar{N}(t)$, we get $\mathbb{E}[\bar{S}_{\bar{N}(t)+1}] = \mu_M(\bar{m}(t) + 1)$. Therefore

$$\mu_M(\bar{m}(t) + 1) \leq t + M$$
$$\frac{\bar{m}(t) + 1}{t} \leq \frac{1}{\mu_M} + \frac{M}{t\mu_M}$$
$$\frac{\bar{m}(t)}{t} \leq \frac{1}{\mu_M} + \frac{M}{t\mu_M} - \frac{1}{t}$$

Apply $\limsup$ in both sides to get

$$\limsup_{t \to \infty} \frac{\bar{m}(t)}{t} \leq \frac{1}{\mu_M}$$

which implies

$$\limsup_{t \to \infty} \frac{m(t)}{t} \leq \limsup_{t \to \infty} \frac{\bar{m}(t)}{t} \leq \frac{1}{\mu_M}.$$

But this inequality is true for any $M > 0$ (M was arbitrary). Therefore, we may take $M \to \infty$, and using that $\lim_{M \to \infty} \frac{1}{\mu_M} = \frac{1}{\mu}$, we conclude

$$\limsup_{t \to \infty} \frac{m(t)}{t} = \frac{1}{\mu}.$$

Combining the two claims above immediately completes the proof of the theorem. ∎

To end this section, we conclude with a convergence in distribution result similar to the regular Central Limit Theorem.

Theorem 11.10 (Central Limit Theorem for Renewal Processes) *Let $X_1, X_2 \ldots$ be i.i.d., positive, $\mu = \mathbb{E}[X_k]$, $\sigma^2 = Var(X_k) < \infty$. Define as usual the renewal process elements*

$$S_n = \sum_{i=1}^{n} X_i, \quad N(t) = \sup\{n : S_n \leq t\}$$

Then

$$P\left(\frac{N(t) - \frac{t}{\mu}}{\sigma\sqrt{\frac{t}{\mu^3}}} < y\right) \xrightarrow{t\to\infty} \Phi(y)$$

where $\Phi(y)$ is the cdf of $N(0,1)$

In other words, when t is large, $N(t)$, the number of renewals up to time t, is approximately distributed as a $N\left(\frac{t}{\mu}, \frac{\sigma^2 t}{\mu^3}\right)$ random variable.

Proof: Fix $y \in \mathbb{R}$.

$$P\left(\frac{N(t) - \frac{t}{\mu}}{\sigma\sqrt{\frac{t}{\mu^3}}} < y\right) = P\left(N(t) < \underbrace{\frac{t}{\mu} + y\sigma\sqrt{\frac{t}{\mu^3}}}_{=r_t}\right)$$

$$= P(N(t) < r_t)$$
$$= P(S_{r_t} > t)$$

where we defined r_t in the expression above. Note that the last equality works only when r_t is an integer. If r_t is not an integer, take $\tilde{r}_t = [r_t] + 1$. In that case, we have the following:

$$P(N(t) < r_t) = P(N(t) < \tilde{r}_t) = P(S_{\tilde{r}_t} > t)$$

and we continue the proof in this more general case.

We may use the classic Central Limit Theorem to finish the proof since $S_{\tilde{r}_t}$ is a sum of i.i.d. random variables with finite variance. Therefore, we have

$$P(S_{\tilde{r}_t} > t) = P\left(\frac{S_{\tilde{r}_t} - \tilde{r}_t\mu}{\sigma\sqrt{\tilde{r}_t}} > \frac{t - \tilde{r}_t\mu}{\sigma\sqrt{\tilde{r}_t}}\right)$$

$$= \lim_{t\to\infty} \Phi\left(\frac{t - \tilde{r}_t\mu}{\sigma\sqrt{\tilde{r}_t}}\right)$$

where Φ is the distribution function of an $N(0,1)$ random variable.

The proof of the theorem will end if we show that $\frac{t - \tilde{r}_t\mu}{\sigma\sqrt{\tilde{r}_t}} \to -y$

We have that

$$\tilde{r}_t = r_t + \underbrace{\{1 - \{r_t\}\}}_{=\Delta_t \in [0,1)}$$

where we used the notation $\{x\}$ for the fractional part of x. This implies

$$\frac{t - \tilde{r}_t\mu}{\sigma\sqrt{\tilde{r}_t}} = \frac{t - r_t\mu - \Delta_t\mu}{\sigma\sqrt{r_t + \Delta_t}}$$

Recall that $r_t = \frac{t}{\mu} + y\sigma\sqrt{\frac{t}{\mu^3}}$. Therefore, we continue:

$$= \frac{t - \Delta_t\mu - \left(t + y\sigma\sqrt{\frac{t}{\mu}}\right)}{\sigma\sqrt{\Delta_t + \frac{t}{\mu} + \sigma y\sqrt{\frac{t}{\mu^3}}}} \xrightarrow{t\to\infty} \frac{-y\sigma}{\sigma} = -y$$

∎

11.2 Discrete Renewal Theory. Blackwell Theorem

Recall that in the definition of renewal processes the only condition about lifetimes was that they be i.i.d. However, what if the lifetimes are discrete random variables? We next give a motivating example of a situation when such a case is encountered.

■ EXAMPLE 11.3 Block Replacement Policy

Consider a type of light bulb with lifetime X, a random variable. Due to economic reasons, it might be cheaper on a per-bulb basis to replace all the bulbs instead of just the one that breaks. A block replacement policy does just that by fixing a time period K and replacing bulbs as they fail at times $1, 2 \ldots K - 1$ and at time K replacing everything regardless of the condition of the bulb. Let

$$c_1 = \text{replacement cost per bulb (block replacement)}$$
$$c_2 = \text{replacement cost per bulb (failure replacement)}$$

where obviously $c_1 < c_2$. Let $N(n)$ be the number of replacements up to time n for *a certain bulb*, and let $m(n) = \mathbb{E}[N(n)]$

For one bulb, the replacement cost up to time K is clearly $c_2 N(K - 1) + c_1$ since we pay c_2 for every time the bulb burns before K and we pay c_1 at time K regardless. Therefore, the expected replacement cost is $c_2 m(K - 1) + c_1$ and the mean cost per unit of time is

$$\frac{\text{expected total cost}}{\text{time}} = \frac{c_2 m(K - 1) + c_1}{K}$$

Suppose furthermore that replacing failed bulbs takes place only at the beginning of the day. Thus, we are only interested in discrete variables to describe the lifetime of a light bulb. So let us assume that X has the arbitrary discrete distribution $P(X = k) = p_k, k = 1, 2 \ldots$.

Fix an $n \leq K$. Look at X_1 – the first lifetime of the light bulb. Clearly, if $X_1 > n$, there was no replacement by time n. If $X_1 = k \leq n$, then we will have $m(n - k)$ expected replacements in the latter time period. Therefore, we can write conditioning on the lifetime of the first bulb

$$m(n) = \mathbf{E}[N(n)] = \sum_{k=1}^{\infty} \mathbf{E}[N(n)|X_1 = k]\mathbf{P}(X_1 = k)$$

$$= \sum_{k=1}^{n}[1 + m(n-k)]p_k + \sum_{k=n+1}^{\infty} 0 \cdot p_k$$

$$= \sum_{k=1}^{n} p_k[1 + m(n-k)]$$

$$= F_X(n) + \sum_{k=1}^{n-1} p_k m(n-k),$$

where $F_X(\cdot)$ is the cdf of X. Then we obtain recursively

$$m(0) = 0$$
$$m(1) = F_X(1) + p_1 m(0) = p_1$$
$$m(2) = F_X(2) + p_1 m(1) + p_2 m(0) = p_1 + p_2 + p_1^2$$

$\vdots$ etc.

■ **EXAMPLE 11.4   continues the example above**

Let us consider a numerical example of the problem above. Suppose that X can only take values $\{1, 2, 3, 4\}$ with $p_1 = 0.1$, $p_2 = 0.4$, $p_3 = 0.3$, and $p_4 = 0.2$; furthermore, the costs are $c_1 = 2$, $c_2 = 3$. Find the optimal replacement policy.
   Using the formulas above, we may calculate

$$m(1) = 0.1, \quad m(2) = 0.51, \quad m(3) = 0.891, \quad m(4) = 1.3231$$

Using these numbers we will try to minimize the expect cost as

$$\text{cost} = \frac{c_1 + c_2 m(K-1)}{K} \quad \leftarrow \text{We will try different K's to get the minimum}$$

We will obtain a table of cost as a function of K as follows:
   Hence the optimal replacement policy is at $K = 2$. We can also continue the calculation of m's:

$$m(5) = 1.6617, m(6) = 2.0647, m(7) = 2.4463, m(8) = 2.8336,$$
$$m(9) = 3.2136, m(10) = 3.6016, \ldots$$

   Now we can calculate u_n, the probability that a replacement occurs in period n, as

$$u_n = m(n) - m(n-1).$$

Table 11.1 Cost evolution with block value K

K	cost
1	2.00
2	1.15
3	1.17
4	1.16
5	1.19

By calculating u_n's for the values given, we can see that pretty quickly we have

$$u_n \approx \frac{1}{\mu} = 0.3846.$$

The limiting value will be explained by the next theorem (Blackwell renewal theorem).

Consider a renewal process with nonnegative integer-valued lifetimes, $X_1, X_2, \ldots$ with $P(X_i = k) = p_k$, $k = 0, 1, 2 \ldots$. Let us consider two examples. In the first example, each lifetime can take values 2 and 3 with probabilities 0.5 and 0.5. Then, the possible renewal times are at 2, 3, $4 = 2 + 2$, $5 = 2 + 3$ or $3 + 2$, and so on. We can see that for any integer $n \geq 2$ there is a chance (positive probability) that a renewal occurs at n. In a second example, consider lifetimes that take values 2 and 4 with probability 0.5 each. We can see that in this case renewals may occur only at even times. Thus for any time $n = 2k + 1$, an odd integer, there is no chance (probability 0) of a renewal happening at that time. The two examples are distinct and they require separate treatment in renewal theory. The next definition formalizes the latter case.

Definition 11.11 X, *an integer random variable, is called a* lattice *if there $\exists d \geq 0$ such that $p_k > 0, \forall k$ not a multiple of d. The largest d with the property that $\sum_{n=1}^{\infty} p_{nd} = 1$ is called the* period *of X. In effect:*

$$d = gcd\{k : p_k > 0\}^6$$

If $gcd\{k : p_k > 0\} = 1$, then X is called a non-lattice-random variable. Also if X, a lattice random variable, has cdf F, then F is called a lattice.

---

[6]The greatest common divisor (gcd) of the set of integers. The greatest common divisor of a set of integers A is defined as the largest integer that is a divisor of all integers in the set A.

■ **EXAMPLE 11.5**

Consider the two simple examples below:

- $p_2 = p_4 = p_6 = \frac{1}{3}$: this gives a lattice distribution, since $gcd\{2, 4, 6\} = 2$.

- $p_3 = p_7 = \frac{1}{2}$: this gives a non-lattice distribution, since $gcd\{3, 7\} = 1$.

In the previous example, we have seen how to establish the equation

$$m(n) = F_X(n) + \sum_{k=1}^{n-1} p_k m(n - k).$$

(Note that, if lifetimes are allowed to be zero, the equation is a little different.)

However, this equation constitutes a particular example of a *renewal equation* (discrete case). In general, **a discrete renewal equation** looks like

$$v_n = b_n + \sum_{k=0}^{n} p_k v_{n-k}, \tag{11.5}$$

where v_i's are unknowns and p_i's are given (known) probabilities. Note that this form of equation has a unique solution, found by recurrently solving the equation above: $v_0 = \frac{b_0}{1-p_0}$, $v_1 = \frac{b_1 + p_1 v_0}{1-p_0}$, etc.

Let u_n be the expected number of renewals that take place in period n. We have seen in the example that $u_n = m(n) - m(n-1)$. This is true only if the lifetimes are nonzero and therefore at most one renewal occurs in any one time period. This is easy to show.

$$\begin{aligned} u_n &= \mathbf{P}\{\text{One renewal occurred at } n\} \\ &= \mathbf{E}[\mathbf{1}_{\{\text{One renewal occurred at } n\}}] \\ &= \mathbf{E}[N(n) - N(n-1)] = m(n) - m(n-1) \end{aligned}$$

We have seen in the previous example that this u_n got closer and closer to $1/\mu$. The next theorem formalizes this fact and generalizes it.

Theorem 11.12 (Blackwell renewal theorem) *Using the notations above, let X_i be i.i.d. discrete random variables defining the lifetimes of a renewal process $N(t)$. Let u_n denote the probability that a renewal takes place at time $t = n$. Then*

1. $u_n \to \frac{1}{\mu}$ as $n \to \infty$.

2. If $X_0 \geq 0$ is a "delay" variable, and $X_1, X_2, \ldots \geq 0$ are i.i.d. lifetimes independent of X_0 with $\mathbf{E}X_1 = \mu$ and non-lattice distribution, then

$$m(t + a) - m(t) \to \frac{a}{\mu}, \quad \text{as } t \to \infty.$$

Note that $m(t + a) - m(t)$ is the expected number of renewals in the interval $[t, t + a]$.

3. *If X_i's are lattice random variables with period d, and $X_0 = 0$, then*

$$\mathbf{E}[\textit{Number of renewals at nd}] \to \frac{d}{n} \quad , n \to \infty$$

Remark 11.13 *About the theorem.*

1. *Even though the section is about renewal processes with discrete inter-arrival time distribution, Part (2) of the Blackwell theorem applies to **any** non-lattice distribution. This includes any inter-arrival times distributions, for example, continuous.*

2. *All the parts of the theorem apply if the mean inter-arrival time is $\mu = \infty$, with the convention $1/\infty = 0$.*

3. *If $\mathbf{P}(X_i = 0) = 0$, part (3) is equivalent to $\mathbf{P}\{\textit{Renewal at nd}\} \to d/\mu$*

We skip the proof of this theorem. We refer the reader to Karlin and Taylor (1975) for a detailed proof.

We write for an infinitesimal increment dy

$$m(dy) \overset{\text{alternative notation}}{=} dm(y) = m(y + dy) - m(y)$$
$$= \mathbf{E}\left[\textit{Number of renewals in the interval } (y, y + dy]\right]$$

This is the *renewal measure*. The Blackwell renewal theorem says that, approximately,

$$m(dy) \simeq \frac{1}{\mu} dy.$$

Lemma 11.14 *We have*

$$m(dy) = \sum_{n=0}^{\infty} \mathbf{P}\left(S_n \in (y, y + dy]\right) \tag{11.6}$$

Proof: The proof is straightforward (here we use a delay, therefore the sum starts from $n = 0$):

$$m(dy) = \mathbf{E}[N(y + dy) - N(y)] = \mathbf{E}[N(y + dy)] - \mathbf{E}[N(y)] =$$
$$= \sum_{n=0}^{\infty} \mathbf{P}\left(N(y + dy) \geq n\right) - \sum_{n=0}^{\infty} \mathbf{P}\left(N(y) \geq n\right)$$

$$= \sum_{n=0}^{\infty} \mathbf{P}\left(S_n \le y + dy\right) - \sum_{n=0}^{\infty} \mathbf{P}\left(S_n \le y\right)$$

$$= \sum_{n=0}^{\infty} \mathbf{P}\left(S_n \in (y, y + dy]\right)$$

∎

Many applications of the renewal theorem are concerned with the behavior of the process near a large time t. We need a final key of the puzzle before we can proceed with the study of such applications, and this key is provided in the next section.

11.3 The Key Renewal Theorem

This is the main result used in applications of renewal processes. We will start with a definition.

Definition 11.15 (Directly Riemann Integrable function) *A function* $h : [0, \infty) \to \mathbb{R}$ *is called a directly Riemann integrable (DRI) function if the upper and lower mesh δ Darboux sums are finite and have the same limit as $\delta \to 0$.*

Reminder of lower (and upper) Darboux sum LDS (and UDS):
Let $\pi = (t_0 = 0 < t_1 < t_2 < \dots)$ be a partition of $[0, \infty)$, with $max_i(t_i - t_{i-1}) \le \delta$. Define

$$LDS(h, \pi, \delta) = \sum_{n=1}^{\infty} \inf_{t \in [t_{n-1}, t_n]} h(t)(t_n - t_{n-1})$$

$$UDS(h, \pi, \delta) = \sum_{n=1}^{\infty} \sup_{t \in [t_{n-1}, t_n]} h(t)(t_n - t_{n-1})$$

EXAMPLE 11.6   Example of Riemann integrable function which is not DRI

Let

$$h(s) = \sum_{k=1}^{\infty} \mathbf{1}_{\{k \le s < k + \frac{1}{k^2}\}}$$

To understand this example, plot it from $s = 0$ to some large number. We have that

$$\int_0^{\infty} h(s)ds = 1 + \frac{1}{4} + \frac{1}{9} + \dots = \sum_{k=1}^{\infty} \frac{1}{k^2} < \infty,$$

so this function is Riemann integrable. However, it is not DRI. Take for some δ the partition $\pi = (t_0 = 0, t_1 = \delta, t_2 = 2\delta, \ldots, t_n = n\delta, \ldots)$. Then

$$UDS(h, \pi, \delta) = \sum_{n=1}^{\infty} \sup_{t \in [(n-1)\delta, n\delta]} h(t)(n\delta - (n-1)\delta)$$

$$= \delta \sum_{n=1}^{\infty} \sup_{t \in [(n-1)\delta, n\delta]} h(t),$$

which holds for any $\delta > 0$ no matter how small. The last term is an infinite sum of 1's, which is infinite.

Proposition 11.16 *The following are sufficient conditions for a function to be DRI:*

1. h is a monotone function (either increasing or decreasing),

2. $\int_0^{\infty} |h(t)| dt < \infty$.

The type of functions which are DRI and we will use will be the ones presented in the previous proposition.

We are now in the position to be able to state the main theorem of this section.

Theorem 11.17 (The Key Renewal Theorem) *For non-lattice $X_1, X_2, \ldots$ (any X_0 "delay" is fine) and if h is a DRI function, we have*

$$\lim_{t \to \infty} \int_0^t h(t - y) m(dy) = \frac{1}{\mu} \int_0^{\infty} h(t) dt$$

The proof of this theorem is standard and follows the definition of the directly Riemann integrable function, or more precisely their approximation with step functions:

$$\sum_{n=1}^{\infty} \inf_{t \in [t_{n-1}, t_n]} h(t) \mathbf{1}_{[t_{n-1}, t_n]}(t) \leq h(t) \leq \sum_{n=1}^{\infty} \sup_{t \in [t_{n-1}, t_n]} h(t) \mathbf{1}_{[t_{n-1}, t_n]}(t)$$

This is a very powerful theorem. We shall see its application in the next section.

11.4   Applications of the Renewal Theorems

Refer back to Figure 11.1 on page 335. We can see there the current age at time t and the remaining lifetime at t. Recall our definitions

$$A(t) = \text{Age of the current item in use at time } t$$
$$Y(t) = \text{Remaining lifetime of the item in use at time } t$$

Applications are concerned with these quantities when t is large. The question is: can we obtain distributions for these random variables? For the remainder of this section, we shall try to express

(a) $\mathbf{P}($Age at time t of the current item $> x) = \mathbf{P}(A(t) > x)$,

(b) $\mathbf{P}($Remaining lifetime of the item in use at $t > x) = \mathbf{P}(Y(t) > x)$,

(c) $\mathbf{P}($Total age of the item in use at $t > x) = \mathbf{P}(X_{N(t)+1} > x)$.

Recall that the distribution completely characterizes a random variable, thus these quantities will completely characterize the state of the renewal process for some large t.

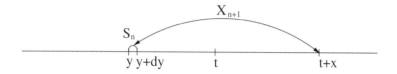

Figure 11.3 Deduction of the formula

We first look at the residual life process $Y(t)$. For exemplification, see Figure 11.3. Recall that X_0 is the delay, and we use the convention $S_0 = X_0$. Note that this renewal is counted in the renewal process $N(t)$. We have

$$\mathbf{P}(Y(t) > x) = \mathbf{P}(S_{N(t)+1} - l > x) = \mathbf{P}(N(t) = 0, X_0 > t + x)$$
$$+ \sum_{n=1}^{\infty} \int_0^t \mathbf{P}(N(t) = n, S_{n-1} \in (y, y + dy], X_n > t + x - y)$$

$$= \mathbf{P}(X_0 > t + x) + \sum_{n=1}^{\infty} \int_0^t \mathbf{P}(S_{n-1} \in (y, y + dy], X_n > t + x - y)$$

$$= (1 - F_0(t + x)) + \sum_{n=0}^{\infty} \int_0^t \mathbf{P}(S_n \in (y, y + dy]) \mathbf{P}(X_{n+1} > t + x - y)$$

Using the notation
$$\overline{F}(x) = 1 - F(x),$$

we continue:

$$\mathbf{P}(Y(t) > x) = \overline{F}_0(t + x) + \int_0^t \overline{F}(t + x - y) \sum_{n=0}^{\infty} \mathbf{P}(S_n \in (y, y + dy])$$

$$= \overline{F}_0(t + x) + \int_0^t \overline{F}(t + x - y) m(dy)$$

$$= \overline{F}_0(t + x) + \int_0^t h(t - y) m(dy),$$

where we used Lemma 11.14 and the notation $h(s) = \overline{F}(s+x)$. Now, $\overline{F}_0(t+x) \xrightarrow{t \to \infty} 0$ (since F_0 is a distribution function) and a direct application of the key renewal theorem (KRT) 11.17 yields

$$\mathbf{P}(Y(t) > x) \xrightarrow{t \to \infty} \frac{1}{\mu} \int_0^\infty \overline{F}(s + x)ds = \frac{1}{\mu} \int_x^\infty \overline{F}(y)dy$$

This result is significant enough to make it a proposition.

Proposition 11.18 *Let $A(t)$ be the age at t of the item and let $Y(t)$ be the residual life of the item alive at t. If F is the cdf of the lifetimes with expected lifetime μ, then the distributions of $A(t)$ and $Y(t)$ for t large have densities proportional, with*

$$f(y) = \frac{\overline{F}(y)}{\mu}$$

Proof: For $Y(t)$, the result is clear since from above we have

$$\mathbf{P}(Y(t) \leq x) \xrightarrow{t \to \infty} = \int_{-\infty}^x \frac{\overline{F}(y)}{\mu}dy$$

For $A(t)$, note that we have

$$\{A(t) > x\} \Leftrightarrow \{Y(t - x) > x\} \quad \text{(No renewal in } [t - x, t]),$$

therefore

$$\lim_{t \to \infty} \mathbf{P}\{A(t) > x\} = \lim_{t \to \infty} \mathbf{P}\{Y(t - x) > x\} = \frac{1}{\mu} \int_x^\infty \overline{F}(y)dy$$

Thus, the cdf of $A(t)$ has the stated property. ∎

Remark 11.19 *If the distribution of the delay has the special form $\mathbf{P}(X_0 > x) = \frac{1}{\mu} \int_x^\infty \overline{F}(y)dy$ (i.e., same as the long-term distribution), then $m(t) = \frac{t}{\mu}$ and the process is stationary (meaning that it looks the same regardless of the time when you start observing it).*

11.5 Special cases of Renewal Processes. Alternating Renewal process. Renewal Reward process.

In the case of Poisson processes, we found some of the best applications using Type I and Type II events and the compound Poisson process. It is worth investigating similar notions for the renewal process.

11.5.1 The alternating renewal process

Let $\{(Z_n, Y_n)\}_{n=1}^{\infty}$ be i.i.d. pairs of random variables[7]. Note that the pairs for $i \neq j$ are independent but Z_n and Y_n may be dependent.

Let $X_n = Z_n + Y_n$ define the lifetimes of the renewal process, and let $S_n = \sum_{i=1}^{n} X_i$ be the corresponding renewal times.

The story The Z_i times represent the light bulb lifetimes or the times when the system is in the ON state; the Y_i times may represent the replacement times or the times when the system is OFF.

Let us denote the cdf of Y_i's with G, the cdf of Z_i's with H, and the cdf of X_i's with F.

Theorem 11.20 *If* $\mathbf{E}[X_n] < \infty$ *and* F *is non-lattice, we have*

$$\mathbf{P}(\text{The system is ON at time } t) \overset{t \to \infty}{\longrightarrow} \frac{\mathbf{E}(Z_n)}{\mathbf{E}(X_n)} = \frac{\mathbf{E}(Z_n)}{\mathbf{E}(Z_n) + \mathbf{E}(Y_n)}$$

Proof:

$$\mathbf{P}(\text{ON at time } t) = \mathbf{P}(Z_1 > t) + \sum_{n=0}^{\infty} \int_0^t \mathbf{P}(S_n \in (y, y+dy], Z_{n+1} > t - y)$$

$$= \overline{H}_1(t) + \sum_{n=0}^{\infty} \int_0^t \mathbf{P}(S_n \in (y, y+dy]) \mathbf{P}(Z_{n+1} > t - y)$$

$$= \overline{H}_1(t) + \int_0^t \overline{H}(t - y) \sum_{n=0}^{\infty} \mathbf{P}(S_n \in (y, y+dy])$$

$$= \overline{H}_1(t) + \int_0^t \overline{H}(t - y) m(dy)$$

$$\overset{t \to \infty}{\longrightarrow} \overline{H}_1(\infty) + \frac{1}{\mu} \int_0^{\infty} \overline{H}(t) dt$$

However, $\mathbf{E}[Z] = \int_0^{\infty} \mathbf{P}(Z > z) dz = \int_0^{\infty} \overline{H}(z) dz$ and $\mathbf{E}[X] = \mu$, so we are done. ∎

▣ EXAMPLE 11.7

We have already seen that the distribution of $A(t)$ has density $\overline{F}(y)/\mu$. We will obtain this distribution again using the previous theorem about alternating renewal processes. Read the next derivations since these provide examples of applying this most useful theorem.

---

[7]Here, (Z_1, Y_1) (delay) is allowed to have a different distribution than the rest.

Once again we will calculate $\mathbf{P}(A(t) > x)$. Fix $x > 0$. Say that the system is ON during the first x units of each lifetime and OFF the rest of the time. Mathematically, using the notation of the alternating renewal processes

$$Z_k := X_k \wedge x = \min(X_k, x)$$
$$Y_k = X_k - Z_k$$

Then the theorem says

$$\mathbf{P}(\text{System is ON at time t}) = \mathbf{P}(A(t) < x) \rightarrow \frac{\mathbf{E}(Z_n)}{\mu}$$

But we can calculate the limit since

$$\mathbf{E}(Z_n) = \int_0^\infty \mathbf{P}(Z_n > y)dy = \int_0^x \mathbf{P}(Z_n > y)dy + \int_x^\infty \mathbf{P}(Z_n > y)dy$$
$$= \int_0^x \mathbf{P}(X_n > y)dy = \int_0^x \overline{F}(y)dy,$$

which will give the density and finish the example.

EXAMPLE 11.8   Limiting distribution of the current lifetime $X_{N(t)+1}$

We want to calculate $\mathbf{P}(X_{N(t)+1} > x)$. Fix x. Construct an alternating renewal process using:

$$Z_n = X_n \mathbf{1}_{\{X_n > x\}}, \quad Y_n = X_n \mathbf{1}_{\{X_n \le x\}}$$

Then

$$\mathbf{P}(\text{System is ON at time t}) = \mathbf{P}(X_{N(t)+1} > x) \rightarrow \frac{\mathbf{E}(Z_n)}{\mu}$$

Again, we can calculate

$$\mathbf{E}(Z_n) = \int_0^\infty \mathbf{P}(Z_n > y)dy = \int_0^x \mathbf{P}(Z_n > y)dy + \int_x^\infty \mathbf{P}(Z_n > y)dy$$
$$= \int_0^x \mathbf{P}(X_n > x)dy + \int_x^\infty \mathbf{P}(X_n > y)dy$$
$$= x\mathbf{P}(X_n > x) + \int_x^\infty \overline{F}(y)dy = \int_x^\infty ydF(y)$$

The last equality is obtained by recalling (Corollary 5.2 on page 159) that for a positive random variable X_n we can write

$$\mathbf{E}[X_n] = \int_0^\infty ydF(y) = \int_0^\infty \overline{F}(y)dy.$$

Using this and then integrating by parts, we get

$$\int_x^\infty \overline{F}(y)dy = \int_0^\infty ydF(y) - \int_0^x \overline{F}(y)dy$$

$$= \int_0^\infty ydF(y) - \left(y\overline{F}(y)\Big|_{y=0}^{y=x} - \int_0^x yd\overline{F}(y) \right)$$

$$= \int_x^\infty ydF(y) - x\overline{F}(x)$$

using that $d\overline{F}(y) = d(1 - F(y)) = -dF(y)$. This gives the limiting distribution

$$\mathbf{P}(X_{N(t)+1} > x) \rightarrow \frac{\int_x^\infty ydF(y)}{\mu}$$

or

$$\mathbf{P}(X_{N(t)+1} \le x) \rightarrow \int_0^x \frac{1}{\mu} ydF(y)$$

and so the density of the current lifetime is $\frac{xf(x)}{\mu}$, where $f(x)$ (if it exists) is the density of the lifetime distribution. Note that the distribution of the current lifetime at t is not the same as any lifetime distribution. This is because in a more general context it is the "stopped" lifetime.

Recall that, if we denote $Y(t)$ the excess or residual lifetime, we have already found its limiting distribution:

$$P(Y(t) > x) \rightarrow \frac{1}{\mu} \int_x^\infty \overline{F}(t)dy$$

We would like to find its expectation, or the limiting expected excess life, $\mathbf{E}[Y(t)]$. A first guess would be obviously the expectation of the previous distribution

$$\mathbf{E}[Y(t)] = \int_0^\infty P(Y(t) > x)dx \rightarrow \frac{1}{\mu} \int_0^\infty \int_x^\infty \overline{F}(y)dydx$$

This of course is correct, but we shall also prove it using the key renewal theorem.

Proposition 11.21 *If X is non-lattice with $\mathbf{E}[X^2] < \infty$, then*

$$\lim_{t \to \infty} \mathbf{E}[Y(t)] = \frac{\mathbf{E}[X^2]}{2\mu}$$

Note that one can show that $\frac{\mathbf{E}[X^2]}{2\mu}$ and $\frac{1}{\mu}\int_0^\infty \int_x^\infty \bar{F}(y)dydx$ are the same quantities by using a change in the order of integration and then integrating by parts.

Proof:

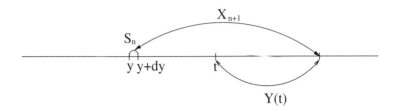

Figure 11.4   Relationship between X_{n+1} and $Y(t)$

We can go ahead and calculate

$$\mathbf{E}[Y(t)] = \sum_{n=0}^\infty \mathbf{E}[Y(t)\mathbf{1}_{\{N(t)=n\}}]$$

$$= \mathbf{E}[Y(t)\mathbf{1}_{\{N(t)=0\}}] + \sum_{n=1}^\infty \mathbf{E}[Y(t)\mathbf{1}_{\{N(t)=n\}}]$$

$$= \mathbf{E}[(X_1 - t)\mathbf{1}_{\{X_1>t\}}] + \sum_{n=1}^\infty \int_0^t \mathbf{E}\left[\underbrace{Y(t)}_{=X_{n+1}-(t-y)} \mathbf{1}_{\{S_n\in(y,y+dy],N(t)=n\}}\right]$$

$$= \mathbf{E}[(X_1 - t)\mathbf{1}_{\{X_1>t\}}] + \sum_{n=1}^\infty \int_0^t \mathbf{E}\left[(X_{n+1}-(t-y))\,\mathbf{1}_{\{S_n\in(y,y+dy]\}}\mathbf{1}_{\{X_{n+1}>t-y\}}\right]$$

$$= \mathbf{E}[(X_1 - t)\mathbf{1}_{\{X_1>t\}}] + \int_0^t \mathbf{E}\left[X-(t-y)\mathbf{1}_{\{X>t-y\}}\right]\underbrace{\sum_{n=1}^\infty \mathbf{E}\left[\mathbf{1}_{\{S_n\in(y,y+dy]\}}\right]}_{=m(dy)}$$

$$= \mathbf{E}[(X_1 - t)\mathbf{1}_{\{X_1>t\}}] + \int_0^t \mathbf{E}\left[X-(t-y)\mathbf{1}_{\{X>t-y\}}\right]m(dy)$$

The first term in the above sum converges to 0 as $t \to \infty$ since $\mathbf{E}[X_1]$ is finite. We can write $h(t-y) = X - (t-y)\mathbf{1}_{\{X>t-y\}}$ and use the KRT for the second term. If we do that, we obtain the limit as

$$\mathbf{E}[Y(t)] \xrightarrow{t \to \infty} = \frac{1}{\mu} \int_0^\infty h(s) ds = \frac{1}{\mu} \int_0^\infty \mathbf{E}[(X - s) \mathbf{1}_{\{X > s\}}] ds$$

$$= \frac{1}{\mu} \int_0^\infty \left[\int_s^\infty (x - s) dF(x) \right] ds$$

$$\text{Fubini} = \frac{1}{\mu} \int_0^\infty \int_0^x (x - s) ds \, dF(x)$$

$$= \frac{1}{\mu} \int_0^\infty -\frac{(x - s)^2}{2} \Big|_0^x dF(x)$$

$$= \frac{1}{\mu} \int_0^\infty \frac{x^2}{2} dF(x) = \frac{\mathbf{E}[X^2]}{2\mu}$$

∎

Corollary 11.22 *If $\mathbf{E}[X^2] < \infty$ and F non-lattice, then (for the undelayed renewal process)*

$$m(t) - \frac{t}{\mu} \xrightarrow{t \to \infty} \frac{\mathbf{E}[X^2]}{2\mu^2} - 1$$

Proof: Note that we have shown $\mathbf{E}[S_{N(t)+1}] = \mu \cdot (m(t) + 1)$
However,

$$\mathbf{E}[t + Y(t)] = t + \mathbf{E}[Y(t)] \to t + \frac{\mathbf{E}[X^2]}{2\mu}$$

Since $S_{N(t)+1} = t + Y(t)$, we obtain

$$m(t) + 1 \to \frac{t}{\mu} + \frac{\mathbf{E}[X^2]}{2\mu^2} \Rightarrow m(t) - \frac{t}{\mu} \to \frac{\mathbf{E}[X^2]}{2\mu^2} - 1$$

∎

◨ **EXAMPLE 11.9**

Let $X_1, X_2 \ldots$ be i.i.d. $U[0, 1]$. Then, $\mu = \frac{1}{2}$, $\mathbf{E}[X^2] = \frac{1}{3}$

Then the corollary says that for $t = 100$

$$m(100) \sim \frac{100}{m} + \frac{\mathbf{E}[X^2]}{2\mu^2} - 1$$

$$= \frac{100}{\frac{1}{2}} + \frac{\frac{1}{3}}{2 \cdot (\frac{1}{2})^2} - 1 \quad \leftarrow \text{better approximation}$$

$$= 199\frac{1}{3} \quad \text{(probably very accurate)}$$

11.5.2   Renewal reward process

This would be the equivalent of the compound Poisson process. Each event comes with a random variable attached to it. Specifically, consider the i.i.d. pairs: (X_1, R_1), $(X_2, R_2) \ldots$

Story: At each of the times $S_n = \sum_{i=1}^{n} X_i$, you get a reward R_n. Assume that $X_i \geq 0$, $\mathbf{E}[X_i] = \mu < \infty$, $\mathbf{E}[R_i] = \mathbf{E}[R] < \infty$.

Let $R(t) = \sum_{i=1}^{N(t)} R_i$ is the total reward up to time t.

Theorem 11.23 *With the notations above, the total reward $R(t)$*

1.

$$\frac{R(t)}{t} \xrightarrow{a.s.} \frac{\mathbf{E}[R]}{\mu} \qquad as \ t \to \infty$$

2.

$$\frac{\mathbf{E}[R(t)]}{t} \to \frac{\mathbf{E}[R]}{\mu} \qquad as \ t \to \infty$$

Proof: **Part (1)** We have

$$\frac{R(t)}{t} = \frac{1}{t} \sum_{i=1}^{N(t)} R_i = \frac{\sum_{i=1}^{N(t)} R_i}{N(t)} \frac{N(t)}{t}$$

The first term in the product above converges to $\mathbf{E}[R]$ using the strong law of large numbers (SLLN), and the second term converges to $1/\mu$ by the renewal SLLN. Therefore, we get the result in part (1).

Part (2) We have
using Wald's theorem for $N(t) + 1$ which is a stopping time,

$$\mathbf{E}[R(t)] = \mathbf{E}[\sum_{i=1}^{N(t)} R_i] = \mathbf{E}[\sum_{i=1}^{N(t)+1} R_i] - \mathbf{E}[R_{N(t)+1}]$$

$$= \mathbf{E}[N(t) + 1]\mathbf{E}[R_n] - \mathbf{E}[R_{N(t)+1}] = (m(t) + 1)\,\mathbf{E}(R) - \mathbf{E}[R_{N(t)+1}]$$

This implies dividing with t and taking the limit as $t \to \infty$:

$$\frac{\mathbf{E}[R(t)]}{t} = \frac{(m(t) + 1)}{t}\mathbf{E}(R) - \frac{\mathbf{E}[R_{N(t)+1}]}{t} \xrightarrow{t \to \infty} \frac{\mathbf{E}(R)}{\mu} - \lim_{t \to \infty} \frac{\mathbf{E}[R_{N(t)+1}]}{t},$$

where we used the elementary renewal theorem for the first term. To complete the proof, we have to show that $\lim_{t \to \infty} \mathbf{E}[R_{N(t)+1}]/t = 0$. We have

$$\mathbf{E}[R_{N(t)+1}] = \mathbf{E}[R_1 \mathbf{1}_{\{X_1>t\}}] + \sum_{n=1}^{\infty} \int_0^t \mathbf{E}\left[R_{n+1}\mathbf{1}_{\{X_{n+1}>t-y, S_n \in (y, y+dy], N(t)=n\}}\right]$$

$$= \mathbf{E}[R_1 \mathbf{1}_{\{X_1>t\}}] + \int_0^t \sum_{n=1}^{\infty} \mathbf{E}\left[R_{n+1}\mathbf{1}_{\{X_{n+1}>t-y\}}\right] \mathbf{E}\left[\mathbf{1}_{\{S_n \in (y, y+dy]\}}\right]$$

$$= \mathbf{E}[R_1 \mathbf{1}_{\{X_1>t\}}] + \int_0^t \mathbf{E}\left[R_2 \mathbf{1}_{\{X_2>t-y\}}\right] \sum_{n=1}^{\infty} \mathbf{E}\left[\mathbf{1}_{\{S_n \in (y, y+dy]\}}\right]$$

$$= \mathbf{E}[R_1 \mathbf{1}_{\{X_1>t\}}] + \int_0^t h(t-y)m(dy),$$

where we denoted $h(t-y) = \mathbf{E}\left[R_2 \mathbf{1}_{\{X_2>t-y\}}\right]$, to apply the KRT. The first term converges to 0 as $t \to \infty$ (justify), and we obtain the limit

$$\lim_{t\to\infty} \mathbf{E}[R_{N(t)+1}] = \frac{1}{\mu}\int_0^\infty h(t)dt = \frac{1}{\mu}\int_0^\infty \mathbf{E}\left[R_2 \mathbf{1}_{\{X_2>t\}}\right] < \frac{\mathbf{E}(R)}{\mu} < \infty$$

Thus, dividing with t and taking the limit, we obtain 0, which completes the proof. ∎

11.6   A generalized approach. The Renewal Equation. Convolutions.

Often, the quantity of interest in renewal theory $Z(t)$ satisfies an equation of the form

$$Z(t) = z(t) + \int_0^t Z(t-y)F(dy),$$

where $F(t) = $ cdf of inter-arriaval time, and $z(t)$ is some known function with the properties

- $z(t) = 0$ if $t < 0$,

- z is bounded on a finite interval.

An equation of this type is called a **renewal equation**.

EXAMPLE 11.10

$m(t)$ satisfies

$$m(t) = F(t) + \int_0^t m(t-y)F(dy)$$

■ **EXAMPLE 11.11**

$P(Y(t) > x)$:

$$P(Y(t) > x) = \bar{F}(t+x) + \int_0^t P(Y(t-y) > x)F(dy)$$

■ **EXAMPLE 11.12**

$\mathbf{E}[Y(t)]$:

$$\mathbf{E}[Y(t)] = \mathbf{E}[X_1 - t]I_{\{X_1 > t\}} + \int_0^t \mathbf{E}[Y(t-y)]dF(y)$$

The next theorem will provide a way to solve the renewal equation.

Theorem 11.24 *If $F(0-) = 0$, $F(0) < 1$, $z(t)$ is bounded on finite intervals, and $z(t) = 0$ for $t < 0$, then the renewal equation*

$$Z(t) = z(t) + \int_0^t Z(t-s)dF(s)$$

has a unique solution, bounded on finite intervals given by

$$Z(t) = z(t) * m_0(t) = \int_0^t z(t-y)m_0(dy) = \sum_{n=0}^\infty \int_0^t z(t-y)dF_n(y),$$

where

$$m_0(t) = \sum_{n=0}^\infty F_n(t) = \sum_{n=0}^\infty P(S_n \le t)$$

$$F_n(t) = \underbrace{F * F \ldots * F}_{n\ times}, \qquad with\ S_0 = 0$$

Recall the convolution and its properties (Chapter 6).

Proof (Proof of the theorem on renewal equation):
  Part 1. Existence of the solution.

$$z * m_0(t) = \sum_{n=0}^\infty z * F_n(t)$$

$$= z * F_0(t) + \sum_{n=1}^\infty z * F_n(t)$$

$$= z(t) * F_0 + \left[\sum_{n=0}^{\infty} z * F_n(t)\right] * F(t)$$

$$= z(t) + (z * m_0) * F(t)$$

$$= z(t) + \int_0^t (z * m_0)(t - s)dF(s)$$

Note that we used $F_0(t) = P(S_0 \leq t) = \mathbf{1}_{\{t \geq 0\}}$.

This shows that $z * m_0$ is a solution for the renewal equation.

Part 2. Uniqueness.

Assume that there exist $Z_1(t)$ and $Z_2(t)$ which are two solutions of the renewal equation. Let $V(t) = (Z_1 - Z_2)(t)$. Since both Z_1, Z_2 are roots, then $V(t)$ should also solve the renewal equation, that is,

$$V(t) = (Z_1 - Z_2)(t)$$

$$= z(t) + \int_0^t Z_1(t - s)dF(s) - z(t) - \int_0^t Z_2(t - s)dF(s)$$

$$= \int_0^t V(t - s)dF(s) = V * F(t)$$

Repeating the argument, we have

$$V(t) = V * F(t) = V * F_2(t) = \ldots = V * F_k(t), \quad \forall k$$

which implies

$$V(t) = \int_0^t V(t - y)F_k(dy)$$

$$\leq \sup_{0 \leq s \leq t} V(s) \int_0^t dF_k(s)$$

$$= \sup_{0 \leq s \leq t} V(s)F_k(t) \xrightarrow{k \to \infty} 0$$

Because $F_k(t) = P(X_1 + X_2 + \ldots + X_k \leq t) \xrightarrow{k \to \infty} 0$, $\forall t$ fixed (CLT or SLLN). ∎

Theorem 11.25 *True for both lattice and non-lattice cases) X_1 has distribution*

$$P(X_1 > x) = \int_0^\infty \frac{1}{\mu}\overline{F}(y)dy \stackrel{def}{=} F_e(x)$$

This is called the equilibrium distribution; *the process with delay X_1 having this distribution is called the* equilibrium renewal process. *Let*

$$m_D(t) \stackrel{def}{=} \sum_{n=1}^{\infty} P(S_n \le t) = \sum_{n=0}^{\infty} F_e * F_n(t),$$

and $Y_D(t)$ be the residual lifetime at t for the delayed process. Then

1. $m_D(t) = \frac{t}{\mu}$

2. $P(Y_D(t) > x) = \overline{F}_e(x)$ *for all $t > 0$*

3. $\{N_D(t)\}_t$ *has stationary increments.*

Proof: **Part 1.**

$$m_D(t) = F_e(t) + \left(F_e * \sum_{n=1}^{\infty} F_n \right)(t)$$

$$= F_e(t) + \left(F_e * \sum_{n=0}^{\infty} F_n \right) * F(t)$$

$$= F_e(t) + m_D(t) * F(t)$$

which implies that $m_D(t)$ solves a renewal equation with $z(t) = F_e(t)$.

If we show that $\frac{t}{\mu}$ also solves the renewal equation with the same $z(t)$, we are done.

Check yourself that $h(t) = \frac{t}{\mu}\mathbf{1}_{\{t>0\}}$ also solves the same renewal equation. By uniqueness of the solution, we are done.

Part 2. Using the usual renewal argument:

$$P(Y_D(t) > x) = \mathbf{P}(X_1 > t + x) + \int_0^t \overline{F}(t - y + x)m_D(dy)$$

$$(\text{From (i)} \Longrightarrow) = \overline{F}_e(t + x) + \int_0^t \overline{F}(t - y + x)\frac{dy}{\mu}$$

$$= \int_{t+x}^{\infty} \frac{1}{\mu}\overline{F}(y)dy + \int_0^t \frac{1}{\mu}\overline{F}(t - y + x)dy$$

$$(\text{c.v. } v = t - y + x) = \int_{t+x}^{\infty} \frac{1}{\mu}\overline{F}(y)dy - \int_{t+x}^{x} \frac{1}{\mu}\overline{F}(v)dv$$

$$= \int_x^{\infty} \frac{1}{\mu}\overline{F}(y)dy$$

$$= \overline{F}_e(x) \quad \Longrightarrow . \text{ DONE.}$$

Part 3. This part follows from part (2) using the fact that $N_D(t + s) - N_D(s)$ is the number of renewals in a time interval of length t for a delayed renewal process. ∎

11.7 Age-Dependent Branching processes

We conclude this chapter with an important class of processes that may be solved as a direct application of the renewal theory.

Story: Suppose each individual in a population lives a certain time which is a random variable with distribution F, with $F(0) = 0$ (lifetime is positive).
Let P_j be the probability that at the death of each such individual we obtain exactly j offspring, $j = 0, 1, 2 \ldots$.
Each offspring then acts independently of others and produce their own offspring according to P_j, and so on and so forth.

Let $X(t)$ denote the number of organisms alive at time t. $\{X\}_{t>0}$ is called an *age-dependent branching process* with $X(0) = 1$.

Quantity of interest $M(t) = \mathbb{E}[X(t)]$ is the expected number of individuals living at time t

$$\text{Let } m = \sum_{j=0}^{\infty} j P_j = \text{expected number of offspring} = \text{basic reproductive rate}$$

Galton–Watson process If lifetimes are identically equal to 1, then we obtain very easily

$$M(k) = m^k$$

Furthermore, the actual process may be expressed as

$$X(k) = \sum_{i=1}^{X(k-1)} \xi_i^{(k)},$$

where $\xi_i^{(k)}$ is the number of offspring of individual j in the previous population $X(n-1)$, which is of course distributed with the discrete probabilities P_j above.

In this particular situation, $X(k)$ is called a *Galton–Watson process* (which is also a Markov chain, as we will see later). This process has a very interesting history. In Victorian England at the end of the nineteenth century, there was concern that the aristocratic surnames were becoming extinct. Francis Galton originally posed the question regarding the probability of such an event in the *Educational Times* in 1873. Reverend Henry William Watson replied with a solution in the same venue. Together, they then wrote a paper entitled "*On the probability of extinction of families*" in 1874.

The probability of extinction may be calculated as a particular case of the more general theorem at the end of this section. There is trivially a case when population never dies out – when each parent produces exactly one offspring. Outside this case, if $m \le 1$, the probability of extinction is equal to 1 and if $m > 1$ this probability is strictly less than 1; that is, there is a positive probability that the population will survive indefinitely.

Remark Usually, one may prove a result of the form

$$\frac{X(k)}{m^k} \to Z$$

with Z a random variable finite a.s.

Theorem 11.26 *If $X(0) = 1$, $m \neq 1$, and F is a non-lattice distribution, then*

$$\lim_{t \to \infty} e^{-\alpha t} M(t) = \frac{m-1}{m^2 \alpha \int_0^\infty x e^{-\alpha x} dF(x)},$$

where α is the solution of the equation $\int_0^\infty e^{-\alpha t} dF(t) = \frac{1}{m}$

Remark 11.27 *The theorem simply says that $M(t) \sim constant \cdot e^{-\alpha t}$*

Proof: Using $ds = (s, s + ds]$ and the renewal argument

$$M(t) = \mathbf{E}[X(t)]$$

$$= \mathbf{E}[X(t)\mathbf{1}_{\{\text{1st life} > t\}}] + \mathbf{E}\left[X(t)\mathbf{1}_{\{\text{1st life} \leq t\}}\right]$$

$$= \mathbf{E}[\mathbf{11}_{\{T_1 > t\}}] + \int_0^t \mathbf{E}\left[\sum_{i=1}^N X_i(t-s)\mathbf{1}_{\{T_1 \in ds\}}\right],$$

where N is the number of offspring that the first individual produces and $X_i(t - s)$ is the number of individuals in the population as a direct result of offspring i. Now, using the Wald's theorem and the fact that all the $X_i(t)$ have the same distribution as the original process, we may continue (with $m = \mathbf{E}[N]$):

$$= P(T_1 > t) + \int_0^t m\mathbf{E}[X(t-s)\mathbf{1}_{\{T_1 \in ds\}}]$$

$$= P(T_1 > t) + \int_0^t m\mathbf{E}[X(t-s)]\mathbf{E}[\mathbf{1}_{\{T_1 \in ds\}}]$$

$$= \overline{F}(t) + \int_0^t mM(t-s)dF(s)$$

This looks a lot like a renewal equation except for the m. In order to eliminate it, we will try to incorporate it into the distribution function. We multiply both sides by $e^{-\alpha t}$,

$$\Rightarrow M(t)e^{-\alpha t} = \overline{F}(t)e^{-\alpha t} + \int_0^t e^{-\alpha t} mM(t-s)dF(s)$$

$$= \overline{F}(t)e^{-\alpha t} + \int_0^t e^{-\alpha(t-s)} M(t-s) \underbrace{me^{-\alpha s}dF(s)}_{=dG(s)}$$

We denoted $dG(s) = me^{-\alpha s}dF(s)$, or $G(t) = \int_0^t me^{-\alpha s}dF(s)$. Now to be able to apply the key renewal theorem we need G to be a proper distribution function. G is clearly an increasing function (the integral of a positive function), right continuous, and $G(0) = 0$. We only need $G(\infty) = 1$ and this is ensured by α being a solution of the equation

$$\int_0^\infty e^{-\alpha t} dF(t) = \frac{1}{m}.$$

Note that, if the expected number of offspring $m > 1$, then the resulting α must be positive. If $m < 1$, then the resulting alpha is negative. The case $m = 1$ has to be treated separately; the resulting alpha is zero, and in fact in this case we can apply the key renewal theorem directly without any modification.

Thus, we obtain a renewal equation $Z(t) = z(t) + \int_0^t Z(t - s) dG(s)$, where

$$Z(t) = e^{-\alpha t} M(t)$$
$$z(t) = e^{-\alpha t} \overline{F}(t)$$

Recall that the unique solution of this equation is given by

$$Z(t) = z * m_0(t) = \int_0^t z(t - s) m_0(ds) \qquad \left(\text{with } m_0 \text{ given by } \sum_{n=0}^\infty G_n(s)\right)$$

$$\xrightarrow{\text{KRT}} \frac{1}{\mu_G} \int_0^\infty z(t) dt,$$

provided that we can apply the KRT. Therefore, we see that

$$Z(t) \to \frac{1}{\mu_G} \int_0^\infty e^{-\alpha t} \overline{F}(t) dt.$$

Now let us calculate the two quantities in the limit while at the same time showing that $z(t)$ is directly Riemann integrable (so that we can apply the KRT). Using $\overline{F}(t) = \int_t^\infty dF(x)$, we have

$$\int_0^\infty z(t) dt = \int_0^\infty \left(\int_t^\infty dF(x) \right) e^{-\alpha t} dt$$

$$= \int_0^\infty \left(\int_0^x e^{-\alpha t} dt \right) dF(x) \qquad \text{(Fubini)}$$

$$= \int_0^\infty \frac{1}{\alpha} (1 - e^{-\alpha x}) dF(x)$$

$$= \frac{1}{\alpha} \left[\underbrace{\int_0^\infty dF(x)}_{=1} - \underbrace{\int_0^\infty e^{-\alpha x} dF(x)}_{=\frac{1}{m} \text{ (def. of } \alpha)} \right]$$

$$= \frac{1}{\alpha} \left[1 - \frac{1}{m} \right] < \infty$$

Thus, $z(t)$ is Riemann integrable, decreasing, and positive, and therefore it is DRI. All that remains is to calculate μ_G.

$$\mu_G = \int_0^\infty x dG(x) = \int_0^\infty x m e^{-\alpha x} dF(x) = m \int_0^\infty x e^{-\alpha x} dF(x)$$

Hence

$$Z(t) = e^{-\alpha t} M(t) \to \frac{\frac{1}{\alpha}\left(1 - \frac{1}{m}\right)}{m \int_0^\infty \alpha e^{-\alpha x} dF(x)}$$

And a little algebra shows that this is exactly the formula we need to prove. ∎

Remark 11.28 *Note that the limit of the theorem works for $m < 1$ and the corresponding $\alpha < 0$. In this case, $z(t) = e^{-\alpha t}\overline{F}(t)$ is going to be also a DRI function. The theorem implies that in this case the population decays to 0.*

In either case $m > 1$ or $m < 1$, the mean lifetime for the G distribution $\mu_G = \infty$ is possible and it will not change the answers.

In the case when $m = 1$, the argument in the theorem leads to

$$M(t) = \overline{F}(t) + \int_0^t M(t - s) dF(s),$$

which is a proper renewal equation and the limit of the solution may be found directly as

$$M(t) = \int_0^t \overline{F}(t - s) m_0(ds) \to \frac{1}{\mu_F} \int_0^\infty \overline{F}(t) dt = 1$$

Problems

11.1   For a branching process, what is the probability that $X(t) = 0$ eventually? (Population dies out). Show the following:

For $m < 1$, $P(\text{Population dies out}) = 1$.

If $m = 1$, then $P(\text{Population dies out}) = 1$ except when the number of offspring is exactly 1.

If $m > 1$? $P(\text{Population dies out}) > 0$

Note that these are exactly the same results as in the case of the Galton–Watson process.

11.2   Show that for a renewal process

$$S_{N(t)-1} \le t \le S_{N(t)},$$

as long as $N(t) \ge 1$.

11.3   Circle all statements that are true. $N(t)$ is a renewal process and S_n is the time of the nth renewal:

    a) $N(t) < n$ if and only if $S_n > t$,

b) $N(t) \leq n$ if and only if $S_n \geq t$,

c) $N(t) > n$ if and only if $S_n < t$.

11.4   Let X_t, the lifetimes for a renewal process, be i.i.d. $Geometric(p)$. Give the distribution of the age of the current item in use at t: $A(t)$, the residual lifetime of the item: $Y(t)$, and the total age of the item in use at t: $X_{N(t)+1}$.

11.5   Let X be an exponential random variable with mean 1. For any $\lambda > 0$, show that the random variable X/λ is an exponential random variable with rate λ.

11.6   Denote by $L(t) = X_{N(t)+1}$ the random variable measuring the current lifetime at t. Suppose the renewal process has discrete lifetimes. Let $n \geq 1$, and set $z(n) = \mathbf{P}(L_t = n)$. Show that $z(n)$ satisfies the renewal equation (11.5) with

$$b_n = \begin{cases} f(t), & \text{for } n = 0, 1, 2, \dots, t-1 \\ 0, & \text{for } n \geq t \end{cases}$$

.

11.7   Suppose we have a type of light bulb with lifetime uniformly distributed on the interval $(0, 10)$ days. You walk into the room which has this type of light bulb after 1 year. We know that the maintenance replaced light bulbs the moment they burned out, and the light bulbs are distributed according to their equilibrium distribution.

  a) What is the average lifetime of a light bulb in this equilibrium distribution?

  b) If a light bulb has been working for 4 days, what is the probability that it will burn at least 2 more days?

11.8   A certain component of a machine has a lifetime distribution $F_1(x)$. When this component fails, there is a probability p that it is instantaneously replaced ($0 < p < 1$). If the component is not replaced, the machine operates for a random amount of time (with distribution $F_2(x)$) and then shuts down. After the system shuts down, the machine is repaired (and the component is replaced), which takes another random time length with distribution $F_3(x)$. Assume that the distributions have expectations μ_1, μ_2, and μ_3 and they are all finite. We also assume that all the random variables in question are independent of each other. Find the asymptotic proportion of time when the machine is operational and the component in question is also working.

11.9   Suppose a high-performance computer (HPC) processes jobs which arrive according to a Poisson process with rate $\lambda = 6$ per day. We assume that all jobs require some time to complete and this time is uniformly distributed between 2 and 3 h. The HPC will process the jobs in the order they are received, and the jobs will be processed continuously as long as there are any in the queue. Calculate the expected proportion of time the HPC is busy.

11.10   Consider an insurance company which receives claims according to a Poisson process with rate $\lambda = 400$/year. Suppose the size of claims are random variables R_n which are distributed with an exponential distribution with mean \$1000. Calculate the expected total amount of the claims during a year. Assuming that the insurance company has n clients, how much should the monthly insurance premium be to make sure that the company has a yearly profit. Provide numerical values if the company has $n = 1,000, 10,000$ and $100,000$ clients. Now assume that the company has 10 people on the staff with a total of $500,000 salary budget and it has to produce profit of $500,000 at the end of the year. How much should the monthly premium be now?

11.11   Suppose I want to cross a road with no stopping sign. On that road, cars pass according to a Poisson process with rate λ. I need at least a τ second gap to cross the road safely. If I observe a gap of at least τ seconds, I start crossing immediately. Let T the amount of time I need to wait until I start crossing.

    a) Set up a renewal equation for $\mathbf{P}(T < x)$.

    b) Find $\mathbf{E}[T] + \tau$, the expected time to cross the road.

11.12   Suppose a light bulb in a renewal process has a lifetime distributed with density $f(x) = \frac{3}{(x+1)^4}$, for $x \geq 0$. Let $Y(t)$ be the residual lifetime for some large t. Use the results in this chapter to calculate the expected value and variance for the residual life $Y(t)$.

11.13   A particular driver commutes between Hoboken (H) and Atlantic City (AC). Every time the driver goes from H to AC, he drives using cruise control at a fixed speed, which is uniformly distributed between 55 and 65 miles per hour (mph). Every time he drives back from AC to H, he drives at a speed which is either 55 mph with probability $1/2$ or 65 mph with probability $1/2$.

    a) In the long run, what proportion of his driving time is spent going from H to AC?

    b) In the long run, what proportion of his driving time is spent at a speed of 65 mph?

11.14   Let $N(t)$ be a renewal process and let $m(t) = \mathbf{E}[N(t)]$ be the renewal function. Suppose that the inter-arrival times are strictly positive, that is, $P(X \leq 0) = F_X(0) = 0$.

    a) For all $x > 0$ and $t > x$, show that

$$\mathbf{E}[N(t) \mid X_1 = x] = \mathbf{E}[N(t - x)] + 1$$

    b) Use part a) to derive the renewal equation for $m(t)$ (see page 359).

11.15   Suppose two independent renewal processes are denoted with $N_1(t)$ and $N_2(t)$. Assume that each inter-arrival time for both processes has the same distribution $F(x)$ and density $f(x)$ exists.

    a) Is the process $N_1(t) + N_2(t)$ a renewal process? Explain.

    b) Let $Y(t)$ denote the time until the first arrival after time t from either of the two processes. Find an expression for the distribution function of $Y(t)$ as $t \to \infty$.

11.16   Suppose a branching process starts with one individual. Suppose each individual has exactly three children each of whom survives until reproductive age with probability p and dies with probability $1-p$, independent of his siblings. The children reaching the reproductive age reproduce according to the same rules.

a) Write down the distribution of offspring reaching the reproductive age.

b) Using Theorem 11.26, calculate the limit of the mean number of individuals in the population as $t \to \infty$.

c) Derive the values of p that will lead to extinction with probability 1.

11.17   A population begins with a single individual. In each generation, each individual dies with probability $1/2$ or divides into two identical individuals with probability $1/2$. Let $N(n)$ denote the number of individuals in the population in the nth generation. Find the mean and variance of $N(n)$. Repeat the problem if the individual dies with probability $2/3$ and divides with probability $1/3$. Calculate the probability of extinction for the second case.

CHAPTER 12

MARKOV CHAINS

In this chapter, we start the study of some the most popular class of models suitable for real-life situations.

12.1 Basic Concepts for Markov Chains

12.1.1 Definition

Consider a set of outcomes S which is finite or countable. S is called the states space. It is convenient to represent the set S as the nonnegative integers $\{0, 1, \ldots\}$ (any discrete set may be put into a bijection with this set). Consider a process $X = (X_1, X_2, \ldots, X_n, \ldots)$ whose components X_n take values in this set S. We will say that the process X is in state $i \in S$ at time n if $X_n = i$.

We next consider a matrix

$$P = \begin{pmatrix} P_{0,0} & P_{0,1} & P_{0,2} & \cdots \\ P_{1,0} & P_{1,1} & P_{1,2} & \cdots \\ \vdots & \vdots & \vdots & \\ P_{i,0} & P_{i,1} & P_{i,2} & \cdots \\ \vdots & \vdots & \vdots & \end{pmatrix},$$

Probability and Stochastic Processes, First Edition. Ionuţ Florescu
© 2015 John Wiley & Sons, Inc. Published 2015 by John Wiley & Sons, Inc.

with the elements

$$\begin{cases} P_{i,j} \geq 0, \forall i, j \\ \sum_{j=0}^{\infty} P_{i,j} = 1, \forall i \text{ (i.e., rows sum to 1).} \end{cases}$$

Such a matrix is often called a *stochastic matrix*.

Furthermore, let $\pi_0 = \{\pi_0(i)\}_{i=0}^{\infty}$ be some initial probability mass function on the elements of S (i.e., $\pi_0(i) \geq 0$ and $\sum_i \pi_0(i) = 1$).

Definition 12.1 *A discrete-time Markov chain[1] on the state space S with initial distribution π_0 and and transition matrix P is a stochastic process $X = \{X_n\}_{n=0}^{\infty}$ with*

$$\mathbf{P}\left(X_0 = i_0, \ldots, X_n = i_n\right) = \pi_0(i_0) \prod_{k=0}^{n-1} P_{i_k, i_{k+1}}$$

Note that with this definition the elements P_{ij} may be interpreted as the transition probability of jumping from state i to state j in one step. Accordingly, the matrix P is called the *transition probability matrix*.

Theorem 12.2 (Markov Property) *If $X = \{X_n\}_{n=0}^{\infty}$ is a discrete-time Markov chain defined as above, then*

$$\mathbf{P}\left(X_{n+1} = i_{n+1} | X_0 = i_0, \ldots, X_n = i_n\right) = \mathbf{P}\left(X_{n+1} = i_{n+1} | X_n = i_n\right)$$
$$= P_{i_n, i_{n+1}}$$

Proof: Exercise. Verify the theorem using Definition 12.1.   ∎

12.1.2   Examples of Markov chains

This section presents several examples of Markov chains.

■ EXAMPLE 12.1

This is a very simple example. Let $X_1, X_2, \ldots, X_n, \ldots$ be i.i.d. random variables on $\mathbb{Z}$. Take the transition probabilities $P_{i,j} = \mathbf{P}(X_k = j)$, for any k and no matter what starting state i is. Then, this forms a (rather trivial) Markov chain on

---

[1] This is in fact the definition of what is called a *homogeneous* Markov chain. That is because the transition probability matrix P does not depend on n, the current step number. A Markov chain with transition matrix $P = P_n$ dependent on the step number is called a *nonhomogeneous* Markov chain. The study of nonhomogeneous Markov chains is subject of active research at the time of writing this book.

$\mathbb{Z}$ with transition probabilities not dependent on the current state (i.e., all rows in the transition matrix are identical and equal to the distribution of X_1).

EXAMPLE 12.2   Generalized random walk on $\mathbb{Z}$

Let $\Delta_0, \Delta_1, \Delta_2, \ldots, \Delta_n, \ldots$ be again i.i.d. random variables on $\mathbb{Z}$ with some probability mass function $p_m = \mathbf{P}(\Delta_k = m)$. Let the state of the process X at time n be defined by

$$X_n = \sum_{k=0}^{n} \Delta_k.$$

Unlike the simple random walk, this more general one may have jumps of any size. It is easy to see that X is a Markov chain on $\mathbb{Z}$ with initial distribution $\pi_0(m) = p_m$ and with transition probability matrix having elements

$$P_{i,j} = \mathbf{P}(\Delta_k = j - i) = p_{j-i}.$$

As a special case, if all the jumps are positive, $\Delta_k \geq 0$, we obtain a renewal process.

EXAMPLE 12.3   Ehrenfest chain

Suppose we have two boxes, box 1 with x balls and box 2 with $d - x$ balls (a total of d balls in both). At each step (time) we pick one of the balls at random and we transfer that ball from one box to another (see Figure 12.1). It is easy to see that the same model may be constructed using one urn containing a total of d balls of two types of color. At every step, we pick one ball at random and change its color to the opposite color.

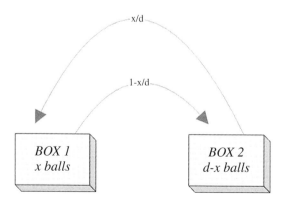

Figure 12.1   A graphical description of the Ehrenfest chain

We consider the process $X = \{X_n\}_{n=0}^{\infty}$ with X_n given by the number of balls in box 1 at time n. We obviously have $X_n \in S = \{0, 1, .., d\}$. Then X is a Markov chain on S with some initial distribution. The transition probabilities for the chain are

$$\mathbf{P}(X_n = y | X_{n-1} = x) = P_{x,y} = \begin{cases} \frac{x}{d}, & \text{if } y = x - 1, \\ \frac{d-x}{d}, & \text{if } y = x + 1, \\ 0, & \text{else .} \end{cases}$$

Paul Ehrenfest (January 18, 1880–September 25, 1933) used this model to study the exchange of air molecules in two chambers connected by a small hole. He created this simple model of diffusion to explain the Second Law of thermodynamics.

We also note that this is an example of a periodic chain (the proper definition to come) since the number of balls in box 1 alternates between odd and even.

EXAMPLE 12.4   Gambler's ruin problem

We have seen this example several times already. This is an example of a Markov chain as well. Recall that we deal with the case of some compulsive gambler who starts with some initial wealth K. At every step, the gambler bets one dollar on some game that has probability p of winning, and if he/she wins his wealth goes up by one dollar. Furthermore, assume that the gambler stops betting only when his/her wealth reaches either 0 or some upper limit fixed in advance $d > K$ (which may be ∞). Let $X_0 = K$ and $X_n =$ Gambler's wealth after n such games. Then X_n is a Markov chain on $S = \{1, 2, \ldots, d\}$, with initial distribution $\pi_0(i) = \mathbf{1}_{\{K\}}(i)$ and transition probabilities

$$\text{If } 0 < i < d \text{ then } \begin{cases} P_{i,i+1} = p \\ P_{i,i-1} = 1 - p \\ P_{i,j} = 0, & \text{for any other } j \end{cases}$$
$$P_{0,j} = \mathbf{1}_{\{0\}}(j); P_{d,j} = \mathbf{1}_{\{d\}}(j);$$

Note that this process can be interpreted as a simple (± 1) random walk with 0 and d absorption states (see diagram in Figure 12.2).

An interesting question (as with all random walks) is: what is the probability of eventually hitting 0 (that is not coming back to the starting state), which is interpreted as the probability of eventual ruin in this particular case?

EXAMPLE 12.5   Birth and Death chain

This is a generalization of both the previous examples. We have absorption states as in Example 12.4 and the probabilities of transition are state dependent as in Example 12.3. Furthermore, the process may remain in the same state (Figure 12.3).

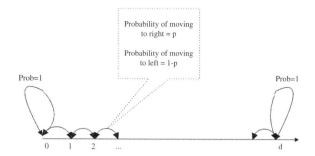

Figure 12.2   A graphical description of the Gambler's ruin problem

The story associated with the name is that we start with a certain population type. We view the time between events as unimportant, so we take that the time step is 1 (thus creating a Markov chain). Think about the steps as being steps between events and not actual time between events. At any time step, one event takes place: either a birth or a death or nothing occurs. If X is the process denoting the size of the population, then it is a Markov chain on $\mathcal{S} = \{0, 1, 2, \ldots\}$ with some initial distribution. We can see a scheme depicting it in Figure 12.3, assuming that the population dies out if it reaches size 0 (no immigration from outside). We could also put an upper cap on the population by creating another (semi) absorbing upper state.

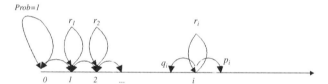

Figure 12.3   A graphical description of the birth and death chain

We can write down the transition probabilities of this chain as follows:

$$\text{If } 0 < i \text{ then} \begin{cases} P_{i,i+1} = p_i \\ P_{i,i-1} = q_i \\ P_{i,i} = r_i \\ P_{i,j} = 0, \quad \text{for any other } j \end{cases}$$
$$P_{0,j} = \mathbf{1}_{\{0\}}(j),$$

assuming that for any i we have $p_i + q_i + r_i = 1$. In general, they should depend on the particular state (e.g., there should be a greater birth/death chance if the actual size of the population is 1,000,000 vs the case when the population size is say 100 individuals).

■ EXAMPLE 12.6   A simple queuing process (G/G/1)

Assume that we have customers who arrive at a service facility that has one or more processing units (a bank with tellers, a gas station with pumps, etc.). The classical example is phone calls that arrive at a telephone company. This is due to Agner Krarup Erlang (January 1, 1878–February 3, 1929), a Danish engineer who worked for the Copenhagen Telephone Exchange in 1909 an who invented the process to model the incoming and processing of calls and to help design an automatic switching board.

Notation   In queuing theory, **A/B/n** is a notation introduced by David G Kendall (15 January 1918–23 October 2007) in 1953 to simplify and classify the multitude of papers and results available at the time. The first letter **A** denotes the distribution of time between arrivals, the second letter **B** stands for the distribution of the service time, and the last letter is a number **n** that gives the number of servers in the queuing system. Possible notations for the distribution of inter-arrival and service include

- M exponential distribution,

- D deterministic (non-random distribution),

- Ek Erlang distribution,

- G general (any distribution other than the ones above).

Note that inter-arrival times are still independent of any of these distributions.

    In fact, the notation is longer **A/B/C/K/N/D**. In this notation, the last three, often omitted, letters represent the following:

- **C** the capacity of the system (think total number of people fitting in the bank building),

- **P** total population size,

- **D** the service discipline (first come first served, last come first served, random order, priority service, etc.).

    Coming from a communist country (where queues were encountered everywhere), growing up I found queuing theory a fascinating area. Their peak of development was in the 1960s and they saw a re-emergence in the 1990s with the development of networks and network analysis. Today, complex networks are simulated rather than analyzed mathematically due to their extremely complex structure. However, blocks of queues could be simulated much faster if only today's engineers would have the patience and will of Erlang.

    In this example we analyze a **G/G/1** queue

    Let Δ_n denote the number of customers arriving during the nth service. Let $p_m = \mathbf{P}(\Delta_n = m)$, $m = 0, 1, 2, \ldots$. In general, this distribution will depend on the length of the nth service and we will obtain a so called nonhomogeneous Markov chain. For this example, let us make the (unrealistic) assumption that the distribution is the same for all n (symbolized above by the fact that p_m does not have an index n).

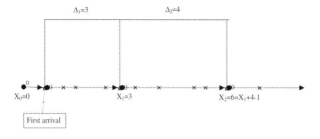

Figure 12.4   A realization of the simple queue system

Let X_{n+1} be the number of customers in queue after the nth service. A typical realization is depicted in Figure 12.4. Then we can write, in general (think why)

$$X_{n+1} = X_n + \Delta_n - \mathbf{1}_{\{X_n > 0\}}.$$

Therefore, $\{X_n\}$ is a Markov chain with state space $\mathbb{N} = \{0, 1, 2, \ldots\}$, with initial distribution $\pi_0(i) = \mathbf{1}_{\{0\}}(i)$, and with transition probabilities given by

$$P_{i,j} = \mathbf{P}(\Delta_n = j - i + 1) = p_{j-i+1}, \text{ if } i \geq 1$$
$$P_{0,j} = p_j$$

EXAMPLE 12.7   Birth and catastrophe chain

This is similar to the birth and death chain in Example 12.5 with the difference that there are no natural deaths but instead at any moment there is a probability that the entire population is wiped out. Of course, since there is a nonzero probability that a wipe-out will eventually happen, to avoid getting stuck at population size 0, we assume that once at zero there is a certain probability for spontaneous life. We can easily show that X, the process of population size after step n, is a Markov chain and its transition probabilities are given by

$$P_{i,j} = \begin{cases} p_i, & \text{if } j = i + 1 \\ q_i, & \text{if } j = 0 \\ 0, & \text{else} \end{cases} , \forall i \geq 0.$$

As an example of such a process. One we have already seen is the current age process $A(t)$ for a renewal process restricted to inter-arrival times in $\mathbb{Z}$.

■ EXAMPLE 12.8   Number of Mutant Genes in a population

Simplified model.

Story: Suppose the existence of i mutant genes (from a total of d genes) in the nth generation of a certain population. To get to the next generation

1. all genes duplicate, therefore there will be $2i$ mutant and $2d - 2i$ normal genes, and

2. d genes are randomly selected from the above possibilities.

If X_n denotes the number of mutant genes in the population after the nth generation, then $X = \{X_n\}_{n=1}^{\infty}$ is a Markov chain with state space $\mathcal{S} = \{1, 2, \ldots, d\}$ and with transition probabilities given by

$$P_{i,j} = \frac{\binom{2i}{j}\binom{2d-2i}{d-j}}{\binom{2d}{d}} \qquad , \forall i, j \in \{1, 2, \ldots, d\}.$$

12.1.3   The Chapman– Kolmogorov equation

Theorem 12.3 *Let $X = \{X_n\}_{n=1}^{\infty}$ be a Markov chain on $\mathcal{S} = \{1, 2, \ldots\}$, with some initial distribution and one-step transition probability matrix P. Denote the n-step transition probability of going from state i to state j by $P_{i,j}^n$, i.e.,*

$$P_{i,j}^n = \mathbf{P}\left(X_{m+n} = j | X_m = i\right).$$

Then we have the Chapman–Kolmogorov relation

$$P_{i,j}^{n+m} = \sum_{k \in \mathcal{S}} P_{i,k}^n P_{k,j}^m. \qquad (12.1)$$

Proof: The proof is simple:

$$P_{i,j}^{n+m} = \mathbf{P}\left(X_{m+n} = j | X_0 = i\right) = \sum_{k \in \mathcal{S}} \mathbf{P}\left(X_{m+n} = j, X_n = k | X_0 = i\right)$$

$$= \sum_{k \in \mathcal{S}} \mathbf{P}\left(X_{m+n} = j | X_n = k, X_0 = i\right) \mathbf{P}\left(X_n = k | X_0 = i\right)$$

$$= \sum_{k \in \mathcal{S}} P_{k,j}^m P_{i,k}^n,$$

where the first equality is the definition, for the second we used the law of total probability (exhaustive events: $\cup_{k \in \mathcal{S}} \{X_n = k\}$), third equality is just the multiplicative rule, and finally the last follows from the Markov property (Theorem 12.2).   ■

Remark 12.4 *This theorem also tells us that, if $P^{(n)}$ is the n-step transition matrix, then $P^{(n)} = P^n$ the nth power of the one-step transition matrix.*

12.1.4 Communicating classes and class properties

Definition 12.5 *We say that state i communicates with state j, and we write $i \longleftrightarrow j$ if there exist $n, m \geq 0$ such that $P_{ij}^n > 0$ and $P_{ji}^m > 0$.*

Simply put, two states communicate if it is possible to reach one from the other. Note that any class communicates with itself. By definition, $P_{ii}^0 = 1$ and $P_{ij}^0 = 0$. This is because the reflexivity property below is valid.

Proposition 12.6 *"$\longleftrightarrow$" is an equivalence relationship.*

Proof: Reflexivity ($i \longleftrightarrow i$) and symmetry ($i \longleftrightarrow j$ implies $j \longleftrightarrow i$) properties are trivial. The proof of the transitivity property ($i \longleftrightarrow j$ and $j \longleftrightarrow k$ implies $i \longleftrightarrow k$) is not hard either, using the Chapman–Kolmogorov equations. For example, having a $P_{ij}^{n_1} > 0$ and a $P_{jk}^{n_2} > 0$ implies using Equation (12.1) that

$$P_{ik}^{n_1 + n_2} \geq P_{ij}^{n_1} P_{jk}^{n_2} > 0,$$

and thus one can get from i to k. Reaching i from k is similar. ∎

Proposition 12.6 is important since it means that the state space S can be partitioned into communicating classes. As we will see, this means that we only have to analyze one member of each class to deduce the properties of the entire class. Therefore, the first step when analyzing a Markov chain is always to group the states into equivalent classes, and then study the properties that will be defined below for each class.

12.1.5 Periodicity

Definition 12.7 (Periodicity) *We call the period of the state $i \in S$ the greatest common divisor of the set $\{n \geq 1 | P_{ii}^n > 0\}$. We denote the period of the state i with $d(i)$.*

We can see an example of calculating these periods in Figure 12.5. Each nonzero probability in the transition matrix corresponds to an arrow in the image.

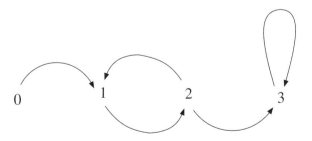

Figure 12.5 A simple Markov chain. The periods for each state are $d(0) = \infty$, $d(1) = d(2) = 2$, $d(3) = 1$

Proposition 12.8 *Periodicity is a class property. More specifically, if $i, j \in S$ are two states with $i \longleftrightarrow j$, then they have the same period $d(i) = d(j)$.*

Proof: Let $d(i) = d$. Since $i \longleftrightarrow j$, there exist $n, m \geq 1$ with $P_{ij}^n > 0, P_{ji}^n > 0$. Thus using the Chapman–Kolmogorov equation again, we have

$$P_{ii}^{n+m} > P_{ij}^n P_{ji}^n > 0 \quad \Rightarrow \quad d|m+n.$$

Assume that for some integer $q \geq 1$ we have $P_{jj}^q > 0$. This q must exist, otherwise we could not communicate from i to j. Then again

$$P_{ii}^{n+m+q} > P_{ij}^n P_{jj}^q P_{ji}^n > 0 \quad \Rightarrow \quad d|m+n+q.$$

Thus, since $d|m+n$ and $d|m+n+q$, then $d|q$, which means that d is a common divisor of $\{n \geq 1 | P_{jj}^n > 0\}$, which means that it must divide the greatest common divisor of the set, or $d(i)|d(j)$.

Repeating the argument with i and j reversed, we obtain that $d(j)|d(i)$, and thus finally $d(i) = d(j)$.   ∎

12.1.6   Recurrence property

Let f_{ij}^n denote the probability of hitting j for the first time in exactly n steps starting at state i, that is,

$$f_{ij}^n = \mathbf{P}\{X_n = j, X_k \neq j, \ \forall 0 < k < n | X_0 = i\}$$

Also let f_{ij} be the probability that the Markov chain *ever* hits j starting at state i. We can write by a simple conditioning argument

$$f_{ij} = \sum_{k=1}^{\infty} f_{ij}^n$$

Definition 12.9 *The state j is called **recurrent** if and only if $f_{jj} = 1$. The state j is called **transient** if and only if $f_{jj} < 1$.*

So, a transient state, as the name suggests, is a state to which we may not come back. Recurrent is a state to which we will come back with probability 1. Next, we will characterize the recurrent states further.

Proposition 12.10 *The state $j \in S$ is recurrent if and only if $\sum_{n=0}^{\infty} P_{jj}^n = \infty$.*

Note that the terms of the sum are the probabilities of going from j to j in exactly n steps. Furthermore, we may also write

$$\mathbf{E}\left[\text{Number of visits to } j | X_0 = j\right] = \sum_{n=0}^{\infty} \mathbf{E}\left[\mathbf{1}_{\{X_n=j\}} | X_0 = j\right] = \sum_{n=0}^{\infty} P_{jj}^n,$$

or, a state is recurrent iff the expected number of visits to it is infinite.

Proof: The proof is drawn using the expectation above. We note that for any state j the number of visits to j starting with itself is a geometric random variable (number of failures to obtain the first success) with probability of success $p = 1 - f_{jj}$. In the definition of the geometric random variable, a success is not coming back to j. Thus

$$\mathbf{E}\left[\text{Number of visits to } j | X_0 = j\right] = \frac{1}{p} - 1 = \frac{f_{jj}}{1 - f_{jj}}$$

Therefore, this expectation is infinite (and equivalently the sum of probabilities is infinite) if and only if $f_{jj} = 1$. ∎

Corollary 12.11 *If $i \in S$ is a recurrent state and $i \longleftrightarrow j$, then j is recurrent. Furthermore, $f_{ij} = 1$ as well.*

Proof: Proving $f_{jj} = 1$. If $P_{ji}^n > 0$ and $P_{ij}^n > 0$ (since $i \longleftrightarrow j$), then by Chapman-Kolmogorov equation $P_{jj}^{n+n+r} \geq P_{ji}^n P_{ii}^r P_{ij}^n$, for all r. Thus

$$\sum_{s=0}^{\infty} P_{jj}^s \geq P_{ji}^n \sum_{r=0}^{\infty} P_{ii}^r P_{ij}^n = \infty,$$

and so j is recurrent by Proposition 12.10.

Proving $f_{ij} = 1$. Let T_k be the time of the kth return to i. Let τ denote the time of the first visit to j. We want to show that $\tau < \infty$ a.s. From the hypothesis, we have $i \longrightarrow j$, therefore there exists $n > 0$ such that

$$\mathbf{P}\{X_n = j, X_k \neq i \text{ and } j, \forall 0 < k < n | X_0 = i\} > \varepsilon > 0$$

.

Since every return to i can be considered a cycle, for any $k > 1$ we have

$$\mathbf{P}_i\{\tau \leq T_k | \tau > T_{k-1}\} > \varepsilon, \text{ or } \mathbf{P}_i\{\tau > T_k | \tau > T_{k-1}\} \leq 1 - \varepsilon,$$

where we denoted with $\mathbf{P}_i$ the conditional probability on the event $\{X_0 = i\}$. However, for all k, we can write

$$\begin{aligned}
\mathbf{P}_i\{\tau > T_k\} &= \mathbf{P}_i\{\tau > T_k | \tau > T_{k-1}\}\mathbf{P}_i\{\tau > T_{k-1}\} = \dots \\
&= \mathbf{P}_i\{\tau > T_k | \tau > T_{k-1}\} \dots \mathbf{P}_i\{\tau > T_2 | \tau > T_1\}\mathbf{P}_i\{\tau > T_1\} \\
&\leq (1 - \varepsilon)^k,
\end{aligned}$$

which implies immediately

$$\mathbf{P}_i\{\tau = \infty\} \leq \mathbf{P}_i\{\tau > T_k\} \leq (1 - \varepsilon)^k, \forall k,$$

and thus by taking $k \longrightarrow \infty$ we are done. ∎

This corollary shows that recurrence is a class property. Furthermore, the next remark shows that, once we get outside a class, things are not as simple as described above.

Remark 12.12 *Proposition 12.10 and its corollary seem to indicate that a condition for $f_{ij} = 1$ would be that $\sum_{n=0}^{\infty} P_{ij}^n = \infty$. That is not so. Recall that the hypothesis of the proposition requires that i and j are in the same class. If this condition is violated, the condition is not true anymore. For example, look at the situations depicted in Figure 12.6.*

- $\sum_{n=0}^{\infty} P_{ij}^n = \infty \not\Rightarrow f_{ij} = 1$.

  In Figure (12.6a, we see an example of a chain where $f_{ij} = p$ and at the same time $\sum_{n=0}^{\infty} P_{ij}^n = \infty$.

- $\sum_{n=0}^{\infty} P_{ij}^n = \infty \not\Leftarrow f_{ij} = 1$.

  In Figure (12.6b, we see an example of a chain where $f_{ij} = 1$ but $\sum_{n=0}^{\infty} P_{ij}^n = 1$; in fact, $P_{ij}^1 = 1$ and $P_{ij}^n = 0$ for all $n \geq 2$.

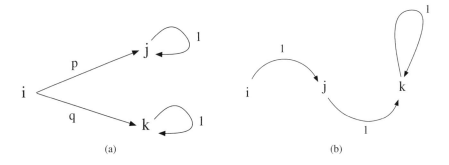

(a)   (b)

Figure 12.6   Counterexamples for $\sum_{n=0}^{\infty} P_{ij}^n = \infty \not\Leftrightarrow f_{ij} = 1$

A very important question is: why should we care about recurrence? The answer to this question is that the recurrence property is very much related with the stationarity of the process. If a process is stationary, then its long-time behavior is easy to understand and describe. Thus, the current notions are just tools we need to describe a stochastic process.

12.1.7   Types of recurrence

It turns out that we can classify recurrence further depending on how long it takes on average to come back to the original starting state. This is important since the stationarity behavior is affected by the type of recurrence the process possesses.

Definition 12.13 *Let μ_{jj} be the mean time to return to j starting at j. More precisely, we define*

$$\mu_{jj} = \begin{cases} \sum_{n=1}^{\infty} n f_{jj}^n, & \text{if } f_{jj} = 1 \quad (j \text{ recurrent}) \\ \infty, & \text{if } f_{jj} < 1 \quad (j \text{ transient}) \end{cases} \tag{12.2}$$

With this definition, we call a recurrent state j

- *positive recurrent iff $\mu_{jj} < \infty$,*

- *null recurrent iff $\mu_{jj} = \infty$ (but $f_{jj} = 1$).*

It is easy to understand that for the transient states the expected number of states to come back is infinite. After all, we have used this before. However, even for the recurrent states the sum in the definition may be infinite. In conclusion, the null recurrent states are those where the Markov chain does come back with probability 1 but the expected amount of time to do so is infinite. One such strange case is presented in the next section (the random walk in one or two dimensions).

12.2 Simple Random Walk on Integers in d Dimensions

In this section, we present a special Markov chain, the random walk, a process used extensively in applications.

Let $\mathbf{X}_k = (X_k^{(1)}, X_k^{(2)} \dots X_k^{(d)}) \in \mathbb{R}^d$ be a d-dimensional vector. Each component variable $X_k^{(i)}$ is independent of the others and takes values ± 1 with probability $\frac{1}{2}$.

Then $\mathbf{S}_n = \sum_{k=1}^{n} \mathbf{X}_k$ is called a d-dimensional simple random walk.

Remark 12.14 *The sum above is done component-wise. We shall use the bold notation $\mathbf{X}, \mathbf{S}$ to denote d-dimensional vectors throughout this section. The random walk above is called "simple" due to the simple structure of the jumps $\mathbf{X}_k$. A more general random walk may have any distribution for the jump components $X_k^{(i)}$ with the property that the distribution has finite mean and variance and it is not concentrated at 0 or on a particular side of 0 (i.e., there exists an $\varepsilon > 0$ such that $\mathbf{P}\{X_k^{(i)} > \varepsilon\} > 0$ and $\mathbf{P}\{X_k^{(i)} < -\varepsilon\} > 0$). For a more complete and general description, consult the forthcoming book Lyons and Peres (2010).*

Note that the probabilities for S_n are only determined by where S_{n-1} is, thus the process is a Markov chain.

Let us study the recurrence of this process. We first note that each state in $\mathbb{Z}^d$ communicates with any other state, thus the process is irreducible and the random walk contains only one class. Since recurrence is a class property, it is enough to study this property for an arbitrary state.

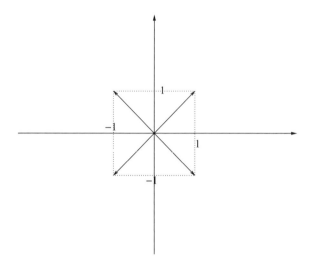

Figure 12.7   Random walk in two dimensions. Possible values for the first jump each with probability 1/4

Question: If we start the process from the origin $\mathbf{0} = (0, 0, \ldots, 0)$, what is the probability that $\mathbf{S}_n = (0, 0, \ldots, 0)$ for some n?

Answer: Let us calculate the probability. For n odd, clearly $P(\mathbf{S}_n = \mathbf{0}) = 0$ (we need a forward step and we also need a backward step to return). For n even, let us denote $n = 2k$, and we have

$$P(\mathbf{S}_{2k} = \mathbf{0}) = P\left(S_{2k}^{(1)} = 0, S_{2k}^{(2)} = 0, \ldots S_{2k}^{(d)} = 0\right)$$
$$= \left[P\left(S_{2k}^{(i)} = 0\right)\right]^d$$

since all components are independent.

Claim: The probability above is $\left[P\left(S_{2k}^{(i)} = 0\right)\right]^d = \left(\dfrac{\text{constant}}{\sqrt{k}}\right)^d$

Justification   Note that we have a total of $2k$ steps and $S_{2k}^{(i)}$ is now one dimensional. To get back to 0 once you start from it, you need k steps in one direction (values of 1) and k steps in the opposite direction (values of -1). But then the number of such paths is

$$2k \text{ total steps} \Rightarrow \begin{cases} k \text{ forward steps} \\ k \text{ back steps} \end{cases} \rightarrow \binom{2k}{k}$$

The probability of any such path is $(\frac{1}{2})^{2k}$, which implies

$$P\left(S_{2k}^{(i)} = 0\right) = \binom{2k}{k}\left(\frac{1}{2}\right)^{2k}$$

We use Stirling's formula for the combinatorial term (i.e., $n! \sim \sqrt{2\pi n}\, n^n e^{-n}$), giving

$$P\left(S_{2k}^{(i)} = 0\right) = \frac{(2k)!}{(k!)^2}\left(\frac{1}{2}\right)^{2k}$$

$$\approx \frac{\sqrt{2\pi 2k}(2k)^{2k}e^{-2k}}{(\sqrt{2\pi k}k^k e^{-k})^2}\left(\frac{1}{2}\right)^{2k}$$

$$= \frac{2\sqrt{\pi k}2^{2k}(k^{2k}e^{-2k})}{2\pi k(k^{2k}e^{-2k})}\left(\frac{1}{2}\right)^{2k}$$

$$= \frac{1}{\sqrt{\pi k}} = \frac{\text{constant}}{\sqrt{k}}$$

which proves the claim.

Theorem 12.15 (Polya) *If $d = 1, 2$, then the d-dimensional simple random walk comes back to 0 infinitely often. If $d \geq 3$, eventually the random walk stops coming back to 0. In other words, if $d = 1, 2$, the chain is recurrent and if $d \geq 3$ the chain is transient.*

Proof: Let $p_d = \mathbf{P}\{$The random walk never comes back to 0 **at all** in $\mathbb{R}^d\}$. This is the same as $1 - f_{00}$ with f_{00} introduced to study the recurrence of the Markov chain.

If we introduce the sum $\sum_{n=1}^{\infty} \mathbf{1}_{\{S_n=0\}}$, this counts the number of times the random walk returns to 0. This sum $\sum_{n=1}^{\infty} \mathbf{1}_{\{S_n=0\}}$ is distributed as a Geometric(p_d) (number of failures) random variable. This is clear if you consider coming back to 0 a failure and not coming back to a success. Therefore we can write

$$\mathbf{E}\left[\sum_{n=1}^{\infty} \mathbf{1}_{\{S_n=0\}}\right] = \frac{1}{p_d} - 1$$

Using the Fubini equation and the previous claim

$$\mathbf{E}\left[\sum_{n=1}^{\infty} \mathbf{1}_{\{S_n=0\}}\right] = \sum_{n=1}^{\infty}\left(\mathbf{E}[\mathbf{1}_{\{S_n=0\}}]\right) = \sum_{n=1}^{\infty} P(\mathbf{S}_n = 0) = \sum_{\substack{n=2k \\ k=1}}^{\infty} \frac{c^d}{k^{\frac{d}{2}}}$$

$$= c^d \sum_{k=1}^{\infty} \frac{1}{k^{\frac{d}{2}}}$$

Therefore, equating the two expressions for the expectation and using that $\sum_k \frac{1}{k^n}$ is convergent if and only if $n > 1$ we obtain

$$\frac{1 - p_d}{p_d} = \begin{cases} < \infty & \text{if } d > 2 \\ \infty & \text{if } d \leq 2 \end{cases}$$

which implies

$$p_d > 0 \Leftrightarrow d \geq 3$$
$$p_d = 0 \Leftrightarrow d = 1, 2$$

As a conclusion, when $d = 1, 2$, the number of visits to 0 is ∞. When $d \geq 3$, the number of visits is finite a.s., which means that eventually you will drift to infinity. ∎

Remark 12.16 *Considering a renewal event, the event that the random walk returns to the origin $\vec{0}$, we may view the random walk as producing a renewal process as well.*

Remark 12.17 *I have been made aware of this result at a conference where the speaker was presenting a related topic and started to tell us that the previous night after leaving the pub he had no idea where the hotel was. But he trusted in the Polya result and started walking at random in two dimensions (on the streets) confident that eventually he will reach the original starting point (the hotel room). However, as we shall see next, the expected time to reach the origin is infinite so this might not have been a good idea at all. Of course, he was in a town, so not quite the same process (boundary). And no memory might be violated, though he claimed that the beer brought him very close to a no-memory process.*

Proposition 12.18 *A simple random walk in $\mathbb{R}^d$ has all the states null recurrent if $d = 1$ or 2.*

Proof: We have already seen that it is possible not to return to state 0 ($p > 0$ or $f_{00} < 1$) when $d \geq 3$, and therefore we are looking at a transient Markov chain.

Suppose that the dimension is $d = 1$ or 2. We have seen that the chain is recurrent ($f_{00} = 1$). However, is it null or positive recurrent? We can use Wald's theorem to calculate the expected return time. We will do so only for the dimension $d = 1$; for dimension 2 the proof is similar. Let N denote the time of first return to 0.

Assume by absurd that $\mathbf{E}[N] < \infty$. Then by Wald's theorem

$$\mathbf{E}\left[\sum_{k=2}^{N} X_k \,\middle|\, X_1 = 1\right] \overset{\text{Wald}}{=} \mathbf{E}[N|X_1 = 1]\,\mathbf{E}[X_2|X_1 = 1] = 0,$$

since X_2 and X_1 are independent. However, given that $X_1 = 1$, we need to have $\sum_2^N X_k = -1$ to return to 0, and therefore the expectation above should also be -1 (a contradiction).

Thus, $\mathbb{E}[N] = \infty$ and the chain is null recurrent. ∎

12.3 Limit Theorems

The following theorem is a consequence of the renewal theorems presented in Chapter 11 when applied to Markov chains. That chapter is useful not in itself for applications but by the strength of the results that may be applied to many other processes. We present these results here since we need them for the Markov chain characterization theorem in the next section.

Theorem 12.19 (Limit Theorems) *Let X_n define a Markov chain with state space S. Let $i, j \in S$ be any two states such that $i \longleftrightarrow j$. Denote by*

$$N_j(t) = \sum_{k \leq t} \mathbf{1}_{\{X_k = j\}}$$

the number of visit to j by time t. If we ever reach j, then the number of successive visits to j defines a renewal process N_j.

1. *Using the Strong Law of Large Numbers for renewal processes, we have*

$$P \left\{ \lim_{t \to \infty} \frac{N_j(t)}{t} = \frac{1}{\mu_{jj}} \, \middle| \, X_0 = i \right\} = 1 \tag{12.3}$$

2. *Using the elementary renewal theorem, we have*

$$\lim_{n \to \infty} \frac{\sum_{k=1}^{n} P_{ij}^k}{n} = \frac{1}{\mu_{jj}}$$

3. *Using Blackwell's renewal theorem, if $d(j) = 1$, we have*

$$\lim_{n \to \infty} P_{ij}^n = \frac{1}{\mu_{jj}}$$

4. *Using Blackwell's renewal theorem (lattice case) if $d(j) = d$, we have*

$$\lim_{n \to \infty} P_{jj}^{nd} = \frac{d}{\mu_{jj}}$$

The theorem, as in the case of the renewal processes, is valid for $\mu_{jj} = \infty$ as well. The proof is left as an exercise. It is a straightforward application of the results in the chapter on renewal processes.

12.4 Characterization of States for a Markov Chain. Stationary Distribution.

Before we give the main theorem of this chapter, we need to introduce several new concepts.

Definition 12.20 *A Markov chain is called* irreducible *if it only contains one class (in other words, all states communicate with each other).*
A Markov chain is called aperiodic *if all states have period $d(i) = 1$.*

A reducible chain may be split into two or more classes.

Definition 12.21 *A probability density* $\{\Pi_j\}_{j\in S}$ *is called* stationary *for the Markov chain if and only if*

$$\Pi_j = \sum_i \Pi_i P_{ij}, \qquad \forall j \in S$$

If we denote $\Pi = \begin{pmatrix} \Pi_1 \\ \Pi_2 \\ \vdots \\ \Pi_n \end{pmatrix}$ we can rewrite the definition as $\Pi^T P = \Pi^T$,

or equivalently, $P^T \Pi = \Pi$, where ^{T} denotes the transposition operation. Thus, if you like the linear algebra language better, Π is an eigenvector of P^T corresponding to the eigenvalue 1.

Note that, if the initial distribution of X_0 is a stationary distribution Π, then X_n will have the same distribution for all n ($\{X_n\}$ is a stationary process):

$$P(X_1 = j) = \sum_{i\in S} \underbrace{P\{X_1 = j | X_0 = i\}}_{=P_{ij}} \underbrace{P\{X_0 = i\}}_{=\Pi_i}$$

$$= \sum_{i\in S} \Pi_i P_{ij} = \Pi_j$$

We finally have all the ingredients to characterize a Markov chain.

Theorem 12.22 *Any irreducible aperiodic Markov chain is of two possible types:*

1. *All the states are either* transient *or* null recurrent, *and in either case*

$$\lim_{n\to\infty} P_{ij}^n = 0 \qquad , \forall i, j \in S$$

and the stationary distribution does not exist. The converse is also true: if the stationary distribution does not exist, then the states are either transient or null recurrent.

2. *All the states are* positive recurrent, *and in this case*

$$\lim_{n\to\infty} P_{ij}^n = \Pi_j > 0 \qquad , \forall i, j \in S$$

where

$$\Pi_j = \frac{1}{\mu_{jj}}$$

and $\Pi = \{\Pi_j\}_{j\in S}$ *defines a unique stationary distribution for the Markov chain.*

Proof: First note that we already showed that transience and recurrence are class properties. This theorem implies that positive recurrence (and implicitly null recurrence) is also a class property.

The key tool of the theorem is the convergence result

$$\lim_{n\to\infty} P_{ij}^n = \frac{1}{\mu_{jj}} \overset{\text{Notation}}{=} \Pi_j$$

from Theorem 12.19. Without loss of generality, let us label the states of the Markov chain $S = \{0, 1, 2, \ldots\}$.

Proof of part (2):

For any state i and for any M fixed integer, we have

$$\sum_{j=0}^{M} P_{ij}^n \le \sum_{j=0}^{\infty} P_{ij}^n = 1.$$

Let $n \to \infty$; using that $P_{ij}^n \longrightarrow \Pi_j$ and that the sum is finite, we get

$$\sum_{j=0}^{M} \Pi_j \le 1, \quad \forall M \in \mathbb{N},$$

and taking $M \to \infty$, we obtain

$$\sum_{j=0}^{\infty} \Pi_j \le 1 \tag{12.4}$$

From Chapman-Kolmogorov equation, we get $P_{ij}^{n+1} = \sum_{k=0}^{\infty} P_{ik}^n P_{kj}$. Taking $\liminf$ over n in the above relation and applying Fatou's lemma, we get

$$\liminf_{n\to\infty} P_{ij}^{n+1} = \liminf_{n\to\infty} \sum_{k=0}^{\infty} P_{ik}^n P_{kj} \ge \sum_{k=0}^{\infty} \liminf_{n\to\infty} P_{ik}^n P_{kj}$$

But $P_{ij}^n \longrightarrow \Pi_j$ as $n \to \infty$, and therefore the $\liminf$ has the same limit; thus

$$\Pi_j \ge \sum_{k=0}^{\infty} \Pi_k P_{kj} \tag{12.5}$$

In (12.5), we sum over j to obtain

$$\sum_{j=0}^{\infty} \Pi_j \ge \sum_{j=0}^{\infty} \sum_{k=0}^{\infty} \Pi_k P_{kj} = \sum_{k=0}^{\infty} \Pi_k \underbrace{\sum_{j=0}^{\infty} P_{kj}}_{=1} = \sum_{k=0}^{\infty} \Pi_k$$

The left-hand side and the right-hand side are identical, and all terms are positive, so we must have the equality $\sum_{j=0}^{\infty} \Pi_j = \sum_{j=0}^{\infty} \sum_{k=0}^{\infty} \Pi_k P_{kj}$. However, from (12.5), all the terms in the sum are positive and the only way a sum of positive terms is zero is if all the terms are zero; therefore, (12.5) holds with equality. The state j was arbitrary, so we have

$$\Pi_j = \sum_{k=0}^{\infty} \Pi_k P_{kj}, \quad \forall j \in \mathcal{S} \tag{12.6}$$

This relation shows that $\{\Pi_j\}$ forms a stationary distribution if it is a distribution (i.e., $\sum \Pi_j = 1$). For our constructed Π_j's, this is not necessarily true, but using (12.4) we can re-normalize the terms to create a probability distribution.

Define

$$\tilde{\Pi}_j = \frac{\Pi_j}{\sum_{j=0}^{\infty} \Pi_j}.$$

Substituting in (12.6) shows that $\{\tilde{\Pi}_j\}_{j \in \mathcal{S}}$ forms a stationary distribution for the original Markov chain.

Uniqueness: Assume that the Markov chain has two stationary distributions Π_j and $\tilde{\Pi}_j$. Then, using the dominated convergence theorem ($\Pi_k < 1$ and $P_{kj} < 1$), we have for all j

$$\Pi_j = \sum_{k=0}^{\infty} \Pi_k P_{kj}^n \overset{n \to \infty}{\longrightarrow} \sum_{k=0}^{\infty} \Pi_k \tilde{\Pi}_j = \tilde{\Pi}_j$$

Proof of part (1):

If we use that $\lim_{n \to \infty} P_{ij}^n = \frac{1}{\mu_{jj}}$, and the fact that for transient states and null recurrent $\mu_{jj} = \infty$ by definition, then the first implication $\lim_{n \to \infty} P_{ij}^n = 0$ is immediate. It remains to show that all states are transient or null recurrent if and only if there exists no stationary distribution.

"$\Longleftarrow$" If not all the states are null recurrent or transient, then there exists a state j with $\mu_{jj} < \infty$, therefore that state is positive recurrent and thus by part (2) there exist a stationary distribution (contradiction).

"$\Longrightarrow$" Assume that there exist a stationary distribution $\{\Pi_j\}_j$. Then by definition and by dominated convergence, again

$$\Pi_j = \sum_{k=0}^{\infty} \Pi_k P_{kj}^n \overset{n \to \infty}{\longrightarrow} \sum_{k=0}^{\infty} \Pi_k \cdot 0 = 0, \quad \forall j.$$

Thus $\Pi_j = 0$ for all j is a contradiction with $\sum_j \Pi_j = 1$. ∎

Definition 12.23 (Ergodic Markov Chain) *An irreducible aperiodic Markov Chain with all states positive recurrent is called* ergodic.

The ergodic Markov chains are the easiest and the simplest chains that one can study. Note that the theorem proves there always exists a unique stationary distribution for an ergodic Markov chain. Thus, any such chain will eventually reach its stationary distribution.

Remark 12.24 *If an irreducible Markov chain admits a stationary distribution* $\{\Pi_j\}_{j \in \mathcal{S}}$, *then necessarily* $\Pi_j > 0$ *for all states j.*

Proof: Without loss of generality, assume that for the state 0 we have $\Pi_0 > 0$ (one of the states has to have nonzero probability since the probabilities sum to 1). Then

$$\Pi_j = \sum_{k=0}^{\infty} \Pi_k P_{kj}^n \geq \Pi_0 P_{0j}^n, \forall n$$

But $0 \longleftrightarrow j$ and so there must exist an m such that $P_{0j}^m > 0$. Since j is arbitrary, done. ∎

Question: Can a Markov chain have **more** than one stationary distribution?

Short answer is yes. Note that the characterization Theorem 12.22 is valid only as long as its hypotheses are satisfied. The main hypotheses are irreducibility and aperiodicity. So, to construct a chain with more than one stationary distribution, try more than one class and/or periodic classes. (See Exercise 12.5 on page 406.)

12.4.1 Examples. Calculating stationary distribution

Let us go back to several of the examples presented earlier. **Birth and disaster chain**

of Example 12.7 on page 377

The Markov chain has the probability matrix with entries

$$P_{i,i+1} = \mathbf{P}(X_{n+1} = i + 1 | X_n = i) = p_i, \forall i \in 0, 1, 2, \dots$$
$$P_{i0} = \mathbf{P}(X_{n+1} = 0 | X_n = i) = q_i = 1 - p_i$$
$$\text{all others are } 0$$

Question: Is the chain irreducible? This has a pretty clear answer: Yes (all states communicate).

Question: Is the chain recurrent?

Let $X_0 = 0$ and let $\tau = \inf\{n \geq 1 | X_n = 0\}$, the time of first return to 0. Assume that all $p_k > 0$. The chain is recurrent if and only if $\mathbf{P}\{\tau < \infty\} = 1$.

Let $n \in \mathbb{N}$. Then

$$
\begin{aligned}
\mathbf{P}(\tau \leq n) &= 1 - \mathbf{P}(\tau > n) \\
&= 1 - \mathbf{P}(\text{hop } n \text{ times to the right}) \\
&= 1 - \prod_{i=0}^{n-1} p_i
\end{aligned}
$$

So the chain is recurrent if and only if

$$\lim_{n \to \infty} \left(1 - \prod_{i=0}^{n-1} p_i \right) = 1 \Leftrightarrow \prod_{i=0}^{\infty} p_i = 0$$

$$\Leftrightarrow \sum_{i=0}^{\infty} (1 - p_i) = \infty \Leftrightarrow \sum_{i=0}^{\infty} q_i = \infty \text{ (using Borel-Cantelli lemma)}.$$

Question: Is the chain *positive* recurrent?

According to theorem 12.22 an irreducible MC is positive recurrent if and only if it has a stationary distribution $\Leftrightarrow \Pi = \Pi \cdot P$, where

$$P = \begin{pmatrix} q_0 & p_0 & 0 & 0 & \cdots \\ q_1 & 0 & p_1 & 0 & \cdots \\ q_2 & 0 & 0 & p_2 & \cdots \\ & & \vdots & & \end{pmatrix}$$

which produces the system of equations

$$\Pi_0 = \sum_{i=0}^{\infty} \Pi_i q_i$$

$$\Pi_1 = p_0 \Pi_0$$

$$\Pi_2 = p_1 \Pi_1$$

$$\vdots$$

$$\Pi_{k+1} = p_k \Pi_k$$

which gives for $k \geq 1$

$$\Pi_k = \Pi_0 \prod_{i=0}^{k-1} p_i$$

Now it is easy to see that the stationary distribution exists iff

$$\sum_k \Pi_k < \infty \qquad \Leftrightarrow \qquad \sum_{k=0}^{\infty} \left(\prod_{i=0}^{k-1} p_i \right) < \infty$$

If that is the case, then we can calculate

$$\Pi_k = \frac{\prod_{i=0}^{k-1} p_i}{\sum_{k=0}^{\infty} \left(\prod_{i=0}^{k-1} p_i \right)}, \quad k \geq 1,$$

and $\Pi_0 = 1 - \sum_{k \geq 1} \Pi_k$

Birth and death chain of example 12.5 on page 374

In this case the transition matrix is

$$P = \begin{pmatrix} r_0 & p_0 & 0 & 0 & \cdots \\ q_1 & r_1 & p_1 & 0 & \cdots \\ 0 & q_2 & r_2 & p_2 & \cdots \\ & & \vdots & & \end{pmatrix}$$

We assume that all $p_k, q_k > 0$ in order to have an irreducible chain.

Question: Is this chain recurrent?

We define $T_b = \inf\{n : X_n = b\}$. Define a function on integers

$$\alpha_b(x) = P(T_0 < T_b | X_0 = x) \qquad x = 1, 2, \ldots, b$$

; this is the probability that the chain reaches 0 before reaching b starting from initial state x. Conditioning on the first step transitions, we get

$$\alpha_b(x) = \underbrace{p_x \cdot \alpha_b(x+1)}_{\text{move to right}} + \underbrace{r_x \cdot \alpha_b(x)}_{\text{stay}} + \underbrace{q_x \cdot \alpha_b(x-1)}_{\text{move to left}} \qquad x = 1, 2 \ldots b,$$

To solve, define

$$\gamma_0 = \frac{1}{p_0}, \qquad \gamma_y = \frac{q_1 q_2 \cdots q_y}{p_0 p_1 \cdots p_y}$$

Then we can directly verify that the following is a solution:

$$\alpha_b(x) = \frac{\sum_{y=x}^{b-1} \gamma_y}{\sum_{y=0}^{b-1} \gamma_y} = 1 - \frac{\sum_{y=0}^{x-1} \gamma_y}{\sum_{y=0}^{b-1} \gamma_y}$$

For more details, we direct the reader to Karlin and Taylor (1975) Section 4.7, which uses a similar argument for a simpler problem.

Studying the recurrence is clearly studying the limit of the above probability $\alpha_b(x)$ as $b \to \infty$. The limit depends on the sum $\sum_{y=0}^{\infty} \gamma_y$.

We have the following:

- If $\sum_{y=0}^{\infty} \gamma_y < \infty$, then $lim_{b \to \infty} \alpha_b(x) < 1$ for any x and thus the chain is *transient*.

- If $\sum_{y=0}^{\infty} \gamma_y = \infty$, then $lim_{b \to \infty} \alpha_b(x) = 1$ for any x and thus the chain is *recurrent*.

Furthermore, one can calculate the expected return time, and it turns out that the following situation is encountered. If the chain is recurrent and in addition

- $\sum_{y=0}^{\infty} \frac{1}{p_y \gamma_y} < \infty$, the chain is positive recurrent,

- $\sum_{y=0}^{\infty} \frac{1}{p_y \gamma_y} = \infty$, the chain is null recurrent.

A special periodic random walk

Consider a special case of the previous example where the transition matrix is

$$P = \begin{pmatrix} 0 & 1 & 0 & 0 & \ldots \\ q_1 & 0 & p_1 & 0 & \ldots \\ 0 & q_2 & 0 & p_2 & \ldots \\ & & \vdots & & \end{pmatrix}$$

This is an irreducible but periodic Markov chain with period 2. Thus we cannot apply directly Theorem 12.22. Nevertheless, we wish to investigate whether the chain has a stationary distribution. Let Π be a potential stationary distribution. If it exists, by definition we should have

$$\Pi_j = \sum_i \Pi_i P_{ij} = \Pi_{j-1} p_{j-1} + \Pi_{j+1} q_{j+1},$$

where $p_{-1} = 0$ and $p_0 = 1$. Thus we get for $j = 0$, $\Pi_0 = q_1 \Pi_1$. Substituting recursively, we obtain the probabilities in terms of Π_0 as

$$\Pi_j = \frac{p_{j-1} p_{j-2} \cdots p_1}{q_j q_{j-1} q_{j-2} \cdots q_1} \Pi_0 = \Pi_0 \prod_{k=0}^{j-1} \frac{p_k}{q_{k+1}},$$

since $p_0 = 1$. Now the stationary distribution exists if and only if we can normalize the probabilities, and we can only do that if $\sum_j \Pi_j < \infty$. Or

$$\Pi_0 + \sum_{j=1}^{\infty} \Pi_0 \prod_{k=0}^{j-1} \frac{p_k}{q_{k+1}} = \Pi_0 \left(1 + \sum_{j=1}^{\infty} \prod_{k=0}^{j-1} \frac{p_k}{q_{k+1}} \right) < \infty.$$

This is clearly happening if and only if

$$\sum_{j=1}^{\infty} \prod_{k=0}^{j-1} \frac{p_k}{q_{k+1}} < \infty$$

If this is the case, then the stationary distribution is uniquely determined using the extra condition $\sum_{i=0}^{\infty} \Pi_i = 1$.

As a particular case, consider a homogeneous random walk such that $p_k = p$ and $q_k = q = 1 - p$ for all $k \geq 1$. In this case, the expression above becomes

$$\sum_{j=1}^{\infty} \frac{1}{p} \prod_{k=1}^{j-1} \frac{p}{q} = \frac{1}{p} \sum_{j=1}^{\infty} \left(\frac{p}{q} \right)^{j-1},$$

which is a geometric series, clearly convergent if and only if $p < q$, or in plain terms, the power of attraction is greater than the power of dispersion.

12.5 Other Issues: Graphs, First-Step Analysis

12.5.1 First-step analysis

Sometimes, if the transition matrix is simple, we can find a relation between states by conditioning on the possible first-step transitions using the law of total probability. An example is the calculation of $\alpha_b(x)$ in the previous subsection. As a general rule, it works well when the transition matrix P is sparse (has many zeroes).

12.5.2 Markov chains and graphs

If the Markov chain is finite and has a small number of states, or if there is a structure in the chain, a better visualization may be obtained by drawing a graph representing the Markov chain. The states are represented by the vertices of the graph. Nonzero one-step transitions are represented by directed edges with weights.

Visualization of such a graph helps in classifying the states of the chain.

- The Markov chain is irreducible if the corresponding graph is connected (from each vertex one can reach any other vertex).

- The Markov chain is periodic if there exists $N > 0$ (which is the period) such that all cycles are multiples of N. If $N = 1$ or there is no such cycle, the chain is aperiodic. A cycle consists of the vertices connected on a trajectory in the graph that starts and ends in the same vertex.

EXAMPLE 12.9

Classify the states of a finite state space Markov chain with transition probability given by

$$P = \begin{pmatrix} 0.4 & 0 & 0.1 & 0 & 0.3 & 0 & 0.2 \\ 0 & 1 & 0 & 0 & 0 & 0 & 0 \\ 0 & 0 & 0.4 & 0.6 & 0 & 0 & 0 \\ 0 & 0 & 0.8 & 0.2 & 0 & 0 & 0 \\ 0 & 0.3 & 0 & 0.2 & 0.4 & 0.1 & 0 \\ 0 & 0 & 0 & 0 & 0.1 & 0 & 0.9 \\ 0 & 0 & 0 & 0.6 & 0 & 0.2 & 0.2 \end{pmatrix}$$

Proof (Solution): If we draw the graph, we see immediately that there are seven states, and if label them in the order of rows in the matrix above, we have four classes:

- States 3 and 4 positive recurrent,

- State 2 positive recurrent,

- State 1 transient,

- States 5, 6, and 7 transient.

■

Remark 12.25 *A finite state space Markov chain cannot have null recurrent states but only transient and positive recurrent states.*

12.6   A general Treatment of the Markov Chains

Suppose we are given a Markov chain with the possibly infinite state space S. The first step in its analysis is to separate it into classes. The treatment presented here is more pertinent to a state space S which is finite, but the same treatment may be extended to countable state spaces. We will point out the differences at appropriate times.

We first reorder the states $X_i^{(j)} \in S$ of the chain in the following way:

$$\underbrace{X_1^{(1)}, X_2^{(1)} \ldots}_{\text{class 1}}, \underbrace{X_1^{(2)}, X_2^{(2)} \ldots}_{\text{class 2}}, \ldots, \underbrace{X_1^{(t)}, X_2^{(t)} \ldots}_{S_T},$$

$$\underbrace{\phantom{X_1^{(1)}, X_2^{(1)} \ldots, X_1^{(2)}, X_2^{(2)} \ldots}}_{S_R}$$

where S_R denotes the recurrent classes and S_T gathers all the transient states regardless of the class.

With this reordering, the transition probability matrix is rewritten as a block matrix:

$$P = \begin{pmatrix} K & 0 \\ L & Q \end{pmatrix}$$

where the states in the rows corresponding to K are in S_R and the rows in the bottom matrices correspond to the states in S_T. K and Q are square matrices. In fact

$$K = \begin{pmatrix} K_1 & 0 & 0 & 0 & \ldots \\ 0 & K_2 & 0 & 0 & \ldots \\ 0 & 0 & K_3 & 0 & \ldots \\ & & & \vdots & \end{pmatrix}$$

where K_i are the transition matrices within each recurrent class.

The upper-right entries in P are all zero; otherwise we could transition from a recurrent state into a transient state (this would contradict the definition of the transient state).

Let

$$G(x, y) = \sum_{n=0}^{\infty} P_{x,y}^n = \mathbf{E}\left[\sum_{n=0}^{\infty} \mathbf{1}_{\{X_n = y | X_0 = x\}}\right],$$

the expected number of visits to y starting at x. Since we started the definition with $n = 0$, we can write using matrix notation

$$G = [G(x, y)]_{(x,y)}$$
$$= I + P + P^2 + P^3 \ldots + P^n + \ldots$$
$$= \begin{pmatrix} \sum K^n & 0 \\ \sum L_n & \sum Q^n \end{pmatrix}$$

Note that, due to the particular form of the matrix P,

$$P^n = \begin{pmatrix} K^n & 0 \\ L_n & Q^n \end{pmatrix}$$

where K^n and Q^n are the original matrices at the power n, and L_n is some matrix. In fact, L_n may be determined using a recurrence relation

$$L_n = L_{n-1}K + Q^{n-1}L$$

but this is irrelevant for us as explained bellow.

If we look at the matrix G, the left blocks in this matrix are not interesting to us. First, for the entries corresponding to K, since the states are recurrent, we know that the entries are all either ∞ or 0 depending on whether x and y are in the same recurrent class or not[2]. The same conclusion applies to the entries corresponding to the L block. There will be an expected infinite number of visits to a recurrent state y if it can be reached from the corresponding transient state x[3].

The interesting entries are the ones corresponding to the block Q since they give the expected number of visits of the Markov chain to transient states before it leaves these states.

Let

$$M = \sum_{n=0}^{\infty} Q^n. \tag{12.7}$$

This matrix converges for every entry (otherwise the entries would correspond to recurrent states). Sometimes M is referred to as the "fundamental matrix associated with Q." Let the entries in this matrix be denoted with m_{ij}. Recall that both i, j are transient states and

$$m_{ij} = \text{expected number of visits to } j \text{ starting at } i$$

Note that, if we multiply both sides of (12.7) by $(I - Q)$, we obtain

$$(I - Q)M = (I - Q)\sum_{k=0}^{\infty} Q^k = \sum_{k=0}^{\infty} (Q^k - Q^{k+1}) = I$$

Multiplying to the right is similar; therefore, if $(I-Q)^{-1}$ exists, then $M = (I-Q)^{-1}$.

---

[2]We note that, if the chain has an infinite number of states, then there may be null recurrent states in which case the study of the corresponding K entries could be interesting.
[3]The same issue applies here as well.

Observation: We may obtain the same equation by following a first-step analysis. Let $i, j \in S_T$. Then

$$m_{ij} = \mathbf{E}[\text{number of visit to } j | X_0 = i]$$

$$= \mathbf{E}\left[\sum_{n=0}^{\infty} \mathbf{1}_{\{X_n=j\}} \,\Big|\, X_0 = i\right]$$

$$= \underbrace{\mathbf{1}_{\{i=j\}}}_{n=0} + \sum_{k \in S_T} \underbrace{\mathbf{E}\left[\sum_{n=1}^{\infty} \mathbf{1}_{\{X_n=j\}} \,\Big|\, X_1 = k, X_0 = i\right]}_{m_{kj}} P_{ik}$$

$$+ \sum_{k \in S_R} \mathbf{E}\left[\sum_{n=1}^{\infty} \mathbf{1}_{\{X_n=j\}} \,\Big|\, X_1 = k, X_0 = i\right] P_{ik}$$

In the last term, the expectation is zero if $k \in S_R$ and thus rewriting everything in matrix notation we regain the same equation:

$$M = I + QM$$

Remark 12.26 *If S_T is an infinite set ($|S_T| = \infty$), the equation $M = I + QM$ might have multiple solutions. Under this circumstance, we take the minimum of all possible such solutions. The minimum here is in terms of rank.*

Regardless how we calculate it, this matrix M is very important since it shows the behavior of the transient states. Recall that the behavior of the recurrent states is completely described by the stationary distribution.

Theorem 12.27 *If $i \neq j$ are two transient states in S_T*

$$f_{ij} = \frac{m_{ij}}{m_{jj}} \qquad \text{and} \qquad f_{ii} = 1 - \frac{1}{m_{ii}}$$

Proof: We have

$$m_{ii} = \mathbf{E}[\text{\# of visits to } i \text{ starting at } i]$$
$$= \mathbf{E}[\text{Geometric(success probability = prob the MC does not come back to } i]$$
$$= \mathbf{E}[\text{Geometric}(1 - f_{ii})]$$
$$= \frac{1}{1 - f_{ii}} < \infty$$

Note that $f_{ii} < 1$ for any transient state i.

We can use the same trick on other states, to obtain

$$m_{ij} = f_{ij} \cdot m_{jj}$$

where f_{ij} is the total transition probability from i to j. ∎

12.6.1 Time of absorption

Suppose we want to calculate the expected number of states visited before leaving the transient states. Let

$$\tau = \inf\{n : X_n \notin S_T\}$$

τ is also referred to as *the time of absorption*. The next three results completely describe this variable.

Proposition 12.28 *For $i \in S_T$, with τ the time of absorption,*

$$\mathbf{E}[\tau | X_0 = i] = \sum_{k \in S_T} m_{ik}$$

Proposition 12.29 *For $i \in S_T$, $j \in S_T^c$ (j any recurrent state)*

$$P\{X_\tau = j | X_0 = i\} = \sum_{k \in S_T} m_{ik} P_{kj}$$

Proof:

$$P(X_\tau = j | X_0 = i) = \sum_{k \in S_T} \sum_{n=0}^{\infty} P_{kj} P(X_n = k | X_0 = i)$$

$$= \sum_{k \in S_T} P_{kj} \underbrace{\sum_{n=0}^{\infty} P(X_n = k | X_0 = i)}_{= \sum_{n=0}^{\infty} \mathbb{E}[\mathbf{1}_{\{X_n = k | X_0 = i\}}]}$$

$$= \sum_{k \in S_T} P_{kj} m_{ik}$$

∎

Proposition 12.30 *For $i \in S_T$, $j \in S_T^c$, $j \in K_a$, K_a is some recurrent class (closed irreducible recurrent class)*

$$f_{ij} = \sum_{j \in k_a} P\{X_\tau = j | X_0 = i\}$$

12.6.2 An example

Suppose we have a Markov chain with eight states where the transition matrix after rearranging the states is

$$P = \begin{pmatrix} \begin{pmatrix} 0.4 & 0.3 & 0.3 \\ 0 & 0.6 & 0.4 \\ 0.5 & 0.5 & 0 \end{pmatrix} & \mathbf{0} & & \mathbf{0} \\ \mathbf{0} & \begin{pmatrix} 0 & 1 \\ 0.8 & 0.2 \end{pmatrix} & \mathbf{0} \\ \begin{pmatrix} 0 & 0 & 0 \\ 0.4 & 0.1 & 0 \\ 0.1 & 0 & 0.1 \end{pmatrix} & \begin{pmatrix} 0 & 0 \\ 0 & 0.3 \\ 0.2 & 0 \end{pmatrix} & \begin{pmatrix} 0.4 & 0.6 & 0 \\ 0 & 0 & 0.2 \\ 0.6 & 0 & 0 \end{pmatrix} \end{pmatrix}$$

In this example, we can easily see that

$$Q = \begin{pmatrix} 0.4 & 0.6 & 0 \\ 0 & 0 & 0.2 \\ 0.6 & 0 & 0 \end{pmatrix}$$

$$M = (I - Q)^{-1} = \begin{pmatrix} 0.6 & 0.60 & 0 \\ 0 & 1 & -0.2 \\ -0.6 & 0 & 1 \end{pmatrix}^{-1} = \frac{1}{66} \begin{pmatrix} 126 & 75 & 15 \\ 15 & 75 & 15 \\ 75 & 45 & 75 \end{pmatrix}$$

Using the results above, we do the following:
a) Calculate $f_8 7$.

$$f_{87} = \frac{m_{87}}{m_{77}} = \frac{45}{75} = \frac{3}{5}$$

Note we can directly calculate this probability from Q as well:

$$f_{87} = (0.6)(0.6) + (0.6)(0.4)(0.6) + (0.6)(0.4)^2(0.6) \ldots$$
$$= (0.6)^2 (1 + 0.4 + (0.4)^2 + \ldots)$$
$$= (0.6)^2 \cdot \frac{1}{1 - 0.4} = (0.6)^2 \frac{1}{0.6} = 0.6$$

b) Find f_{67}.

$$f_{67} = \frac{m_{67}}{m_{77}} = \frac{75}{75} = 1$$

Again, direct calculation from Q gives

$$f_{67} = 0.6 + (0.4)(0.6) + (0.4)^2(0.6) + \ldots$$
$$= 0.6 \cdot \frac{1}{1 - 0.4} = 1$$

c) Find f_{78}. Use theorem

$$f_{78} = \frac{m_{78}}{m_{88}} = \frac{15}{75} = \frac{1}{5}$$

Direct Calculation from Q

$$f_{78} = 0.2 \qquad \text{(given)}$$

d) Find $\mathbb{E}\{\tau|X_0 = 6\}$.

$$\mathbb{E}\{\tau|X_0 = 6\} = m_{66} + m_{67} + m_{68}$$
$$= \frac{216}{66} = 3.257$$

e) Find $P\{X_\tau = 1|X_0 = 6\}$.

$$P(X_\tau = 1|X_0 = 6) = m_{66}P_{61} + m_{67}P_{71} + m_{68}P_{81}$$
$$= \frac{1}{66}(125 \cdot 0 + 75 \cdot 0.4 + 15 \cdot 0.1)$$
$$= 0.47$$

f) Find f_{61}.

$$f_{61} = \sum_{j \in K_1} P(X_\tau = j|X_0 = 6) = \sum_{j \in \{1,2,3\}} \sum_{k \in S_T} m_{6k}P_{kj}$$
$$= \sum_{k \in S_T} m_{6k} \sum_{j \in \{1,2,3\}} P_{kj} = \text{(calculate)}$$

■ EXAMPLE 12.10   Phenotype evolution as a Markov chain

In this example, we will discuss a model of genetic evolution based on a simple dominant recessive relationship. This model was postulated by Gregor Johann Mendel (July 20, 1822–January 6, 1884) who basically invented genetics 50 years before anybody recognized it and in fact rediscovered and republished it at the beginning of 1900s by Thomas Hunt Morgan and others. We follow his flies example here. This particular type of fly, a fruit fly (Drosophila), was chosen due to its fast reproductive cycle as well as a very visible recessive trait.

Suppose that the color of eyes for a particular fruit fly type is determined by a particular gene. In this example, we assume that this gene has two alleles B and b, where B is dominant and b is recessive. Therefore, each fly has exactly one of the possible types: BB, Bb, bB, and bb. Since for all intents and purposes Bb and bB have identical results, the possibilities are just three: BB, Bb, and bb. This set of possibilities is called the **genotype** of the individual.

Since the B allele is dominant, a fly with one of the allele being the dominant one (BB, Bb, or bB) has red eyes. A fly with both recessive alleles (bb) has white eyes. The collection of traits exhibited as a result of the genotype is called the **phenotype** of that gene.

Every offspring takes one allele at random from each parent's gene. Therefore, the offspring's eye color is heavily dependent on the genotype of the parents. In the original study, Mendel and later Thomas Morgan crossed successive generations of offspring with a particular individual type. Let us look at this type of evolution and create a Markov chain to study it.

We will assume that genes inherited from each parent are selected at random, independently of each other. It is easy to see that

- The offspring of two BB parents is always BB,

- The offspring of two bb parents is always bb,

- The offspring of one BB and one bb is always Bb.

All the other cases must be analyzed individually. If we follow the original model hypothesized by Mendel, we can construct a Markov chain that will describe the evolution depending on what we crossbreed.

Suppose, for example, generations are crossbred with the genotype BB (no recessive gene). The resulting transition probability matrix is presented in Table 12.1. We can similarly construct transition probability tables for crossbreeding with genotype Bb (Table 12.2) and genotype bb (Table 12.3).

Table 12.1   Transition probability for crossbreeding with genotype BB

	BB	Bb	bb
BB	1	0	0
Bb	1/2	1/2	0
bb	0	1	0

Table 12.2   Transition probability for crossbreeding with genotype Bb

	BB	Bb	bb
BB	1/2	1/2	0
Bb	1/4	1/2	1/4
bb	0	1/2	1/2

When we crossbreed with the white-eyed individual, it is pretty clear we are dealing with Table 12.3. However, when we crossbreed with a red-eyed fly, we don't know which genotype that red fly is (BB or Bb). So suppose that we have

Table 12.3   Transition probability for crossbreeding with genotype bb

	BB	Bb	bb
BB	0	1	0
Bb	0	1/2	1/2
bb	0	0	1

controlled breeding with a red-eyed fly and assume that both allele are equally likely to be in the population. Thus any particular genotype BB, Bb, bB, and bb is equally likely. This makes the probabilities of a randomly selected fly from the population:

$$\mathbf{P}(BB) = \frac{1}{4}; \quad \mathbf{P}(Bb) = \frac{1}{2}; \quad \mathbf{P}(bb) = \frac{1}{4}.$$

Based on these assumptions, we have the following probabilities of the particular genotype for a red-eyed fly:

$$\mathbf{P}(\text{genotype is } BB \mid \text{fly has red eyes}) = \frac{\mathbf{P}(BB \text{ and red eyes})}{\mathbf{P}(\text{fly has red eyes})}$$

$$= \frac{\mathbf{P}(BB)}{\mathbf{P}(BB) + \mathbf{P}(Bb)} = \frac{1}{3}$$

and of course $\mathbf{P}(\text{genotype is } Bb \mid \text{fly has red eyes}) = 2/3$. With these assumptions, we can calculate the probabilities of the genotype for the next generation after breeding with a red-eyed fly. These are presented in Table 12.4. The calculation for each transition probability requires conditioning on the other parent genotype. For example

$$\mathbf{P}(\text{offspring is } BB \mid \text{parent 1 is } BB)$$
$$= \mathbf{P}(\text{offspring is } BB \text{ and parent 2 is } BB \mid \text{parent 1 is } BB)$$
$$\quad + \mathbf{P}(\text{offspring is } BB \text{ and parent 2 is } Bb \mid \text{parent 1 is } BB)$$
$$= \mathbf{P}(\text{offspring } BB) \mid \text{parents } BB \times BB)\mathbf{P}(\text{parent 2 is } BB)$$
$$\quad + \mathbf{P}(\text{offspring } BB \mid \text{parents } BB \times Bb)\mathbf{P}(\text{parent 2 is } Bb)$$
$$= 1\frac{1}{3} + \frac{1}{2}\frac{2}{3} = \frac{2}{3}$$

Similarly

$$\mathbf{P}(Bb \mid BB) = \mathbf{P}(Bb \mid BB \times BB)\mathbf{P}(BB) + \mathbf{P}(Bb \mid BB \times Bb)\mathbf{P}(Bb)$$
$$= 0\frac{1}{3} + \frac{1}{2}\frac{2}{3} = \frac{1}{3}$$

and so on, for the rest of the table.

Table 12.4 Transition probability for crossbreeding with red-eyed flies

	BB	Bb	bb
BB	2/3	1/3	0
Bb	1/3	1/2	1/6
bb	0	2/3	1/3

Note that the process described above needs controlled reproductions, and this will not describe the gene evolution in the population. However, if we assume the same distribution of basic genes in the population following the model of crossbreeding with red-eyed flies, we can calculate transition probabilities for this Markov chain. Calculate this matrix in Problem 12.17.

Thomas Hunt Morgan discovered that the results obtained by using this simple model did not agree with the experimental evidence. This led to the discovery of sex linkage and inheritance patterns and eventually to the discovery of the X and Y chromosomes. We now know that for the particular example (Drosophila) the genetic material responsible for the color of the eye is in fact located on the X chromosome.

■ EXAMPLE 12.11 Genetics inbreeding

In genetics literature, there exists a different approach for a type of population inbreeding (without any individuals from the outside world). This is applicable, for example, in a situation of a hybridized mono-culture where the seeds for the next season are selected from the current plants.

Specifically, in generation 0 we select two individuals of opposite sex from the general population. We thus obtain generation 1. We then select two (and only two) individuals at random from this offspring generation. We mate them and their offspring makes generation 2. We continue in this way, every time selecting two individuals at random from the current generation to produce the next generation.

We present this approach because it leads to a Markov chain with a larger number of states. We describe the genetic evolution using a Markov chain where X_n represents the genotypes of BOTH parents of one individual. If we do this, the possible states of the Markov process are presented in Table 12.5.

Let us determine the transition probabilities for this chain. We present the results in Table 12.6. We next show how we obtained those numbers.

Suppose that a pair of parents is selected at random from the population to form generation 0. Clearly, if the individuals selected in generation 0 happen to be BB×BB, then all offspring for all generations will have genotype BB. Similarly, if the parents are bb×bb, all the offspring will have this genotype. Thus the first and last row of the transition probability are easy to calculate.

Table 12.5   Possible genotypes of parents

X_n	Genotype
1	BB×BB
2	BB×Bb
3	BB×bb
4	Bb×Bb
5	Bb×bb
6	bb×bb

Table 12.6   Transition probability for the interbreeding model

	BB×BB	BB×Bb	BB×bb	Bb×Bb	Bb×bb	bb×bb
BB×BB	1	0	0	0	0	0
BB×Bb	1/4	1/2	1/4	0	0	0
BB×bb	0	0	0	1	0	0
Bb×Bb	1/16	1/4	1/8	1/4	1/4	1/16
Bb×bb	0	0	0	1/4	1/2	1/4
bb×bb	0	0	0	0	0	1

Next, let the individuals selected be BB×Bb (second row). Since each off-spring gets one allele from parent 1 and the other allele from parent 2, it is easy to see that the proportion of the offspring genotypes will be

$$\mathbf{P}(BB) = \frac{1}{2}; \quad \mathbf{P}(Bb) = \frac{1}{2}; \quad \mathbf{P}(bb) = 0$$

Since we now pick two individuals at random (and independently) from the population, the probability that the pair is BB×BB is

$$\mathbf{P}(BB \times BB) = \mathbf{P}(BB)\mathbf{P}(BB) = \frac{1}{4}$$

and, for example

$$\mathbf{P}(BB \times Bb) = \binom{2}{1}\mathbf{P}(BB)\mathbf{P}(Bb) = \frac{1}{2}$$

In a very similar way, we can fill the entire probability matrix. We do one more example of calculation when parents are Bb×Bb. In this case, the offspring distribution is

$$\mathbf{P}(BB) = \frac{1}{4}; \quad \mathbf{P}(Bb) = \frac{1}{2}; \quad \mathbf{P}(bb) = \frac{1}{4}.$$

Therefore,

$$\mathbf{P}(BB \times Bb) = \binom{2}{1}\mathbf{P}(BB)\mathbf{P}(Bb) = \frac{1}{4}$$

, and so on.

Problems

12.1 For each example in Section 12.1.2, write down the transition probability matrix.

12.2 Give the proof of the Markov property (Theorem 12.2 on page 372).

12.3 Show that the Remark 12.4 is true.

12.4 Draw the "arrows diagram" of a communicating class with period d, for $d = 2, 3, 4, 5$. Generalize.

12.5 Give an example of a Markov chain with more than one stationary distribution. Draw its associated graph.

12.6 Let X_n be a birth and catastrophy chain. Its transition matrix is

$$P = \begin{pmatrix} q_0 & p_0 & 0 & 0 & \cdots \\ q_1 & 0 & p_1 & 0 & \cdots \\ q_2 & 0 & 0 & p_2 & \cdots \\ & & \vdots & & \end{pmatrix}$$

Let $X_0 = 0$. Assume $p_k > 0 \, \forall k$. Define $T_b = \inf\{n : X_n = b\}$ as the time of first entry in b. Calculate $\mathbf{E}_0[T_b] = \mathbf{E}[T_b|X_0 = 0]$.

12.7 Using the matrix exemplified in Subsection 12.6.2, calculate the following: f_{68}, f_{76}, $\mathbf{E}[X_\tau = 4|X_0 = 6]$, f_{81}, f_{74}.

12.8 Consider a Markov chain with the following transition probability matrix:

$$P = \begin{pmatrix} 0 & 0 & 1/2 & 1/2 \\ 0 & 0 & 1/2 & 1/2 \\ 1/2 & 1/2 & 0 & 0 \\ 1/2 & 1/2 & 0 & 0 \end{pmatrix}$$

Show that the chain is ergodic but not irreducible.

12.9 Consider the gambler's ruin problem where probability of winning is $p = 1/2$ and the initial wealth is $100. Suppose the gambler stops if he reaches $200 or if he goes bankrupt. Let X_n the total wealth after n games. Show that X_n is a Markov chain, and write the probability transition matrix for the wealth. Classify the states in the chain as recurrent or transient. Does the chain have a stationary distribution? Is it unique?

12.10 Consider the gene mutating Example 12.8 on page 378. Show that the process forms a Markov chain, and write its probability transition matrix. Classify the states in the chain as recurrent or transient. Does the chain have a stationary distribution? Is it unique?

12.11 Consider a random walk with $n + 1$ states. The transition probability from state i to state j is

$$P_{ij} = \begin{cases} 1 - p & \text{if } i = j = 0 \\ p & \text{if } j = i + 1, \text{ and } i = 0, \ldots, n - 1 \\ p & \text{if } i = j = n \\ 1 - p & \text{if } j = i - 1, \text{ and } j = 1, \ldots, n \\ 0 & \text{otherwise.} \end{cases}$$

Draw the states and transitions of the Markov chain. Classify its states and calculate its stationary distribution.

12.12 Consider an irreducible Markov chain. Show that, if it is periodic, then $P_{ii} = 0$ for all its states i. Is the reciprocal implication true, that is, if $P_{ii} = 0$ for all i, is the chain periodic? Give a proof or counterexample.

12.13 Suppose we have an asymmetric random walk on integers with $P_{i,i+1} = p$ and $P_{i,i-1} = 1 - p$ for all $i \geq 1$. For $i = 0$, we have $P_{0,1} = p$ and $P_{0,0} = 1 - p$. Classify the states of the Markov chain and find its stationary distribution. Discuss the existence of the stationary distribution depending on the value of p.

12.14 In a network queue transmission packets go through two consecutive processes. Router A processes the packets in a time distributed with a geometric distribution with mean 10 ms. Router B routs the packets in a time distributed as a *Geometric*(20) random variable (also in milliseconds). All the packets need to go through router A and then router B, and we assume there is no shortage of packets for the first queue (immediately after router A processes a packet there is another one in line entering service). Give a graph of the Markov chain describing the states of the two routers.

    a) What is the probability for each router being in use?

    b) What is the probability that both routers are in use simultaneously?

   Hint: look at the stationary distribution.

12.15 A baseball player is very emotional. Every time he gets a hit, his confidence grows and the probability of making a hit at the next try increases. Similarly, if he strikes out or fails to make a hit, his confidence drops and his probability of getting a hit decreases. Assume that before the first ball thrown his probability of getting a hit is 0.3. Every time he gets a hit, the probability increases by 0.05 and every time he fails to get a hit his probability drops by 0.05. Assume that the probability of getting a hit cannot go above 0.6 and cannot go below 0.1 regardless of the sequence of hits or misses. Graph a Markov chain for the baseball player with the states being probabilities. Specifically, let state 0 represent the initial state before any tries, positive states representing higher probabilities, and negative numbers representing lower probabilities. Over a long sequence of balls thrown to this player, what is the probability of making a hit in the seventh ball thrown to him?

12.16   In this problem, we will analyze the model presented in Example 12.10 on page 401. Read and understand this example first.

- a) Consider a Markov chain described by the transition probabilities in Table 12.1. Calculate the stationary distribution if it exists. Suppose we start with a single fly and in generation 1000 we have 50 individuals. What proportion of these is expected to have white eyes.
- b) Repeat part a) for the Markov chain described by the transition probabilities in Table 12.2.
- c) Repeat part a) for the Markov chain described by the transition probabilities in Table 12.3.
- d) Suppose we know that the fly we crossbred with has the same genotype but we do not know which one it is. We observe after 1000 generations 17 flies with white eyes out of total 50 in that generation. What is most likely the genotype of the crossbreeding fly.

12.17   We refer again to the Example 12.10 on page 401. Suppose that the individuals we choose for crossbreeding are always coming from a sample which contains 25% individuals with genotype BB, 50% individuals with genotype Bb, and 25% individuals with genotype bb. Calculate offspring transition probability based on these assumptions. Give the stationary distribution of the resulting Markov chain.

12.18   In this problem, we look at the model presented in Example 12.11 and the resulting transition probability presented in Table 12.6. Identify the transient and recurrent states of the Markov chain. Reorder the states so that the recurrent states are ordered first, and write down the resulting transition matrix. Follow the general treatment in Section 12.6 and calculate the relevant matrices. Then answer the following questions:

- a) What are the eventual limiting states of the Markov chain?
- b) If the initial generation 0 is BB×Bb, calculate the probabilities of absorption into the relevant recurrent states. Calculate the expected number of generations until the population becomes uniform.
- c) For the initial generation Bb×Bb, calculate the probabilities of absorption into the relevant recurrent states. Calculate the expected number of generations until the population becomes uniform.

d) Repeat the above part for initial pair BB×bb. Compare the results obtained with the previous part. Is there an intuitive explanation of the connection between answers?

12.19 Assume it is known that the color of eyes of humans is determined by a particular gene. We assume that this gene has three alleles B, b, and g, where B is dominant and both b and g are recessive. A person with genotype containing the dominant allele B has brown eyes. A person with both recessive alleles containing allele b (bb or bg) has blue eyes. Finally, a person with genotype gg has green eyes. Every child takes one allele at random from each of the parents.

Suppose, in a remote village where all the villagers have blue eyes (genotype bb), one day an escaped fugitive with brown eyes (genotype BB) takes refuge. He marries one of the villagers and they have one child. This child marries further in the village and the story continues every time having exactly one descendant. Write down the transition probability matrix for the successive generations of children. Calculate the expected number of generations until a descendant with blue eyes appears.

12.20 Refer to the genetic transition presented in the previous problem. Suppose that I know that my parents have brown eyes (genotype Bg) and blue eyes (genotype bg). I know that I have brown eyes and my wife has blue eyes. Calculate the probability that my son has blue eyes.

12.21 Suppose the color of rats is determined by three different alleles B, G, and b. Suppose B and G are codominant alleles. There are four different types of individual rat colors even though there are six genotypes:

Color	Genotype
Black	BB or Bb
Grey	GG or Gb
Half black, half gray	BG
Brown	bb

Suppose we create an interbreeding model as in Example 12.11. We select a pair of rats at random and all offspring are interbred. Write down the transition probability model for this situation. Write a computer program that will simulate the color of the pair selected for further breeding. Start with a random pair where each individual in the pair is selected from the six genotypes equally likely. Suppose that each pair has 100 offspring and each time we proceed with the selection from this 100 offspring (do not simulate gender). Perform 100,000 simulations of this process and report the estimated expected number of brown rats in generation 1000.

CHAPTER 13

SEMI-MARKOV PROCESSES AND CONTINUOUS TIME MARKOV PROCESSES

In many situations, we wish to model the time between transitions as well as the nature of the transitions between states. Often, we have a pretty good idea about the distribution of the time spent in a particular state. The semi-Markov process is an example of a process that tries to capture both time and state transitions. However, this process is not Markov in general, in the sense that next transitions may depend on the previous states besides the current one. In a special case where the transitions only depend on the current states, we obtain the Markov process as a particular case of the process presented in this chapter. We decided to include the semi-Markov process in this book since the mathematical treatment of the two processes is very similar. Naturally, since the Markov process is a particular case, it will have extra properties, but the mathematical framework is identical.

Definition 13.1 *Define the state space as $S = \{0, 1, 2, \ldots\}$. Suppose that we have a process which jumps between these states according to a Markov chain with transition probability matrix $\{P_{ij}\}_{ij}$.*

However, unlike in the discrete-time Markov chain case of the previous chapter, we may stay in the state i for a random amount of time which may be dependent on

Probability and Stochastic Processes, First Edition. Ionuţ Florescu
© 2015 John Wiley & Sons, Inc. Published 2015 by John Wiley & Sons, Inc.

where we jump <u>next</u>. Mathematically, let

$F_{ij} =$ *the distribution of time T_{ij} spent in state i given that the chain jumps to j next.*

*Let Z_t denote the state of the process at time t. Then Z_t defines a **semi-Markov** process.*

Remark 13.2 *We make the following notes about the process we have just defined:*

- *A semi-Markov process in general does not have the Markov property. This is easy to see since the length of the stay depends on where the next jumps goes.*

- *If the distribution F_{ij} is the same for all j and only depends on the state i, then the process has the Markov property.*

- *In the **special case** when F_{ij} is the same for all j **and** additionally is the distribution function of an exponential random variable with rate v_i, the process is called a* continuous-time Markov chain. *We will analyze this special process in detail in Section 13.2.*

Let $H_i(t)$ denote the distribution function of the time spent in state i before making a transition[1]. We have a formula that describes this cdf:

$$H_i(t) = \sum_{j \in \mathcal{S}} P_{ij} F_{ij}(t)$$

by conditioning on where the semi-Markov process jumps next.
Let

$$\mu_i = \mathbf{E}[\text{time spend in state } i] = \int_0^\infty x \, dH(x)$$

If we let X_n denote the nth state visited by the process, then X_n is a Markov chain with transition matrix $\{P_{ij}\}_{ij}$. This X_n is called *the embedded Markov chain*. Note that only the time duration spent in state i is affected by the destination, not the actual state.

We say that the semi-Markov process is *irreducible* if and only if its *embedded Markov chain* is irreducible. We also note that transitions to the same state are irrelevant since we could easily change the distribution F_{ii} to contain the time distribution where we can go back to the state. Therefore, we can consider the embedded Markov chain having the property $P_{ii} = 0$.

Remark 13.3 *Let us make a separate remark on why the process is called semi-Markov. Let S_i denote the time of the ith jump. Let $X_i = Z_{S_i}$ denote the state of the semi-Markov process at the time S_i of the ith jump. If we denote by $T_i = S_i - S_{i-1}$ the time spent in state i, then the two-dimensional continuous-time process (X_i, T_i) has the Markov property. This is why the process is called semi-Markov. By itself, Z_t does not have the Markov property. However, a simple modification (augmentation) transforms it into a process that has the Markov property.*

---

[1]Note that if $H_i(t)$ is the distribution of the constant 1, we obtain the discrete-time Markov chain of the previous chapter as a special case.

13.1 Characterization Theorems for the General semi- Markov Process

Define T_{ii} to be the length of an i-to-i cycle of the semi-Markov process and let $\mu_{ii} = \mathbb{E}[T_{ii}]$ be the expected time length of an i-to-i cycle.

Proposition 13.4 *If the semi-Markov process is irreducible, T_{ij} is a non-lattice random variable, and $\mu_{ii} < \infty$, $\forall i$, then*

$$P_i \stackrel{def}{=} \lim_{t \to \infty} P(Z(t) = i | Z(0) = j) = \frac{\mu_i}{\mu_{ii}},$$

where μ_i is the mean stay in state i, and μ_{ii} is the expected length of an i-to-i cycle.

Proof: If the semi-Markov process is irreducible, all states communicate. If N_{ij} is the number of times the process transition to j given that the initial state is i, then is it easy to see that the process $N_{ij}(t)$ is a delayed renewal process with inter-arrival times T_{ii} and the delay T_{ij}. Thus, we can simply apply the alternating renewal theorem (Theorem 11.20 on page 353). Let us denote

- the time in state i as ON time, and

- the time in any other state $j \neq i$ as OFF time.

Apply the theorem and we are immediately done! ∎

We note that to be in the hypotheses of the Theorem 11.20 we need the delay time to be a.s. finite.

Corollary 13.5 *We have*

$$\frac{Time \; in \; state \; i \; during \; [0, t]}{t} \to \frac{\mu_i}{\mu_{ii}} \quad a.s.$$

Proof: Let

$$R_i(t) = \int_0^t \mathbf{1}_{\{Z_s = i\}} ds = \text{Amount of time spent in state } i \text{ before time } t$$

If we denote by $N_i(t)$ the number of visits to i before t, we have

$$\frac{R_i(t)}{t} = \frac{\frac{R_i(t)}{N_i(t)} \to \text{SLLN regular} \to \mu_i}{\frac{t}{N_i(t)} \to \text{SLLN for renewal} \to \mu_{ii}} \quad \text{Done!}$$

∎

In the case of the semi-Markov process, we can relate the proportion of time spent in one state with the stationary distribution if it exists. This relationship is detailed in the next theorem. Before that, we need to introduce the concept of stationary measure.

Definition 13.6 *Let $X = \{X_n\}$ be an irreducible Markov chain with countable state space $S = \{1, 2, 3, \dots\}$ and transition matrix P. A measure η on S is called a* stationary measure *for the Markov chain X if*

$$\eta_j = \sum_{i \in S} \eta_i P_{ij}.$$

Note that the expression above mimics the definition of a stationary distribution. Indeed, it is the same mathematical expression; however, for a stationary measure the η_i's do not have to sum to 1.

Theorem 13.7 *If T_{ij} have a non-lattice distribution, if $\mu_{ii} < \infty$, and if the embedded Markov chain defined by $\{P_{ij}\}$ is irreducible, recurrent, with a stationary measure*

$$\eta_i = \mathbf{E}[\text{number of visits to } i \text{ in a 0-to-0 cycle}],$$

then the limiting probability defined in Proposition 13.4 is

$$P_i = \frac{\eta_i \mu_i}{\sum_j \eta_j \mu_j}.$$

Proof: First, let us show that the η_i's defined above determine a stationary measure for the embedded Markov chain. For that Markov chain, we let $N_{00} = \inf\{n : X_n = 0, X_{n-1} \neq 0, \dots, X_1 \neq 0 | X_0 = 0\}$ the length of a 0-to-0 cycle. Then we have for N_i, the number of visits to i in a 0-to-0 cycle,

$$N_i = \sum_{t=1}^{N_{00}} \mathbf{1}_{\{X_t = i\}}.$$

If the chain is irreducible and if we start it from the stationary distribution, then the chain is stationary and thus all X_t have the same distribution. In this case, we must have (applying expectations and Wald's theorem)

$$\eta_i = \eta_{00} \pi_i.$$

Therefore, the η_i's are equal to a constant multiplying the stationary distribution at i. It is easy to check using this expression that the η_i's define a stationary measure according to the definition.

Now let us go back to the semi-Markov process and use the following notation:

$$\mu_{00} = \text{average length of an 0-to-0 cycle,}$$
$$\mu_j = \text{expected time spent in state } j \text{ during a visit to } j.$$

We can derive in a similar way by writing in terms of variables, applying expectations, and then Wald's theorem

$$\mu_{00} = \sum_j \eta_j \mu_j$$

If we now apply Proposition 13.4 for $i = 0$, we obtain

$$P_0 = \frac{\mu_0}{\mu_{00}} = \frac{\mu_0}{\sum_j \eta_j \mu_j} = \frac{\overbrace{\eta_0}^{=1} \mu_0}{\sum_j \eta_j \mu_j}$$

Thus we prove the theorem for $i = 0$. For a general i, we denote N_i, the random variable measuring the number of visits to i during a 0-to-0 cycle. Note that if we visit state i N_i times during one $0 - to - 0$ cycle, we in fact cycle through state i N_i times. Formally,

$$\text{Time}_{\text{0-to-0}} = \text{Time}_{\text{0-to-i}} + \text{Time}_{\text{i-to-i cycle 1}} + \cdots + \text{Time}_{\text{i-to-i cycle } N_i - 1} + \text{Time}_{\text{i-to-0}}.$$

Note that $\text{Time}_{\text{0-to-i}} + \text{Time}_{\text{i-to-0}} = \text{Time}_{\text{i-to-0}} + \text{Time}_{\text{0-to-i}}$ forms an i-to-i cycle. Also note that we have only $N_i - 1$ cycles since the number of visits to state i has to be N_i. Thus, we may write

$$\text{Time}_{\text{0-to-0}} = \sum_{k=1}^{N_i} \text{Time}_{\text{i-to-i cycle } k},$$

and applying Wald's lemma once again we obtain

$$\mu_{00} = \eta_i \mu_{ii}. \tag{13.1}$$

Thus, we finally get using Proposition 13.4 and Equation (13.1)

$$P_i = \frac{\mu_i}{\mu_{ii}} = \frac{\mu_i}{\mu_{00}/\eta_i} = \frac{\eta_i \mu_i}{\sum_j \eta_j \mu_j}$$

which concludes the proof of the theorem.   ∎

EXAMPLE 13.1   Continuous-time random walk

Consider a random walk with jumps equal to ± 1 with probability $\frac{1}{2}$ on $\mathbb{Z}$. We have already seen in the previous chapter that if the process jumps at integer times, it is a null recurrent Markov chain. Therefore, according to the main theorem of the Markov chain (Theorem 12.22) the probability that we are in state i at time t goes to zero as $t \to \infty$.

We will change this process and embed it into a semi-Markov process (in fact a continuous-time Markov chain). To do so, assume that the time we spend into any state is not fixed equal to 1 anymore, but it is a random variable which has a certain distribution.

By changing the time at which the process moves, we can make the resulting semi-Markov process positive recurrent. To do so, we set the expected amount of time spend in state j smaller and smaller if j is further away from the origin.

To give a physical analogy, the process is heated up (moves faster) the further the state j is from the origin. In states close to the origin, the process cools down (slows down). As a specific example, we can construct distributions such that the expectations of the times spent in respective states are as follows:

$$\mu_0 = 2$$

$$\mu_1 = \mu_{-1} = \frac{1}{2^2}$$

$$\mu_2 = \mu_{-2} = \frac{1}{2^3}$$

$$\vdots$$

$$\mu_n = \mu_{-n} = \frac{1}{(n+1)^2}$$

$$\vdots$$

We want to see if there exists a stationary measure. Take a candidate measure $\eta = \{\eta_i\}_{i \in \mathbb{Z}}$, and due to the simple form of the transition matrix, we have

$$\eta_j = \sum_{i \in \mathbb{Z}} \eta_i P_{ij},$$

or,

$$\eta_j = \eta_{j-1} \left(\frac{1}{2}\right) + \eta_{j+1} \left(\frac{1}{2}\right)$$

The system above is trivially verified by $\eta_j \equiv 1, \forall j$.

However, this stationary measure is not a stationary distribution since $\sum_j \eta_j = \infty$. Despite this, the semi-Markov process is in fact positive recurrent and we can find the stationary distribution by applying the theorem:

$$P_i = \frac{\mu_i}{\sum_{i \in \mathbb{Z}} \mu_i}$$

Now you can see why we picked the expectations of the times spent in each state the way we did. Specifically, since we have $\sum_{i \in \mathbb{Z}} \mu_i = 2 + 2\frac{1}{2^2} + 2\frac{1}{3^2} \ldots = 2\frac{\pi^2}{6} = \frac{\pi^2}{3}$, the stationary measure defined by P_i's is in fact a stationary distribution:

$$P_i = \frac{3}{((i+1)\pi)^2} \quad \text{if } i \neq 0$$

and

$$P_0 = \frac{6}{\pi^2}$$

13.2   Continuous-Time Markov Processes

In the case where a semi-Markov process with the distribution of the time spent in state i is exponential and does not depend on the next state j, we obtain what is called the Markov process.

Specifically, let $\{X(t)\}_{t \geq 0}$ be a stochastic process with values in the set of non-negative integers.

Definition 13.8 *We say that $\{X(t)\}_t$ is a (homogeneous) continuous-time Markov chain if and only if it is a semi-Markov process and the time of stay in state i is distributed as an exponential random variable with rate v_i.*

This definition implies the following behavior of the process each time it enters state i:

- It stays in state i an amount of time exponentially distributed with rate[2] v_i, which does not depend on the next state.

- After that time, it jumps to some other state j with probability P_{ij}.

Remark

- Without loss of generality, we can assume that the embedded Markov chain has $P_{ii} = 0$.

- The amounts of time spent in different states are independent.

Theorem 13.9 $\{X(t)\}$ *has the Markov property.*

Proof: This is immediate due to the memoryless property of the exponential.

$$\begin{aligned}
P\{X(t+s) &= j | X(s) = i, X(r) = k, 0 \leq r \leq s\} \\
&= P\{X(t+s) = j | X(s) = i\} \\
&= P_{ij}(t) \qquad \forall s, t > 0 \quad i, j, k \in \mathbb{N}
\end{aligned}$$

∎

Definition 13.10 (Regular Markov Chain) *A continuous Markov chain is called* regular *if almost surely the number of transitions in finite time is finite.*

EXAMPLE 13.2   Nonregular Markov Chain

Consider the case when the chain always jumps to a higher state $P_{i,i+1} = 1$, and the rate $v_i = i^2$. This chain is nonregular. It has an infinite expected number of jumps in a finite time interval (note that the expected time to have an infinite number of jumps: $\sum_i \frac{1}{v_i} = \sum_i \frac{1}{i^2} < \infty$).

---

[2]Note that the expected length of time spent in state i is $\frac{1}{v_i}$

Proposition 13.11 (Criteria for regularity of the continuous-time Markov chains) *If the following, two conditions hold then the continuous-time Markov chain is regular:*

  a) *There exists some constant M such that $v_i \leq M < \infty \; \forall i$.*

  b) *The embedded discrete-time Markov chain is irreducible and recurrent.*

Having a regular Markov process avoids dealing with the possibility of an infinite number of jumps in a finite interval. Since this possibility complicates the analysis, we will make the following assumption:

For the remainder of this chapter, we assume that all the continuous-time Markov chains encountered are regular.

Definition 13.12 *For a continuous-time Markov process, we define the quantity*

$$q_{ij} = v_i P_{ij}, \text{ for } i \neq j.$$

q_{ij} *is called* the transition rate from i to j.

Note that v_i is the rate at which the process leaves state i, and P_{ij} is the probability of the embedded Markov chain going from i to j. Thus the definition of the rate q_{ij} is quite natural.

Definition 13.13 *We define the transition probability of the Markov process from state i at time s to state j in t units of time as*

$$P_{ij}(s,t) = P\{X(t+s) = j | X(s) = i\}.$$

In the general case of a semi-Markov process, this transition probability depends on both time s and the time interval t. However, in the case of the continuous-time Markov chain, we have

$$P_{ij}(s,t) = P\{X(t+s) = j | X(s) = i\} = P\{X(t) = j | X(0) = i\} = P_{ij}(t)$$

due to the memoryless property of the exponential distribution. Thus, for a Markov process the probability depends only on the time elapsed between the two moments in time (s and $s+t$). This is the reason why the Markov process as defined in Definition 13.8 was called homogeneous.

These transition probabilities completely describe the evolution of the Markov chain. *The main problem when studying continuous-time Markov chains is finding these probabilities.* Unfortunately, only in very simple cases can they actually be explicitly found. We shall detail several such cases in this chapter.

In general, we will have to work with Kolmogorov-type differential equations (Theorems 13.16 and 13.17) to find them. These equations oftentimes have to be solved numerically. The next lemma presents the instantaneous behavior of these probabilities.

Lemma 13.14 (Limiting probabilities) *For a continuous-time Markov chain as defined in Definition 13.8 and with the transition probabilities as in Definition 13.12, the following limits hold:*

$$\lim_{t \to 0} \frac{1 - P_{ii}(t)}{t} = v_i \tag{13.2}$$

$$\lim_{t \to 0} \frac{P_{ij}(t)}{t} = q_{ij} \tag{13.3}$$

Proof: We shall prove (13.3) first. Equation (13.2) will then be an immediate implication of (13.3).

Let

$$T_k = \text{The time of the } k\text{th jump.}$$

The idea is to show that two or more jumps are impossible in a very short interval.

$$\mathbf{P}\{T_2 < t | X(0) = i\} = \mathbf{P}_i\{T_2 < t\}$$

$$= \sum_{k \neq i} P_i\{T_2 < t | T_1 < t, X(t_1) = k\} P_i\{ \overbrace{T_1 < t}^{(1 - e^{-v_i t})}, \overbrace{X(t_1) = k}^{P_{ik}} \}$$

$$\leq \sum_{k \neq i} (1 - e^{-v_k t}) \cdot \underbrace{(1 - e^{-v_i t})}_{\text{not dependent on } k} P_{ik}$$

Thus

$$\frac{P_i(T_2 < t)}{t} \leq \underbrace{\frac{1 - e^{-v_i t}}{t}}_{\text{converges to } v_i \text{ as } t \to 0} \sum_{k \neq i} (1 - e^{-v_k t}) P_{ik}$$

Using the dominated convergence theorem with the sum in the right-hand side (each term is bounded by 1), we get

$$\lim_{t \to 0} \frac{P_i(T_2 < t)}{t} = 0 \quad \Rightarrow P_i(T_2 < t) \sim o(t)$$

Thus, the chances of two jumps in an infinitesimal interval are small. This is not surprising since the underlying process counting the number of jumps occurring by time t is a Poisson process.

Furthermore, we can continue as

$$|P_{ij}(t) - P_i\{T_1 < t, X(T_1) = j\}| \leq P_i(T_2 < t)$$

Dividing both sides by t, and then taking the limit as $t \to 0$, we get

$$\left| \frac{P_{ij}(t)}{t} - \frac{P_i(T_1 < t, X(T_1) = j)}{t} \right| \leq \frac{P_i(T_2 < t)}{t} = o(t),$$

and so

$$\lim_{t \to 0} \frac{P_{ij}(t)}{t} = \lim_{t \to 0} \frac{P_i(T_1 < t, X(T_1) = j)}{t} = \lim_{t \to 0} \frac{(1 - e^{-v_i t}) P_{ij}}{t} = v_i P_{ij} = q_{ij}$$

Finally, we show that (13.3) implies (13.2). We know that

$$1 - P_{ii}(t) = P\{X(t) \neq i | X(0) = i\}$$
$$= \sum_{j \neq i} P\{X(t) = j | X(0) = i\} = \sum_{j \neq i} P_{ij}(t)$$

Dividing both sides by t and using (13.3), we get

$$\frac{1 - P_{ii}(t)}{t} = \sum_{j \neq i} \frac{P_{ij}(t)}{t} \to \sum_{j \neq i} q_{ij} = \sum_{j \neq i} v_i P_{ij} = v_i,$$

since $P_{ii} = 0$ ∎

Lemma 13.15 (Continuous-Time Chapman–Kolmogorov Equation) *For all $s, t > 0$, we may write*

$$P_{ij}(t + s) = \sum_k P_{ik}(t) P_{kj}(s)$$

This lemma is an immediate consequence of the definition of the transition probabilities.

13.3 The Kolmogorov Differential Equations

The goal of this section is to give a general methodology of finding the transition probabilities $P_{ij}(t)$ which we claimed completely characterize a continuous-time Markov chain. Note that these probabilities are functions of the time t elapsed between the two states. They will be expressed as solutions of a system of differential equations in t.

Theorem 13.16 (Kolmogorov's Backward Equation) *For all $i, j \in S$, and $t \geq 0$,*

$$\frac{dP_{ij}}{dt}(t) = \sum_{k \neq i} q_{ik} P_{kj}(t) - v_i P_{ij}(t)$$

where q_{ik} is the transition rate from i to k and $P_{kj}(t)$ is the probability that the chain moves from k to j in t units of time.

If we denote a matrix $R = (R_{ik})_{ik}$ by

$$R_{ik} = \begin{cases} q_{ik} & \text{if } i \neq k \\ -v_i & \text{if } i = k \end{cases}$$

then we can write the backward Kolmogorov equation in matrix form as

$$P'(t) = RP(t)$$

where $P'(t) = \left[\frac{dP_{ij}}{dt}(t)\right]_{i,j}$ is a matrix notation.

Theorem 13.17 (Kolmogorov's Forward Equation) *Under certain regularity conditions*[3]

$$P'_{ij}(t) = \sum_{k \neq j} q_{kj} P_{ik}(t) - v_j P_{ij}(t),$$

or in matrix notation

$$P'(t) = P(t)R$$

Proof (Proof of the Kolmogorov theorems): Both proofs are similar. The idea is to apply the Chapman-Kolmogorov equations for an interval $[0, t + h]$ made of a tiny interval of length h and a long interval of length t. The difference in the two results is where the tiny interval is situated: at the beginning for the backward differential equation and at the end for the forward differential equation.

For the backward version we have

$$P_{ij}(t + h) = \sum_k P_{ik}(h) P_{kj}(t) \tag{13.4}$$

Thus

$$P_{ij}(t + h) - P_{ij}(t) = \sum_{k \neq i} P_{ik}(h) P_{kj}(t) - (1 - P_{ii}(h)) P_{ij}(t)$$

Dividing by h and taking the limit as $h \to 0$, we have

$$P'_{ij}(t) = \lim_{h \to 0} \sum_{k \neq i} \frac{P_{ik}(h)}{h} P_{kj}(t) - \lim_{h \to 0} \frac{(1 - P_{ii}(h))}{h} P_{ij}(t)$$

The second term is no problem, and it produces exactly the needed result. However, we need to change the order of the limit and summation for the first term and that is not always possible. In this case where we can follow the argument in Lemma 13.14, and we have for an N large enough

$$\sum_{k > N} \frac{P_{ik}(h)}{h} \leq \frac{P_i\{T_1 < h, X(T_1) > N\}}{h} + \frac{P_i\{T_2 < h\}}{h}$$

The second term goes to zero and the first term is less than

$$\frac{1 - e^{-v_i h}}{h} \sum_{k > N} P_{ik}.$$

---

[3] The precise statement of these conditions is beyond the purpose of this chapter. Their exact place where these conditions are needed is pointed out in the proof. We also note that for us having a chain with a finite number of states is sufficient.

The first term converges to v_i, and the sum can be made arbitrarily small (recall that $\sum P_i k = 1$) and thus we can apply the bounded convergence theorem to conclude the proof of the theorem.

It is at this precise point where the proof of the forward Kolmogorov equation breaks down and this is the reason why we need some regularity conditions for the forward equation to ensure that the limit exists and is finite.

Specifically for the forward equation, we repeat the arguments above and we have in this case

$$P_{ij}(t+h) = \sum_k P_{ik}(t)P_{kj}(h) \tag{13.5}$$

Thus

$$P_{ij}(t+h) - P_{ij}(t) = \sum_{k \neq j} P_{ik}(t)P_{kj}(h) - P_{ij}(t)(1 - P_{jj}(h))$$

Dividing by h and taking the limit as $h \to 0$, we obtain

$$P'_{ij}(t) = \lim_{h \to 0} \sum_{k \neq i} \frac{P_{kj}(h)}{h} P_{ik}(t) - P_{ij}(t) \lim_{h \to 0} \frac{(1 - P_{jj}(h))}{h}$$

If the limit and the sum in the first term can be interchanged, then we obtain the forward equation. However, this is not always permitted. If we introduce regularity conditions to make sure that we can always change the limit and the sum, then the equation exists.

We do note, however, that in the case when the state space S is a finite set (e.g., $\{1, 2, \ldots, n\}$), the sum contains a finite number of elements and therefore exchanging the order of the summation and the limit is possible.   ∎

▣ EXAMPLE 13.3

Consider a simple system consisting of only two states 0 and 1, where the distribution of stay in state 0 is exponential with rate λ, and the distribution of the time spent in state 1 is also exponential with rate μ.

$$\underbrace{0}_{Exp(\lambda)} \longleftrightarrow \underbrace{1}_{Exp(\mu)}$$

For example, this represents a queuing system where the system alternates between empty (state 0) and busy (state 1).

We want to find the transition probabilities $P_{ij}(t)$ for this chain.

Proof (Solution): Let us calculate the transition probabilities by solving the Kolmogorov equations. For example, we use the forward equations. Note we can do this since in this case the state space $S = \{0, 1\}$ is finite.

Forward Equation

$$P'_{11}(t) = q_{01}P_{10}(t) - \underbrace{v_1}_{=\mu} P_{11}(t)$$

$$q_{01} = v_0 P_{01} = \lambda P_{01} = \lambda \cdot 1 = \lambda$$

$$\Rightarrow P'_{11}(t) = \lambda P_{10}(t) - \mu P_{11}(t)$$

Now, since $P_{10}(t) + P_{11}(t) = 1$,

$$P'_{11}(t) = \lambda(1 - P_{11}(t)) - \mu P_{11}(t)$$
$$= \lambda - (\lambda + \mu)P_{11}(t)$$

Denote $P_{11}(t) = y(t)$, then multiply both side by $e^{(\lambda+\mu)t}$. After reordering, we obtain the equation

$$y'(t)e^{(\lambda+\mu)t} + y(t)(\lambda + \mu)e^{(\lambda+\mu)t} = \lambda e^{(\lambda+\mu)t}$$

$$\frac{d}{dt}\left(e^{(\lambda+\mu)t}y(t)\right) = \lambda e^{(\lambda+\mu)t}$$

$$e^{(\lambda+\mu)t}y(t) = \frac{\lambda}{\lambda+\mu}e^{(\lambda+\mu)t} + c$$

$$y(t) = \frac{\lambda}{\lambda+\mu} + ce^{-(\lambda+\mu)t}$$

Note that

$$1 = P_{11}(0) = y(0) = \frac{\lambda}{\lambda+\mu} + c \quad \Rightarrow c = \frac{\mu}{\lambda+\mu}$$

which gives the solution

$$P_{11}(t) = \frac{\lambda}{\lambda+\mu} + \frac{\mu}{\lambda+\mu}e^{-(\lambda+\mu)t}$$

We can find P_{10} next using the fact that $P_{10}(t) + P_{11}(t) = 1$.

If we use of the backward Kolmogorov equations, we can find P_{01} and P_{00}. See Exercise 13.7 on page 433. ∎

The next three examples are quite classical. We only present them here, and we give the Kolmogorov equations. They are similar to the discrete-time Markov chains but the transitions are now being made at random times. For in depth details about their application, just use any book which describes population dynamics, for example, Renshaw (2011).

◨ EXAMPLE 13.4   The Birth–death Process

Let the state space $S = \{0, 1, 2 \ldots\}$. We define

$$\begin{cases} q_{i,i+1} = \lambda_i & \text{"a birth"} \\ q_{i,i-1} = \mu_i & \text{"a death"} \\ v_i = \lambda_i + \mu_i \end{cases}$$

In addition, we define $v_0 = \lambda_0$, $q_{0,-1} = \mu_0 = 0$.

Note that in this case the birth and the death rates depend on the population size i. This is very realistic, but complicates the differential equations presented next.

Backward Equation

$$P'_{ij}(t) = \mu_i P_{i-1,j}(t) + \lambda_i P_{i+1,j}(t) - (\lambda_i + \mu_i) P_{ij}(t)$$

where $P_{-1,0}(t) = 0$. (Special case $i = 0$)

Forward Equation

$$P'_{ij}(t) = P_{i,j-1}(t)\lambda_{j-1} + P_{i,i+1}(t)\mu_{j+1} - P_{ij}(t)(\mu_j + \lambda_j)$$

where $P_{0,-1}(t) = 0$. (Special case $j = 0$)

In general, there are no analytic solutions to these equations. Also, we need to solve systems of ordinary differential equations (ODEs) numerically. However, in the particular cases presented next, we can find analytic solutions.

▦ EXAMPLE 13.5   Pure birth process

This is a particular case of the previous example of a population where only births occur and no individual dies. More precisely, $\mu_i = 0$, $\forall i$. The forward equations become

$$P'_{ij}(t) = P_{i,j-1}(t)\lambda_{j-1} - P_{ij}(t)\lambda_j$$

If $i = j$, we obtain an equation that can be solved analytically.

$$P'_{ii}(t) = \underbrace{P_{i,i-1}(t)}_{=0 \text{ (pure birth)}} \cdot \lambda_{i-1} - P_{ii}(t)\lambda_i$$

$$= -P_{ii}(t)\lambda_i \qquad \Rightarrow P_{ii}(t) = e^{-\lambda_i t}$$

This is quite easy to reason as well, since we never come back to i again and the amount of time we spend in state i is exponential with the rate λ_i.

■ **EXAMPLE 13.6   Yule Process**

This is an even more specialized process: a pure birth process where $\lambda_j = j \cdot \lambda$. This particular birth rate appears if we assume that each individual reproduces at the same rate and independently of all others. In this case, all the i individuals present in the population at time t have the same probability λdt of producing one offspring in the next dt time. This will result in the particular choice of the birth rate.

13.4   Calculating Transition Probabilities for a Markov Process. General Approach

Recall: We defined the matrix $R = (R_{ij})_{ij}$, which contains the transition rates as

$$R_{ij} = \begin{cases} q_{ij} & \text{if } i \neq j \\ -v_i & \text{if } i = j \end{cases}$$

Using this notation, we rewrite

- Backward equation: $P'(t) = R \cdot P(t)$,

- Forward equation: $P'(t) = P(t) \cdot R$.

Note that either of these is a matrix notation for a system of first-order ODE's. Solving such systems of ODE's has strong connections with linear algebra.

Suppose we use a first-order approximation for backward equation:

$$P(t + h) - P(t) \sim RP(t)h$$
$$P(t + h) \sim (I + Rh)P(t)$$

Similarly, for the forward equation we obtain

$$P(t + h) \sim P(t)(I + Rh)$$

; in both expressions I denotes the Identity matrix. Therefore, $P(t)$ is the inverse matrix of $I + Rh$, and we can use the same reasoning as in Section 12.6 on page 396 but in the reverse direction.

We also note that

$$\lim_{n \to \infty} \left(I + R\frac{t}{n} \right)^n = \sum_{i=0}^{\infty} \frac{(Rt)^i}{i!} = e^{Rt}$$

Skipping linear algebra details, when t is large (or h is small), we can write $h = t/n$ for some n large and thus obtain

$$P(t) \sim \left(I + R\frac{t}{n} \right)^n$$

This expression provides an approximation for the transition probability matrix when n is large. However, this is not useful if t is small, and that is in fact the interesting case. Most of the times we want to know what will happen with our Markov process in a short time span.

Fortunately, in the case where the state space $S = \{1, 2, \ldots, N\}$ is finite, we can reduce the formulae above further. We first find the eigenvalues $\lambda_1, \ldots, \lambda_N$ of the matrix R and their corresponding eigenvectors $u_1, \ldots, u_N$. Then, we construct the following matrices:

$$U = \begin{pmatrix} u_1 & \cdots & u_N \end{pmatrix}$$

where the columns are the eigenvectors, and

$$D = \begin{pmatrix} e^{\lambda_1 t} & 0 & 0 \\ 0 & e^{\lambda_2 t} & 0 \\ & & \vdots \\ 0 & \cdots & e^{\lambda_n t} \end{pmatrix}$$

Then we can write exactly

$$P(t) = UDU^{-1} = UDU^T$$

Therefore we have a potential expression for the transition probabilities of the process in time t.

13.5   Limiting Probabilities for the Continuous-Time Markov Chain

As we have seen, it is possible to obtain exact probabilities in only a few cases. However, can we at least find limiting values of these probabilities? If we can, then we at least can say something about the behavior of the chain at some time in the distant future.

This question has an immediate first answer. Since a continuous-time Markov chain is a special case of the semi-Markov process, we can certainly apply the limiting theorem we have for that case. This is presented in the next proposition.

Proposition 13.18 *If μ_{00}, the expected length of a 0-to-0 cycle, is finite for a continuous-time Markov chain, then*

$$P_{ij}(t) \to P_j = \frac{\eta_j / v_j}{\sum_k \eta_k / v_k},$$

where η_j denotes the expected number of visits to j during a 0-to-0 cycle.

However, recall that the continuous-time Markov chain is a more specialized process than the semi-Markov process. Due to this, we have a more powerful result when the process is positive recurrent.

Proposition 13.19 (Balance equations) *Suppose that a continuous-time, irreducible Markov chain is positive recurrent ($\mu_{00} < \infty$). Then it has a unique stationary probability distribution P_j solving*

$$v_i P_i = \sum_{j \neq i} v_j P_j P_{ji} = \sum_{j \neq i} P_j q_{ji}. \tag{13.6}$$

Furthermore,

$$\lim_{t \to \infty} P_{ij}(t) = P_j.$$

Note that $v_i P_i$ represents the long-term rate of leaving the state i. Also note that $P_j q_{ji}$ represents the long-term rate of going from j to i. Therefore, the sum $\sum_{j \neq i} P_j q_{ji}$ represents the long-term rate of entering the state i. This is the reason why the equation is called the balance equation since it equates the two rates. Intuitively since the chain is positive recurrent i is visited infinitely often and for every transition out of i there must be a subsequent transition into i with probability 1.

Proof:

To prove the theorem, recall that an irreducible *discrete-time* Markov chain has a unique probability measure η given by $\eta_i = \sum_j \eta_j P_{ji}$. Therefore, by Proposition 13.18, the limit of $P_{ij}(t)$ exists for all i, j, when $t \to \infty$ and

$$P_{ij}(t) \to P_j = \frac{\eta_j / v_j}{\sum_k \eta_k / v_k},$$

and the P_j's sum to 1. Using the Kolmogorov forward equations, we have

$$P'_{ji}(t) = \sum_{k \neq i} q_{ki} P_{jk}(t) - v_i P_{ji}(t)$$

But since the limit $P_{ji}(t)$ exists when $t \to \infty$, then we must have $P'_{ji}(t)$ going to 0 as $t \to \infty$. The function $P_{ji}(t)$ converges to a constant, and therefore the derivative of the function must converge to 0. Taking the limit at $t \to \infty$ in the Kolmogorov forward equations, and changing the limit and summation, we obtain

$$0 = \sum_{k \neq i} q_{ki} P_k - v_i P_i.$$

Rearranging the terms finalizes the proof. ∎

Theorem 13.20 *If $P_j = P_j^{(k)} = \lim_{t \to \infty} P_{kj}(t)$ exists **for any** j, then*

$$P_j = \sum_i P_i P_{ij}(s), \quad \forall s > 0.$$

Remark 13.21 *Even though the limit exists in the theorem above, it is not always true that $\sum_j P_j = 1$. Furthermore, applying renewal theory in the case where the expected renewal times is infinite ($\mu = \infty$), we may see that $\lim_{t \to \infty} P_{ij}(t) = 0$ for* *any irreducible null recurrent or transient continuous-time Markov chain (thus the stationary probability does not exist for these chains).*

Definition 13.22 (Ergodic Markov process) *When the continuous-time Markov chain is irreducible and has limiting probabilities $P_j = \lim_{t \to \infty} P_{kj}(t)$ which are strictly positive, $P_j > 0$, and $\sum_j P_j = 1$, the Markov process is called ergodic.*

EXAMPLE 13.7   The Birth–death Process – continued

Let us continue with Example 13.4. Recall that the state space of the process is $S = \{0, 1, 2 \ldots\}$, and

$$\begin{cases} q_{i,i+1} = \lambda_i & \text{"a birth"} \\ q_{i,i-1} = \mu_i & \text{"a death"} \\ v_i = \lambda_i + \mu_i \end{cases}$$

In addition, we have $v_0 = \lambda_0$, $q_{0,-1} = \mu_0 = 0$. Let us calculate the limiting probabilities for this process. Applying Proposition 13.19, the chain is positive recurrent if we have for state 0

$$\lambda_0 P_0 = \mu_1 P_1,$$

and in general for state i

$$(\lambda_i + \mu_i) P_i = \mu_{i+1} P_{i+1} + \lambda_{i-1} P_{i-1}.$$

Expressing everything in terms of P_0 gives

$$P_1 = \frac{\lambda_0}{\mu_1} P_0$$

$$P_2 = \frac{1}{\mu_2} (\lambda_1 P_1 + \mu_1 P_1 - \lambda_0 P_0) = \frac{\lambda_1}{\mu_2} P_1 = \frac{\lambda_1 \lambda_0}{\mu_2 \mu_1} P_0$$

$$P_3 = \cdots = \frac{\lambda_2}{\mu_3} P_2 = \frac{\lambda_2 \lambda_1 \lambda_0}{\mu_3 \mu_2 \mu_1} P_0$$

$$\ldots \ldots$$

Using the sum of P_i's equaling to 1, we obtain

$$P_0 = \frac{1}{1 + \sum_{n=1}^{\infty} \frac{\lambda_0 \lambda_1 \ldots \lambda_{n-1}}{\mu_1 \mu_2 \ldots \mu_n}},$$

and

$$P_i = \frac{\lambda_0 \lambda_1 \ldots \lambda_{i-1}}{\mu_1 \mu_2 \ldots \mu_i} \frac{1}{1 + \sum_{n=1}^{\infty} \frac{\lambda_0 \lambda_1 \ldots \lambda_{n-1}}{\mu_1 \mu_2 \ldots \mu_n}}.$$

These expressions also give the condition for the limiting probabilities to exist. Specifically, we must have

$$\sum_{n=1}^{\infty} \frac{\lambda_0 \lambda_1 \dots \lambda_{n-1}}{\mu_1 \mu_2 \dots \mu_n} < \infty$$

EXAMPLE 13.8   An M/M/1 queue

In this type of queuing process, the customers arrive according to a Poisson process with constant rate λ and there is one server which takes an exponential time with constant rate μ to service each client. Thus this process can be written as a Markov chain. More specifically, this is a birth–death chain where the transition rates are $\lambda_n = \lambda$ and $\mu_n = \mu$. Simply applying the result in the previous exercise, we obtain

$$P_i = \left(\frac{\lambda}{\mu}\right)^i \frac{1}{1 + \sum_{n=1}^{\infty} \left(\frac{\lambda}{\mu}\right)^n}.$$

Clearly, this probability exists only if $\lambda < \mu$, and in this case we have

$$P_i = \left(\frac{\lambda}{\mu}\right)^i \left(1 - \frac{\lambda}{\mu}\right) = l^i(1 - l).$$

In the expression above, we denoted $\frac{\lambda}{\mu}$ by l the load of the system. This is easily understood if we think about the system's evolution. If the customers arrive at a rate λ larger than the service rate μ, then the system will overload and the queue will become infinite.

13.6 Reversible Markov Process

Reversibility is a general property, as explained next. We attach this notion to the Markov processes since in this case it leads to easy calculation for transition probabilities.

Definition 13.23 *A stochastic process X_t is time-reversible if the joint distribution of $(X_{t_1}, X_{t_2} \dots, X_{t_n})$ is the same as the joint distribution of $(X_{t-t_1}, X_{t-t_2} \dots, X_{t-t_n})$, for all times $t_1, \dots, t_n$, and t.*

Since the choice of t can be anything, we can easily prove the following lemma.

Lemma 13.24 *Any reversible process is stationary.*

In the case of the Markov processes, we can give simple conditions on the transition rates to ensure that the resulting process is reversible.

Theorem 13.25 (Reversible Markov process) *A stationary Markov chain is reversible if and only if there exist positive numbers P_i for all states i in the state space of the process such that the detailed balance equations hold:*

$$P_j q_{ji} = P_i q_{ij} \tag{13.7}$$

Such a collection of P_i's form a stationary distribution for the Markov process.

Proof: Assume first that the process is reversible. Since the process is stationary, $\mathbf{P}(X_t = i)$ does not depend on t, and let P_i denote this probability. We clearly have this probabilities sum to 1, and we can write by the reversibility

$$\mathbf{P}(X_t = i, X_{t+\tau} = j) = \mathbf{P}(X_{\tau+2t-t} = i, X_{\tau+2t-(t+\tau)} = j)$$
$$= \mathbf{P}(X_{t+\tau} = i, X_t = j)$$

Therefore,

$$\mathbf{P}(X_{t+\tau} = j \mid X_t = i)\mathbf{P}(X_t = i) = \mathbf{P}(X_{t+\tau} = i \mid X_t = j)\mathbf{P}(X_t = j)$$
$$\frac{\mathbf{P}(X_{t+\tau} = j \mid X_t = i)}{\tau} P_i = \frac{\mathbf{P}(X_{t+\tau} = i \mid X_t = j)}{\tau} P_j$$

and taking the limit as $\tau \to 0$, we obtain the relationship stated in the theorem.

Conversely, suppose that Equation (13.7) holds for some numbers P_i. Summing this equation over $j \neq i$, we will obtain the balance Equation (13.6), therefore the P_i's are a stationary distribution. To show reversibility, consider the behavior of the process for some interval $[-T, T]$. Suppose we start at $-T$ in state i_1, stay there a time h_1, then jump to state i_2, stay there a time h_2, then keep jumping until time T when we are in state i_m and have been there for some period h_m. Since the process is Markov, we can write the density of this probability as

$$P_{i_1} e^{-v_{i_1} h_1} q_{i_1 i_2} e^{-v_{i_2} h_2} \cdots q_{i_{m-1} i_m} e^{-v_{i_m} h_m},$$

where the density is a function of the h_j's.

To obtain the actual probability, we need to integrate over $h_1 + \ldots h_m = 2T$. The hypothesis equation (13.7) applied successively implies that

$$P_{i_1} q_{i_1 i_2} \cdots q_{i_{m-1} i_m} = P_{i_m} q_{i_m i_{m-1}} \cdots q_{i_2 i_1},$$

and therefore the density is equal to the probability density that the system starts in state i_m at time $t = -T$, remains for a time h_m there, then jumps to i_{m-1}, and so on. Therefore, the probabilistic behavior of X_t is the same as the behavior of X_{-t} on the interval $[-T, T]$ for all T. Since the t_i's and h_i's are arbitrary, the distribution of $(X_{t_1}, \ldots, X_{t_m})$ is the same as the distribution of $(X_{-t_1}, \ldots, X_{-t_m})$. However, by stationarity, the latter vector has the same distribution as $(X_{\tau-t_1}, \ldots, X_{\tau-t_m})$, which concludes the proof that the process is reversible. ∎

The balance equations can be extended to any collection of states.

Lemma 13.26 *For a reversible Markov process with state space S, let $\mathcal{A}$ be a subset of states. Then we have*

$$\sum_{i \in \mathcal{A}} \sum_{j \in \mathcal{A}^c} P_i q_{ij} = \sum_{i \in \mathcal{A}} \sum_{j \in \mathcal{A}^c} P_j q_{ji},$$

where $\mathcal{A}^c$ is the complement of $\mathcal{A}$ with respect to the set S, i.e., $\mathcal{A}^c = S \backslash \mathcal{A}$. Therefore, the probability of flow into subset $\mathcal{A}$ is the same as that of going out of $\mathcal{A}$.

Proof: The proof is immediate by summing the balance equations (13.6) for all $i \in \mathcal{A}$ and then subtracting

$$\sum_{i \in \mathcal{A}} \sum_{j \in \mathcal{A}} P_j q_{ji} = \sum_{i \in \mathcal{A}} \sum_{j \in \mathcal{A}} P_i q_{ij},$$

which is a consequence of the reversibility condition (13.7). ∎

◼ EXAMPLE 13.9   Birth–Death process

Any ergodic birth–death process is in a steady state time-reversible.

To show this, we need to prove that the rate at which the process goes from i to $i+1$ is equal to the rate at which it goes from $i+1$ to i. But in Example 13.7 we already calculated the stationary probabilities P_i and found the condition under which they exist. Even if the condition does not hold, we can take

$$P_i = \frac{\lambda_0 \lambda_1 \ldots \lambda_{i-1}}{\mu_1 \mu_2 \ldots \mu_i}$$

and show that for these numbers the reversibility condition (13.7) holds:

$$P_{i+1} q_{i+1,i} = P_i q_{i,i+1}$$

or

$$\frac{\lambda_0 \lambda_1 \ldots \lambda_i}{\mu_1 \mu_2 \ldots \mu_{i+1}} \mu_{i+1} = \frac{\lambda_0 \lambda_1 \ldots \lambda_{i-1}}{\mu_1 \mu_2 \ldots \mu_i} \lambda_i,$$

Therefore, the birth–death process is always time reversible.

Lemma 13.27 (Truncated Markov process) *Consider a time-reversible Markov process with limiting probabilities P_i for all $i \in S$. We truncate the process to a set of states $\mathcal{A} \in S$ by setting all transition rates $q_{ij} = 0$ for all $i \in \mathcal{A}$ and $j \in S \backslash \mathcal{A}$. All other transition rates stay the same. In other words, once the Markov process enters any state in $\mathcal{A}$, it cannot leave set $\mathcal{A}$. The resulting process restricted to the states in $\mathcal{A}$ is also time reversible and has limiting probabilities*

$$P_i^{\mathcal{A}} = \frac{P_i}{\sum_{i \in \mathcal{A}} P_i},$$

or what one would expect as conditional probabilities.

The reversibility conditions (13.7) hold for the probabilities of the truncated process. This is immediate since they hold for the original process.

■ **EXAMPLE 13.10   A queue system with limited queue size**

Suppose we have an $M/M/1/N$ queue. That is, the customers arrive according to a Poisson process with rate λ, there is one server with processing time exponential with rate μ, and there are only N places in the waiting queue. Any customer arriving when there are already N customers in the queue leaves rather than wait. Let X_t be the number of customers in the system at time t.

This is an example of a truncated Markov process where states $\{1, 2, \dots, N\}$ are absorbing. Recall that in Example 13.8 we calculated the limiting probabilities for an M/M/1 queue. We can use the lemma above to calculate the limiting probabilities as

$$P_i = \frac{\left(\frac{\lambda}{\mu}\right)^i}{\sum_{i=0}^{N} \left(\frac{\lambda}{\mu}\right)^i}$$

· for all i in $\{1, \dots, N\}$ and 0 otherwise.

Problems

13.1   Consider a sequence of Bernoulli(1/2) random variables $X_0, X_1, \dots, X_n, \dots$. Construct a process $Y_n = X_{n-1} + 2X_n + 3X_{n+1}$.

1. Calculate $\mathbf{P}(Y_0 = 1, Y_1 = 3, Y_2 = 2)$ and $\mathbf{P}(Y_1 = 3, Y_2 = 2)$.

2. Calculate $\mathbf{P}(Y_2 = 2 \mid Y_0 = 1, Y_1 = 3)$ and $\mathbf{P}(Y_2 = 2 \mid Y_1 = 3)$.

3. Does the process Y_n have the Markov property? Argue by using the answers to the previous two parts.

13.2   Consider a sequence of Bernoulli(1/2) random variables $X_0, X_1, \dots, X_n, \dots$. Construct a process $Y_n = X_n - X_{n-1}$.

1. Does the process Y_n have the Markov property?

2. Define the two-dimensional process $Z_n = (Y_n, X_{n-1})$. Show that this two-dimensional process has the Markov property.

13.3   Suppose a semi-Markov process has two states $S = \{1, 2\}$. Assume that the distributions of times spent in each state are the cdfs of the following distributions:

$$F_{11}(t) \sim Uniform[0, 1]; \quad F_{12}(t) \sim Uniform[0, 3];$$
$$F_{21}(t) \sim Uniform[1, 3]; \quad F_{22}(t) \sim Uniform[0, 2].$$

Calculate the transition rates for this continuous-time process. Calculate the expected time spent in state 1 and state 2.

This semi-Markov process can be made equivalent to a semi-Markov process that transitions from 1 to 2 and 2 to 1 only. Calculate the cdf of the transition times for such a Markov process.

13.4 Show that the continuous-time Markov chain defined in Example 13.2 on page 417 is nonregular.

13.5 Prove Proposition 13.11 on page 418.

13.6 Prove Lemma 13.15.

13.7 Use the backward Kolmogorov equations to find to find P_{01} and P_{00} in Example 13.3.

13.8 Consider the Yule process presented in Example 13.6 on page 425. Calculate the transition probabilities $P_{ii}(t)$ and $P_{37}(0.5)$.

13.9 Consider a Markov process with state space $S = \{0, 1, 2, \dots\}$ and transition rates $q_{i,i+1} = a^i$, $q_{i,0} = b$, and $q_{i,k} = 0$ in all other cases. Show the following:
  a) When $a > 1$, this process is not regular.
  b) Find the stationary distribution when $a \leq 1$ and $b > 0$.
  c) Show that the stationary distribution does not exist when $a > 0$ and $b = 0$.

13.10 Consider an M/M/1/c queue. This is a queuing process where arrivals are according to a Poisson process with rate λ the service is exponentially distributed with rate μ, there is one server, and the queue capacity is c. Write this queue as a Markov chain in continuous time. Calculate the stationary distribution for the chain and write the probabilities in term of the load $l = \lambda/\mu$.

13.11 Suppose a shop has three machines and a repairman. Machines are breaking down independently with rate $\lambda = 3$ per week and the rate at which the repairman fixes any given machine is constant $\mu = 1$ per day. Construct a Markov process X_t which counts the number of machines broken down at time t. Show that this system can be modeled as a birth–death process. Calculate the parameters of the birth and death process. Use Proposition 13.19 to calculate the limiting probability that i machines will not be in use for i equals to 1, 2, and 3. Calculate the proportion of time for which all three machines are functioning at the same time.

13.12 Everyone has seen this situation. You chose one line and see that the other lines next to the line you chose are moving faster. Yet other times, lines are combined into one and the next customer in line can go to any server. In this exercise, we will compare the two situations using a very simple example. Suppose customers are arriving according to a $Poisson(\lambda)$ process and there are two servers each acting independently and the time to process a customer is exponential with rate μ.
  a) Suppose that there are two separate lines. Each person arriving according to the Poisson process chooses a line at random (with probability $1/2$). Once the customer chooses a line, he/she cannot change the line. What is the expected total number of people waiting in the line. Express the result in terms of the load $l = \lambda/\mu$.

b) Suppose that there is one common line and, as soon as one of the servers become available, the next person in the common line goes to that server. Again, calculate the expected number of people in the line in terms of l.

c) Compare the two answers. Which solution is better?

13.13   Consider a population in which each individual independently gives birth at an exponential rate λ and dies at an exponential rate μ. What kind of process is this? What are its parameters? Find $\mathbf{E}[X_t \mid X_0 = n]$.

13.14   Consider a population as in the previous problem, but in addition new citizens immigrate according to a Poisson process with a rate θ. This process is called a *birth and death process with immigration*. Since each immigrant arrives alone, show that this process is equivalent to a birth–death process. Calculate $\mathbf{E}[X_t \mid X_0 = n]$.

13.15   Consider a birth and death model with immigration as in the previous problem. However, assume that once the population has reached a size N immigration is forbidden. Set this process as a birth–death model. Suppose $N = 6$, $\lambda = \theta = 2$, and $\mu = 4$. Calculate the proportion of time that immigration is not allowed.

13.16   Let X_t denote a Markov process with two states 1 and 2, and let $q_{12} = 1$ and $q_{21} = 1/2$. Show that the Markov process is reversible. Calculate the limiting probabilities P_1 and P_2.

13.17   Consider a discrete-time Markov chain with transition times in integers, with the following transition probability matrix:

$$P = \begin{pmatrix} 0 & 1 & 0 \\ \frac{3}{4} & 0 & \frac{1}{4} \\ 1 & 0 & 0 \end{pmatrix}$$

Show that the Markov chain is positive recurrent, and find its stationary distribution. Show that the Markov chain is not reversible. The equivalent reversibility equation to (13.7) for a discrete-time Markov chain is the existence of numbers π_i:

$$\pi_i P_{ij} = \pi_j P_{ji}.$$

13.18   Consider an M/M/s queue with s servers, where the customers arrive with rate λ and each server processes with rate μ independently of others. Show that the number of customers in the queue is a birth and death process. Show that its stationary distribution can be calculated using

$$P_i = P_0 \left(\frac{\lambda}{\mu}\right)^i \frac{1}{i!}, \quad \text{for } i \in \{1, 2, \ldots, s\}$$

$$P_i = P_s \left(\frac{\lambda}{s\mu}\right)^{i-s}, \quad \text{for } i \geq s$$

if $\lambda < s\mu$. Here, the ratio $l = \frac{\lambda}{s\mu}$ is the load of the queue. Show that, if a customer arrives when all s servers are busy, his waiting time is exponentially distributed with mean $\frac{1}{s\mu - \lambda}$

13.19 Suppose orders are coming to a market exchange according to a Poisson process with rate λ from high-frequency traders. However, the exchange can only process K orders at the same time, so an order sent when the system is processing K orders is lost. Furthermore, assume that the order processing time is an exponential random variable with rate μ (mean $1/\mu$). Show that the number of orders being processed at time t is a birth and death process with transition rates

$$q_{i,i-1} = i\mu, \text{ for } i \in \{1, 2, \ldots, K\}$$
$$q_{i,i+1} = \lambda, \text{ for } i \in \{0, 1, \ldots, K-1\}$$

Calculate the stationary distribution and the probability that an order is going to be lost (thus the long-term proportion of orders lost). This is called the *Erlang formula*.

13.20 Refer to the previous problem and assume that in fact the number of traders submitting orders is finite of size M. Thus, instead of growing at a constant rate, it makes more sense for the resulting chain to have

$$q_{i,i+1} = (M-i)\lambda, \text{ for } i \in \{0, 1, \ldots, K-1\}$$

so that each trader has at most one order in the queue waiting to be processed. Calculate the stationary distribution and the probability that an order is going to be lost in this situation.

CHAPTER 14

MARTINGALES

A martingale originally referred to a betting strategy. In eighteenth century France, especially due to Monaco's Monte Carlo and its casinos, it was not unusual for gamblers with a sure winning strategy (a martingale) to speak with probabilists and ask their opinion. The word itself was first mentioned in the thesis of Jean Ville (Ville, 1939), where it refers to a "gaming strategy" or martingale (for an entire etymology of the word, see Mansuy (2009)). Joseph Doob, who wrote a classic Doob (1953) in which martingales were established as one of the important groups of stochastic processes, later explains (Snell, 1997) that after reviewing the book by Ville in 1939 he was intrigued by the work and after formalizing the definition he kept the name.

Martingales may be defined as processes in continuous time or in discrete time. In this chapter, we consider only discrete-time martingales. In general, continuous-time martingales have the same basic properties as the discrete-time ones. The differences between discrete-time and continuous-time martingales come from analyzing the situation when two times s and t are very close to each other. We shall talk more about the differences born out of this issue later in this chapter. For a complete and wonderful reference, consult Karatzas and Shreve (1991), and for the notions of local martingales and semimartingales, look at Protter (2003).

Probability and Stochastic Processes, First Edition. Ionuţ Florescu

14.1 Definition and Examples

Definition 14.1 (Discrete-time martingale) *Let* $(\Omega, \mathscr{F}, \{\mathscr{F}_n\}, \mathbf{P})$ *be a filtered probability space. Recall the definition of the filtration (Definition 9.2 on page 294). A process* $X = \{X_n\}$ *defined on this space is called a* **martingale** *with respect to the filtration* $\{\mathscr{F}_n\}$ *and probability measure* $\mathbf{P}$ *if and only if*

(i) $X = \{X_n\}$ *is adapted with respect to the filtration* $\{\mathscr{F}_n\}$,

(ii) $\mathbf{E}[|X_n|] < \infty$ *for all* n,

(iii) $\mathbf{E}[X_{n+1} \mid \mathscr{F}_n] = X_n$, *a.s. for all* $n \geq 0$.

Definition 14.2 (Submartingale, Supermartingale) *A process* $X = \{X_n\}$ *is called a* **submartingale** *relative to* $(\{\mathscr{F}_n\}, \mathbf{P})$ *if the property* (iii) *above is replaced with*

$$\mathbf{E}[X_{n+1} \mid \mathscr{F}_n] \geq X_n, \ a.s., \ \forall n \geq 0.$$

A process $X = \{X_n\}$ *is called a* **supermartingale** *relative to* $(\{\mathscr{F}_n\}, \mathbf{P})$ *if the property* (iii) *above is replaced with*

$$\mathbf{E}[X_{n+1} \mid \mathscr{F}_n] \leq X_n, \ a.s., \ \forall n \geq 0.$$

Note Since we can apply expectation to both sides in the inequalities above, note that for a submartingale the expectations increase ($\mathbf{E}[X_{n+1}] \geq \mathbf{E}[X_n]$) while for a supermartingale the expectations decrease ($\mathbf{E}[X_{n+1}] \leq \mathbf{E}[X_n]$). Thus a supermaringale decreases in value on average, while a submartingale increases. This is totally counterintuitive if we think that the word "martingale" was initially used for a game strategy. However, if we look at the strategy from the casino's perspective, then the names make sense.

Second note In cases when the filtration is not defined specifically and only the process is specified, we always use the standard filtration defined by the martingale in the sense of Definition 9.6 on page 295 (Chapter 9).

Third note Note that if X is a supermartingale, it is very easy to show that the process $-X = \{-X_n\}_n$ is a submartingale, and vice versa. Using this, results pertaining to supermartingales will also hold for submartingales with obvious changes (mostly the inequalities will change direction). Thus, we will concentrate on only one type for all results stated.

Furthermore, the process $X_n - X_0$ has the same properties as the process X_n and thus we will assume without loss of generality that all martingale-type processes start from $X_0 = 0$.

Finally, observe that $\mathbf{E}[X_n \mid \mathscr{F}_m] = X_m$ for any $m < n$ and not just consecutive ones.

$$\mathbf{E}[X_n \mid \mathscr{F}_m] = \mathbf{E}[\mathbf{E}[X_n \mid \mathscr{F}_{n-1}] \mid \mathscr{F}_m] = \mathbf{E}[X_{n-1} \mid \mathscr{F}_m] = \cdots = X_m$$

14.1.1 Examples of martingales

▣ EXAMPLE 14.1   Martingale sum of i.i.d. random variables

Consider $X_1, X_2 \ldots X_n \ldots$ i.i.d. with $\mathbf{E}[X_i] = 0, \forall i$. Define

$$Z_k = \sum_{i=1}^{k} X_i$$

Then $\{Z_n\}$ is a martingale.

The first two properties in Definition 14.1 are easily verified; for the third, we have

$$\mathbf{E}[Z_{n+1} \mid \mathscr{F}_n] = \mathbf{E}[X_{n+1} + Z_n \mid \mathscr{F}_n] = \mathbf{E}[X_{n+1} \mid \mathscr{F}_n] + \mathbf{E}[Z_n \mid \mathscr{F}_n] = Z_n$$

by independence of X_{n+1} of the sigma algebra $\mathscr{F}_n = \sigma(X_1, \ldots, X_n)$ and $\mathbf{E}[X_n] = 0$ as well as the measurability of Z_n with respect to the same $\mathscr{F}_n$.

▣ EXAMPLE 14.2   Martingale product of i.i.d. random variables

Consider $X_1, X_2 \ldots X_n$ i.i.d. random variables with $\mathbf{E}[X_i] = 1$. Let

$$Z_n = \prod_{i=1}^{n} X_i$$

Then, $\{Z_n\}$ is a martingale.

Again, the first two properties are easy to verify. To show the third, we use the elementary properties of conditional expectation. For example

$$\mathbf{E}[Z_{n+1} \mid \mathscr{F}_n] = \mathbf{E}\left[\prod_{i=1}^{n+1} X_n \,\middle|\, \mathscr{F}_n\right] = \mathbf{E}\left[\prod_{i=1}^{n} X_i \cdot X_{n+1} \,\middle|\, \mathscr{F}_n\right]$$

$$= \prod_{i=1}^{n} X_i \underbrace{\mathbf{E}[X_{n+1} \mid \mathscr{F}_n]}_{=\mathbf{E}[X_{n+1}]=1} = \prod_{i=1}^{n} X_i = Z_n$$

where we used measurability and independence properties, respectively

▣ EXAMPLE 14.3   Doob martingale

Consider $Y_1, Y_2 \ldots Y_n$ random variables, and let X be a random variable with $\mathbf{E}[|X|] < \infty$. Define $\mathscr{F}_n = \sigma(Y_1, \ldots, Y_n)$, the generated filtration, and let

$$Z_n = \mathbf{E}[X \mid \mathscr{F}_n]$$

Z_n is a martingale with respect to the filtration generated by Y_n (sometimes called the *Doob martingale* or the *Doob process*).

14.2   Martingales and Markov Chains

It is important to note that the two types of stochastic processes are very different. A martingale in essence has the same future conditional expectations given the present and the past. A Markov chain is a process whose entire future distribution depends only on the present (not the past). Also, its expectation may be completely different from the value today. Note, for example, that the martingales in Examples 14.1 and 14.2 do not have the Markov property. The process in Example 14.3 is both a Markov process and a martingale, but to change the martingale property just modify the definition of Z_n slightly, for example, $Z_n = \mathbf{E}[X + n \mid \mathscr{F}_n]$.

However, there are tight relations between the two notions of stochastic processes. Specifically, given a Markov chain, we can define a martingale process that depends on the original Markov chain.

14.2.1   Martingales induced by Markov chains

Suppose X_n is a Markov chain with state space $\mathcal{S} = \{1, 2, \ldots\}$ and transition probability matrix $P = \{P_{ij}\}_{i,j \in \{1,\ldots,n\}}$. Let $\mathscr{F} = \{\mathscr{F}_n\}_n$ denote the standard filtration induced by the Markov chain X_n.

14.2.1.1   *Martingale induced by right regular sequences.*   Suppose $f : \mathbb{Z} \to \mathbb{R}$ is some nonnegative, bounded function with the property

$$f(i) = \sum_j P_{ij} f(j) \tag{14.1}$$

Note that this is not the defining property of a stationary distribution. f would be the stationary distribution if P is replaced by its transpose.

Define $Y = \{Y_n\}_n$ the process with components

$$Y_n = f(X_n).$$

Then the process Y_n is a martingale. It is also a Markov chain but this is trivial.

Since X_n is generating $\mathscr{F}_n$, then clearly Y is adapted to this filtration. We have $\mathbf{E}[Y_n] = \mathbf{E}[f(X_n)] < \infty$ since f is bounded, so the first two defining properties for martingales (definition 14.1) are verified. We have

$$\mathbf{E}[Y_{n+1} \mid \mathcal{F}_n] = \mathbf{E}[f(X_{n+1}) \mid \mathcal{F}_n] = \mathbf{E}[f(X_{n+1}) \mid X_n]$$
$$= \sum_j f(j) P_{X_n j} \quad \text{(by definition of expectation)}$$
$$= f(X_n) \quad \text{(using (14.1))}$$
$$= Y_n$$

Definition (14.1) defines the function $f(\cdot)$ similar to the way a stationary distribution is defined. The function f is calculated in a similar way with the stationary distribution calculation. Many martingales appear in this way. In the next example, we give a more general construction.

14.2.1.2 *Martingale induced by eigenvectors of the transition matrix.* Review the eigenvector and eigenvalue decomposition of matrices. A brief summary of these and other linear algebra notions is given in Appendix A.2.

The characteristic equation to calculate the eigenvectors corresponding to eigenvalue λ may be written as

$$\sum_{j=1}^{n} P_{ij} f(j) = \lambda f(i), \quad \text{for all } i.$$

Observe that the function defined in equation (14.1) corresponds to the eigenvalue $\lambda = 1$ of the transition matrix P. Also note that since all the rows of the transition probability matrix sum to 1, $\lambda = 1$ is always an eigenvalue for any probability matrix.

Suppose that $\mathbf{f}$ is an eigenvector of the matrix P corresponding to the eigenvalue λ, that is,

$$\mathbf{f} = \begin{pmatrix} f(1) \\ f(2) \\ \vdots \\ f(n) \end{pmatrix}$$

Furthermore, suppose that $\mathbf{E}|f(X_n)| < \infty$ for all n. If the chain is finite dimensional, this is always going to be true. Finally, define

$$Y_n = \frac{f(X_n)}{\lambda^n}.$$

This process is a martingale. The proof is identical to that given for the eigenvalue 1 in the previous subsection.

■ EXAMPLE 14.4 Branching process as a martingale

As a special case of the example above, we may consider the branching process introduced in Section 11.7 on page 363. Using the same notation, let X_n denote the size of the nth generation in the branching process, and let m denote the mean number of offspring for each individual in the population. Then,

$$Z_n = \frac{X_n}{m^n}$$

is a martingale.

Again, the first two properties in the definition of a martingale are satisfied; for the last property, we have

$$\mathbf{E}[X_{n+1} \mid \mathscr{F}_n] = \mathbf{E}[X_{n+1} \mid X_n] = \mathbf{E}\left[\sum_{i=1}^{X_n} Y_i \mid X_n\right] = \sum_{i=1}^{X_n} \mathbf{E}[Y_i] = mX_n,$$

where Y_i is the random variable denoting the number of offspring of individual i in the population size X_n at time n. Dividing both ends with m^{n+1} shows that Z_n is a martingale.

🖿 EXAMPLE 14.5   Wald's martingale

Suppose that $X_1, X_2, \ldots$ are i.i.d. random variables with the moment generating function

$$M(\theta) = \mathbf{E}[e^{\theta X_n}].$$

Assume that for some $\theta \neq 0$ the moment generating function is finite. If we define the partial sums process

$$S_n = \sum_{i=1}^{n} X_i$$

with $S_0 = 0$, then we can construct the process

$$M_n = \frac{e^{\theta S_n}}{(M(\theta))^n},$$

which is called the Wald martingale. The process is a martingale with respect to the filtration generated by $\{X_n\}_n$ for any such θ.

This is easy to show, since the process M_t is a product of terms involving X_i's and the product increment has expectation 1 (see Example 14.2 above).

The example above also works with the characteristic function replacing the moment generating function. That would be more general since any random variable has a characteristic function.

As an application of such a construction, suppose the original X_n's are i.i.d. normal variables with mean 0 and some variance σ^2. Then, the moment generating function of X_i's is

$$M(\theta) = \mathbf{E}\left[e^{\theta X_1}\right] = e^{\frac{\theta^2 \sigma^2}{2}},$$

and this function is finite for any finite θ.

Therefore, the process

$$M_n = e^{\theta(X_1 + \cdots + X_n) - \frac{n}{2}\theta^2\sigma^2}$$

is a martingale for any θ that is real. This is a very useful martingale, called the *exponential martingale*. We shall observe this process again when we talk about Brownian motion.

14.3   Previsible Process. The Martingale Transform

We now introduce processes which at time n only depend on the information available at a previous time $n-1$.

Definition 14.3 (Previsible process) *A process $G = \{G_n\}_n$ is called previsible with respect to a filtration $\mathcal{F} = \{\mathcal{F}_n\}$ if*

$$G_n \text{ is } \mathcal{F}_{n-1} \text{ measurable, } \forall n \geq 1$$

A previsible process G is used in stochastic processes, and will be used extensively when we will talk about stochastic integration. As a simple example, consider a gambling strategy where G_n is the stake you have in the nth game. This stake has to be decided based on the history of the outcomes up to $n-1$ and not including the unknown gambling outcome n before the nth game is played. This is why the process is called previsible; the strategy has to be thought in advance. Also note that at time 0 the process is not defined, since there is no gaming history at that time.

Now suppose that the game outcome at every step is modeled by a stochastic process $X = X_n$ adapted to the filtration. For example, in a game with two outcomes, X would be a $Bernoulli(p)$ sequence of 1's and 0, with p the probability of 1 shown at step n, that is, the probability of winning the respective game at step n. This can get very complicated for realistic games because in a betting game one can gain a partial stake and many other possibilities, in general modeled by the stochastic process X.

Assume that the gain obtained at every step n depends on the game outcome $X_n - X_{n-1}$ multiplied by the respective stake G_n that the player has in the game n. The total gain up to time n is a process denoted $G \cdot X$, where

$$(G \cdot X)_n = \sum_{i=1}^n G_i(X_i - X_{i-1})$$

By convention, $(G \cdot X)_0 = 0$.

This newly defined process $G \cdot X$ is called the **martingale transform** of X by G. This process is *the discrete-time analog of the stochastic integral*. The stochastic integral is defined using continuous-time processes.

Proposition 14.4 *Suppose that $G = \{G_n\}$ is a previsible process as before. If either*

(i) *G is a bounded process (i.e., $|G_n(\omega)| < M$, for some finite constant M, for any n and every ω), or*

(ii) *G_n and X_n are both L^1 random variables for any n,*

then we have the following:

- *If the process X is a martingale, then the transform $G \cdot X$ is a martingale started from 0.*

- *If the process X is a supermartingale and in addition G is positive ($G_n(\omega) \geq 0$ for all n and ω), then the transform $G \cdot X$ is a supermartingale.*

This result is important for gambling. If the casino is fair, in other words the game X_n is a martingale, then any realistic gambling strategy will be fair (a martingale). Note, here realistic means one can gamble with only positive and finite amounts of money. If, on the other hand, the casino is unfair (as all casinos are), then there is no gambling strategy that depends only on previously seen outcomes that will make money in the long run. The only way to beat the casino is to cheat (have the gambling strategy G_n not be previsible and depend on the current outcome X_n).

Proof: Conditions (i) and (ii) are the needed conditions for the conditional expectation to exist. To prove the result, we write for example in the supermartingale case

$$\mathbf{E}[(G \cdot X)_n - (G \cdot X)_{n-1} \mid \mathcal{F}_{n-1}] = \mathbf{E}[G_n(X_n - X_{n-1}) \mid \mathcal{F}_{n-1}]$$
$$= G_n \mathbf{E}[X_n - X_{n-1} \mid \mathcal{F}_{n-1}] \leq 0$$

using, in order, G_n is $\mathcal{F}_{n-1}$ measurable, X_n is supermartingale, and G_n is positive. For the martingale proof, we do not need positivity since any a.s. finite number multiplied with 0 is 0. Also, note that for the submartingale case we need G positive as well. ∎

14.4   Stopping Time. Stopped Process

As we mentioned already, martingales have been introduced as betting strategies. Any betting strategy (or any strategy for that matter) needs to have a stopping strategy. The concept of stopping is very important since one needs to define a set of criteria for stopping and these criteria should only use information which is available at the moment when the strategy ends.

We presented the definition of a discrete stopping time in Definition 11.7 on page 339). We recall it here as well.

Definition 14.5 (Stopping Time) *The random variable T is a stopping time for the filtration $\mathcal{F} = \{\mathcal{F}_n\}$ if and only if*

1. *T is adapted to the filtration $\mathcal{F}$, that is, the set $\{T \leq n\} \in \mathcal{F}_n$, and*

2. *$P(T < \infty) = 1$*

Definition 14.6 (Stopped process) *Let T be a stopping time for the stochastic process $\{Z_n\}$ with the filtration $\mathcal{F} = \{\mathcal{F}_n\}$.*
Define the stopped process $Z^T = \{Z_n^T\}_n$, where

$$Z_n^T(\omega) = Z_{T(\omega) \wedge n}(\omega) = \begin{cases} Z_n(\omega) & \text{if } n \leq T(\omega) \\ Z_T(\omega) & \text{if } n > T(\omega) \end{cases}$$

Thus a particular path of the process would stop at some $T(\omega)$ and be a particular value ($= Z_{T(\omega)}(\omega)$) afterwards: $Z^T = (Z_1, Z_2, Z_3 \ldots, ., Z_T, Z_T, Z_T, Z_T, \ldots)$

Since T is a random variable, the precise moment when the process stops is unknown before the process Z is actually observed. In general, the final value $Z_{T(\omega)}(\omega)$ is unknown as well. We recall the definition of the stopping time, which says that the moment when the process is stopped is precisely determined by knowing only the information available at that respective moment in time.

One of the most commonly encountered stopping times (and therefore strategies) is the first entry time in some set. The payer may say: "I will stop when I make 60%

more than what I had when entering the game." This translates to having a stopping time as the first time when the total gain process enters the set $[0.6, \infty)$. We plot such a strategy in Figure 14.1. Since every path is stochastic, the actual value of the random variable T stopping time is random as well, as we see in the two simulated paths presented. Mathematically, the stopping time is

$$T(\omega) = \inf\{t : Z_t(\omega) \geq 0.6\}$$

We shall see that any stopping time defined as a first entry in some set is a true stopping time.

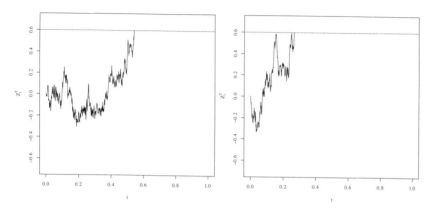

Figure 14.1   A Z_t^T process which stops once it reaches 0.6

Lemma 14.7 *Let X_n be an adapted process with respect to some filtration $\mathscr{F} = \{\mathscr{F}_n\}$. Let A be any Borel set, and define a stopping time*

$$T = \inf\{n : X_n \in A\},$$

which is the time of first entry in the set A and sometimes called the hitting time of A. Then T is a stopping time.

Proof: By convention, we have $\inf \emptyset = \infty$, so $T = \infty$ if the process never enters the set A. We have

$$\{T \leq n\} = \bigcup_{i \leq n}\{X_k \in A\} \in \mathscr{F}_n,$$

since $\{T \leq n\}$ means that at some time before n the process was in A, and since the process was adapted, the set is in $\mathscr{F}$. Thus T is a stopping time.   ∎

Note the time in Figure 14.1 is the first entry in the set $A = [0.6, \infty)$.

14.4.1 Properties of stopping time

1. Any constant $T(\omega) = C$ is a stopping time.

2. $S \wedge T$ and $S \vee T$ are stopping times, where we used the notation

$$S \wedge T(\omega) = \max\{S(\omega), T(\omega)\},$$

and

$$S \vee T(\omega) = \min\{S(\omega), T(\omega)\}.$$

3. If $S \leq T$ a.s., then $\mathscr{F}_S \subset \mathscr{F}_T$.

4. If $\{T_n\}$ is a sequence of stopping times, then

$$\sup_k T_k$$

   is a stopping time. This property of stopping times is very important for local martingale processes.

5. If, in addition, the filtration $\{\mathscr{F}_n\}$ is right-continuous, then $\inf_k T_k$, $\limsup T_k$, $\liminf T_k$ are all stopping times.

A right-continuous filtration has the property that

$$\mathscr{F}_t = \bigcap_{u>t} \mathscr{F}_u$$

This particular property is applicable only when the time is continuous; for discrete-time processes it is automatically true.

The next theorem is the reason why we reintroduced stopping times.

Theorem 14.8 (Stopped martingale) *Suppose T is a stopping time for the stochastic process $\{X_n\}_n$. If X_n is a martingale, then the stopped process $\{Z_n^T\}_n$ is also a martingale, and in particular*

$$\mathbf{E}[X_{T \wedge n}] = \mathbf{E}[X_0].$$

If X_n is a supermartingale (or submartingale), then the stopped process $\{Z_n^T\}_n$ is also a supermartingale (submartingale), and in particular

$$\mathbf{E}[X_{T \wedge n}] \leq \mathbf{E}[X_0], \quad (\text{respectively } \geq \mathbf{E}[X_0]).$$

Proof: Given the stopping time T, construct a martingale transform using the previsible process G^T:

$$G_n^T(\omega) = \mathbf{1}_{[0,T(\omega)]}(n) = \begin{cases} 1, & \text{if } T(\omega) \geq n \\ 0, & \text{else} \end{cases}$$

This process is always positive and bounded; in fact, it can only be 0 or 1. Since the event

$$\{G_n^T = 0\} = \{T < n\} = \{T \le n - 1\} \in \mathscr{F}_{n-1},$$

and the event

$$\{G_n^T = 1\} = \{T \ge n\} = \{T \le n - 1\}^C \in \mathscr{F}_{n-1},$$

the process G^T is previsible. Therefore, we can apply Proposition 14.4 about martingale transforms to conclude that the martingale transform is a martingale or supermartingale or submartingale depending on the type of process X. However, the martingale transform with our choice of the process G is

$$(G^T \cdot X)_n = \sum_{i=1}^{n} G_i^T (X_i - X_{i-1}) = \begin{cases} X_n - X_0, & \text{if } T \ge n \\ X_T - X_0, & \text{if } T < n \end{cases} = X_{T \wedge n} - X_0$$

But this is just the stopped process. Therefore, the stopped process has the same type as the original process X. ∎

Note that, as a consequence of the stopped process being a martingale, we must have $\mathbf{E}[X_{T \wedge n}] = \mathbf{E}[X_0]$. However, as the next example shows, this does not necessarily imply that $\mathbf{E}[X_T] = \mathbf{E}[X_0]$.

▣ EXAMPLE 14.6

Consider a simple random walk X_n on the integers starting at $X_0 = 0$. We know from Example 14.1 that the process is a martingale, since the increment of the process has mean 0.

Define T as the first entry time in the set $[1, \infty)$ for the random walk

$$T = \inf\{n \mid X_n = 1\}$$

We will show later that $\mathbf{P}(T < \infty) = 1$, or in other words T is finite almost surely. Since T is a first entry random variable and is finite a.s., it is a stopping time and we must have

$$\mathbf{E}[X_{T \wedge n}] = \mathbf{E}[X_0], \ \forall n.$$

However, $\mathbf{E}[X_T] = 1$ using the definition of the stopping time T and $\mathbf{E}[X_0] = 0$ since the process starts from 0. Therefore, we have

$$\mathbf{E}[X_T] \ne \mathbf{E}[X_{T \wedge n}] = \mathbf{E}[X_0].$$

The example above presents a problem to us, since in practical applications we want to have the terminal expectation at the time when the stopping time is exercised somehow predictable, specifically in the martingale case the same as the initial expectation at time 0. In this way, we can analyze the future behavior of the process. If we take n to infinity, the stopped process should go to the same X_T. There is an entire theory of convergence dealing with this. For us, it is more interesting to know conditions under which we have $\mathbf{E}[X_T] = \mathbf{E}[X_0]$.

So, what are the conditions needed for the equality $\mathbf{E}[X_T] = \mathbf{E}[X_0]$? The next theorem gives the *sufficient* conditions.

Theorem 14.9 (Doob's Optional-Sampling theorem) *Let $X = \{X_n\}$ be a martingale (supermartingale) with respect to the filtration $\mathscr{F} = \{\mathscr{F}_n\}_n$. Let T be a stopping time for the filtration $\mathscr{F}$. If any one of the following conditions is true:*

1. *The stopping time T is bounded, (i.e., there exists some constant N and $T(\omega) \leq N$ for all ω), or*

2. *The martingale process X is bounded and the stopping time T is finite a.s. (i.e., there exists some constant M and $X_n(\omega) \leq M$, $\forall n, \omega$, and $\mathbf{P}(T < \infty) = 1$)), or*

3. $\mathbf{E}[T] < \infty$ *and there exists a constant M such that*

$$|X_n(\omega) - X_{n-1}(\omega)| < M, \ \forall n, \omega,$$

then $\mathbf{E}[X_T] = \mathbf{E}[X_0]$ if X is a martingale. If X is a supermartingale, $\mathbf{E}[X_T] \leq \mathbf{E}[X_0]$.

Note that the conditions of the theorem are not satisfied for Example 14.6. Specifically, a finite a.s. variable is not the same as a bounded random variable. For example, an $N(0, 1)$ random variable is finite a.s. since the probability that the value is finite is 1, but it is not bounded since there does not exist a real number N so that all the values of the random variable are less than N.

Proof: Suppose, first, that X is a supermartingale. We have already seen that for any stopping time X^T is a supermartingale. Therefore

$$\mathbf{E}[X_{T \wedge n} - X_0] \leq 0, \text{ for any } n \in \mathbb{N}. \tag{14.2}$$

If we can take n to infinity in this expression and exchange the order of limit and expectation, the theorem would be proven. But wait, we have specific theorems that allow us to change the order of the limit and integration. In fact, each of the three sufficient conditions stated in the Doob's theorem corresponds to one such specific theorem on changing the order of limit and integration.

Under condition 1) the stopping time T is bounded by N so we always have $T(\omega) \wedge N = T(\omega)$, $\forall \omega$. So we just take $n = N$ in (14.2) and are done without going to the limit.

Under condition 2), we can use the bounded convergence theorem (Theorem 4.10) with (14.2) to conclude $\mathbf{E}[X_T] \leq \mathbf{E}[X_0]$.

Under condition 3), we have

$$|X_{T \wedge n} - X_0| = \left| \sum_{i=1}^{T \wedge n} (X_i - X_{i-1}) \right| \leq M(T \wedge n) \leq MT.$$

But condition 3) has that $\mathbf{E}[T] < \infty$ and thus $MT \in L^1$, and applying the dominated convergence theorem (Theorem 4.9), we obtain the result once again.

To prove the result for martingales, consider the process X as a supermartingale. We can do this since the definition of supermartingales include the equality case. Applying the above, we obtain

$$\mathbf{E}\left[X_T\right] \leq \mathbf{E}[X_0].$$

Then we consider the process $-X$ also as a supermartingale to obtain

$$\mathbf{E}\left[-X_T\right] \leq \mathbf{E}[-X_0].$$

Combining the two inequalities gives the equality for martingales.   ∎

Stochastic processes which are martingales are some of the most powerful in the entire probability theory. The stopping time reasoning is unique among any and all branches of mathematics. However, to apply this reasoning correctly, we need to be careful, particularly to verify the hypotheses of the optional sampling theorem.

14.5   Classical Examples of Martingale Reasoning

In this section, we present examples of using martingales to provide elegant solutions to seemingly complicated problems.

14.5.1   The expected number of tosses until a binary pattern occurs

Suppose we toss an unbalanced coin which has probability p of landing Head and probability $q = 1 - p$ of landing Tail. What is the expected number of tosses until a certain pattern appears for the first time?

To make the reasoning more accessible, we consider a specific pattern, say, for example, $HHHTTTHH$. This pattern may be any sequence of heads and tails. There are other ways to find the expected number of tosses until the pattern occurs for the first time, but by far the most elegant and interesting reasoning is using martingale theory.

To perform the martingale reasoning, we need to construct a martingale. To this end, consider a game where at each step we toss this coin which has probability p of landing H. Consider a sequence of gamblers. Gambler i starts playing at toss number i, and every gambler bets on the exact pattern we want to calculate the probability for.

Specifically, in our example ($HHHTTTHH$), gambler i bets \$1 on H at toss i. If the gambler wins, then he/she bets all the winnings on H at the next toss $i + 1$. If the gambler wins again, he/she bets all the money on the next outcome in the pattern (H), and so on and so forth. If the gambler loses at any point, he/she has really lost the initial dollar that he/she started with. The game stops at the moment when one gambler wins the entire pattern, in our example $HHHTTTHH$.

Let N be the time (toss) when the pattern ($HHHTTTHH$) appears for the first time. Let X_n denote the bank total wealth after n tosses (the bank starts with 0 wealth and can hold a negative position).

We need to make sure that X_n is a martingale. This is accomplished by determining the right payoff for each successful bet that will make X_n a martingale. Since H has probability p and T has probability $q = 1 - p$, the right payout for an x amount bet is

$$Payout = \begin{cases} x\frac{1}{p}, & \text{if the outcome is } H \text{ and the bet is for } \$x \text{ on } H, \\ x\frac{1}{q}, & \text{if the outcome is } T \text{ and the bet is for } \$x \text{ on } T. \end{cases}$$

With this payout after each game, the bank receives x from a certain player and pays back x on average ($x\frac{1}{p}p + x\frac{1}{q}q$), which is a zero net expected game. With this particular payoff structure, X_n is a martingale since the expected wealth at step $n+1$, given X_n and the amount bet at step n, is exactly equal to X_n. Furthermore, it is easy to see that N is a stopping time for the X_n sequence since the event $\{N \leq n\}$ is determined by knowing the outcome of the first n games only.

Note also that we are in condition 3) of the Doob's optional sampling theorem, since $\mathbf{E}[T] < \infty$, and clearly the difference in bank's wealth at consecutive steps is bounded (not proven but easy to prove). Thus, we can use Doob's optional sampling theorem to conclude that the wealth process has the property

$$\mathbf{E}[X_N] = \mathbf{E}[X_0].$$

Since $X_0 = 0$, this clearly implies that $\mathbf{E}[X_N] = 0$. But what is X_N? We claim that it is pretty obvious that for our particular pattern $HHHTTTHH$ this is

$$X_N = N - \frac{1\,1\,1\,1\,1\,1\,1\,1}{p\,p\,p\,q\,q\,q\,p\,p} - \frac{1\,1}{p\,p} - \frac{1}{p}$$

Why? Well, there have been N games and at each of these N games the bank gained the dollar that each gambler had. The bank also needs to pay everyone who gained at any game. Clearly, with the exception of the gambler starting at toss $N - 7$ who actually won, all the previous ones lost (it is the first time the pattern appears). However, we need to take into account all the people who started after toss $N - 7$ and may have received a gain. For example, the gambler starting at toss $N - 6$ bets on HHH but we know because of our winner that the pattern this gambler got is HHT, so he/she lost everything on the third toss. The reasoning is repeated for all the others, and now we can see that the last two will actually win their bets. This is the reason for the last two terms in X_N; they represent the payoff to these individuals. Now, applying expectation in both sides above, it is easy to see that

$$\mathbf{E}[N] = \left(\frac{1}{p}\right)^5 \left(\frac{1}{q}\right)^3 + \left(\frac{1}{p}\right)^2 + \frac{1}{p}$$

As a more particular pattern, we can answer the following question: what is the first time until we see n heads in a row? Repeating the argument above, we notice that we have to pay all the people starting to play after our winner. This makes the expected value equal to

$$\mathbf{E}[N] = \left(\frac{1}{p}\right)^n + \left(\frac{1}{p}\right)^{n-1} + \ldots + \frac{1}{p} = \frac{1}{p}\frac{1 - \left(\frac{1}{p}\right)^n}{1 - \frac{1}{p}} = \frac{1 - p^n}{p^n(1-p)}$$

To give an even more specific example, consider a fair coin (probability of H is 0.5). Then the expected number of tosses to obtain three heads in a row is 14.

Here is an even funnier application. How many times have you seen a basketball announcer saying that he has scored 8 (or 15, or 20) free throws in a row, and he is due a miss by the law of averages. So let us play a bit with this.

Suppose a player has an excellent, true free throw percentage 0.9 for the season. The expected number of free throws until the first miss is not a big deal, since it is the expected value of a geometric random variable ($1/0.1 = 10$ free throws). But using the formula, we can calculate the expected number of throws until two misses in a row (110) or until three misses in a row (1, 110). This of course calculates an expected value (not the actual number) and assumes that each throw is independent (obviously not the case or we would not talk about hot & cold streaks in basketball).

14.5.2   Expected number of attempts until a general pattern occurs

The reasoning presented in the previous question for a binary random variable which takes just two outcomes at every step can be extended to more general discrete random variables. David Williams in his excellent book *Probability with Martingales* (Williams, 1991) presents an example of a monkey typing a sequence of capital letters at random on a typewriter. He then wonders (and gives the answer) on how many letters the monkey types until the pattern

$$ABRACADABRA$$

appears for the first time. The idea is identical to the one presented in the previous section except that the gamblers have more choices at every step and the bank payoff needs to be adjusted accordingly. Specifically, if we assume that the gamblers in the construction above can bet on any one of the 26 capital letters, and since the letters are random, the probability of any one letter being the winning bet is $1/26$; then, repeating the reasoning, the number of letters typed until the pattern appears for the first time is

$$\mathbf{E}[N] = 26^{11} + 26^4 + 26 = 3.670344 \times 10^{15}$$

which is an astronomical number. Note that the last two terms are due to the patterns *ABRA* and *A* – winning patterns at the moment of stopping. If we assume a fast typing monkey who types at a speed of 1 letter per second, then it would need an average of 3.670344×10^{15} s or 1.019×10^{12} h, or if we assume the monkey types nonstop day and night, 116.3 million years.

It is also easy to think about the complete works of Shakespeare as being a long sequence of words. Using again the martingale reasoning, a monkey would type a finite number of letters until the complete work of Shakespeare appears randomly but obviously since the simple pattern $ABRACADABRA$ takes more time than the history of mammals: for all intents and purposes, the task is impossible.

14.5.3   Gambler's ruin probability – revisited

Recall the gambler's ruin example. In this example, a gambler starts playing a two-outcome betting game starting with initial wealth x. Each time the gambler bets \$1, he wins with probability p and will stop when the total wealth reaches either 0 or a fixed wealth amount $M > x$. Of course, the resulting process is a random walk with absorbing states at 0 and M.

Let X_n be the result of the nth game. These X_n's are i.i.d., each taking values $+1$ or -1 with probabilities p and $q = 1 - p$, respectively.

The process

$$W_n = \sum_{i=0}^{n} X_i$$

with $X_0 = x$, which stops when either $W_T = 0$ or $W_T = M$ is the amount of wealth after n games. It is easy to see that the process

$$S_n = W_n - x$$

is an equivalent process but one that is now starting from 0 and stopping when reaching either $-x$ or $M - x$. The S_n process represents the profit after n games. We look at S rather than W since the martingales in this chapter are starting from 0.

The case when the game is fair   Let us first assume that $p = q = 1/2$. It is clear that in this case $\mathbf{E}[X_n] = 0$ and thus the process S_n is a martingale without any modifications.

Define a stopping time by the first time when the gambler is ruined or reaches his desired wealth:

$$T = \min\{n \mid S_n = -x \text{ or } S_n = M - x\}$$

This is a stopping time since it is the first time S_n enters the set $(-\infty, 0] \cup [x, \infty)$. It is also easy to show that the expected value of T is finite, so we are in the hypotheses of the Doob's optional sampling theorem.

Let p_0 be the probability of ruin or, put differently, the probability that S_n reaches $-x$ before reaching $M - x$. Then, by the optional sampling theorem

$$\mathbf{E}[S_0] = 0 = \mathbf{E}[S_T] = (-x)p_0 + (M - x)(1 - p_0)$$

Solving for p_0 gives the ruin probability

$$p_0 = \frac{M - x}{M} = 1 - \frac{x}{M},$$

which is the same probability we reached earlier in this book using different arguments.

Let us further calculate the expected time to ruin. Consider the following process:

$$Z_n = S_n^2 - n.$$

The Z_n process is also a martingale. To prove this

$$\begin{aligned}
\mathbf{E}[Z_{n+1} \mid \mathcal{F}_n] &= \mathbf{E}[Z_{n+1} \mid Z_n] = \mathbf{E}[(S_n + X_{n+1})^2 - (n+1) \mid S_n] \\
&= \mathbf{E}[S_n^2 \mid S_n] + 2\mathbf{E}[S_n X_{n+1} \mid S_n] + \mathbf{E}[X_{n+1}^2 \mid S_n] - n - 1 \\
&= S_n^2 - n = Z_n
\end{aligned}$$

using that X_{n+1} has expected 0, variance 1, and is independent of S_n. Now, applying the optional sampling theorem for the martingale Z_n we obtain

$$0 = \mathbf{E}[Z_T] = \mathbf{E}[S_T^2] - \mathbf{E}[T] = (-x)^2 p_0 + (M-x)^2(1 - p_0) - \mathbf{E}[T]$$

Substituting the formula we just found for p_0, we obtain

$$\mathbf{E}[T] = x(M - x).$$

The case when the game is NOT fair   Suppose that the game is not fair, in other words $p \neq 1/2$. We need to construct a martingale to apply the martingale reasoning. To do so, we need to find a transformation so that the expected value of that transformation applied to the wealth increment is going to be zero. In the pattern example, we modified the game probabilities so that the expectation of the increment is 0. Here, the game probabilities are fixed and by making a transformation we modify the payoff. Recall that the values of the game n are $X_n = 1$ with probability p and $X_n = -1$ with probability $q = 1 - p$. Observe

$$1 = p + q = q\frac{p}{q} + p\frac{q}{p} = \left(\frac{q}{p}\right)^1 p + \left(\frac{q}{p}\right)^{-1} q = \mathbf{E}[X_n].$$

This is the idea when constructing the martingale process next.

We consider the process

$$Z_n = \left(\frac{q}{p}\right)^{S_n}$$

which we can show is a martingale. Simply note that $Z_{n+1} = \left(\frac{q}{p}\right)^{X_{n+1}} Z_n$, and that

$$\mathbf{E}\left[\left(\frac{q}{p}\right)^{X_{n+1}}\right] = \left(\frac{q}{p}\right)^1 p + \left(\frac{q}{p}\right)^{-1} q = 1$$

Therefore, by the same optional sampling theorem

$$1 = \mathbf{E}[Z_T] = p_0 \left(\frac{q}{p}\right)^{-x} + (1 - p_0) \left(\frac{q}{p}\right)^{M-x}$$

which gives the probability of ruin in this biased case as well:

$$p_0 = \frac{1 - \left(\frac{q}{p}\right)^{M-x}}{\left(\frac{q}{p}\right)^{-x} - \left(\frac{q}{p}\right)^{M-x}}$$

Note that the gambler's ruin problem is the same as a random walk stopped at 0 or M. In the next section, we will look specifically at random walks but keep in mind that the results in this section can be applied as well.

14.5.3.1 *The distribution of hitting time for random walks* This is a really interesting derivation using martingales. We consider the symmetric random walk with $p = 1/2$ started at 0, and we let

$$T = \inf\{n \mid S_n = 1\}.$$

T, the first time when the random walk hits 1, is a stopping time for the filtration generated by S_n. Now, as we have done earlier, we need to come up with a martingale process that will help us to find the distribution of T. Since the random walk is a Markov chain, we will follow Wald's martingale example. We calculate the moment generating function. We have

$$\mathbf{E}[e^{\theta X_1}] = \frac{1}{2}e^{\theta} + \frac{1}{2}e^{-\theta} = \cosh \theta$$

Therefore, we calculate

$$\mathbf{E}[\operatorname{sech} \theta e^{\theta X_1}] = \frac{1}{\cosh \theta} \mathbf{E}[e^{\theta X_1}] = 1,$$

where the hyperbolic secant is defined as $\operatorname{sech}\theta = \frac{1}{\cosh\theta}$. The angle θ is chosen so that the generating function calculated at θ exits. Thus, we construct Wald's martingale as

$$M_n = (\operatorname{sech} \theta)^n e^{\theta S_n}.$$

Since T is a stopping time for the associated filtration, we must have

$$\mathbf{E}[M_{T \wedge n}] = \mathbf{E}\left[(\operatorname{sech} \theta)^{T \wedge n} e^{\theta S_{T \wedge n}}\right] = 1, \qquad \forall n$$

We cannot directly apply the optional sampling theorem to this problem since none of the hypotheses is verified (nether the martingale nor the stopping time is bounded).

Instead, *pay attention to the following argument*, as it shows how to tackle the case of possibly unbounded stopping times.

Let us take $\theta > 0$. Such a theta exists since the moment generating function is finite at 0 and the radius of convergence is strictly positive. Now note that $\exp(\theta S_{T \wedge n})$ is bounded by e^{θ} by the definition of T and the fact that theta is positive. Since e^{θ} is integrable, we can apply the bounded convergence theorem to conclude that, as $n \to \infty$, the limit and integral can be exchanged and we have

$$\mathbf{E}[M_T] = \mathbf{E}\left[(\operatorname{sech}\theta)^T e^{\theta S_T}\right] = \mathbf{E}\left[(\operatorname{sech}\theta)^T e^{\theta}\right] = 1$$

Thus

$$\mathbf{E}\left[(\operatorname{sech}\theta)^T\right] = e^{-\theta}, \text{ for some } \theta > 0 \tag{14.3}$$

If $T = \infty$, the equality does not hold, so we need to make sure this does not happen. Either we use the argument we used previously when we showed that a one-dimensional random walk is recurrent and thus $\mathbf{P}(T < \infty) = 1$, or we take $\theta \downarrow 0$ in the previous expectation and use

$$\mathbf{E}\left[(\operatorname{sech}\theta)^T\right] = \mathbf{E}\left[(\operatorname{sech}\theta)^T \mathbf{1}_{\{T < \infty\}}\right] + \mathbf{E}\left[(\operatorname{sech}\theta)^T \mathbf{1}_{\{T = \infty\}}\right]$$

When $T < \infty$, we have $(\operatorname{sech}\theta)^T \uparrow 1$, and when $T = \infty$ we have $(\operatorname{sech}\theta)^T \uparrow 0$ as $\theta \to 0$. Since the functions of θ are either increasing or decreasing, we can apply monotone convergence in the expression in both terms to end up with

$$\mathbf{E}[\mathbf{1}_{\{T < \infty\}}] = \mathbf{P}\{T < \infty\} = 1$$

Now, to calculate the distribution of T, let us use Equation (14.3). Let $\alpha = \operatorname{sech}\theta$. Then we have

$$\mathbf{E}\left[\alpha^T\right] = e^{-\theta}$$

Recall that $\alpha = \operatorname{sech}\theta = \frac{2}{e^{-\theta} + e^{\theta}}$. From this last expression, we denote $y = e^{-\theta}$ and solve the resulting equation in terms of y:

$$\alpha y^2 - 2y + \alpha = 0$$

Recall that $y = e^{-\theta} < 1$ since $\theta > 0$, so the only valid solution is

$$e^{-\theta} = y = \frac{1}{\alpha}(1 - \sqrt{1 - \alpha^2})$$

But we must have

$$\mathbf{E}\left[\alpha^T\right] = \sum_{n=1}^{\infty} \alpha^n \mathbf{P}\{T = n\} = e^{-\theta}$$

By expressing the function of α as a power series

$$\frac{1}{\alpha}(1 - \sqrt{1 - \alpha^2}) = \frac{1}{\alpha}\left(1 - \sum_{k=0}^{\infty} \binom{\frac{1}{2}}{k}(-1)^k \alpha^{2k}\right) = \sum_{k=1}^{\infty} \binom{\frac{1}{2}}{k}(-1)^{k+1} \alpha^{2k-1}$$

we are able to identify coefficients and get the exact distribution of the stopping time.

$$\mathbf{P}\{T = 2k - 1\} = (-1)^{k+1}\binom{\frac{1}{2}}{k} = \binom{2k+1}{k}\frac{k+1}{2^{2k}(2k-1)(2k+1)}$$

In the last formula, we used an identity for the binomial coefficient $\binom{n}{k}$ when $n = 1/2$ to see that the probabilities are in fact positive. The binomial coefficient when n is a fractional or in fact any complex number z is defined as

$$\binom{z}{k} = \frac{z(z-1)(z-2)\ldots(z-k+1)}{k!}$$

With this notation, we have the generalized binomial expansion (which we used above)

$$(1+\alpha)^z = \sum_{k=0}^{\infty} \binom{z}{k} \alpha^k$$

14.6 Convergence Theorems. L^1 Convergence. Bounded Martingales in L^2

This section is concerned with the long-time behavior of the martingale: $\lim_{n\to\infty} X_n$. The next theorem is the classical Doob convergence theorem. I will just state the theorem without proving it. The proof requires the classical upcrossing argument. While this argument is indeed very interesting probabilistically, I believe it is a very local proof and never used beyond this result.

Theorem 14.10 (Doob's convergence theorem) *Let X be a supermartingale such that either*

1. *X is bounded in L^1 (i.e., $\sup_n \mathbf{E}[|X_n|] < \infty$), or*

2. *X is nonnegative (i.e., $X_n(\omega) \geq 0$, $\forall n, \omega$).*

Then there exists a random variable X_∞ a.s. finite such that

$$\lim_{n\to\infty} X_n = X_\infty, \quad a.s.$$

In the hypothesis of the theorem, we need the martingale bounded in L^1. Rather than showing that the process is bounded in L^1, sometimes it is easier to show that it is bounded in L^2 (i.e., $\sup_n \mathbf{E}[X_n^2] < \infty$) even though this is a stronger results. This is because for martingales the following orthogonality result holds.

Let X_n be a martingale with respect to filtration $\mathcal{F}_n$ and assume that each X_n is in L^2 ($\mathbf{E}[X_n^2] < \infty$). Then we have by definition

$$\mathbf{E}[X_n - X_m \mid \mathcal{F}_m] = 0, \forall m < n$$

which means that $X_n - X_m$ is orthogonal (perpendicular) on the information up to time m and, furthermore, because of this

$$\mathbf{E}[X_n^2] = \mathbf{E}\left[\left(X_0 + \sum_{k=1}^{n}(X_k - X_{k-1})\right)^2\right]$$

$$= \mathbf{E}[X_0^2] + \sum_{k=1}^{n}\mathbf{E}\left[(X_k - X_{k-1})^2\right] \tag{14.4}$$

The last equality holds because the expectation of each cross-product term is zero. For example, with $m < n$

$$\mathbf{E}\left[(X_m - X_{m-1})(X_n - X_{n-1})\right] = \mathbf{E}\left[\mathbf{E}[(X_m - X_{m-1})(X_n - X_{n-1}) \mid \mathcal{F}_m]\right]$$
$$= \mathbf{E}\left[(X_m - X_{m-1})\mathbf{E}[(X_n - X_{n-1}) \mid \mathcal{F}_m]\right]$$
$$= \mathbf{E}\left[(X_m - X_{m-1})(\mathbf{E}[X_n \mid \mathcal{F}_m] - \mathbf{E}[X_{n-1} \mid \mathcal{F}_m])\right]$$
$$= \mathbf{E}\left[(X_m - X_{m-1})(X_m - X_m)\right] = 0$$

Thus we only need to calculate the second moment of the increments, which are typically easier to calculate. If the second moment exists, then the convergence holds, as the next theorem shows.

Theorem 14.11 *Let X be a martingale such that $X_n \in L^2$ for all n. If X is bounded in L^2 or, equivalently,*

$$\sum_{k=1}^{\infty}\mathbf{E}\left[(X_k - X_{k-1})^2\right] < \infty,$$

then there exists the random variable X_∞ and

$$X_n \to X_\infty, \text{ almost surely and in } L^2.$$

The random variable X_∞ may be defined for each $\omega \in \Omega$ as

$$X_\infty(\omega) = \limsup_{n \to \infty} X_n(\omega)$$

The $\limsup$ always exists for any sequence, although it could be ∞. The point of the theorem is that, for a martingale as in the theorem, the $\limsup$ is actually finite a.s. and exists a.s. The proof is fairly simple. Due to (14.4), the boundness in L^2 is evidently equivalent to the expression in the hypothesis of the theorem. Furthermore, we know that, if a variable is in L^2 is also in L^1, the variable X_∞ exists by the Doob convergence theorem and the convergence is almost sure. The only thing remaining is to show that the convergence also holds in L^2. We have (using a similar argument with (14.4)) that

$$\mathbf{E}\left[(X_n - X_m)^2\right] = \sum_{k=m+1}^{n}\mathbf{E}\left[(X_{k+1} - X_k)^2\right]$$

Now, take $\liminf_{n\to\infty}$ and apply Fatou's lemma to obtain

$$\mathbf{E}\left[\liminf_n (X_n - X_m)^2\right] \le \liminf_n \mathbf{E}\left[(X_n - X_m)^2\right]$$

$$= \liminf_n \sum_{k=m+1}^{n} \mathbf{E}\left[(X_{k+1} - X_k)^2\right]$$

or

$$\mathbf{E}\left[(X_\infty - X_m)^2\right] \le \sum_{k=m+1}^{\infty} \mathbf{E}\left[(X_{k+1} - X_k)^2\right]$$

But the right-hand side is the tail of a convergent series (by the L^2 boundedness), so it converges to 0 as $m \to \infty$. Therefore, we must have

$$\lim_{m\to\infty} \mathbf{E}\left[(X_\infty - X_m)^2\right] = 0,$$

or the L^2 convergence of the martingale.

Finally, we conclude the chapter on martingales with an inequality which bounds the supremum for continuous martingales.

Theorem 14.12 (Doob's martingale inequality) *Suppose M_t is a martingale with an associated filtration and assume that its paths are continuous a.s. That is, the function $t \to M_t(\omega)$ is continuous for all omega except for some in a set of probability 0. Then for all $p \ge 1$, $T \ge 0$, and all $\lambda > 0$ we have*

$$\mathbf{P}\left(\sup_{0\le t\le T} |M_t| \ge \lambda\right) \le \frac{1}{\lambda^p}\mathbf{E}[|M_T|^p]$$

For a proof, we refer to (Stroock and Varadhan, 1979, Theorem 1.2.3) or (Revuz and Yor, 2004, Chapter II, Theorem 1.7).

Problems

14.1 Let $X = \{X_n\}_n$ a stochastic process so that X_i's are squared integrable and let $\mathscr{F}_n$ be its standard filtration. Suppose that

$$Z_n = \sum_{i=1}^{n} X_n$$

is a martingale. Show that $\mathbf{E}[X_i X_j] = 0$ for all $i \ne j$.

14.2 Let $X = \{X_n\}_n$ be a martingale with $X_0 = 0$ and let $\mathscr{F}_n$ be its standard filtration. Show that

$$\mathbf{E}[X_n]^2 = \sum_{i=1}^{n} \mathbf{E}[(X_i - X_{i-1})^2].$$

14.3   [Not a stopping time] We saw the first entry defined as an infimum is a stoping time. How about the supremum? Let X_t be a martingale and let

$$S = \sup\{n \mid n \le 100, X_n \in B\}$$

for some interval B in $\mathbb{R}$ with the convention $\sup(\emptyset) = 0$. Give an argument to show that S is **not** a stopping time.

14.4   Let X_n be in L^1 and $\mathscr{F}_n = \sigma(X_1, X_2, \dots, X_n)$ its standard filtration. Suppose that we have

$$\mathbf{E}[X_{n+1}] = aX_n + bX_{n-1},$$

for some $a, b \in (0, 1)$ with $a + b = 1$. Find α so that the process

$$Z_n = \alpha X_n + X_{n-1}$$

is a martingale with respect to $\mathscr{F}_n$.

14.5   Let $X_1, X_2, \dots, X_n$ be i.i.d. with mean 0 and variance σ^2 finite. Suppose T is a stopping time for the standard filtration generated by the X_i's with a finite expectation. Show that

$$Var\left(\sum_{i=1}^{T} X_i\right) = \sigma^2 \mathbf{E}[T].$$

14.6   For a simple random walk S_n, which of the following are stopping times?
   a) $T_1 = \min\{n \ge 3 \mid S_n = S_{n-3} + 3\}$,
   b) $T_2 = \min\{n \ge 3 \mid S_n = S_{n-3} + 2\}$,
   c) $T_3 = T_1 + 3$,
   d) $T_4 = T_2 \wedge T_3$,
   e) $T_5 = T_1 \vee 5$.

14.7   [De Mere's martingale] We have a fair coin $p = 1/2$ and a fair game i.e., twice the amount gambled if the toss is won, and lose all amount if the toss is lost. Suppose we adopt a doubling strategy, that is, we start with x dollars, if we lose, we next bet $2x$, and so on. If we lose n bets, the amount bet on the $n+1$ toss is $2^{n+1}x$. We stop playing the moment we win (or equivalently we bet 0 dollars for all subsequent tosses). Let Z_n the net profit after the nth toss occurred.
   a) Show that Z_n is a martingale.
   b) Show that the game ends in a finite time almost surely.
   c) Calculate the expected time until the gambler stops betting.
   d) Calculate the expected net profit at the time when the gambler stops betting.
   e) Calculate the expected value of the gambler's maximum loss during the game. Comment.

14.8   Consider De Mere's martingale in the previous problem. To eliminate such strategies, most casinos impose a maximum amount that can be bet at any one game.

Suppose that in some casino like this the minimum amount bet is $1 and the maximum is 10,000. Calculate the expected net profit for a doubling strategy starting with $1 that will end when either the player wins or the maximum bet limit is reached.

14.9 [Polya's urn] Suppose an urn contains one white ball and one green ball. At time 1, a ball is selected randomly and a ball of the same color is added to the urn so that there are now three balls in the urn. Then at time 2, a ball from the urn is selected randomly and a new ball of the same color is added. The process continues like this. Note that after n selections are completed, the urn contains $n + 2$ balls. These $n + 2$ balls are made of $Z_n + 1$ white balls and $n - Z_n + 1$ green balls, where W_n is the number of new white balls that have been added to the urn. Similarly, $n - Z_n$ is the number of new green balls.

   a) Show that $\mathbf{P}(Z_n = i) = \frac{1}{n+1}$ for all $i \in \{0, 1, \ldots, n\}$.
   b) Show that the process

   $$M_n = \frac{Z_n + 1}{n + 2}$$

   is a martingale. Note M_n is the proportion of white balls in the urn after the nth selection.
   c) Show that M_∞ exists almost surely and calculate it.
   d) Show that the process

   $$N_n = \frac{(n + 1)!}{Z_n!(n - Z_n)!} \theta^{Z_n} (1 - \theta)^{n - Z_n}$$

   is a martingale for all $\theta \in (0, 1)$.

14.10 [Wright–Fisher population model] In a population we have two types of individuals, type A and type B. Type A is purely dominant and type B carries a particular type of recessive allele. Let X_n be the proportion of population of type A in generation n. Suppose the population is always constant of size N. Each individual in generation $n + 1$ chooses its ancestor at random from one of the N individuals existing at time n. The parent genetic type is always preserved. At time $t = 0$, the initial proportion X_0 is 0.5.

   a) Find the distribution of $NX_{n+1} \mid X_n$, or the number of individuals of type A in generation $n + 1$, given the proportion of individuals of type A in generation n.
   b) Show that X_n is a martingale.
   c) Show that the process

   $$Z_n = \frac{X_n(1 - X_n)}{(1 - N)^n}$$

   is a martingale.

14.11 Suppose that random variables X_n are such that the distribution of X_{n+1} given X_n is $Binomial(10, \frac{X_n}{10})$.

a) Show that X_n is a martingale.
b) We define

$$T = \inf\{n \geq 0 \mid X_n = 0 \text{ or } X_n = 10\}.$$

Show that T is a stopping time for the filtration generated by X_n.
c) Show that X_n converges almost surely to a limiting random variable X_∞, for any initial value X_0.
d) Derive the distribution of $X_\infty \mid X_0 = a$ for $a \in \{0, 1, 2, \ldots, 10\}$.
e) Show that

$$Z_n = \frac{X_n(10 - X_n)}{\left(1 - \frac{1}{10}\right)^n}$$

is a martingale.

14.12   Consider a simple random walk with $p = 1/2$ started from $S_0 = 0$, and let

$$T_\alpha = \inf\{n \in \mathbb{N} \mid S_n = \alpha \text{ or } S_n = -\alpha\},$$

the first time the random walk exits from $(-\alpha, \alpha)$, or equivalently first hitting time of $(-\infty, \alpha] \cup [\alpha, \infty)$. Using the methodology presented in the examples, find an estimate for the expected value of T_α and for the moment generating function of T_α.

14.13   Consider a branching process with m the expected offspring size. We have seen that the process X_n/m^n is a martingale where X_n is the size of generation n. Show that if $m > 1$, then $X_n \to \infty$ almost surely.

14.14   Suppose that $X_1, \ldots, X_n, \ldots$ are i.i.d. random variables. Construct the partial sums process:

$$S_n = \sum_{i=1}^{n} X_i,$$

and the standard filtration $\mathscr{F}_n = \sigma(X_1, \ldots, X_n)$. Suppose that there exists an $M > 0$ such that the moment generating function for the X_i's is finite on $(-M, M)$, that is,

$$M(\theta) = \mathbf{E}\left[e^{\theta X_1}\right] < \infty, \quad \forall \theta \in (-M, M).$$

Denote $\phi(\theta) = \log(M(\theta))$.
a) Show that X_n^θ defined as

$$X_n^\theta = e^{\theta S_n - n\phi(\theta)}$$

is a martingale with respect to $\mathscr{F}_n$.
b) Now assume that $X_i \sim N(0, 1)$. Calculate $M(\theta)$ and $\phi(\theta)$. Calculate the martingale X_n^θ. Show that for all $\theta \neq 0$ we have $X_n^\theta \to 0$ almost surely. Do we have $\mathbf{E}[X_n^\theta] \to 0$?

14.15   Recall Section 14.5.3.1 on page 454 and the definition of the hitting time T given there. Let $M(\theta)$ be the moment generating function of the X_i's, $\phi(\theta) = \log(M(\theta))$, and

$$X_n^\theta = e^{\theta S_n - n\phi(\theta)}.$$

a) Calculate $M(\theta)$ and $\phi(\theta)$ for $\theta \in \mathbb{R}$, and show that $\phi(\theta) \geq 0$ for all $\theta > 0$.
b) For $\theta > 0$, show that

$$\mathbf{E}\left[M(\theta)^{-T}\right] = e^{-\theta}.$$

c) Let $\gamma = M(\theta)^{-1}$; then let $y = e^{-\theta}$ to obtain

$$\gamma(1 - p)y^2 - y + \gamma p = 0.$$

Solve the equation and deduce that for $p < 1$

$$\mathbf{E}\left[\gamma^T\right] = \frac{1 - \sqrt{1 - 4p(1 - p)\gamma^2}}{2(1 - p)\gamma}.$$

What is the value of $\mathbf{E}\left[\gamma^T\right]$ when $p = 1$?
d) Derive from the previous question that $\mathbf{E}[T] < \infty$ if and only if $p > 1/2$.

14.16   [Doob's inequalities] For a stochastic process $X = \{X_n\}_n$ with a filtration $\mathcal{F} = \{\mathcal{F}_n\}_n$, we define the maximum process

$$\bar{X}_n(\omega) = \max_{0 \leq i \leq n} X_i(\omega).$$

a) Suppose the original process $X = \{X_n\}_n$ is a submartingale. Show that

$$\mathbf{P}(\bar{X}_n \geq \lambda) \leq \frac{1}{\lambda}\mathbf{E}\left[X_n \mathbf{1}_{\bar{X}_n \geq \lambda}\right] \leq \frac{1}{\lambda}\mathbf{E}\left[X_n \mathbf{1}_{\bar{X}_n \geq 0}\right] = \mathbf{E}\left[X_n^+\right]$$

where X_n^+ is the positive part of the process.
   Hint: For the first inequality, take

$$N = \inf\{i : X_i \geq \lambda\} \wedge n$$

and show that $\mathbf{E}[X_0] \leq \mathbf{E}[X_N] \leq \mathbf{E}[X_n]$.
b) [Doob's inequality for martingales] Let X_n be a squared-integrable martingale. Derive from part a) that

$$\mathbf{P}\left(\max_{0 \leq i \leq n} |X_i| \geq \lambda\right) \leq \frac{1}{\lambda^2}\mathbf{E}[X_n^2].$$

c) For $p > 1$, assume that $X = \{X_n\}_n$ is a submartingale with $X_n \in L^p$ for all n. Show that

$$\mathbf{E}[\bar{X}_n^p] \le \left(\frac{p}{p-1}\right)^p \mathbf{E}[(X_n^+)^p].$$

Hint: Use part a) and work with $\bar{X}_n \wedge M$ for some $M > 0$, and use the expectation formula for positive random variables:

$$\mathbf{E}\left[(\bar{X}_n \wedge M)^p\right] = \int_0^\infty px^{p-1}\mathbf{P}(\bar{X}_n \wedge M \ge x)dx.$$

d) **[Doob's inequality for submartingales]** Assume $p > 1$, and let X_n be a submartingale with $X_n \in L^p$ for all n. Show that

$$\mathbf{E}\left[\max_{0\le i\le n} |X_i|^p\right] \le \left(\frac{p}{p-1}\right)^p \mathbf{E}[|X_n|^p].$$

CHAPTER 15

BROWNIAN MOTION

In this chapter we introduce – without a doubt – the most important stochastic process. This particular process is a Markov process, is a martingale, and, due to the martingale representation theorem, one of the most important in the theory of continuous-time–continuous-state space processes.

15.1 History

In 1827, the biologist, Robert Brown, looking through a microscope at pollen grains in water, noted that the grains moved through the water even though the stand was perfectly stationary. Repeating the experiment with inorganic matter, he disproved the hypothesis that the movement is caused by life matter; however, he was not able to determine the exact cause of the movement. This discovery led to speculation about atoms and molecules long before they were discovered. Of course, now we know that in fact the movement is caused by bombardment from the water molecules. Even though a pollen grain is 10,000 times larger than a water molecule, the cumulated effect of them is the driving force of the grain. The direction of the force of atomic

Probability and Stochastic Processes, First Edition. Ionuţ Florescu
© 2015 John Wiley & Sons, Inc. Published 2015 by John Wiley & Sons, Inc.

bombardment is constantly changing, and at different times the pollen grain is hit more on one side than another, leading to the seemingly random nature of the motion. It is the fact that, even though the chances are that the pollen is hit equally from all sides, due to the random nature of life, this never happens in reality and this is the great discovery behind stochastic processes. This type of phenomenon is named in Brown's honor.

The first person trying to describe some mathematics behind the Brownian motion was Thorvald N. Thiele in a paper on the method of least squares published in 1880. In 1890, Louis Bachelier, coming from a completely different direction, presented for the first time a mathematical description of the phenomenon in his PhD thesis "The theory of speculation." This is also the first ever stochastic analysis of the stock and option markets (no option actually existed at the time). His notation is cumbersome by modern standards; nevertheless, he created a complete market model some 70 years before the famous Black-Scholes–Merton model. True, the stock prices can be negative in his model, but at a time when there was no knowledge of formal probability, even his thesis was remarkable.

Albert Einstein (in one of his 1905 papers) is typically credited with the formal introduction of the Brownian motion as a two-dimensional process; however, he allegedly reviewed and rejected the Marian Smolukovski paper (published in 1906) on exactly the same subject. Nevertheless, these papers brought the solution of the problem to the attention of physicists, and presented it as a way to indirectly confirm the existence of atoms and molecules. Their equations describing Brownian motion were subsequently verified by the experimental work of Jean Baptiste Perrin in 1913. Using Brownian motion as a model of particles in suspension, Einstein estimated the Avogadro's number $6 \times 10^2 3$ based on the diffusion coefficient D in the Einstein relation. Both Einstein and Smolukovsky reached a solution by solving the associated heat equation, which will be described in this chapter. Their approach, while describing the probability distribution in a Markov process, does not show the existence or construction of such a process. A process with continuous paths which are nowhere differentiable was unheard of or unimaginable at that time.

In 1923, the existence and construction of such a one-dimensional process in time was established by Norbert Wiener. This Wiener process became synonymous with Brownian motion and the measure he used to establish the construction was called the Wiener measure in his honor. The proof he gave is fairly technical even by modern standards, and we will not see it here. He even created the first stochastic integral with respect to the Brownian motion - however, since the integrand considered was deterministic, this integral behaves exactly like a Riemann–Stieltjes integral. This type of stochastic integral is named the *Wiener integral* in his honor (and to distinguish it from the "real" stochastic one, the Itô integral).

In recent times, the Brownian motion has been used everywhere in science. In mathematics, Mark Yor is one of the current researchers on anything related to Brownian motion and much more. For an easier textbook dedicated to Brownian motion, we cite here Mörters and Peres (2010) and for a more technical and in-depth one Revuz and Yor (2004).

15.2 Definition

We work in a probability space $(\Omega, \mathscr{F}, \mathbf{P})$ endowed with a complete filtration $\{\mathscr{F}_t\}$. We shall give two definitions of this process. The reason for two and not one definition is the historical perspective presented above. The two processes in the definitions are identical.

Definition 15.1 (Wiener Process) *We say that the process W_t is a standard (one-dimensional) Wiener process if it is adapted to the filtration $\{\mathscr{F}_t\}$ and has the following properties:*

1. *$W_0 = 0$ (it starts from 0);*

2. *The process has stationary, independent increments;*

3. *The increment $W_{t+h} - W_t$ is distributed as an $N(0, h)$ random variable;*

4. *The function $t \to W_t$ is continuous almost surely.*

The last property says that almost surely the paths of the process are continuous. If you recall, one may look at a process either at a fixed t as a random variable, or pathwise. The continuity property is a path property.

If the filtration is not specified, then we consider the standard filtration generated by the process $\mathscr{F}_t = \sigma(W_s, s \le t)$.

The problem with this definition is that it arises from the perceived properties of the physical phenomenon. As such, it is not very clear if they do not contradict themselves and if such a process in fact exists. The main issue is the fourth property (4): it is not very clear whether or this property contradicts the other three, and therefore a process with all these properties may not in fact exist. Norbert Wiener proved in 1923 that such a process in fact exists, and the main idea of his proof is to restrict his search to the space of continuous functions where he was able to construct a process with the first three properties thus proving that such a process (later called Wiener process) exists.

Definition 15.2 (Brownian Motion) *For $x, y \in \mathbb{R}$ and $t > 0$, define*

$$p(t, x, y) = \frac{1}{2\pi t} e^{-\frac{(x-y)^2}{2t}}.$$

For a probability space $(\Omega, \mathscr{F}, \mathbf{P})$, define a stochastic process with finite dimensional distribution

$$\mathbf{P}(B_{t_1} \in A_1, \dots, B_{t_n} \in A_n) =$$

$$\int_{A_1 \times \cdots \times A_n} p(t_1, 0, x_1) p(t_2 - t_1, x_1, x_2) \dots p(t_n - t_{n-1}, x_{n-1}, x_n) dx_1 dx_2 \dots dx_n$$

for any $n \in \mathbb{N}$, $0 < t_1 < t_2 \cdots < t_n$, and any A_i's Borel sets in $\mathbb{R}$. We also use the convention $p(0, 0, y)dy = \delta_{\{0\}}(y)$, the Dirac function at 0.

A stochastic process B_t with these properties is called a Brownian motion starting at zero.

A Brownian motion with the properties stated above and starting at 0 is called a *standard Brownian motion*. Either of the definitions defines the same process. The first definition concentrates on path properties of the process, while the second defines the *transition probability* of the stochastic process. From the second definition, we clearly see that the joint density for $B_{t_1}, \ldots, B_{t_n}$ is

$$f(x_1, \ldots, x_n) = p(t_1, 0, x_1)p(t_2 - t_1, x_1, x_2) \ldots p(t_n - t_{n-1}, x_{n-1}, x_n)$$

Using this density, we can calculate any desired probabilities associated with the Brownian motion.

◼ EXAMPLE 15.1   **Conditional on the future**

Suppose we observe the Brownian motion at t and its value is a particular number, say β. We want to look at where the process was in between 0 and t; for example, we look at the distribution of B_s conditional of $B_t = \beta$ for some $0 < s < t$. Since we know the joint density of B_s, B_t, which is

$$f_{s,t}(x, y) = p(s, 0, x)p(t - s, x, y),$$

we can easily calculate the density of $B_s | B_t = \beta$ as

$$f_{s|t}(x|\beta) = \frac{f_{s,t}(x, \beta)}{f_t(\beta)} = \frac{p(s, 0, x)p(t - s, x, \beta)}{p(t, 0, \beta)}$$

Substituting the transition probability after simplifications, we can express

$$f_{s|t}(x|\beta) = Ke^{-\frac{t(x - \beta s/t)^2}{2s(t-s)}}, \quad x \in \mathbb{R}$$

This is the density of a normal with mean $\frac{\beta s}{t}$ and variance $\frac{s(t-s)}{t}$.

◼ EXAMPLE 15.2   **Brownian motion is a Martingale**

Let $0 \le s \le t$. Suppose the filtration $\{\mathscr{F}_t\}_t$ is either given and the Brownian motion B_t is adapted to it, or it is generated by the Brownian motion. We can write

$$\begin{aligned}
\mathbf{E}[B_t \mid \mathscr{F}_s] &= \mathbf{E}[B_t - B_s + B_s \mid \mathscr{F}_s] \\
&= \mathbf{E}[B_t - B_s \mid \mathscr{F}_s] + \mathbf{E}[B_s \mid \mathscr{F}_s] \\
&= \mathbf{E}[B_t - B_s] + B_s \\
&= B_s,
\end{aligned}$$

where we used the independence of increments in the first expectation and measurability of B_s in the second expectation. Since we also have $\mathbf{E}[B_t^2] = t$, the Brownian motion is an L^2 martingale.

15.2.1 Brownian motion as a Gaussian process

Brownian motion falls into a wider category of stochastic processes; it is a special case of a Gaussian process.

Recall, if $X = (X_{t_1}, ..., X_{t_d})$ is a random vector, then

$$\mathbf{E}X = (\mathbf{E}X_{t_1}, ..., \mathbf{E}X_{t_d}) \in \mathbb{R}^d$$

and the covariance matrix of X, denoted by $\Sigma_X = (\Sigma_X(i, j))_{i,j=1,...,d}$, is defined by

$$\Sigma_X(i, j) = Cov(X_{t_i}, X_{t_j})$$

for every $i, j = 1, ..., d$.

Recall the notation of a scalar product of two vectors:

$$\langle x, y \rangle = x^T y = \sum_{i=1}^{d} x_i y_i$$

if $x = (x_1, ..., x_d), y = (y_1, ..., y_d) \in \mathbb{R}^d$, and we denoted by x^T the transpose of the $d \times 1$ matrix x.

With this, we recall the definition of a multivariate normal distribution with mean $\mu \in \mathbb{R}^d$ and covariance matrix Σ, denoted $N(\mu, \Sigma)$. This is the distribution of a d-dimensional vector with joint density

$$f(\mathbf{x}) = f(x_1, \dots, x_d) = \frac{1}{\sqrt{(2\pi|\Sigma|)^d}} e^{-(\mathbf{x}-\mu)^T \Sigma^{-1}(\mathbf{x}-\mu)}, \qquad (15.1)$$

where $|\Sigma|$ and Σ^{-1} are, respectively, the determinant and the inverse of the matrix Σ.

Definition 15.3 (Gaussian process) *A stochastic process X_t on a filtered probability space $(\Omega, \mathscr{F}, \mathbf{P}, \{\mathscr{F}_t\}_t)$ is called a Gaussian process if and only if*

1. *X_t is adapted to the filtration $\{\mathscr{F}_t\}_t$, and*

2. *For any $d \in \mathbb{N}$ and all $0 \leq t_1 \leq t_2 \leq \cdots \leq t_d$, the vector $(X_{t_1}, \dots, X_{t_d})$ has a multivariate normal distribution and the corresponding covariance matrix Σ is invertible.*

*The vector $(X_{t_1}, \dots, X_{t_d})$ with these properties is called **a Gaussian vector**.*

For a Gaussian process (and vector), we can calculate its characteristic function easily.

Theorem 15.4 *Let $X = (X_1, ..., X_d)$ be a Gaussian vector, and denote by μ its expectation vector and by Σ_X its covariance matrix. Then the characteristic function of X is*

$$\varphi_X(u) = e^{i\langle \mu, u \rangle - \frac{1}{2}\langle u, \Sigma_X u \rangle} \qquad (15.2)$$

for every $u \in \mathbb{R}^d$.

We pose the proof of the theorem as an exercise.

Remark 15.5 *We present the theorem here since, in the literature, oftentimes a Gaussian vector is defined through its characteristic function given in Theorem 15.4. That is, a random vector X is called a Gaussian random vector if its characteristic function is given by*

$$\varphi_X(u) = e^{i\langle \mu, u\rangle - \frac{1}{2}\langle u, \Sigma u\rangle}$$

for some vector μ and some symmetric positive definite matrix Σ.

The next two theorems allow us to define a general version of the Brownian motion.

Theorem 15.6 *If X is a d-dimensional Gaussian vector, then there exist a vector $m \in \mathbb{R}^d$ and a d-dimensional square matrix $A \in M_d(\mathbb{R})$ such that*

$$\mathbf{X} =^{(d)} m + A\mathbf{Z}$$

(where $=^{(d)}$ stands for the equality in distribution) and $\mathbf{Z} \sim N(0, I_d)$.

Proof: Let $A \in M_d(\mathbb{R})$ be a d-dimensional square matrix with real elements and define

$$\mathbf{Y} = m + A\mathbf{Z}$$

where $\mathbf{Z}$ is an $N(0, I_d)$ Gaussian vector. For every $u \in \mathbb{R}^d$, we have

$$
\begin{aligned}
\varphi_{\mathbf{Y}}(u) &= \mathbf{E}e^{i\langle u, \mathbf{Y}\rangle} \\
&= e^{i\langle m, u\rangle}\mathbf{E}e^{i\langle u, A\mathbf{Z}\rangle} \\
&= e^{i\langle m, u\rangle}\mathbf{E}e^{i\langle A^T u, \mathbf{Z}\rangle} \\
&= e^{i\langle m, u\rangle}\varphi_{\mathbf{Z}}(A^T u).
\end{aligned}
$$

Now suppose that $\mathbf{X}$ is a given Gaussian vector. Since Σ_X is symmetric and positive definite, there exists $A \in M_d(\mathbb{R})$ such that

$$\Lambda_X = AA^T.$$

Let $\mathbf{Z}_1 \sim N(0, I_d)$. We apply Theorem 15.4 and the beginning of this proof. It follows that $\mathbf{X}$ and $m + A\mathbf{Z}_1$ have the same characteristic function and therefore the same law. ∎

The converse of Theorem 15.6 is also true.

Theorem 15.7 *If*

$$\mathbf{X} = m + A\mathbf{Z}$$

with $m \in \mathbb{R}^d$, $A \in M_d(\mathbb{R})$, and $\mathbf{Z} \sim N(0, 1)$, then $\mathbf{X}$ is a Gaussian vector.

Proof: We just use the theorem 15.4 again.   ∎

Definition 15.8 (Generalized Brownian motions) *Suppose on a probability space* $(\Omega, \mathscr{F}, \mathbf{P})$ *we are given a Brownian motion* B_t. *Then*

- *for any* x, $B_t + x$ *is a Brownian motion* **started from** x;

- *for any* $\sigma > 0$, *the process* σB_t *is a* **scaled** *Brownian motion, or a Brownian motion with* **volatility** σ;

- *for any* $\mu \in \mathbf{R}$, *the process* $\mu t + B_t$ *is called a Brownian motion* **with drift**.

15.3 Properties of Brownian Motion

The Brownian motion is the stochastic process most studied and best understood. The properties it possesses are numerous, and books have been published which only state these properties. For instance, Borodin and Salminen (2002) is the best reference existing today about Brownian motion. In this section, we only present the fundamental properties.

Proposition 15.9 *Suppose* B_t *is a Brownian motion. Then for all* $0 \leq t_1 \leq \cdots \leq t_d$, $(B_{t_1}, \ldots, B_{t_d})$ *is a Gaussian vector with expectation* 0 *and covariance matrix* $\Sigma = (\Sigma_{i,j})$ *given by*

$$\Sigma_{i,j} = t_i \wedge t_j = \min(t_i, t_j).$$

Proof: We just need to calculate the covariance between two values of the Brownian motion. We have, for some $t_i < t_j$

$$
\begin{aligned}
Cov(B_{t_i}, B_{t_j}) = \mathbf{E}[B_{t_i} B_{t_j}] &= \mathbf{E}[B_{t_i}(B_{t_j} - B_{t_i} + B_{t_i})] \\
&= \mathbf{E}[B_{t_i}(B_{t_j} - B_{t_i})] + \mathbf{E}[B_{t_i}^2] \\
&= \mathbf{E}[B_{t_i}]\mathbf{E}[(B_{t_j} - B_{t_i})] + t_i \\
&= t_i
\end{aligned}
$$

by the independence of the increments and since the increment $B_{t_i} = B_{t_i} - B_0$ is $N(0, t_i)$. Repeating the argument for $t_j < t_i$, we get

$$Cov(B_{t_i}, B_{t_j}) = t_j$$

Thus we obtain the result stated.   ∎

Lemma 15.10 (Scaling invariance property) *Suppose a standard Brownian motion* B_t *is defined on a probability space* $(\Omega, \mathscr{F}, \mathbf{P})$. *Then the process*

$$X_t = \frac{1}{a} B_{a^2 t}$$

is also a standard Brownian motion, for any $a > 0$.

Proof: The proof is very simple; one just needs to verify the properties in Definition 15.1.   ∎

Lemma 15.11 (Time invariance property) *Suppose a standard Brownian motion B_t is defined on a probability space $(\Omega, \mathscr{F}, \mathbf{P})$. Then the process*

$$X_t = tB_{\frac{1}{t}}, \text{ for all } t > 0,$$

with $X_0 = 0$, is also a standard Brownian motion.

Proof: Suppose $0 \le t_1 \le \cdots \le t_d$, be real times. For each of these, the resulting $X_{t_i} = t_i B_{1/t_i}$ is a Gaussian random variable and the expectations are all zero. To show that X_t is a Brownian motion, it is then enough to check that the covariance structure is the same. For t_i, t_j, we have

$$Cov(X_{t_i}, X_{t_j}) = t_i t_j Cov(B_{1/t_i}, B_{1/t_j}) = t_i t_j \left(\frac{1}{t_i} \wedge \frac{1}{t_j} \right) = \begin{cases} t_i, & \text{if } t_i \le t_j \\ t_j, & \text{if } t_i > t_j \end{cases},$$

so the process has the same covariance structure. The paths of X_t are a.s. continuous if $t > 0$. At $t = 0$, it is a bit tricky but we can use the law of iterated logarithm (stated next) to show that $B_{1/t}$ is dominated by t as $t \to 0$. Therefore, the limit of X_t is 0 and thus the paths are a.s. continuous. This concludes the proof.   ∎

In Definition 15.1, we explicitly stated that the Brownian motion has continuous paths a.s., while in Definition 15.2 we did not. The next definition and theorem explain why we will get the continuity property for the Brownian motion automatically.

Definition 15.12 *Suppose that X_t and Y_t are stochastic processes on some probability space $(\Omega, \mathscr{F}, \mathbf{P})$. We say that X_t is a version of Y_t if*

$$\mathbf{P}(\{\omega : X_t(\omega) = Y_t(\omega)\}) = 1, \text{ for all } t.$$

Essentially, from a distribution perspective, the processes are the same but their path properties may be different.

Theorem 15.13 (Kolmogorov continuity theorem) *Suppose the stochastic process X_t on $(\Omega, \mathscr{F}, \mathbf{P})$ has the following property: For all $T > 0$, there exist positive constants $\alpha, \beta,$ and C such that*

$$\mathbf{E}[|X_t - X_s|^\alpha] \le C|t - s|^{1+\beta}, \text{ for all } s, t \in [0, T].$$

Then there exists a continuous version of X_t.

We will not prove this theorem here; for a proof, see (Stroock and Varadhan, 1979, p. 51). We can attempt to apply the theorem for a Brownian motion. For $\alpha = 2$, we will obtain

$$\mathbf{E}[|B_t - B_s|^2] = |t - s|,$$

which does not work because β has to be strictly positive. However, we can show (see Problem 15.5) that

$$\mathbf{E}[|B_t - B_s|^4] = 4|t - s|^2$$

so the inequality holds with $\alpha = 4$, $\beta = 1$, and $C = 4$, and therefore the Brownian motion always has a version which is continuous. From now on, we assume that B_t is always the continuous version of the Brownian motion.

■ EXAMPLE 15.3   Continuous version to a totally discontinuous process

Consider the following probability space $([0, \infty), \mathscr{B}([0, \infty), \mathbf{P})$, where the probability measure $\mathbf{P}$ defined on the Borel sets has the property that $\mathbf{P}(\{x\}) = 0$ for any $x \in [0, \infty)$. Now let the stochastic processes X_t and Y_t be defined in the following way:

$$X_t(\omega) = \begin{cases} 1, & \text{if } t = \omega \\ 0, & \text{otherwise}, \end{cases}$$

and,

$$Y_t(\omega) = 0, \ \forall (t, \omega) \in [0, \infty) \times [0, \infty).$$

Then we can show that Y_t is a version of X_t. Note that X_t is discontinuous for all paths ω, while Y_t is continuous for all ω.

To show that Y_t is a version of X_t, let us look at the one-dimensional distribution. We have

$$F_{X_t}(x) = \mathbf{P}(X_t \le x) = \mathbf{P}(X_t^{-1}(-\infty, x]) = \begin{cases} \mathbf{P}(\emptyset), & \text{if } x < 0 \\ \mathbf{P}([0, \infty) \setminus \{t\}), & \text{if } 0 \le x < t \\ \mathbf{P}([0, \infty)), & \text{if } x \ge t \end{cases}$$

$$= \begin{cases} 0, & \text{if } x < 0 \\ 1 - \mathbf{P}(\{t\}), & \text{if } 0 \le x < t \\ 1, & \text{if } x \ge t \end{cases} = \begin{cases} 0, & \text{if } x < 0 \\ 1, & \text{if } x \ge 0 \end{cases}$$

since the probability of a singleton is zero ($\mathbf{P}(\{x\}) = 0$). Furthermore, the distribution for Y_t is easily calculated as

$$F_{Y_t}(x) = \mathbf{P}(Y_t \le x) = \mathbf{P}(Y_t^{-1}(-\infty, x]) = \begin{cases} 0, & \text{if } x < 0 \\ 1, & \text{if } x \ge 0 \end{cases}$$

It is easy to see that a similar derivation will hold for $F_{X_{t_1}, \ldots, X_{t_n}}$. Therefore, this will imply that Y_t is a version of X_t, that is, $\mathbf{P}\{\omega : X_t(\omega) = Y_t(\omega)\} = 1$ for any t.

Definition 15.14 (Multidimensional Brownian motion) *We say that a d-dimensional process* $\mathbf{B}_t = (B_1(t), \ldots, B_d(t))$ *is a Brownian motion on* $\mathbb{R}^d$ *if the one-dimensional components* $B_i(t)$ *are all independent standard Brownian motions on* $\mathbb{R}$.

15.3.1 Hitting times. Reflection principle. Maximum value

Recall that a standard Brownian motion B_t starts from 0. Since the process is symmetric around 0, all results concerning hitting a positive value $a > 0$ are identical for hitting its symmetric value $-a$. Let τ_a be the first time the Brownian motion process hits a. This is clearly a stopping time and we want to calculate its distribution. We start with a very important principle that will help us to calculate this distribution.

Lemma 15.15 (Reflection principle) *Let B_t a Brownian motion, and for a number $a > 0$, let τ_a the first hitting time of a. Then for any number $z \le a$, we have*

$$\mathbf{P}(\tau_a \le t, B_t \le z) = \mathbf{P}(B_t \ge 2a - z).$$

Proof: To prove this result, as soon as the Brownian motion hits a, we reflect it around a. Figure 15.1 on page 474 shows one such reflected path.

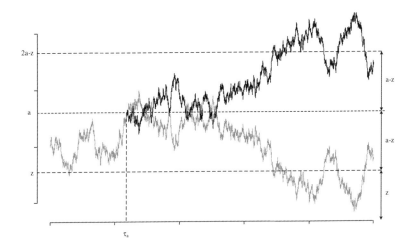

Figure 15.1   The initial Brownian motion is in gray. At the moment when a is hit for the first time, we reflect the process around a in black.

The reasoning notices that the event $\{\omega \mid \tau_a(\omega) \le t, B_t(\omega) \le z\}$ contains every path ω such that a was reached before time t, and the Brownian motion ends below z at time t. However, for each such path we reflect starting at the moment τ_a, which we know is happening before t, and we find the black path in Figure 15.1. It is easy to see that each and every such path will end above $2a - z$ and each and every such path is a path in the event $\{\omega \mid B_t(\omega) \ge 2a - z\}$. Therefore, we must have

$$\{\tau_a \le t, B_t \le z\} \subseteq \{B_t \ge 2a - z\}$$

Conversely, take a path in $\{\omega \mid B_t(\omega) \ge 2a - z\}$. This is the black path in Figure 15.1. Since the B_t process is continuous and at time t it is above a (since

$2a - z = a + a - z > a$), it must have reached a at some earlier time. Call τ_a the first time a is reached, and reflect from that time forward. This will produce the gray path in Figure 15.1 and we can see immediately that the corresponding B_t value is below z. Thus we showed that

$$\{B_t \geq 2a - z\} \subseteq \{\tau_a \leq t, B_t \leq z\}$$

By double inclusion, the two sets must be the same and therefore they have the same probability. ∎

We can write using the reflection principle for $z = a$

$$\mathbf{P}(\tau_a \leq t, B_t \leq a) = \mathbf{P}(B_t \geq a)$$

Also note that, since B_t is continuous, a path w in $B_t(w) \geq a$ will need to have $\tau_a(w)$ happen before t, therefore $\{B_t \geq a\} \subseteq \{\tau_a \leq t\}$, and so

$$\mathbf{P}(\tau_a \leq t, B_t \geq a) = \mathbf{P}(\{\tau_a \leq t\} \cap \{B_t \geq a\}) = \mathbf{P}(B_t \geq a).$$

But we can write

$$\begin{aligned}
\mathbf{P}(\tau_a \leq t) &= \mathbf{P}(\tau_a \leq t, B_t \leq a) + \mathbf{P}(\tau_a \leq t, B_t > a) \\
&= \mathbf{P}(B_t \geq a) + \mathbf{P}(\tau_a \leq t, B_t \geq a) \\
&= 2\mathbf{P}(B_t \geq a) \\
&= \frac{2}{\sqrt{2\pi t}} \int_a^\infty e^{-\frac{x^2}{2t}} \, dx
\end{aligned}$$

Making a change of variables $y = x/\sqrt{t}$ will allow us to write the distribution in terms of the standard normal cdf; more specifically

$$\mathbf{P}(\tau_a \leq t) = 2\left(1 - \Phi\left(\frac{a}{\sqrt{t}}\right)\right)$$

We can also look at the case when a is negative, treating the paths in the same way, and we can also calculate the density of the hitting time by taking derivatives. We state the distribution in the following theorem.

Theorem 15.16 (Distribution of the hitting time for a Brownian motion) *Let B_t be a Brownian motion, and for an $a \neq 0$, let τ_a be the first hitting time of a. Then the random variable τ_a has the distribution function*

$$F_{\tau_a}(t) = 2\left(1 - \Phi\left(\frac{|a|}{\sqrt{t}}\right)\right) = \frac{2}{\sqrt{2\pi t}} \int_{\frac{|a|}{\sqrt{t}}}^\infty e^{-\frac{x^2}{2}} \, dx,$$

and the corresponding density

$$f_{\tau_a}(t) = \frac{|a|}{t\sqrt{2\pi t}} e^{-\frac{a^2}{2t}}.$$

In the expression above, $\Phi(x)$ is the cdf of an $N(0,1)$ random variable.

Using this distribution, we can also calculate, since τ_a is a positive random variable,

$$\mathbf{E}[\tau_a] = \int_0^\infty \mathbf{P}(\tau_a > t)dt = \int_0^\infty \left(1 - \frac{2}{\sqrt{2\pi t}} \int_{\frac{|a|}{\sqrt{t}}}^\infty e^{-\frac{x^2}{2}} dx\right) dt$$

$$= \frac{2}{\sqrt{2\pi t}} \int_0^\infty \int_0^{\frac{|a|}{\sqrt{t}}} e^{-\frac{x^2}{2}} dxdt = \frac{2}{\sqrt{2\pi t}} \int_0^\infty \int_0^{\frac{a^2}{x^2}} e^{-\frac{x^2}{2}} dtdx$$

$$= \frac{2a^2}{\sqrt{2\pi t}} \int_0^\infty \frac{1}{x^2} e^{-\frac{x^2}{2}} dx \geq \frac{2a^2 e^{-\frac{1}{2}}}{\sqrt{2\pi t}} \int_0^1 \frac{1}{x^2} dx = \infty.$$

Thus, τ_a, the time to hit a for the first time, has $\mathbf{P}(\tau_a < \infty) = 1$, yet it has infinite expectation. This is similar to the case of the random walk.

Finally, we can also look at the running maximum random variable. Specifically, consider

$$M_t = \max_{0 \leq s \leq t} B_s,$$

where B_t is a standard Brownian motion. Note that $M_t \geq a$ if and only if $\tau_a \leq t$ since the Brownian motion is continuous. This observation allows us to write

$$\mathbf{P}(M_t \geq a) = \mathbf{P}(\tau_a \leq t) = \frac{2}{\sqrt{2\pi t}} \int_a^\infty e^{-\frac{x^2}{2t}} dx,$$

which will produce the density of the maximum of the Brownian motion. Note that, in fact, this is the same density that we found earlier. Thus the expected value of the maximum on any interval $[0, t]$ is infinite.

15.3.2 Quadratic variation

We now proceed with notions which are going to be extremely important for the next chapter dedicated to stochastic calculus. These notions are born from deterministic equivalents, so let us give the proper definitions first.

Definition 15.17 (Variation for deterministic functions) *Let $f : [0, \infty) \to \mathbb{R}$ be a deterministic function. Let $\pi_n = (0 = t_0 < t_1 < \ldots t_n = t)$ be a partition of the interval $[0, t]$ with n subintervals. Let $\|\pi_n\| = \max_i(t_{i+1} - t_i)$ be the length of the largest subinterval in the partition. We define the first-order variation as*

$$FV_t(f) = \lim_{\|\pi_n\| \to 0} \sum_{i=0}^{n-1} |f(t_{i+1}) - f(t_i)|.$$

We define the quadratic variation as

$$[f, f]_t = \lim_{\|\pi_n\| \to 0} \sum_{i=0}^{n-1} |f(t_{i+1}) - f(t_i)|^2.$$

In general, the d-order variation is defined as

$$\lim_{||\pi_n||\to 0} \sum_{i=0}^{n-1} |f(t_{i+1}) - f(t_i)|^d.$$

We shall see in a moment why we have not attached a notation to higher order variations.

Lemma 15.18 *The first-order variation at t of a differentiable function f with continuous derivative is the length of the curve from 0 to t, that is*

$$FV_t(f) = \int_0^t |f'(s)| ds$$

Proof: This lemma is easy to prove using the mean value theorem. Recall that for any differentiable function f with continuous derivative ($f \in \mathscr{C}^1([0, \infty))$), we have

$$f(t_{i+1}) - f(t_i) = f'(t_i^*)(t_{i+1} - t_i),$$

where t_i^* is some point between t_i and t_{i+1}. Using this, we obtain

$$FV_t(f) = \lim_{||\pi_n||\to 0} \sum_{i=0}^{n-1} |f(t_{i+1}) - f(t_i)|$$

$$= \lim_{||\pi_n||\to 0} \sum_{i=0}^{n-1} |f'(t_i^*)|(t_{i+1} - t_i)$$

$$= \int_0^t |f'(s)| ds,$$

recognizing that the sum is just a Darboux sum corresponding to the integral. ∎

Lemma 15.19 *For a deterministic function $f \in \mathscr{C}^1([0, \infty))$, all d-order variations with $d \geq 2$ are zero.*

Proof: This lemma is the reason why we do not need to talk about higher order variations, since they are all 0. To prove the lemma, we look at the quadratic variation. All higher orders use the same reasoning. We have

$$[f, f]_t = \lim_{||\pi_n||\to 0} \sum_{i=0}^{n-1} |f(t_{i+1}) - f(t_i)|^2$$

$$= \lim_{||\pi_n||\to 0} \sum_{i=0}^{n-1} |f'(t_i^*)|^2 (t_{i+1} - t_i)^2$$

$$\leq \lim_{||\pi_n||\to 0} ||\pi|| \sum_{i=0}^{n-1} |f'(t_i^*)|^2 (t_{i+1} - t_i)$$

$$= \lim_{||\pi_n||\to 0} ||\pi|| \lim_{||\pi_n||\to 0} \sum_{i=0}^{n-1} |f'(t_i^*)|^2 (t_{i+1} - t_i).$$

Once again the second term is the integral $\int_0^t |f'(s)|^2 ds$ and thus bounded, and the first term converges to 0. Therefore, the product goes to 0. $\blacksquare$

Note that the only way the product in the proof above is not 0 is if the integral is infinite. In order for this to happen, the derivative cannot exist on the interval $[0, t]$. We can, in fact, repeat this argument for any interval $[s, t]$ and conclude that the only way a function (path) has finite quadratic variation is if it is not derivable on any interval $[s, t]$. But now, since s and t are arbitrary, we can conclude that the function is not derivable at any point.

Next we look at the corresponding notions for stochastic processes.

Definition 15.20 (Quadratic Variation for stochastic processes) *Let X_t be a stochastic process on the probability space $(\Omega, \mathscr{F}, \mathbf{P})$ with filtration $\{\mathscr{F}_t\}_t$. Let $\pi_n = (0 = t_0 < t_1 < \ldots t_n = t)$ be a partition of the interval $[0, t]$. We define the quadratic variation process*

$$[X, X]_t = \lim_{||\pi_n|| \to 0} \sum_{i=0}^{n-1} |X_{t_{i+1}} - X_{t_i}|^2,$$

where the limit is defined in probability.

Note that the quadratic variation process is a stochastic process. The quadratic variation may be calculated explicitly only for certain classes of stochastic processes. In the next chapter, we shall see that the quadratic variation process for martingales may be expressed in terms of a stochastic integral. In this chapter, we concentrate on the Brownian motion process.

Theorem 15.21 (Quadratic variation for the Brownian motion) *If B_t denotes a standard Brownian motion, the quadratic variation on $[0, t]$ is always equal to t for all t, that is*

$$[B, B]_t = t$$

.

Proof: This remarkable theorem says that the stochastic process, which is the quadratic variation in the case of the Brownian motion, is not stochastic at all! In this case of the Brownian motion, it is deterministic and very simple to express. To prove the theorem, we first look at the expectation of the quadratic variance process. For a partition $\pi_n = (0 = t_0 < t_1 < \ldots t_n = t)$, we denote

$$V_{\pi_n}(t) = \sum_{i=0}^{n-1} |B_{t_{i+1}} - B_{t_i}|^2$$

The expected value is

$$\mathbf{E}[V_{\pi_n}(t)] = \mathbf{E}\left[\sum_{i=0}^{n-1} |B_{t_{i+1}} - B_{t_i}|^2\right] = \sum_{i=0}^{n-1} \mathbf{E}|B_{t_{i+1}} - B_{t_i}|^2 = \sum_{i=0}^{n-1} t_{i+1} - t_i = t$$

This shows that the expected value is always t for all interval $[0, t]$ and any partition. We will show next that the variance of $V_{\pi_n}(t)$ converges to 0 as the partition norm goes to 0 in probability. This will imply that the limiting random variable has mean t and variance 0, and the only such random variable is the constant t. To proceed

$$Var(V_{\pi_n}(t)) = \mathbf{E}[(V_{\pi_n}(t) - \mathbf{E}[V_{\pi_n}(t)])^2] = \mathbf{E}\left[\left(\sum_{i=0}^{n-1} |B_{t_{i+1}} - B_{t_i}|^2 - t\right)^2\right]$$

We will use the notation $\Delta B_i = B_{t_{i+1}} - B_{t_i}$, and we will take the partition made of equal-length subintervals $t_{i+1} - t_i = h$. Note that $h = t/n$. With this, we have

$$Var(V_{\pi_n}(t)) = \mathbf{E}\left[\sum_{i=0}^{n-1} \Delta B_i^2 \sum_{j=0}^{n-1} \Delta B_j^2\right] - 2t\mathbf{E}\left[\sum_{i=0}^{n-1} \Delta B_i^2\right] + t^2$$

$$= \mathbf{E}\left[\sum_{i=0}^{n-1}\sum_{j=0}^{n-1} \Delta B_i^2 \Delta B_j^2\right] - 2t\sum_{i=0}^{n-1} h + t^2$$

$$= \sum_{i=0}^{n-1} \mathbf{E}[\Delta B_i^4] + 2\sum_{i<j} \mathbf{E}[\Delta B_i^2]\mathbf{E}[\Delta B_j^2] - t^2,$$

where we used that the increments are independent and that all terms containing expectations of odd powers of the increment are zero. Now we shall use the fact that the increments are normally distributed $N(0, h)$. Therefore we have

$$\mathbf{E}[\Delta B_i^2] = h \text{ and } \mathbf{E}[\Delta B_i^4] = 3h^2.$$

Thus

$$Var(V_{\pi_n}(t)) = \sum_{i=0}^{n-1} 3h^2 + 2\sum_{i<j} h^2 - t^2 = 3nh^2 + 2\frac{n(n-1)}{2}h^2 - t^2$$

$$= 2nh^2 + n^2h^2 - (nh)^2 = 2nh^2 = 2th,$$

where we used that $t = nh$. The last term goes to 0 as $h \to 0$ since t is fixed, so the variance converges to 0 along a subsequence. We can also show that the variance is bounded for any partition with n elements, which will imply that the variance indeed converges to 0. Thus the proof is completed. ∎

Remark 15.22 *It is easy to see that all variations of higher order than 2 will be zero. We will just repeat the argument we made for deterministic functions. The first-order variation is infinite. This we already remarked after Lemma 15.19, but it is easy enough to show for the Brownian motion.*

$$FV_t(B) = \lim_{||\pi_n|| \to 0} \sum_{i=0}^{n-1} |B_{t_{i+1}} - B_{t_i}|$$

The density of the absolute value of a normal with mean 0 and variance h is easily calculated as

$$f(x) = \frac{\sqrt{2}}{\sqrt{h\pi}} e^{-\frac{x^2}{2h}}, x \geq 0.$$

The expected value is $\sqrt{\frac{2h}{\pi}}$, so for any partition with n equal interval length h, we have

$$\mathbf{E}\left[\sum_{i=0}^{n-1} |B_{t_{i+1}} - B_{t_i}|\right] = n\sqrt{\frac{2h}{\pi}} = \sqrt{\frac{2nt}{\pi}}.$$

Thus the expectation goes to infinity and thus the limit in L^1 of the random variable is ∞. This will imply that in probability the convergence is also to ∞, and thus first-order variation is unbounded.

15.4 Simulating Brownian Motions

We conclude this chapter with two very important issues from the perspective of applications. However, they are also very simple if we understand the theory of Brownian motion.

15.4.1 Generating a Brownian motion path

To generate a Brownian motion, one needs to remember that the paths have infinite variation and therefore are nowhere differentiable. Such a process is hard to imagine and of course impossible to draw. Thus, we will only be able to provide the sample points of a generated path for some times $0 = t_0 < t_1 < \cdots < t_n$. To do so, we need to remember the properties of Brownian motion. Recall that the increments are independent, and all have normal distribution with mean 0 and variance of the size of the time increment.

For simplicity, the following example we generate a standard Brownian motion from $t_0 = 0$ to $t_n = 1$ and we sample the Brownian motion at each of the times $t_i = i\Delta t$ with $\Delta t = 1/10000$, $i \in \{0, 1, 2, \ldots, 10,000\}$. Thus we need to generate normals with mean 0 and variance 0.0001. The R code below does just that.

```
times=seq(from=0,to=1,by=0.0001)
brownincrements=rnorm(10000,0,sd=sqrt(0.0001))
brownian=cumsum(c(0,brownincrements))
plot(times,brownian,type="l")
```

In the code above, the *brownincrements* vector contains the increments $\Delta B_i = B_{t_i} - B_{t_{i-1}}$. To obtain the path, we need to apply a function called *cumsum* which calculates a cumulative sum of the vector elements. Specifically, the element i of the cumulative sum vector would be

$$\sum_{j=1}^{i} \Delta B_j = B_i - B_0$$

for all i from 1 to n. We need to finally add the element $B_0 = 0$ and this will produce the entire path. The last command plots the path.

To create a Brownian motion with drift μ and volatility σ, we need to remember that such a path can be easily obtained using

$$\mu t + \sigma B_t$$

Thus such a path can be created by operating on the standard Brownian path created *brownian*. For example,

```
mu=0.01
sigma =0.3
xprocess=mu*times+sigma*brownian
plot ( times , xprocess , type=" l " )
```

15.4.2 Estimating parameters for a Brownian motion with drift

The second problem is to estimate the parameters of a Brownian motion. For a standard Brownian motion, there is nothing to estimate since there are no parameters. Suppose, however, that we observe a Brownian motion with drift

$$X_t = \mu t + \sigma B_t$$

Suppose that we observe this process at times $t_1, \ldots, t_n$ with $t_i = i\Delta t$. For example, the time interval Δt could be a day $1/365$ or a trading day $1/252$, and ΔX_i could be the daily return for day i.

If this is the case, let us look at the distribution of the increments $\Delta X_i = X_{t_i} - X_{t_{i-1}}$. We can write these increments as

$$\Delta X_i = \mu\Delta t + \sigma\Delta B_i,$$

and we can easily see that the increments of the process X_t are independent, normally distributed, with mean $\mu\Delta t$ and variance $\sigma^2\Delta t$. Thus we can apply the concepts presented in Chapter 8 to find the MLE's for both as

$$\hat{\mu} = \frac{1}{\Delta t}\overline{\Delta X} = \frac{1}{\Delta t}\frac{1}{n}\sum_{i=1}^{n}\Delta X_i,$$

and

$$\hat{\sigma} = \frac{1}{\sqrt{\Delta t}}S_{\Delta X} = \frac{1}{\sqrt{\Delta t}}\sqrt{\frac{1}{n}\sum_{i=1}^{n}(\Delta X_i - \overline{\Delta X})^2}.$$

Problems

15.1   Show that the standard Brownian motion is a martingale. Show that it is a Markov process.

15.2   Calculate the characteristic function of a Gaussian vector (see Theorem 15.4 on page 469).

15.3   Let B_t a standard Brownian motion. Show that $X_t = 3B_{\frac{t}{9}}$ is also a standard Brownian motion.

15.4   [Ornstein–Uhlenbeck process] Consider the process

$$X_t = e^{-t}B_{e^{2t}},$$

where B_t is a standard Brownian motion and the time corresponding to t in X_t is e^{2t} for the Brownian motion.
   a) Calculate $\mathbf{E}[X_t]$ and $V(X_t)$.
   b) Show that X_t has a normal distribution.
   c) Show that X_t and X_{-t} have the same finite dimensional distribution.

15.5   Let B_t a standard Brownian motion on $\mathbb{R}$.
   a) Show that for all f such that the integral exists we have

$$\mathbf{E}[f(B_t)] = \frac{1}{\sqrt{2\pi t}} \int_{-\infty}^{\infty} f(x)e^{-\frac{x^2}{2t}}\, dx$$

.
   b) Apply the previous part to $f(x) = x^{2k}$ and find a recurrence relation between $\mathbf{E}[B_t^{2k}]$ and $\mathbf{E}[B_t^{2(k-1)}]$.
   c) Use the previous part and induction to show that

$$\mathbf{E}[B_t^{2k}] = \frac{(2k)!}{2^k k!}t^k, \quad k \in \mathbb{N}.$$

15.6   Using your choice of software, simulate B_t a standard Brownian motion on the interval $t \in [0, 2]$. Using the simulated paths and the Central Limit Theorem, estimate

$$\mathbf{E}[B_2^4], \mathbf{E}[B_2^8], \mathbf{E}[B_2^{20}].$$

   Then simulate each of the processes B_t^4, B_t^8, B_t^{20} separately and obtain the previous expectations at $t = 2$ that way as well.

   Compare all the numbers obtained and also compare with the values in the previous problem. Use a minimum of 1 million simulated paths, and for each path use a time increment of $\Delta t = 0.01$.

15.7   Let $X_t, t \geq 0$ be defined as

$$X_t = \{B_t \mid B_t \geq 0\}, \quad \forall t > 0,$$

where B_t is a Brownian motion. That is, the process has the paths of the Brownian motion conditioned by being positive at t.
   a) Show that the pdf of X_t is

$$f_{X_t}(x) = 2f_{B_t}(x), \forall x \geq 0.$$

b) Calculate $\mathbf{E}[X_t]$ and $V(X_t)$.

c) Is X_t a Gaussian process?

d) Is X_t stationary?

e) Are X_t and $|B_t|$ identically distributed?

15.8   Let $err f(x)$ denote the regular error functions, that is,

$$err f(x) = \frac{2}{\sqrt{\pi}} \int_0^x e^{-t^2} dt.$$

Show that

$$erf(x) = 1 - \frac{e^{-x^2}}{\sqrt{\pi}} \left(\frac{1}{x} - \frac{1}{2x^3} + \ldots \right).$$

Furthermore, show that

$$\lim_{\varepsilon \to 0} \frac{2}{\varepsilon} \left(1 - \Phi \left(\frac{\delta}{\sigma\sqrt{\varepsilon}} \right) \right) = 0$$

for all $\sigma > 0$ and $\delta > 0$. Here, $\phi(x)$ denotes the cdf of a $N(0, 1)$ random variable.

15.9   If B_t is a Brownian motion, calculate the distribution of $|B_t|$. Calculate $\mathbf{E}|B_t|$ and $V(|B_t|)$.

15.10   If B_t is a Brownian motion and $X_t = e^{\sigma B_t}$, for some $\sigma > 0$, calculate the pdf of X_t. Calculate $\mathbf{E}[X_t]$ and $V(X_t)$. Calculate the transition probability

$$\mathbf{P}(X_t \leq y \mid X_{t_0} = y_0),$$

and give the density of this transition probability. For this part, use the functions in Definition 15.2.

15.11   If B_t is a Brownian motion on $\mathbb{R}$, show that $\frac{1}{t} B_t^2$ is distributed as a chi-squared random variable.

15.12   Suppose we observe the trajectory of a particle at time $t = 1$ and we see that the position is $X_t = 4$. We need to estimate the distribution of the particle at time $s = 1/2$. To do this, we assume that the position at time t is given by a standard Brownian motion. Calculate the distribution of the position at time $s = 1/2$ given that at time $t = 1$ it was at $X_t = 4$ and we know that at time 0 the position was at $X_0 = 0$. How does this distribution change if we know that at time 0 the position of the particle was at $X_0 = 3$?

15.13   [Brownian bridge] Let B_t be a Brownian motion started from 0. Consider the process B_t conditional on $B_1 = 0$. That is, consider the process $B_t \mid B_1 = 0$.

a) Show that this process is a Gaussian process.

b) Calculate for $s < t$ its covariance function, that is,

$$Cov(B_s, B_t \mid B_1 = 0).$$

c) Define the process
$$Z_t = B_t - tB_1.$$

Show that this process is a Brownian bridge.

15.14   Let B_t be a standard Brownian motion. Define

$$N_t = \min_{0 \le s \le t} B_s.$$

Calculate the pdf of N_t. Calculate its expected value.

15.15   Generate and plot the following processes. Use the time interval $[0, 1]$ and generate the sampled values with frequency $\Delta t = 1/252$.
   a) A standard Brownian motion started from 0.
   b) A Brownian motion with drift $\mu = 0.01$ and volatility $\sigma = 0.3$
   c) The process
$$S_t = 100 e^{\mu t + \sigma B_t},$$

with $S_0 = 100$ and parameters as in the previous part.
   d) The process
$$X_t = B_t - tB_1.$$

What do you observe about this process? B_1 is the value of the Brownian motion at time $t = 1$.

15.16   Generate one path of the following process:

$$X_t = 0.01t + 0.3B_t,$$

for $t \in [0, 1]$ and sampled with frequency $\Delta t = 1/252$. Let the observations be denoted with $x_0, x_1 \ldots, x_{252}$. Estimate the parameters μ and σ of the process

$$X_t = \mu t + \sigma B_t,$$

using these observations. Compare the estimates with the actual parameter values 0.01 and 0.3. Give 95% confidence intervals based on the estimators. Do these intervals contain the true parameters?

15.17   Repeat the previous problem but now generate a path using the process

$$S_t = 100 e^{\mu t + \sigma B_t},$$

with $S_0 = 100$ and the parameter values as in the previous problem. You need to think about how to find confidence intervals for the parameters in this problem.

## CHAPTER 16

STOCHASTIC DIFFERENTIAL EQUATIONS WITH RESPECT TO BROWNIAN MOTION

This is a chapter not normally included in a traditional textbook on stochastic processes. However, it is a crucial chapter for any and all applications of stochastic processes to finance. Since my main area of application is finance, this chapter is absolutely necessary for my students.

I am going to start this chapter with a motivating argument. Consider a regular differential equation

$$\frac{dx(t)}{dt} = b(t, x(t)),$$

where b is just a function of two variables. For example, $b(t, x) = 5x$ will produce a linear ordinary differential equation (ODE)

$$\frac{dx(t)}{dt} = 5x(t),$$

or the way it is commonly written in ODE textbooks

$$\frac{dx}{dt} = 5x.$$

These types of equations were good 50 or 100 years ago when we could not really observe things and when the models used were very simple. However, today we are

Probability and Stochastic Processes, First Edition. Ionuţ Florescu
© 2015 John Wiley & Sons, Inc. Published 2015 by John Wiley & Sons, Inc.

dealing with, can measure, and can observe a very dynamic world. From these observations, we know that we need to construct better models than in a deterministic world. To cope with the random nature of things, at the beginning of the last century Albert Einstein and others postulated principles that helped create such type of dynamics. To keep it a simple story, it seemed evident that to model reality an equation of the following type needs to be defined:

$$\frac{dx(t)}{dt} = b(t, x(t)) + \sigma(t, x(t)) \times \text{"}noise\text{"}.$$

The term "$noise$" needs to be properly defined. The first impulse was to define a stochastic process w_t such that "$noise$" $= w_t$. This will make the equation

$$dx(t) = b(t, x(t))dt + \sigma(t, x(t))w_t dt. \tag{16.1}$$

This process was named the "white noise" process, and from practical applications the following properties were postulated as being necessary for this process:

1. for any times $t_1 \neq t_2$, the random variables w_{t_1} and w_{t_2} are independent;

2. w_t has a stationary distribution;

3. $\mathbf{E}[w_t] = 0$;

4. w_t has continuous paths.

However, as was discovered quickly, such a process does not exist!

Norbert Wiener first introduced a form of this noise term for the case when $\sigma(t, x) = \sigma(t)$ was just a function of time (1933). This was the first time we could write an equation of the type described which we could define and made sense. Later, in 1944, Kiyoshi Itô handled the case when $\sigma(t, x)$ was a general function, and this led to what we now call the *stochastic integral* or the *Itô integral*. This introduction was revolutionary; it led to another way of thinking about stochastic processes and in fact it changed completely one of the fundamental calculus formulas, namely the integration by parts formula. This led to an entirely different type of calculus which is now called *stochastic calculus*.

The key idea is to look at the associated integral form of the ODE. Thus, instead of trying to construct a white noise process, we shall try to construct the integral with respect to this white noise process. The idea is similar to our previous treatment of random variables when the probability density function does not exist. It is amazing to me as a probabilist that today we teach undergraduate courses in the way probability was developed 150 years ago. We only use the modern theory of probability in graduate courses and that too only in mathematics and statistics, yet the material was developed 50–80 years ago. I strongly believe that those pioneers would be extremely unhappy with our inability, even today, of explaining their concepts in simple terms to engineers and practitioners of other branches of science.

16.1   The Construction of the Stochastic Integral

I will start this section continuing the motivating example above. We consider a discrete version of Equation (16.1) for times t_i in a partition $\pi_n = (0 = t_0 < t_1 < \cdots < t_n = t)$. We have, for $i \in \{1, 2, \ldots, n\}$

$$X_{t_{i+1}} - X_{t_i} = b(t_i, X_{t_i})\Delta t_i + \sigma(t_i, X_{t_i})w_{t_i}\Delta t_i, \qquad (16.2)$$

where $\Delta t_i = t_i - t_{i-1}$.

The key idea is to replace the term $w_{t_i}\Delta t_i$ with the increment of a stochastic process $\Delta B_i = \Delta B_{t_i} - \Delta B_{t_i - 1}$. Again, the requirements on the increments this time not on the process are the same as before: independent, stationary, mean 0, and continuous paths for the process. But we already know a stochastic process whose paths have these properties. That is the Brownian motion we introduced in the previous chapter. In fact, there are a multitude of processes with these properties. Within this chapter, we only consider the Brownian motion. If we take B_t to be a standard Brownian motion and sum the increments for all i in (16.2), we obtain

$$X_{t_n} - X_0 = X_t - X_0 = \sum_{i=1}^{n} b(t_i, X_{t_i})\Delta t_i + \sum_{i=1}^{n} \sigma(t_i, X_{t_i})\Delta B_i \qquad (16.3)$$

As we make $n \to \infty$ and consequently the partition norm $||\pi_n|| = \max_i(t_i - t_{i-1}) \to 0$, the first sum in (16.3) is a usual Riemann sum and it goes to a regular integral almost surely (in fact always for every path w):

$$\int_0^t b(t, X_t)dt.$$

However, for the second one we do not know whether the limit exists or even how to define it. This is why we need to construct and define the concept of the stochastic integral.

The actual construction is not very difficult. It actually proceeds in the same manner as we constructed expectations in Chapter 4. We once again start with simple functions. Consider a function of the type

$$f(t, \omega) = \sum_{i=1}^{n} e_i(\omega)\mathbf{1}_{(t_{i-1}, t_i]}(t).$$

Note that e_i's are random variables. To be able to get to the motivating example, the goal is to define

$$\int_s^t f(t, \omega)dB_t(\omega) = e_i(\omega)(B_{t_m}(\omega) - B_s(\omega)) + \sum_{i=m+1}^{n} e_i(\omega)(B_{t_i}(\omega) - B_{t_{i-1}}(\omega)),$$

for some m such that $t_{m-1} < s < t_m < \cdots < t_n = t$.

Note that we will work on the interval $[0, t]$ and will define the integral for this interval. We could work on more general intervals $[s, t]$, but the construction is identical, with obvious changes related to s.

Remark 16.1 *The way the e_j's are defined is very important as the next example shows.*

◼ EXAMPLE 16.1

In the Riemann integral, the e_j's have to be such that they are evaluated at some point in the interval $[t_{j-1}, t_j]$, that is, $e_j(omega) = e(\xi_j, \omega)$, with ξ_j anywhere in the interval $[t_{j-1}, t_j]$. The equivalent concept in the world of stochastic processes is measurability. In the stochastic integral definition, the sigma algebra is crucial.

Consider two simple functions as before:

$$f_1(t, \omega) = \sum_{i=0}^{n} B_{t_{i-1}}(\omega) \mathbf{1}_{(t_{i-1}, t_i]}(t),$$

and

$$f_2(t, \omega) = \sum_{i=0}^{n} B_{t_i}(\omega) \mathbf{1}_{(t_{i-1}, t_i]}(t)$$

Note that the only thing that has changed is the time chosen in each interval for the Brownian term. In particular, for f_1, the term e_i is measurable with respect to $\mathscr{F}_{t_{i-1}}$, that is, the left point on the interval. In f_2, the term e_i is measurable with respect to $\mathscr{F}_{t_i}$. This makes a huge difference. The integrals as defined above are (we drop the omega from the notation)

$$\int_0^t f_1(t) dB_t = \sum_{i=1}^{n} B_{t_{i-1}}(B_{t_i} - B_{t_{i-1}}),$$

and

$$\int_0^t f_2(t) dB_t = \sum_{i=1}^{n} B_{t_i}(B_{t_i} - B_{t_{i-1}}).$$

If there is any hope for the limit to be the same, their expected values must be identical. However, we have

$$\mathbf{E}\left[\int_0^t f_1(t) dB_t\right] = \sum_{i=1}^{n} \mathbf{E}[B_{t_{i-1}}(B_{t_i} - B_{t_{i-1}})]$$

$$= \sum_{i=1}^{n} \mathbf{E}[B_{t_{i-1}}]\mathbf{E}[(B_{t_i} - B_{t_{i-1}})] = 0$$

because $B_{t_{i-1}} = B_{t_{i-1}} - B_0$ and $B_{t_i} - B_{t_{i-1}}$ are increments which are independent. On the other hand

$$\mathbf{E}\left[\int_0^t f_2(t)dB_t\right] = \sum_{i=1}^n \mathbf{E}[B_{t_i}(B_{t_i} - B_{t_{i-1}})]$$

$$= \sum_{i=1}^n \mathbf{E}[(B_{t_{i-1}} + B_{t_i} - B_{t_{i-1}})(B_{t_i} - B_{t_{i-1}})]$$

$$= \sum_{i=1}^n \mathbf{E}[B_{t_{i-1}}(B_{t_i} - B_{t_{i-1}})] + \sum_{i=1}^n \mathbf{E}[(B_{t_i} - B_{t_{i-1}})^2]$$

$$= 0 + \sum_{i=1}^n (t_i - t_{i-1}) = t$$

So, clearly the nature of the integrand makes a big difference.

In general, if one chooses a $t_i^* \in [t_{i-1}, t_i]$, the following are the resulting integrals:

- For $t_i^* = t_{i-1}$, we obtain the Itô integral (covered in this chapter).

- For $t_i^* = \frac{t_{i-1}+t_i}{2}$, we obtain the Fisk–Stratonovich integral. Regular calculus rules apply to this integral.

- For $t_i^* = t_i$, we obtain the forward or anticipative integral sometimes used in control theory.

On a probability space (), let $\mathscr{F}_t$ denote the filtration generated by the Brownian motion. Recall that a stochastic process X_t is adapted with respect to a filtration $\mathscr{F}_t$ if and only if X_t is $\mathscr{F}_t$ measurable for all t.

Definition 16.2 (The class of integrands for stochastic integral) *We will define a stochastic integral $\int_0^t X_t dB_t$ for any stochastic process X_t satisfying the following:*

- *The function$(t, \omega) \rightarrow X_t(\omega)$ is measurable with respect to the combined space sigma algebra $\mathscr{B}([0, \infty)) \times \mathscr{F}$;*

- *The process X_t is $\mathscr{F}_t$ adapted;*

- *$\mathbf{E}[\int_0^T X_t(\omega)^2 dt] < \infty$, for $T > 0$.*

We will denote with ν_T this class of integrands for which we can define a stochastic integral.

Note that the first property is normal and in fact satisfied by pretty much everything. The second is needed to properly define the integral as mentioned above. The third is a technical property; it simplifies the mathematical derivations by allowing us to concentrate on limits in L^2.

16.1.1 Itô integral construction

Suppose that $f(t) \in \nu_T$; specifically

$$f(t, \omega) = \sum_{i=1}^{n} e_{t_{i-1}}(\omega) \mathbf{1}_{(t_{i-1}, t_i]}(t), \tag{16.4}$$

where we used the index to signify that $e_{t_{i-1}}$ is $\mathscr{F}_{t_{i-1}}$ measurable, and $\pi_n = (0 = t_0 < t_1 < \cdots < t_n = T)$. For this type of simple functions, we define

$$\int_0^T f(t) dB_t = \sum_{i=1}^{n} e_{t_{i-1}} (B_{t_i} - B_{t_{i-1}}).$$

Lemma 16.3 (Itô isometry for simple functions) *If f is a simple function in ν_T and bounded, then*

$$\mathbf{E}\left[\left(\int_0^T f(t) dB_t\right)^2\right] = \mathbf{E}\left[\int_0^T f(t)^2 dt\right] = \int_0^T \mathbf{E}[f(t)^2] dt.$$

Proof: Let $\Delta B_i = B_{t_i} - B_{t_{i-1}}$. Then

$$\mathbf{E}[e_{t_{i-1}} e_{t_{j-1}} \Delta B_i \Delta B_j] = \begin{cases} 0, & \text{if } i \neq j \\ \mathbf{E}[e_{t_{i-1}}^2](t_i - t_{i-1}), & \text{if } i = j \end{cases}$$

The expression is easy to prove since $e_{t_{i-1}}$ is $\mathscr{F}_{t_{i-1}}$ measurable and ΔB_i is independent of $\mathscr{F}_{t_{i-1}}$. Using this result, we have

$$\mathbf{E}\left[\left(\int_0^T f(t) dB_t\right)^2\right] = \mathbf{E}\left[\left(\sum_i e_{t_{i-1}} \Delta B_i\right)\left(\sum_j e_{t_{i-j}} \Delta B_j\right)\right]$$

$$= \sum_{i,j} \mathbf{E}\left(e_{t_{i-1}} e_{t_{i-j}} \Delta B_i \Delta B_j\right)$$

$$= \sum_i \mathbf{E}[e_{t_{i-1}}^2](t_i - t_{i-1})$$

$$= \int_0^T \mathbf{E}[f(t)^2] dt$$

∎

For the rest of the construction, we will state the steps. These steps are essentially the standard construction we used in Chapter 4. For the full proof of the steps, we refer the reader to the excellent book by Øksendal (2003).

Step 1. Let $f \in \nu_T$ and bounded such that all the paths are continuous ($f(., \omega)$ is continuous for all ω). Then there exist simple functions f_n of the form (16.4) such that

$$\mathbf{E}\left[\int_0^T (f - f_n)^2 dt\right] \to 0$$

as $n \to \infty$. Step 1 says that any bounded, adapted process with continuous paths may be approximated with these simple functions – L^2 approximation.

Step 2. Let $g \in \nu_T$ and bounded. Then there exist bounded functions in ν_T denoted g_n with continuous paths and

$$\mathbf{E}\left[\int_0^T (g - g_n)^2 dt\right] \to 0$$

as $n \to \infty$. Step 2 is a slight generalization. It says that the bounded (not necessarily continuous) integrands may be approximated with bounded continuous functions and thus with simple functions from step 1. Again, the approximation is in L^2.

Step 2. Let $h \in \nu_T$. Then there exist bounded functions in ν_T denoted h_n such that

$$\mathbf{E}\left[\int_0^T (h - h_n)^2 dt\right] \to 0$$

as $n \to \infty$. It was pretty clear where we were going: any integrand in this ν_T can be approximated with simple functions by steps 1, 2, and 3. Because of these three steps, we can give the following definition:

Definition 16.4 (Itô integral) *Let $f \in \nu_T$. Then the stochastic integral or Itô integral of f with respect to the Brownian motion B_t is defined as*

$$\int_0^T f(t, \omega) dB_t(\omega) = \lim_{n \to \infty} \int_0^T f_n(t, \omega) dB_t(\omega),$$

where f_n's form a sequence of simple functions as in (16.4) such that

$$\mathbf{E}\left[\int_0^T (f - f_n)^2 dt\right] \to 0, \quad \text{as } n \to \infty.$$

Such a sequence exists by the steps above. There remains to show that the limit is unique regardless of the approximating sequence, but for this, one needs to go to the proof of the steps. We refer to Øksendal (2003) once again.

Remark 16.5 *Since the approximation sequence does not matter when we need to approximate an integral, for example*

$$\int_0^T \cos(B_t^2) X_t dB_t,$$

using for a partition of $[0, t]$*:* $\pi_n = (0 = t_0 < t_1 < t_2 < \cdots < t_n = t)$

$$\sum_{i=1}^{n} \cos(B_{t_{i-1}}^2) X_{t_{i-1}} (B_{t_i} - B_{t_{i-1}}),$$

since the only requirement is that the integrand is $\mathscr{F}_{t_{i-1}}$ *measurable and approximate the given function in* L^2*. Note that the expression is the stochastic integral corresponding to the integrand*

$$f(t) = \sum_{i=1}^{n} \cos(B_{t_{i-1}}^2) X_{t_{i-1}} \mathbf{1}_{(t_{i-1}, t_i]}(t),$$

which is in ν_T *and of the form* (16.4).

Corollary 16.6 (Itô isometry) *Let* f *be adapted and measurable. Then*

$$\mathbf{E}\left[\left(\int_0^T f(t) dB_t\right)^2\right] = \int_0^T \mathbf{E}[f(t)^2] dt$$

This is easily proven by using Lemma 16.3 and going to L^2 limits.

16.1.2 An illustrative example

We have seen before that the stochastic integral really depends on the construction. Now we will point out that the stochastic integral is really different from a regular integral.

Suppose we have a deterministic function; let us call it $f(t)$ and suppose we want to calculate the integral

$$\int_0^T f(t) df(t).$$

We, of course, need to explain the object above. This is a Riemann–Stieltjes integral which is constructed in a way very similar to the Riemann integral construction. Specifically, for a partition π_n, we have

$$\lim_{||\pi|| \to 0} \sum_{i=1}^{n} f(\xi_i)(f(t_i) - f(t_{i-1})) = \int_0^T f(t) df(t)$$

for any ξ_i in the interval $[t_{i-1}, t_i]$. For more details, we refer the reader to any real analysis book such as Royden (1988) or Wheeden and Zygmund (1977).

Since we can write the integral as the limit using any point in the interval, we can calculate

$$\sum_{i=1}^{n} f(t_i)(f(t_i) - f(t_{i-1})) = \sum_{i=1}^{n} \left(f(t_i)^2 - f(t_i)f(t_{i-1}) - f(t_{i-1})^2 + f(t_{i-1})^2 \right)$$

$$= \sum_{i=1}^{n} \left(f(t_i)^2 - f(t_{i-1})^2 \right) - \sum_{i=1}^{n} f(t_{i-1})(f(t_i) - f(t_{i-1}))$$

$$= f(T)^2 - f(0)^2 - \sum_{i=1}^{n} f(t_{i-1})(f(t_i) - f(t_{i-1}))$$

Since both end sums go to the same integral by taking $||\pi_n|| \to 0$, we obtain

$$\int_0^T f(t)df(t) = f(T)^2 - f(0)^2 - \int_0^T f(t)df(t).$$

Therefore, in the case when $f(t)$ is deterministic, we have

$$\int_0^T f(t)df(t) = \frac{f(T)^2 - f(0)^2}{2}.$$

To mimic the stochastic case to be presented in a minute, if $f(0) = 0$, we will have

$$\int_0^T f(t)df(t) = \frac{f(T)^2}{2}.$$

Now consider the case of the stochastic integral $\int_0^T B_t dB_t$, where B_t is a Brownian motion. Remembering the definition, we cannot use the argument above since the sum we started with will not converge to the Itô integral. Instead

$$\sum_{i=1}^{n} B_{t_{i-1}}(B_{t_i} - B_{t_{i-1}}) = \sum_{i=1}^{n} \left(B_{t_{i-1}}B_{t_i} - B_{t_{i-1}}^2 + \frac{1}{2}B_{t_i}^2 - \frac{1}{2}B_{t_i}^2 \right)$$

$$= \sum_{i=1}^{n} \left(B_{t_{i-1}}B_{t_i} - \frac{1}{2}B_{t_i}^2 - \frac{1}{2}B_{t_{i-1}}^2 \right) + \sum_{i=1}^{n} \left(\frac{1}{2}B_{t_i}^2 - \frac{1}{2}B_{t_{i-1}}^2 \right)$$

$$= -\sum_{i=1}^{n} \frac{1}{2} \left(B_{t_i} - B_{t_{i-1}} \right)^2 + \frac{1}{2}\sum_{i=1}^{n} \left(B_{t_i}^2 - B_{t_{i-1}}^2 \right)$$

$$= \frac{1}{2}\sum_{i=1}^{n} \left(B_{t_i} - B_{t_{i-1}} \right)^2 - \frac{1}{2}B_T^2$$

Notice now that the remaining sum in the expression converges to the quadratic variation of the Brownian motion which is equal to T (Theorem 15.21 on page 478). Therefore, taking the limit in L^2 as $||\pi_n|| \to 0$, we obtain

$$\int_0^T B_t dB_t = \frac{1}{2}(B_T^2 - T) \tag{16.5}$$

As we can easily see, the stochastic integral is really a different notion from the regular integral.

16.2 Properties of the Stochastic Integral

First, I would like to point out the Wiener integral which was developed by Norbert Wiener and then preceded Itô integral and Itô calculus. The Wiener integral is obtained when the integrand of the stochastic integral is a deterministic function, that is,

$$\int_0^T f(t)dB_t$$

where f is deterministic. For this integral, all the regular calculus rules apply. For example, the integration by parts formula is

$$\int_0^T f(t)dB_t = f(T)B_T - \int_0^T B_t df(t),$$

where the latter is a Riemann–Stiletjes integral. Furthermore, this is a random variable which is normally distributed with mean 0 and variance $\int_0^T f^2(t)dt$.

Theorem 16.7 (Properties of the Itô integral) *Let X_t, Y_t be stochastic processes in ν_T, and let $\mathcal{F}_t$ denote the standard filtration generated by B_t. We have*

1. $\int_0^T X_t dB_t = \int_0^S X_t dB_t + \int_S^T X_t dB_t$, *for all $0 \le S \le T$;*

2. $\int_0^T (aX_t + bY_t)dB_t = a \int_0^T X_t dB_t + b \int_0^T Y_t dB_t$, *for all $a, b \in \mathbb{R}$;*

3. $\int_0^T X_t dB_t$ *is $\mathcal{F}_T$ measurable;*

4. $\mathbf{E}\left[\int_0^T X_t dB_t\right] = 0$;

5. $\mathbf{E}\left[\left(\int_0^T X_t dB_t\right)^2\right] = \int_0^T \mathbf{E}\left[X_t^2\right] dt$, *(Itô isometry).*

The proof of the theorem is skipped. All these properties are proven first for simple functions, and then going to the L^2 limit in the general stochastic integrands in ν_T.

Lemma 16.8 (Quadratic variation of the stochastic integral) *If $X_t \in \nu_T$ and we denote*

$$I_t = \int_0^t X_s dB_s$$

the Itô integral of X_t, then the quadratic variation of the process I_t is

$$[I, I]_t = \int_0^T X_t^2 dt.$$

Theorem 16.9 (Continuity of the stochastic integral) *For $X_t \in \nu_T$, we denote*

$$I_t = \int_0^t X_s dB_s$$

Then the Itô integral above has a continuous version J_t such that

$$\mathbf{P}\left(J_t = \int_0^t X_s dB_s\right) = 1, \quad \text{for all } t, \text{ with } 0 \le t \le T.$$

For the proof of these two results, we refer to (Øksendal, 2003, Chapter 3). The last theorem is especially important. It says that there always exists a version of the stochastic integral which is continuous a.s. regardless of whether the integrand has jumps or is not well behaved. From now on, we will always assume that the Itô integral means a continuous version of the integral.

Proposition 16.10 *The stochastic integral $\int_0^T X_t dB_t$ is an $\mathscr{F}_T$- martingale. Furthermore, we have*

$$\mathbf{P}\left(\sup_{0 \le \le T}\left|\int_0^t X_s dB_s\right| \ge \lambda\right) \le \frac{1}{\lambda^2}\mathbf{E}\left[\int_0^T X_t^2 dt\right].$$

Proof: The proof is very simple. One needs to use the previous theorem, Doob's martingale inequality (Theorem 14.12 on page 458), and Itô isometry. ∎

16.3 Itô lemma

We are now in position to provide the most important tool of stochastic calculus. However, before we introduce the lemma, let us introduce the type of processes that will be the object of the lemma.

Definition 16.11 *Let B_t be a standard one-dimensional Brownian motion on $(\Omega, \mathscr{F}, \mathbf{P})$. Then the process*

$$X_t(\omega) = X_0(\omega) + \int_0^t \mu(s, \omega)ds + \int_0^t \sigma(s, \omega)dB_s(\omega) \tag{16.6}$$

is called an Itô process. The processes μ and σ need to be adapted to the filtration generated by the Brownian motion. σ is in ν_T, and to have everything making sense, we require

$$\mathbf{P}\left(\int_0^t |\mu(s, \omega)|ds < \infty \text{ for all } t > 0\right) = 1$$

Note that the functions $\mu, \sigma \in \nu(0, \infty)$ are random in general. Sometimes for brevity of notation the equation above is written as

$$dX_t = \mu dt + \sigma dB_t.$$

This latter equation has no meaning whatsoever other than providing a symbolic notation for Equation (16.6). We shall use this latter notation at all times, meaning an equation of the type (16.6).

Lemma 16.12 (The one-dimensional Itô formula) *Let X be an Itô process as in Equation (16.6). Suppose that $f(t, x)$ is a function defined on $[0, \infty) \times \mathbb{R}$, twice differentiable in x, and one time differentiable in t. Symbolically, $f \in \mathcal{C}^{1,2}([0, \infty) \times \mathbb{R})$.*
Then, the process $Y_t = f(t, X_t)$ is also an Itô process and

$$df(t, X_t) = \frac{\partial f}{\partial t}(t, X_t)dt + \frac{\partial f}{\partial x}(t, X_t)dX_t + \frac{1}{2}\frac{\partial^2 f}{\partial x^2}(t, X_t)d[X, X]_t,$$

where $[X, X]_t$ is the quadrating variation of X_t. To simplify the calculation, we rewrite the formula as

$$dY_t = \frac{\partial f}{\partial t}dt + \frac{\partial f}{\partial x}dX_t + \frac{1}{2}\frac{\partial^2 f}{\partial x^2}(dX_t)^2.$$

The last term can be calculated as $(dX_t)^2$, and using the rules, we have

$$dtdt = dB_tdt = dtdB_t = 0$$
$$dB_tdB_t = dt.$$

We will skip the proof of the lemma for this edition; instead, we will concentrate on the interpretation and reaching stochastic differential equations.

▦ EXAMPLE 16.2

Recall the derivation in (16.5). Here is the same result using a much shorter derivation. Suppose $X_t = B_t$ (i.e., $dX_t = dB_t$) which is an Itô process. Let the function $f(t, x) = \frac{1}{2}x^2$. Then $Y_t = \frac{1}{2}B_t^2$. We can calculate easily $\frac{\partial f}{\partial t} = 0$, $\frac{\partial f}{\partial x} = x$, and $\frac{\partial^2 f}{\partial x^2} = 1$. We apply Itô formula to this function:

$$dY_t = \frac{\partial f}{\partial t}dt + \frac{\partial f}{\partial x}dX_t + \frac{1}{2}\frac{\partial^2 f}{\partial x^2}(dX_t)^2$$
$$d\left(\frac{1}{2}B_t^2\right) = B_tdB_t + \frac{1}{2}(dB_t)^2 = B_tdB_t + \frac{1}{2}dt,$$

or in the proper notation

$$\frac{1}{2}B_t^2 - \frac{1}{2}B_0^2 = \int_0^t B_sdB_s + \frac{1}{2}t$$

Rewriting this expression and using that $B_0 = 0$, we obtain the integral of the Brownian motion with respect to itself:

$$\int_0^t B_sdB_s = \frac{1}{2}\left(B_t^2 - t\right)$$

■ EXAMPLE 16.3   The geometric Brownian motion

In the Black–Scholes–Merton model the stock price follows the equation

$$dS_t = \mu S_t dt + \sigma S_t dB_t, \tag{16.7}$$

where μ and σ are constants and B_t is a regular one-dimensional Brownian motion. This is one of the simplest stochastic models in finance. The process in (16.7) is also called a *geometric Brownian motion*.

Note that the process is an Itô process since it has the form in the formula (16.6). However, since the process contains S_t on the right side, what we are looking at is a stochastic differential equation (SDE). We can use Itô's lemma to calculate the solution of this equation.

Consider the process $R_t = \log S_t$. This is called the *log process*. Notice that $R_t - R_{t-\Delta t} = \log S_t - \log S_{t-\Delta t} = \log(S_t/S_{t-\Delta t})$ for any interval $[t-\Delta t, t]$. This particular difference is known in finance as the continuously compounded return of the stock S_t over the interval $[t-\Delta t, t]$.

We can obtain an equation for the return by applying the Itô lemma to the function $f(t, x) = \log x$. We obtain

$$
\begin{aligned}
dR_t &= \frac{\partial f}{\partial t}(t, S_t)dt + \frac{\partial f}{\partial x}(t, S_t)dS_t + \frac{1}{2}\frac{\partial^2 f}{\partial x^2}(t, S_t)(dS_t)^2 \\
&= \frac{1}{S_t}(\mu S_t dt + \sigma S_t dB_t) + \frac{1}{2}\left(-\frac{1}{S_t^2}\right)(\mu S_t dt + \sigma S_t dB_t)^2 \\
&= \mu dt + \sigma dB_t - \frac{1}{2S_t^2}\left(\mu^2 S_t^2 dt^2 + 2\mu\sigma S_t^2 dt dB_t + \sigma^2 S_t^2 dB_t^2\right) \\
&= \left(\mu - \frac{\sigma^2}{2}\right)dt + \sigma dB_t,
\end{aligned}
$$

or

$$R_t - R_0 = \int_0^t \left(\mu - \frac{\sigma^2}{2}\right)ds + \int_0^t \sigma dB_s = \left(\mu - \frac{\sigma^2}{2}\right)t + \sigma B_t.$$

Recall that $R_t = \log S_t$. Substituting back and solving for S_t, we obtain an explicit formula

$$S_t = S_0 e^{\left(\mu - \frac{\sigma^2}{2}\right)t + \sigma B_t} \tag{16.8}$$

Proposition 16.13 (Product rule) *If X_t, Y_t are two Itô processes of the form* (16.6), *then*

$$d(X_t Y_t) = X_t dY_t + Y_t dX_t + dX_t dY_t,$$

where in the last term the same rules as in the Itô formula apply.

Lemma 16.14 (The general Itô formula) *Suppose* $\mathbf{B}_t = (B_1(t), \ldots, B_d(t))$ *is a d-dimensional Brownian motion. Recall that each component is a standard Brownian motion. Let* $\mathbf{X} = (X_1(t), X_2(t), \ldots, X_d(t))$, *be an n-dimensional Itô process, that is,*

$$dX_i(t) = \mu_i dt + \sigma_{i1} dB_1(t) + \sigma_{i2} dB_2(t) + \cdots + \sigma_{id} dB_d(t)$$

for all i from 1 to n. This expression can be represented in matrix form as

$$d\mathbf{X}_t = U dt + \Sigma d\mathbf{B}_t,$$

where

$$U = \begin{pmatrix} \mu_1 \\ \mu_2 \\ \vdots \\ \mu_n \end{pmatrix}, \; and \; \Sigma = \begin{pmatrix} \sigma_{11} & \sigma_{12} & \cdots & \sigma_{1d} \\ \sigma_{21} & \sigma_{22} & \cdots & \sigma_{2d} \\ & & & \\ \sigma_{n1} & \sigma_{n2} & \cdots & \sigma_{nd} \end{pmatrix}$$

Suppose that $f(t, x) = (f_1(t, \mathbf{x}), f_2(t, \mathbf{x}), \ldots, f_m(t, \mathbf{x}))$ *is a function defined on* $[0, \infty) \times \mathbb{R}^n$ *with values in* $\mathbb{R}^m$ *with* $f \in \mathcal{C}^{1,2}([0, \infty) \times \mathbb{R}^n)$.

Then, the process $\mathbf{Y}_t = f(t, \mathbf{X}_t)$ *is also an Itô process and its component k is given by*

$$dY_k(t) = \frac{\partial f_k}{\partial t} dt + \sum_{i=1}^{n} \frac{\partial f_k}{\partial x_i} dX_i(t) + \frac{1}{2} \sum_{i,j=1}^{n} \frac{\partial^2 f_k}{\partial x_i \partial x_j} (dX_i(t))(dX_j(t)),$$

for all k from 1 to m. The last term is calculated using the following rules:

$$dt dt = dB_i(t) dt = dt dB_i(t) = dB_i(t) dB_j(t) = 0, \quad \forall i \neq j$$
$$dB_i(t) dB_i(t) = dt.$$

EXAMPLE 16.4

Suppose the process $\mathbf{X}_t$ is

$$\mathbf{X}_t = \begin{pmatrix} B_1(t) + B_2(t) + 3t \\ B_1(t) B_2(t) B_3(t) \end{pmatrix}$$

where $\mathbf{B}_t = (B_1(t), B_2(t), B_3(t))$ is a three-dimensional Brownian motion. We want to show whether this process is a general Itô process.

Let

$$f(t, \mathbf{x}) = \begin{pmatrix} x_1 + x_2 + 3t \\ x_1 x_2 x_3 \end{pmatrix}$$

We apply the general Itô's lemma to the process $\mathbf{X}_t = f(t, \mathbf{B}_t)$. We should have

$$dX_1(t) = 3dt + dB_1(t) + dB_2(t),$$

and

$$dX_2(t) = B_2(t)B_3(t)dB_1(t) + B_1(t)B_3(t)dB_2(t) + B_1(t)B_2(t)dB_3(t)$$
$$+ B_3(t)dB_1(t)dB_2(t) + B_2(t)dB_1(t)dB_3(t) + B_1(t)dB_2(t)dB_3(t)$$
$$= B_2(t)B_3(t)dB_1(t) + B_1(t)B_3(t)dB_2(t) + B_1(t)B_2(t)dB_3(t)$$

Therefore, X_t is a general Itô process.

16.4 Stochastic Differential Equations (SDEs)

If the functions μ and σ in (16.6) depend on ω through the process X_t itself, then (16.6) defines an SDE. Specifically, assume that $X_t \in \nu(0, \infty)$. Assume that the functions $\mu = \mu(t, x)$ and $\sigma = \sigma(t, x)$ are twice differentiable with continuous second derivative in both variables. Then

$$X_t = X_0 + \int_0^t \mu(s, X_s)ds + \int_0^t \sigma(s, X_s)dB_s \qquad (16.9)$$

defines a *stochastic differential equation*.

◼ **EXAMPLE 16.5   Brownian motion on the unit circle**

Recall that the point $(\cos x, \sin x)$ belongs to the unit circle. Thus, naturally if we take the process
$$Y_t = (\cos B_t, \sin B_t),$$

this will belong to the unit circle. We wish to obtain an SDE such that its solution will be the process Y_t. To obtain such an SDE, we use the only tool we have, the Itô lemma. Let $f(t, x) = (\cos x, \sin x)$. Note that $Y_t = (Y_1(t), Y_2(t)) = f(t, B_t)$, and applying the general Itô lemma we obtain

$$\begin{cases} dY_1(t) &= -\sin B_t dB_t - \frac{1}{2}\cos B_t dt \\ dY_2(t) &= \cos B_t dB_t - \frac{1}{2}\sin B_t dt \end{cases}$$

Therefore, the Brownian motion on the unit circle is a solution of

$$\begin{cases} dY_1(t) &= -\frac{1}{2}Y_1(t)dt - Y_2(t)dB_t, \\ dY_2(t) &= -\frac{1}{2}Y_2(t)dt + Y_1(t)dB_t \end{cases}$$

From the example above, we see that the Brownian motion on the unit circle must be a solution of the system presented. But is it the only solution? The next theorem gives conditions on the SDE coefficients μ and σ under which a solution exists and it is unique.

Theorem 16.15 (The existence and uniqueness of the solution to an SDE) *Let $T > 0$, and $\mu : [0, T] \times \mathbb{R}^n \to \mathbb{R}^n$, $\sigma : [0, T] \times \mathbb{R}^n \to \mathbb{R}^{n,d}$ are both measurable functions. Let $\mathbf{B}_t$ denote a d-dimensional Brownian motion and consider the stochastic differential equation*

$$d\mathbf{X}_t = \mu(t, \mathbf{X}_t)dt + \sigma(t, \mathbf{X}_t)d\mathbf{B}_t, \tag{16.10}$$

with initial condition $\mathbf{X}_0 = \mathbf{Z}$, some random vector with $\mathbf{E}[|\mathbf{Z}|^2] < \infty$.
  Suppose the coefficients μ and σ have the following properties:

1. ***Linear growth condition:***

$$|\mu(t, \mathbf{x})| + |\sigma(t, \mathbf{x})| \leq C(1 + |\mathbf{x}|),$$

   for some constant C, for all $\mathbf{x} \in \mathbb{R}^n$ and $t \in [0, T]$.

2. ***Lipschitz condition:***

$$|\mu(t, \mathbf{x}) - \mu(t, \mathbf{y})| + |\sigma(t, \mathbf{x}) - \sigma(t, \mathbf{y})| \leq D|\mathbf{x} - \mathbf{y}|,$$

   for some constant D, for all $\mathbf{x}, \mathbf{y} \in \mathbb{R}^n$ and $t \in [0, T]$.

Then the SDE (16.10) has a unique solution $\mathbf{X}_t$ adapted to the filtration generated by Z and B_t and

$$\mathbf{E}\left[\int_0^T |\mathbf{X}_t|\right] < \infty.$$

Remark 16.16 *We chose to give the more general form when the SDE is n dimensional. For the one-dimensional case, the conditions are identical of course.*
  The absolute value in the theorem is the modulus in R^n for μ, that is,

$$|\mathbf{x}|^2 = \sum_{i=1}^n x_i^2,$$

and for the matrix sigma we have

$$|\sigma|^2 = \sum_{i=1}^n \sum_{j=1}^d \sigma_{ij}^2.$$

We will not prove this theorem. For details, consult Øksendal (2003).
  The linear growth condition does not allow the coefficients to grow faster than x. The Lipschitz condition is a regularity condition. The condition is stronger than requiring continuity but weaker than requiring that μ and σ be derivable on their domain. These conditions are actually coming from the ODE theory. In fact, the next two examples are ODE-related. In practice, the Lipschitz condition is rarely verified. Indeed, if the functions μ and σ are differentiable in the x variable with continuous derivative, then the Lipschitz condition is satisfied.

■ **EXAMPLE 16.6   Violating the linear growth condition**

Consider the following SDE:

$$dX_t = X_t^2 dt, \quad X_0 = 1$$

Note that $\mu(t, x) = x^2$, and it does not satisfy the linear growth condition. Since there is no stochastic integral, this SDE can be solved like a regular ODE:

$$\frac{dX_t}{X_t} = dt \quad \Rightarrow \quad -\frac{1}{X_t} = t + k \quad \Rightarrow \quad X_t = -\frac{1}{t + k}$$

$$\text{Since } X_0 = 1 \quad \Rightarrow \quad k = -1 \quad \Rightarrow \quad X_t = \frac{1}{1 - t}$$

Note that $\frac{1}{1-t}$ is the unique solution; however, the solution does not exist at $t = 1$, and therefore there exists no global solution defined on the entire $[0, \infty)$.

■ **EXAMPLE 16.7   Violating the Lipschitz condition**

Consider the following SDE:

$$dX_t = 3\sqrt[3]{X_t^2} dt, \quad X_0 = 0$$

Note that $\mu(t, x) = 3\sqrt[3]{x^2}$ does not satisfy the Lipschitz condition at $x = 0$. This SDE has multiple solutions:

$$X_t = \begin{cases} 0, & \text{if } t \leq a \\ (t - a)^3, & \text{if } t > a \end{cases}$$

is a solution for any value of a.

16.4.1   A discussion of the types of solution for an SDE

Definition 16.17 (Strong and weak solutions) *If the Brownian motion B_t is given and we can find a solution X_t which is adapted with respect to $\mathcal{F}_t$, the filtration generated by the Brownian motion, then X_t is called a **strong solution**.*
   *If only the functions μ and σ are given and we need to find both X_t and B_t that solve the SDE, then the solution is a **weak solution**.*

A strong solution is a weak solution. However, the converse is not necessarily true. The most famous example is the Tanaka equation

$$dX_t = sign(X_t) dB_t, \quad X_0 = 0,$$

where the sign function just gives the sign of the real number x. Note that the function's sign does not satisfy the Lipschitz condition at 0. This equation has no strong solution, but it does have a weak solution. The idea is to play with the fact that $-B_t$ is also a standard Brownian motion. For more details, refer to (Øksendal, 2003, Exercise 4.10 and Chapter 5)

Definition 16.18 (Strong and weak uniqueness) *For an SDE, we say that the solution is **strongly unique** if for any two solutions $X_1(t)$ and $X_2(t)$ we have*

$$X_1(t) = X_2(t), \ a.s. \ for \ all \ t, T.$$

*The solution is called **weakly unique** if for any two solutions $X_1(t)$ and $X_2(t)$ they have the same finite dimensional distribution.*

Lemma 16.19 *If μ and σ satisfy the conditions in Theorem 16.15, then a solution is weakly unique.*

16.5   Examples of SDEs

I strongly believe that the only way one learns stochastic calculus is by practicing exercises. There aren't many tools one needs to use; in fact, the most important one to know is the Itô lemma.

■ EXAMPLE 16.8

Show that the solution to the SDE

$$dX_t = \frac{1}{2}X_t dt + X_t dB_t, \quad X_0 = 1$$

is e^{B_t}.

Note that the coefficients are $\mu(t, x) = \frac{1}{2}x$ and $\sigma(t, x) = x$, which satisfy the linear growth condition and the Lipschitz condition. Therefore, the solution exists and it is weakly unique.

Note that this is an example we have seen before of the geometric Brownian motion. We again take $f(t, x) = \log x$ and $Y_t = \log X_t$. Applying the Itô formula, we obtain

$$dY_t = \frac{1}{X_t}dX_t + \frac{1}{2}\left(-\frac{1}{X_t^2}\right)(dX_t)^2$$

$$= \frac{1}{X_t}\left(\frac{1}{2}X_t dt + X_t dB_t\right) - \frac{1}{2X_t^2}\left(\frac{1}{2}X_t dt + X_t dB_t\right)^2$$

$$= \frac{1}{2}dt + dB_t - \frac{1}{2}dt = dB_t$$

So $Y_t - Y_0 = B_t$, and therefore $X_t = e^{B_t}$.

EXAMPLE 16.9

Let $X_t = \sin B_t$, with $B_0 = \alpha \in (-\frac{\pi}{2}, \frac{\pi}{2})$. Show that X_t is the unique solution of

$$dX_t = -\frac{1}{2}X_t dt + \sqrt{1 - X_t^2}dB_t,$$

for $t < \inf\{s > 0 : B_s \notin (-\frac{\pi}{2}, \frac{\pi}{2})\}$.

We take $f(t, x) = \sin x$ and B_t, and as always apply the Itô formula to the process $X_t = f(t, B_t)$. We obtain

$$dX_t = \cos B_t dB_t + \frac{1}{2}(-\sin B_t)dt$$

$$= \sqrt{1 - X_t^2}dB_t - \frac{1}{2}X_t dt$$

Thus the process X_t is a solution of the equation. To show that it is the unique solution, we need to show that its coefficients are Lipschitz. The coefficient $\mu(t, x) = -\frac{1}{2}x$ clearly is. For $\sigma(t, x) = \sqrt{1 - x^2}$, we need to restrict the time domain to the interval 0 until the first time the argument leaves the interval $(-\frac{\pi}{2}, \frac{\pi}{2})$. Otherwise, we will have multiple solutions. If B_t is in that interval, the corresponding x, y are in $(-1, 1)$ and we have

$$|\sqrt{1 - x^2} - \sqrt{1 - y^2}| = \frac{|\sqrt{1 - x^2} - \sqrt{1 - y^2}||\sqrt{1 - x^2} + \sqrt{1 - y^2}|}{|\sqrt{1 - x^2} + \sqrt{1 - y^2}|}$$

$$= \frac{|1 - x^2 - 1 + y^2|}{\sqrt{1 - x^2} + \sqrt{1 - y^2}} = \frac{|x - y||x + y|}{\sqrt{1 - x^2} + \sqrt{1 - y^2}}$$

$$\leq \frac{2}{\sqrt{1 - x^2} + \sqrt{1 - y^2}}|x - y|$$

Since x and y are bounded away from -1 and 1, there exists a constant c such that

$$\min(\sqrt{1 - x^2}, \sqrt{1 - y^2}) > c,$$

and therefore

$$|\sqrt{1 - x^2} - \sqrt{1 - y^2}| \leq \frac{1}{c}|x - y|$$

for all $x, y \in (-1, 1)$; so the function $\sigma(t, x)$ is Lipschitz on the interval of definition. This shows that the SDE has the unique solution stated for $t < \inf\{s > 0 : B_s \notin (-\frac{\pi}{2}, \frac{\pi}{2})\}$.

EXAMPLE 16.10   Ørnstein–Uhlenbeck (O–U) SDE

In this example, we start the presentation of a technique which applies to linear SDEs. Specifically, we present the idea of using an integrating factor to simplify the equation. The O–U SDE written in the differential form is

$$dX_t = \mu X_t dt + \sigma dB_t$$

Find the solution for this SDE with $X_0 = x \in \mathbb{R}$.

We group the terms

$$dX_t - \mu X_t dt = \sigma dB_t \qquad (16.11)$$

We next note that this is similar to regular ODEs, and so the term on the left looks like what we obtain once we multiply with the integrating factor and take derivatives. Specifically, it looks like the derivative of

$$d(X_t e^{-\mu t})$$

However, these being stochastic processes, we need to verify that this is indeed the case.

$$d(X_t e^{-\mu t}) = e^{-\mu t} dX_t - \mu X_t e^{-\mu t} dt - \mu e^{-\mu t} dt dX_t = e^{-\mu t} dX_t - \mu X_t e^{-\mu t} dt$$

Applying the product rule (Proposition 16.13 on page 497) and noticing that no matter what the stochastic process X_t is, the last term only contains terms of the type $dt dt$ and $dt dB_t$.

So now we see that indeed the left side of Equation (16.11) is the derivative if we make the integrating term appear. So we multiply both sides of (16.11) with the integrating factor $e^{-\mu t}$ to obtain

$$e^{-\mu t} dX_t - \mu e^{-\mu t} X_t dt = \sigma e^{-\mu t} dB_t$$
$$d(X_t e^{-\mu t}) = \sigma e^{-\mu t} dB_t$$

Writing both sides in the actual integral form, we have

$$X_t e^{-\mu t} - X_0 e^{-\mu 0} = \int_0^t \sigma e^{-\mu s} dB_s,$$

or

$$X_t = x e^{\mu t} + e^{\mu t} \int_0^t \sigma e^{-\mu s} dB_s,$$

or even the more compact form

$$X_t = x e^{\mu t} + \int_0^t \sigma e^{\mu(t-s)} dB_s.$$

This guarantees the unique solution since the coefficients of the equations satisfy the conditions in the existence and uniqueness in Theorem 16.15. Furthermore, since given B_t we found the solution explicitly in terms of the B_t, this is a strong solution.

We can also calculate the moments of the process X_t. To do so, it is important to remember that the stochastic integral is a martingale started from 0 and therefore its expectation is always 0. Therefore, we have

$$\mathbf{E}[X_t] = x e^{\mu t}.$$

Thus the expectation of this process goes to 0 or ∞ in the long run depending on the sign of the parameter μ.

For the variance, we use Itô's isometry

$$V(X_t) = \mathbf{E}\left[(X_t - \mathbf{E}[X_t])^2\right] = \mathbf{E}\left[\left(e^{\mu t}\int_0^t \sigma e^{-\mu s}dB_s\right)^2\right]$$

$$= e^{2\mu t}\int_0^t \sigma^2 e^{-2\mu s}ds = \frac{\sigma^2}{2\mu}e^{2\mu t}\left(1 - e^{-2\mu t}\right)$$

$$= \frac{\sigma^2}{2\mu}\left(e^{2\mu t} - 1\right)$$

So, again we reach the same conclusion that the variance becomes either 0 or ∞.

◾ EXAMPLE 16.11   General integrating factor for SDEs

We present this example since it is the most general integrating factor we can show. I have seen this type of approach to solving SDEs in Øksendal (2003). Suppose we have an SDE with $X_0 = x$ of the form

$$dX_t = f(t, X_t)dt + c(t)X_t dB_t. \tag{16.12}$$

Note that in this equation the deterministic integral can contain any term in t and X_t. The stochastic integrand needs to be linear in X_t.

Integrating factor for this equation is

$$F_t = e^{-\int_0^t c(s)dB_s + \frac{1}{2}\int_0^t c^2(s)ds}.$$

Using this integrating factor, we can show that

$$d(X_t F_t) = F_t f(t, X_t)dt.$$

If we now take $Y_t = F_t X_t$, we can see from the above that Y_t satisfies

$$dY_t = F_t f(t, \frac{Y_t}{F_t})dt.$$

Since there is no stochastic integral (so no strange derivation by parts rules), the equation above can be solved just as it is a deterministic type.

First I will show that indeed the equation for $d(X_t F_t)$ does not contain a stochastic integral term. Note that both terms in the product are stochastic, so we need to apply the full Itô's rule. To do so, we need to calculate dF_t. Note that

$$F_t = e^{Z_t},$$

where

$$Z_t = -\int_0^t c(s)dB_s + \frac{1}{2}\int_0^t c^2(s)ds,$$

or in differential form

$$dZ_t = -c(t)dB_t + \frac{1}{2}c^2(t)dt.$$

Thus, applying Itô's rule to Z_t and e^x, we get

$$dF_t = e^{Z_t}dZ_t + \frac{1}{2}e^{Z_t}(dZ_t)^2$$

$$= F_t\left(-c(t)dB_t + \frac{1}{2}c^2(t)dt\right) + \frac{1}{2}F_t\left(-c(t)dB_t + \frac{1}{2}c^2(t)dt\right)^2$$

$$= -c(t)F_tdB_t + \frac{1}{2}c^2(t)F_tdt + \frac{1}{2}F_tc^2(t)dt$$

$$= -c(t)F_tdB_t + c^2(t)F_tdt$$

Finally, we use this expression to calculate

$$d(X_tF_t) = X_tdF_t + F_tdX_t + dX_tdF_t$$

$$= X_t\left(-c(t)F_tdB_t + c^2(t)F_tdt\right) + F_t\left(f(t,X_t)dt + c(t)X_tdB_t\right)$$

$$+ \left(-c(t)F_tdB_t + c^2(t)F_tdt\right)\left(f(t,X_t)dt + c(t)X_tdB_t\right)$$

$$= -c(t)X_tF_tdB_t + c^2(t)X_tF_tdt + f(t,X_t)F_tdt + c(t)X_tF_tdB_t$$

$$- c^2(t)X_tF_tdt$$

$$= f(t,X_t)F_tdt,$$

which shows what we needed to show. Everything else is simple, and depends on the equation

$$dY_t = F_tf(t,\frac{Y_t}{F_t})dt$$

actually providing an easy-to-solve differential equation. We will exemplify on a specific example next.

■ EXAMPLE 16.12

Solve the SDE

$$dX_t = \frac{3}{X_t}dt + \alpha X_tdB_t,$$

with $X_0 = x > 0$.

To solve this, we note that the equation is of the type presented in Example 16.11, and we construct

$$F_t = e^{-\alpha B_t + \frac{1}{2}\alpha^2 t}.$$

We can either re-derive the result, or just use the example to get that we must have

$$d(F_tX_t) = 3\frac{F_t}{X_t}dt.$$

We now substitute $Y_t = F_t X_t$ to obtain the following equation:

$$dY_t = 3\frac{F_t^2}{Y_t}dt$$

This can be solved as a regular ODE with separable variables since regular calculus applies. We have

$$Y_t dY_t = 3F_t^2 dt$$

$$\frac{1}{2}Y_t^2 - \frac{1}{2}x = 3\int_0^t F_s^2 ds$$

$$X_t^2 = F_t\left(x + 6\int_0^t F_s^2 ds\right)$$

$$X_t = e^{-\frac{1}{2}\alpha B_t + \frac{1}{4}\alpha^2 t}\left(x + 6\int_0^t e^{-2\alpha B_s + \alpha^2 s}ds\right)$$

which is a strong solution.

16.5.1 An analysis of Cox– Ingersoll– Ross (CIR) type models

We conclude the examples section with an in-depth analysis of CIR type models. These types of models are heavily used in finance, especially in fixed-income security markets.

The CIR model was first formulated in Cox, Ingersoll, and Ross (1985). Let r_t denote the stochastic process modeled as a CIR-type process:

$$dr_t = \alpha(\mu - r_t)dt + \sigma\sqrt{r_t}dB_t, \tag{16.13}$$

with α, μ, and σ real constants and B_t a standard Brownian motion adapted to $\mathcal{F}_t$. μ, sometimes denoted with $\bar{r}$, as we shall see, is the long-term mean of the process. α is called the rate of return (or speed of return), a name inherited from the simpler mean-reverting O–U process but the meaning is similar here as well. Finally, σ is a constant that controls the variability of the process.

In general, this process does not have an explicit (stochastic) solution. However, its moments may be found by using a standard stochastic technique. It is this technique that we illustrate now. For calculation using a simpler process (the mean reverting O–U process), refer to the Appendix in Florescu and Păsărică (2009), where a similar calculation is performed.

16.5.2 Models similar to CIR

Before we analyze the actual process, let us present some other models that are related.

The Ørnstein-Uhlenbeck process is

$$dr_t = \alpha r_t dt + \sigma dB_t. \tag{16.14}$$

As mentioned in the examples section, it requires an integrating factor to obtain an explicit solution.

The *mean-reverting Ørnstein-Uhlenbeck process* is

$$dr_t = \alpha(\mu - r_t)dt + \sigma dB_t. \tag{16.15}$$

Note that this process is slightly more general than the regular O–U process. This process is called mean-reverting because of the dt term. To understand the name, consider the behavior of the increment for a small time step Δt. We can write

$$r_t - r_{t-\Delta t} = \alpha(\mu - r_{t-\Delta t})\Delta t + \sigma \Delta B_t.$$

Now look at the deterministic term. If $r_{t-\Delta t} < \mu$, then $\mu - r_{t-\Delta t}$ and the whole drift term will be positive. Thus the increment will tend to be positive, which means that r_t will tend to be closer to μ. If $r_{t-\Delta t} > \mu$, then $\mu - r_{t-\Delta t}$ is negative. Thus the increment will tend to be negative and once again r_t tends to get closer to the mean μ. The magnitude of the parameter α can amplify or decrease this mean reversion trend, and thus it is called the speed of reversion. The parameter σ governs the size of random fluctuations.

This model can be solved in exactly the same way as the regular O–U process. In finance, this particular model is known as the *Vasiček model* after the author of the paper that introduced the model Vasicek (1977).

The *Longstaff and Schwartz model*. This model writes the main process as a sum:

$$r_t = \alpha X_t + \beta Y_t,$$

where each of the stochastic processes X_t, Y_t are written as a CIR model:

$$dX_t = (\gamma - \delta X_t)dt + \sqrt{X_t}\, dW_t^1, \tag{16.16}$$
$$dY_t = (\eta - \theta Y_t)dt + \sqrt{Y_t}\, dW_t^2,$$

Note that the form is a bit different but, of course, each of the equations may be written as a CIR by simply factoring the coefficient of the process in the drift (i.e., $\gamma - \delta X_t = \delta\left(\frac{\gamma}{\delta} - X_t\right)$) and taking the parameter σ in CIR equal to 1.

The *Fong and Vasiček model* once again uses a couple of CIR models but this time embedded into each other:

$$dr_t = \alpha(\mu - r_t)dt + \sqrt{\nu_t}dB_1(t)$$
$$d\nu_t = \gamma(\eta - \nu_t)dt + \xi\sqrt{\nu_t}dB_2(t),$$

where the Brownian motions are correlated, that is, $dB_1(t)dB_2(t) = \rho dt$. This model is an example of what is called a *stochastic volatility model*. Note that the coefficient of the Brownian motion driving the main process $B_1(t)$ is stochastic, and the randomness is coming from the outside process ν_t. Any model that has this property is called a stochastic volatility model.

The *Black–Karasinski model*. This model is more and more prevalent in the financial markets. This process is in fact a generalization of the simple mean-reverting

O–U model above. Specifically, if we denote $X_t = \log r_t$, the logarithm of the short rate, then the Black–Karasinski model is

$$dX_t = (\theta(t) - \phi(t)X_t)dt + \sigma(t)dB_t. \tag{16.17}$$

We note that

$$dX_t = \phi(t)\left(\frac{\theta(t)}{\phi(t)} - X_t\right)dt + \sigma(t)dB_t,$$

and thus it is a mean-reverting O–U model where the coefficients are deterministic functions of time. The reason for the time-varying coefficients and their wide use in finance is the inability of the constant coefficient model to fit the yield curve produced by observed bond prices. This problem is common in finance for any of the models with constant coefficients. Having time-dependent coefficients means that one cannot solve the SDE analytically anymore. The reason for the Black–Karasinski model being popular in industry is that one can construct a tree that approximates its dynamics.

16.5.3   Moments calculation for the CIR model

In this subsection we will talk about calculating moments for stochastic processes described by SDEs. This is a very useful methodology which is applicable even in the case where the SDE cannot be solved explicitly. I will skip through tedious calculations and just point out the main points in the derivations.

The CIR process in (16.13) can be written in its integral form as

$$r_t - r_0 = \int_0^t \alpha(\mu - r_s)ds + \int_0^t \sigma\sqrt{r_s}\,dW_s.$$

Next we apply expectations on both sides and use Fubini's theorem (Theorem 5.1 about changing the order of integration) in the first integral:

$$\mathbf{E}[r_t] - r_0 = \mathbf{E}\left[\int_0^t \alpha(\mu - r_s)ds\right] + \mathbf{E}\left[\int_0^t \sigma\sqrt{r_s}\,dW_s\right]$$
$$= \int_0^t \mathbf{E}[\alpha(\mu - r_s)]ds = \int_0^t \alpha(\mu - \mathbf{E}[r_s])ds,$$

where we assume that at time 0 the process is a constant r_0. The second integral disappears since any stochastic integral is a martingale and any martingale has the property that $\mathbf{E}[M_t] = \mathbf{E}[M_0]$. But since at 0 the stochastic integral is equal to zero, then any stochastic integral has an expected value 0.

Next, let us denote $u(t) = \mathbf{E}[r_t]$, which is just a deterministic function. Substituting in the above and going back to the differential notation, we have

$$du(t) = \alpha(\mu - u(t))dt$$

But this is a first-order linear differential equation. We can solve it easily by multi-plying with the integrating factor.

$$du(t) + \alpha u(t)dt = \alpha\mu dt$$

$$e^{\alpha t}du(t) + \alpha e^{\alpha t}u(t)dt = \alpha\mu e^{\alpha t}dt$$

$$d\left(e^{\alpha t}u(t)\right) = \alpha\mu e^{\alpha t}dt$$

and, finally, integration gives

$$e^{\alpha t}u(t) - u(0) = \mu(e^{\alpha t} - 1)$$

Now using that $u(0) = \mathbf{E}[r_0] = r_0$ and after rearranging terms, we finally obtain

$$\mathbf{E}[r_t] = u(t) = \mu + (r_0 - \mu)e^{-\alpha t} \tag{16.18}$$

What about the variance? We have

$$Var(r_t) = \mathbf{E}\left[(r_t - \mathbf{E}[r_t])^2\right] = \mathbf{E}[r_t^2] - (\mathbf{E}[r_t])^2$$

So we will need to calculate the second moment $\mathbf{E}[r_t^2]$. The second term is just the square of expectation calculated earlier. We cannot just square the dynamics of r_t, we need to calculate the proper dynamics of r_t^2 by using Itô's lemma and apply expectations like we did earlier. Take the function $f(x) = x^2$ and apply Itô's lemma to this function and r_t to obtain the dynamics of $r_t^2 = f(r_t)$. We get

$$r_t^2 = 2r_t dr_t + \frac{1}{2}2(dr_t)^2 = 2r_t(\alpha(\mu - r_t)dt + \sigma\sqrt{r_t}dW_t) + \sigma^2 r_t dt$$

$$= (2\alpha\mu + \sigma^2)r_t dt - 2\alpha r_t^2 dt + 2\sigma r_t\sqrt{r_t}dW_t$$

Once again, we write the equation in the integral form, apply expectation in both sides, use Fubini just like we did earlier, and use the fact that the expectation of a stochastic integral is zero.

Next, we denote $y_t = \mathbf{E}[r_t^2]$, and we note that $y_0 = r_0^2$ to obtain

$$dy_t + 2\alpha y_t dt = (2\alpha\mu + \sigma^2)\mathbf{E}[r_t]dt.$$

Once again, we multiply with the integrating factor, which in this case is $e^{2\alpha t}$, to get

$$d\left(e^{2\alpha t}y_t\right) = e^{2\alpha t}(2\alpha\mu + \sigma^2)\mathbf{E}[r_t]dt$$

$$e^{2\alpha t}y_t - y_0 = \int_0^t e^{2\alpha s}(2\alpha\mu + \sigma^2)\mathbf{E}[r_s]ds$$

All that remains is to calculate the integral on the right-hand side:

$$\int_0^t e^{2\alpha s}(2\alpha\mu + \sigma^2)\mathbf{E}[r_s]ds = \int_0^t e^{2\alpha s}(2\alpha\mu + \sigma^2)(\mu + (r_0 - \mu)e^{-\alpha s})ds$$

$$= (2\alpha\mu + \sigma^2)\left(\frac{\mu}{2}(e^{2\alpha t} - 1) + (r_0 - \mu)(e^{\alpha t} - 1)\right)$$

Now going back, we finally obtain the second moment as

$$\mathbf{E}[r_t^2] = r_0^2 e^{-2\alpha t} + (2\alpha\mu + \sigma^2)\left(\frac{\mu}{2}(1 - e^{-2\alpha t}) + (r_0 - \mu)(e^{-\alpha t} - e^{-2\alpha t})\right)$$
(16.19)

To calculate the variance means using Equation (16.19) and the square of the mean in (16.18). After much tedious algebra, we get

$$Var(r_t) = \frac{\sigma^2\mu}{2\alpha} + \frac{\sigma^2}{\alpha}(r_0 - \mu)e^{-\alpha t} + \frac{4\alpha^2 r_0 + \sigma^2(\mu - 2r_0)}{2\alpha}e^{-2\alpha t}$$
(16.20)

Clearly, if one is dedicated enough, one may find expressions for higher moments as well, just by following the same procedure.

16.5.4 Interpretation of the formulas for moments

If we look at the expectation formula in (16.18), we see that the mean of r_t converges to μ and it does so very fast (exponential convergence). If the process starts from $r_0 = \mu$, then in fact the mean at t is constant and equal to μ, and the process is stationary in mean.

Furthermore, looking at the variance formula in (16.20), one sees that the variance converges to $\frac{\sigma^2\mu}{2\alpha}$ and again the convergence is faster with the choice $r_0 = \mu$. However, the variance is not stationary unless either $r_0 = \mu = 0$ (this gives the trivial 0 process) or the following relationship between the parameters hold:

$$r_0 = \mu, \quad 4\alpha^2 - \sigma^2 = 0.$$

Also note that, for the X and Y in the Longstaff and Schwartz model, the formulas for the moments remain the same with the parameter $\sigma = 1$.

16.5.5 Parameter estimation for the CIR model

This section is presented only for the sake of financial applications. Suppose we have a history of the forward rate r_t. Let us denote these observations $r_0, r_1, \ldots, r_n$, equidistant at times Δt apart, and we calculate the increments of the process as $\Delta r_i = r_i - r_{i-1}$ (n total increments). The process r_t in the CIR specification is not Gaussian, nor are its increments. Note that the noise term is the Brownian motion increment but multiplied with the square root of the actual process. The estimation, however, proceeds as if the increments are Gaussian. It is worth noting that this is in fact an approximate estimation. An essential step in the derivation is to realize that the process r_t is Markov.

To continue, let us denote the mean in (16.18) with

$$m(r_0, t) = \mathbf{E}[r_t] = \mu + (r_0 - \mu)e^{-\alpha t}$$
(16.21)

and variance with

$$v(r_0, t) = \frac{\sigma^2\mu}{2\alpha} + \frac{\sigma^2}{\alpha}(r_0 - \mu)e^{-\alpha t} + \frac{4\alpha^2 r_0 + \sigma^2(\mu - 2r_0)}{2\alpha}e^{-2\alpha t}$$
(16.22)

Since the process is Markov, we can calculate the mean and variance of r_t given the value of the process at time s, namely r_s, by just using

$$m(r_s, t - s), \text{ and } v(r_s, t - s).$$

Thus, if we now use a standard Markov argument, the joint density function may be written as

$$f(\Delta r_n, \ldots, \Delta r_1, \theta) = f(\Delta r_n \mid \Delta r_{n-1} \ldots, \Delta r_1, \theta) \cdots f(\Delta r_2 \mid \Delta r_1, \theta)$$

$$= \prod_{i=1}^{n-1} f(\Delta r_{i+1} \mid \Delta r_i, \theta),$$

where $\theta = (\mu, \alpha, \sigma)$ is the vector of parameters. Therefore, the log-likelihood function (see Chapter 8) is

$$L_n(\theta) = \sum_{i=1}^{n-1} \log f(\Delta r_{i+1} \mid \Delta r_i, \theta) \qquad (16.23)$$

and this function needs to be maximized with respect to the vector of parameters $\theta = (\mu, \alpha, \sigma)$.

However, the transition density function f is not known for the CIR process. Here is the point where the approximation with the Gaussian distribution is performed. Let ϕ be the density of the normal distribution with mean μ and variance Δt, that is,

$$\phi(x, \mu, \Delta t) = \frac{1}{\sqrt{2\pi \Delta t}} e^{-\frac{(x-\mu)^2}{2\Delta t}}$$

This function will be substituted for the f in the expression above. The logarithm of this function has a simpler expression:

$$\log \phi(x, \mu, \Delta t) = -\frac{1}{2} \log \Delta t - \frac{(x - \mu)^2}{2\Delta t},$$

where we neglected the constant $-\frac{1}{2} \log 2\pi$ since it is irrelevant for the maximization. With the notation above, the approximate log-likelihood function is

$$L_n(\theta) = \sum_{i=1}^{n-1} \log \phi(\Delta r_{i+1}, m(r_i, \Delta t), v(\Delta r_i, \Delta t)), \qquad (16.24)$$

where Δt is the time between consecutive observations. Note that the expressions $m(r_i, \Delta t)$ and $v(\Delta r_i, \Delta t)$ contain the parameters $\theta = (\mu, \alpha, \sigma)$ and the optimization procedure is with respect to these variables.

Remark 16.20 *We should ask: how good is the approximate estimation? Since we are using an approximate distribution, we need to understand how good the approximation is. In fact, we can show that the estimators are consistent, that is, they converge to the true parameters when the number of observations grows. One may follow the ideas in Bibby and Sørensen (1995) to show this. The derivation shows that the approximate likelihood function in (16.24) is a martingale.*

16.6 Linear Systems of SDEs

In practice, we often need to solve systems of SDEs. Consider the following stochastic system:

$$dX_1(t) = 2tdt + 4dB_1(t)$$
$$dX_2(t) = X_1(t)dB_2(t)$$

We can write this system in a matrix form as

$$d\mathbf{X}_t = Cdt + Dd\mathbf{B}_t$$

where the coefficients are the matrices

$$C = \begin{pmatrix} 2t \\ 0 \end{pmatrix}, \quad D = \begin{pmatrix} 4 & 0 \\ 0 & X_1(t) \end{pmatrix}$$

This particular system can be easily solved by finding the solution for X_1 first and then using that solution to find X_2, but we are looking for a more general theory.

Assume that we have the following system of SDEs:

$$d\mathbf{X}_t = Cdt + Dd\mathbf{B}_t,$$

for an n-dimensional process $\mathbf{X}_t$ and further assume that the vector C can be written as

$$C = A\mathbf{X}_t,$$

for an $n \times n$ matrix A. Also assume that the matrix D is only a function of t (does not contain the process $\mathbf{X}_t$). In this case, the system can be written as

$$d\mathbf{X}_t = A\mathbf{X}_t dt + Dd\mathbf{B}_t, \tag{16.25}$$

and since the matrix D is independent of the variable $\mathbf{X}_t$, we use an idea similar to the integrating factor we arc going to solve the system. But this idea is very similar to the integrating factor idea in the first-order differential equations case. However, here we need to deal with exponential matrices. Review Section A.4 in the Appendix on page 533. If the matrix A is given, we define the exponential matrix:

$$e^A = \sum_{n=0}^{\infty} \frac{1}{n!} A^n,$$

where $A^0 = I$ the identity matrix. If we have this, we can multiply Equation (16.25) to the right with the exponential matrix:

$$e^{-At}d\mathbf{X}_t - e^{-At}A\mathbf{X}_t dt = e^{-At}Dd\mathbf{B}_t.$$

Since the matrix A does not depend on $\mathbf{X}_t$, the regular calculus rules apply and we will obtain

$$d(e^{-At}\mathbf{X}_t) = e^{-At}D d\mathbf{B}_t.$$

Thus we can obtain the solution

$$\mathbf{X}_t = e^{At}\mathbf{X}_0 + e^{At}\int_0^t e^{-As}D d\mathbf{B}_s$$

$$= e^{At}\mathbf{X}_0 + \int_0^t e^{A(t-s)}D d\mathbf{B}_s$$

The issue when calculating the solution is to calculate the exponential. Consult Section A.4.3 on page 534 for details. The idea is to write $A = UDU^{-1}$, where D is diagonal, thus the powers will be

$$A^n = UD^nU^{-1},$$

which is easy to calculate. If we have this, it is then easy to see that

$$e^A = Ue^DU^{-1}.$$

■ **EXAMPLE 16.13**

Consider the system

$$dX_1(t) = X_2 dt + \alpha dB_1(t)$$
$$dX_2(t) = -X_1(t)dt + \beta dB_2(t)$$

with $\mathbf{X}_0 = (0,0)^t$.

This is a system of the type (16.25):

$$d\mathbf{X}_t = A\mathbf{X}_t dt + D d\mathbf{B}_t,$$

where

$$A = \begin{pmatrix} 0 & 1 \\ -1 & 0 \end{pmatrix}, \quad D = \begin{pmatrix} \alpha & 0 \\ 0 & \beta \end{pmatrix}$$

Therefore, the solution is

$$\mathbf{X}_t = \int_0^t e^{A(t-s)}D d\mathbf{B}_s$$

and we just need to calculate e^{At}. To do so, let us use Cholesky decomposition. We need to first calculate the eigenvalues and eigenvectors of A. We have

$$det(I - \lambda A) = 1 + \lambda^2,$$

thus the eigenvalues are i and $-i$. We can calculate the corresponding eigenvectors as

$$u_1 = \begin{pmatrix} i \\ -1 \end{pmatrix}, \quad u_2 = \begin{pmatrix} i \\ 1 \end{pmatrix}$$

We then need to calculate U^{-1}. We skip details and just show the result:

$$U = \begin{pmatrix} i & -1 \\ i & 1 \end{pmatrix}, \quad U^{-1} = \frac{1}{2i}\begin{pmatrix} 1 & -i \\ 1 & i \end{pmatrix}, \quad D = \begin{pmatrix} i & 0 \\ 0 & -i \end{pmatrix}.$$

Therefore,

$$e^{At} = \sum_{i=0}^{\infty} \frac{1}{n!} U(Dt)^n U^{-1} = U \begin{pmatrix} e^{it} & 0 \\ 0 & e^{-it} \end{pmatrix} U^{-1}.$$

After multiplications, we finally obtain

$$e^{At} = \begin{pmatrix} \cos t & \sin t \\ -\sin t & \cos t. \end{pmatrix}$$

Substituting $e^{A(t-s)}$, we can finally find the solution

$$\mathbf{X}_t = \begin{pmatrix} \int_0^t \alpha \cos(t-s)dB_1(s) + \int_0^t \beta \sin(t-s)dB_2(s) \\ -\int_0^t \alpha \sin(t-s)dB_1(s) + \int_0^t \beta \cos(t-s)dB_2(s) \end{pmatrix}$$

16.7 A Simple Relationship between SDEs and Partial Differential Equations (PDEs)

IN this section, we talk about a simple relationship that will allow us to find candidate solutions for SDEs. There is a deeper relationship which involves a famous result: the Feynman–Kaç theorem. We will not explore that result. Consult Karatzas and Shreve (1991) for very general versions of this theorem.

Instead, let us consider the simple SDE in (16.9):

$$X_t = x_0 + \int_0^t \mu(s, X_s)ds + \int_0^t \sigma(s, X_s)dB_s. \tag{16.26}$$

A natural candidate solution would be a function of time and B_t, that is, we may look for a solution of the type

$$X_t = f(t, B_t).$$

If we can determine such a function, then using the existence and uniqueness of the solution (Theorem 16.15) we should be able to conclude that this is the unique solution. But how do we find the function $f(t, x)$? Well, once again, the only tool in this stochastic calculus is the Itô lemma. Applying it to the function f, we must have

$$X_t = f(t, B_t) = f(0, 0) + \int_0^t \left(\frac{\partial f}{\partial s} + \frac{1}{2} \frac{\partial^2 f}{\partial x^2} \right) ds + \int_0^t \frac{\partial f}{\partial x} dB_s \qquad (16.27)$$

This indicates that, if we choose the function f so that it satisfies the following system of PDEs, then the solution is going to provide us with a **candidate** for the solution of the SDE. We still need to verify that the process solves the equation.

The system is obtained by equating terms in (16.26) and (16.27):

$$\frac{\partial f}{\partial t}(t, x) + \frac{1}{2} \frac{\partial^2 f}{\partial x^2}(t, x) = \mu(t, f(t, x))$$

$$\frac{\partial f}{\partial x}(t, x) = \sigma(t, f(t, x)) \qquad (16.28)$$

$$f(0, 0) = x_0$$

Note that this method does not always work. Often, it is the case that the resulting PDE (16.28) is harder to solve than the original SDE (16.26).

▉ EXAMPLE 16.14

Let us look at the Ørnstein-Uhlenbeck SDE (Equation (16.14))

$$dX_t = \alpha X_t dt + \sigma dB_t, \quad X_0 = x.$$

If we follow the technique in this section, we obtain the following PDE system:

$$\frac{\partial f}{\partial t}(t, x) + \frac{1}{2} \frac{\partial^2 f}{\partial x^2}(t, x) = -\alpha f(t, x)$$

$$\frac{\partial f}{\partial x}(t, x) = \sigma$$

$$f(0, 0) = x_0$$

The second equation gives $f(t, x) = \sigma x + c(t)$. Substituting into the first equation gives

$$c'(t) = -\alpha(\sigma x + c(t)),$$

or

$$c'(t) + \alpha c(t) = -\alpha \sigma x.$$

This equation has to hold for all x and t. However, the left-hand side is a function of t only, while the right-hand side is a function of x only. Therefore, the only way this can happen is when both sides are equal to a constant, say λ. However, this will imply that $-\alpha \sigma x = \lambda$, and that happens only if x is a constant, which is absurd.

Thus the system cannot be solved, and we can see that even in this very simple case the PDE method produces nothing.

16.8   Monte Carlo Simulations of SDEs

In this section, we provide a simple pseudo-code to generate sample paths of a process defined by an SDE. We will focus first on the simplest way to do it, which is the Euler's method.

Consider the following SDE:

$$dX_t = \alpha(t, X_t)dt + \beta(t, X_t)dB_t, \quad X_0 = x_0,$$

with B_t a one-dimensional standard Brownian motion.

As you know already, this equation does not mean anything; it is just a notation for the following integral equation:

$$X_t = x_0 + \int_0^t \alpha(s, X_s)ds + \int_0^t \beta(s, X_s)dW_s \qquad (16.29)$$

Therefore, approximating the path of X_t is equivalent to approximating the integral. There are many ways to approximate the first integral. In fact, any sum of the form

$$\sum_{i=1}^{n} \alpha(\xi_i, X_{\xi_i})(t_i - t_{i-1})$$

will converge to the integral when the norm of the partition $\pi_n = (0 = t_0 < t_1 < \cdots < t_n = t)$ goes to 0 and for any point $\xi_i \in [t_{i-1}, t_i]$.

However, the stochastic integral has to be approximated using the leftmost point in the interval (t_{i-1} in the above example).

The Euler method uses a simple rectangular rule. Assume that the interval $[0, t]$ is divided into n subintervals. For simplicity of notation, we take the subintervals to be equal in length, but they do not need to be. This means that the increment is $\Delta t = t/n$ and that the points are $t_0 = 0, t_1 = \Delta t, \ldots, t_i = i\Delta t, \ldots, t_n = n\Delta t = t$. Thus, using X_i to denote X_{t_i}, we have

$$\begin{cases} X_0 &= x_0 \\ X_i &= X_{i-1} + \alpha(t_{i-1}, X_{i-1})\Delta t + \beta(t_{i-1}, X_{i-1})\Delta B_i, \quad \forall i \in \{1, 2, \ldots, n\}, \end{cases}$$

where ΔB_i is the increment of a standard Brownian motion over the interval $[t_{i-1}, t_i]$. Since we know from the basic properties of the Brownian motion that the increments are independent and stationary, it follows that each such increment is independent of all others and is distributed as a normal (Gaussian) random variable with mean 0 and variance the length of the time subinterval (which is Δt for us). Therefore, the standard deviation of the increment is $\sqrt{\Delta t}$.

This is the whole information we need to generate paths of solutions to the SDEs using Euler's method. The following exemplifies the process of creating Monte Carlo paths.

■ EXAMPLE 16.15

Suppose we want to generate a path for a period of one year and we have determined that the stochastic process solves an SDE of the type presented in (16.29). We have estimated the functions $\alpha(t, x) = 15x$ and $\beta(t, x) = 12 \log x$. Specifically, the SDE is

$$dX_t = 15X_t dt + 12 \log X_t dB_t.$$

Suppose also that the starting variable X_0 is a random variable uniformly distributed in $[-15, 15]$.

First, we need to decide on how many points we want for our simulated path. Since this is a path with a length of one year, suppose we want to simulate daily values. In this example, therefore, we pick $t = 1$ and $n = 365$ assuming that there are process movements during weekends and holidays. Thus, $\Delta t = 1/365$. Let us suppose that we have two functions that generate random numbers: *rand* generates a uniform between 0 and 1; and *randn* generates a normal with mean 0 and variance 1.

You should immediately see that the particular equation presented above may not have a solution since the β coefficient contains the $\log$ function which requires a positive argument. Let us continue, assuming that the modeler forgot to add the absolute value so we will use $\beta = 12 \log |x|$.

Algorithm   Generating a sample path using Euler's method.

---

initialize $n, t, \Delta t$
$X[0] = 30 * rand - 15$ {Generate the starting value}
for $i = 1$ to n **do**
    $\Delta W = \sqrt{\Delta t} * randn$ {Generate the increment for BM}
    $X[i] = X[i - 1] + 15 * X[i - 1] * \Delta t + 12 * \log |X[i - 1]| * \Delta W$
    Store $X[i]$
end for
Plot the whole path.

You can extend this algorithm to general α and β functions by creating separate functions and calling these functions in your code. You also work with stochastic volatility models such as the Ho–Lee model which has two equations running in parallel by first generating from one and then using the result in the other at every single step from 1 to n.

If we need to generate correlated Brownian motions, consult Section 5.4 on page 173.

The Euler method gives a first-order approximation for the stochastic integral. The Euler–Milstein method, detailed next, provides an improvement by including second-order terms. The idea in this scheme is to consider expansions on the coefficients μ and σ. We refer to Kloeden and Platen (1992) and Glasserman (2003) for an in-depth analysis of the order of convergence. This scheme is only one of the schemes based on Milstein's work and it is applied when the coefficients of the process are functions of only the main process (do not depend on time). Specifically, the scheme is designed to work with SDEs of the type

$$dX_t = \alpha(X_t)dt + \beta(X_t)dB_t, \quad X_0 = x_0.$$

We consider expansions on the coefficients $\alpha(X_t)$ and $\beta(X_t)$ using Itô's lemma. We then have

$$d\alpha(X_t) = \alpha'(X_t)dX_t + \frac{1}{2}\alpha''(X_t)(dX_t)^2$$

$$d\alpha(X_t) = \left(\alpha'(X_t)\alpha(X_t) + \frac{1}{2}\alpha''(X_t)\beta^2(X_t)\right)dt + \alpha'(X_t)\beta(X_t)dB_t.$$

Proceeding in a similar way for $\beta(X_t)$ and writing in the integral form from t to u for any $u \in (t, t + \Delta t]$, we obtain

$$\alpha_u = \alpha_t + \int_t^u \left(\alpha'_s\alpha_s + \frac{1}{2}\alpha''_s\beta_s^2\right)ds + \int_t^u \alpha'_s\beta_s dB_s$$

$$\beta_u = \beta_t + \int_t^u \left(\beta'_s\alpha_s + \frac{1}{2}\beta''_s\beta_s^2\right)ds + \int_t^u \beta'_s\beta_s dB_s,$$

where we used the notation $\alpha_u = \alpha(X_u)$ and similarly for all other terms but omitted for lack of space. Substituting these expressions in the original SDE, we obtain

$$X_{t+\Delta t} = X_t + \int_t^{t+\Delta t} \left(\alpha_t + \int_t^u \left(\alpha'_s\alpha_s + \frac{1}{2}\alpha''_s\beta_s^2\right)ds + \int_t^u \alpha'_s\beta_s dB_s\right)du$$

$$+ \int_t^{t+\Delta t} \left(\beta_t + \int_t^u \left(\beta'_s\alpha_s + \frac{1}{2}\beta''_s\beta_s^2\right)ds + \int_t^u \beta'_s\beta_s dB_s\right)dB_u$$

In this expression, we eliminate all terms which will produce, after integration, higher orders than Δt. That means eliminating terms of the type $dsdu = O(\Delta_t^2)$ and $dudB_s = O(\Delta t^{\frac{3}{2}})$. The only terms remaining other than simply du and ds are the ones involving $dB_u dB_s$ since they are of the right order. Thus, after eliminating the terms, we are left with

$$X_{t+\Delta t} = X_t + \alpha_t \int_t^{t+\Delta t} du + \beta_t \int_t^{t+\Delta t} dB_u + \int_t^{t+\Delta t}\int_t^u \beta'_s\beta_s dB_s dB_u \quad (16.30)$$

For the last term, we apply Euler discretization in the inner integral:

$$\int_t^{t+\Delta t}\left(\int_t^u \beta'_s\beta_s dB_s\right)dB_u \approx \int_t^{t+\Delta t} \beta'_t\beta_t(B_u - B_t)dB_u$$

$$= \beta'_t\beta_t\left(\int_t^{t+\Delta t} B_u dB_u - B_t\int_t^{t+\Delta t} dB_u\right)$$

$$= \beta'_t\beta_t\left(\int_t^{t+\Delta t} B_u dB_u - B_t B_{t+\Delta t} + B_t^2\right).$$

$$(16.31)$$

For the integral inside, recall that we showed (in a couple of ways)

$$\int_0^v B_u dB_u = \frac{1}{2}(B_t^2 - t).$$

Therefore, applying for t and $t + \Delta t$ and taking the difference, we obtain

$$\int_t^{t+\Delta t} B_u dB_u = \frac{1}{2}(B_{t+\Delta t}^2 - t - \Delta t) - \frac{1}{2}(B_t^2 - t)$$

Therefore, substituting back in (16.31), we have

$$\int_t^{t+\Delta t} \left(\int_t^u \beta_s' \beta_s dB_s \right) dB_u \approx \beta_t' \beta_t \left(\frac{1}{2}(B_{t+\Delta t}^2 - B_t^2 - \Delta t) - B_t B_{t+\Delta t} + B_t^2 \right)$$

$$= \beta_t' \beta_t \left(\frac{1}{2}B_{t+\Delta t}^2 + \frac{1}{2}B_t^2 - B_t B_{t+\Delta t} - \Delta t \right)$$

$$= \beta_t' \beta_t \left(\frac{1}{2}(B_{t+\Delta t} - B_t)^2 - \Delta t \right).$$

Note that $B_{t+\Delta t} - B_t$ is the increment of the Brownian motion which we know is $N(0, \Delta t)$ or $\sqrt{\Delta t}Z$, where $Z \sim N(0, 1)$. This actually concludes the derivation of the Euler–Milstein scheme. To summarize, for the SDE

$$dX_t = \alpha(X_t)dt + \beta(X_t)dB_t, \quad X_0 = x_0,$$

the Euler–Milstein scheme starts with $X_0 = x_0$ and for each successive point, first generates $Z \sim N(0, 1)$ and then calculates the next point as

$$X_{t+\Delta t} = X_t + \alpha(X_t)\Delta t + \beta(X_t)\sqrt{\Delta t}Z + \frac{1}{2}\beta'(X_t)\beta(X_t)\Delta t(Z^2 - 1) \quad (16.32)$$

◼ EXAMPLE 16.16 Euler–Milstein scheme for the Black–Scholes model

In the Black–Scholes model, the stock follows

$$dS_t = rS_t dt + \sigma S_t dB_t,$$

where S_t is the stock price at time t, r is the risk-free rate, and σ is the volatility of the stock. Therefore, the coefficients depend only on S_t, and thus we can apply the Milstein scheme. Specifically

$$\alpha(x) = rx, \quad \beta(x) = \sigma x$$

Therefore, the scheme is

$$S_{t+\Delta t} = S_t + rS_t \Delta t + \sigma S_t \sqrt{\Delta t}Z + \frac{1}{2}\sigma^2 S_t \Delta t(Z^2 - 1).$$

Since the Black–Scholes scheme is none other than the geometric Brownian motion and we have already seen how to solve it by taking $X_t = \log S_t$ and applying the Itô lemma, the equation for X_t is

$$dX_t = \left(r - \frac{\sigma^2}{2} \right) dt + \sigma dB_t.$$

Therefore, in this case

$$X_{t+\Delta t} = X_t + \left(r - \frac{\sigma^2}{2} \right) \Delta t + \sigma \sqrt{\Delta t} Z,$$

since the last term contains the derivative of the $\beta(x) = \sigma$, which is a constant function. Notice that in this case Milstein and Euler are identical, thus we get no improvement by applying the more complex scheme.

■ EXAMPLE 16.17 A stochastic volatility example

Let us finish this chapter by talking about approximating using the Euler–Milstein scheme a more complex model involving two related processes. These processes appear in finance and we shall use the notation which is common there. We will focus on the Heston model (Heston, 1993) which is very popular with practitioners. This model is formulated in differential form as

$$
\begin{aligned}
dS_t &= r S_t dt + \sqrt{Y_t} S_t dB_1(t) \\
dY_t &= \kappa(\bar{Y} - Y_t) dt + \sigma \sqrt{Y_t} dB_2(t)
\end{aligned}
\tag{16.33}
$$

where the Brownian motions are correlated with the correlation coefficient ρ.

The Euler scheme for (16.33) is straightforward:

$$
\begin{aligned}
dS_{t+\Delta t} &= S_t + r S_t \Delta t + \sqrt{Y_t} S_t \sqrt{\Delta t} Z_1 \\
dY_{t+\Delta t} &= Y_t + \kappa(\bar{Y} - Y_t) \Delta t + \sigma \sqrt{Y_t} \sqrt{\Delta t} Z_2
\end{aligned}
$$

where we start from known values S_0 and Y_0 and the correlated increments are created by first generating two independent $N(0, 1)$ variables (call them W_1 and W_2) and taking

$$
\begin{aligned}
Z_1 &= W_1 \\
Z_2 &= \rho W_1 + \sqrt{1 - \rho^2} W_2.
\end{aligned}
$$

The Milstein scheme requires more work. For the Y_t process (which we note is just a CIR-type process), the coefficients are $\alpha(x) = \kappa(\bar{Y} - x)$ and $\beta(x) = \sigma\sqrt{x}$, and therefore a straightforward application of (16.32) produces

$$dY_{t+\Delta t} = Y_t + \kappa(\bar{Y} - Y_t)\Delta t + \sigma\sqrt{Y_t}\sqrt{\Delta t} Z_2 + \frac{1}{4}\sigma^2 \Delta t(Z_2^2 - 1),$$

where Z_1 and Z_2 are generated in the same way as in the Euler scheme. For the S_t process, the coefficients are $\alpha(x) = rx$ and $\beta(x) = \sqrt{Y_t}x$. Therefore, the scheme is

$$dS_{t+\Delta t} = S_t + rS_t\Delta t + \sqrt{Y_t}S_t\sqrt{\Delta t}Z_1 + \frac{1}{2}Y_tS_t\Delta t(Z_2^2 - 1).$$

For the process $X_t = \log S_t$, we can apply Itô's lemma to the two-dimensional function $(\log x, y)$ to obtain the dynamics as

$$dX_t = \left(r - \frac{Y_t}{2}\right)dt + \sqrt{Y_t}dZ_1(t).$$

So the Milstein scheme for this process will be

$$X_{t+\Delta t} = X_t + \left(r - \frac{Y_t}{2}\right)\Delta t + \sqrt{Y_t}\sqrt{\Delta t}Z_1,$$

which, in fact, is the same as the Euler discretization in this case.

Problems

16.1   Calculate

$$\int_0^t sdB_s.$$

To this end, apply Itô lemma to $X_t = B_t$ and $f(t, x) = tx$.

16.2   Calculate the solution to the following SDE:

$$dX_t = \mu X_t dt + \sigma dB_t$$

with $X_0 = x$. The process satisfying this equation is called the Ørnstein–Uhlenbeck process.

16.3   Calculate the solution to the following SDE:

$$dX_t = \alpha(m - X_t)dt + \sigma dB_t$$

with $X_0 = x$. The process satisfying this equation is called the mean-reverting Ørnstein–Uhlenbeck process.

16.4   Using a software of your choice, generate points $(Y_1(i\Delta t), Y_2(i\Delta t))$ solving

$$\begin{cases} dY_1(t) &= -\frac{1}{2}Y_1(t)dt - Y_2(t)dB_t \\ dY_2(t) &= -\frac{1}{2}Y_2(t)dt + Y_1(t)dB_t \end{cases}$$

using an Euler approximation. Use $Y_1(0) = 1$, $Y_2(0) = 0$, $\Delta t = 1/1000$, $i \in \{1, 2, \ldots, 1000\}$. Plot the resulting pairs of points in a two-dimensional space where the first coordinate is the Y_1 value and the second coordinate is the Y_2 value.

16.5   Let B_t be a standard Brownian motion started at 0. Use that for any function f we have:

$$\mathbf{E}[f(B_t)] = \frac{1}{\sqrt{2\pi t}} \int_{-\infty}^{\infty} f(x) e^{-\frac{x^2}{2t}} \, dx,$$

to calculate:

$$\mathbf{E}[B_t^{2k}]$$

for some k, an integer. As a hint, you may want to use integration by parts and induction to come up with a formula for $\mathbf{E}[B_t^{2k}]$.

16.6   Using your choice of software, simulate B_t a standard Brownian motion on the interval $t \in [0, 2]$. Using the simulated paths and the Central Limit Theorem, estimate

$$\mathbf{E}[B_2^4], \mathbf{E}[B_2^8], \mathbf{E}[B_2^{20}].$$

Then simulate each of the processes B_t^4, B_t^8, and B_t^{20} separately and obtain the previous expectations at $t = 2$ that way as well.

Compare all the numbers obtained and also compare with the values in the previous problem. Use a minimum of 1 million simulated paths, and for each path use a time increment of $\Delta t = 0.01$.

16.7   Let X_t, $t \geq 0$, be defined as

$$X_t = \{B_t \mid B_t \geq 0\}, \quad \forall t > 0$$

That is, the process has the paths of the Brownian motion conditioned by the current value being positive.

   a) Show that the pdf of X_t is

$$f_{X_t}(x) = 2 f_{B_t}(x), \forall x \geq 0.$$

   b) Calculate $\mathbf{E}[X_t]$ and $V(X_t)$.
   c) Is X_t a Gaussian process?
   d) Is X_t stationary?
   e) Are X_t and $|B_t|$ identically distributed?

16.8   If $X_t \sim N(0, t)$, calculate the distribution of $|X_t|$. Calculate $\mathbf{E}|X_t|$ and $V(|X_t|)$.

16.9   If $X_t \sim N(0, \sigma^2 t)$ and $Y_t = e^{X_t}$, calculate the pdf of Y_t. Calculate $\mathbf{E}[Y_t]$ and $V(Y_t)$. Calculate the transition probability

$$\mathbf{P}(Y_t = y \mid Y_{t_0} = y_0),$$

and give the density of this transition probability.

16.10   Prove by induction that

$$\int_0^T B_t^k \, dB_t = \frac{B_T^{k+1}}{k+1} - \frac{k}{2} \int_0^T B_t^{k-1} \, dt.$$

16.11 Solve the following SDEs using the general integrating factor method presented in Example 16.11, on page 505, with $X_0 = 0$:

    a) $dX_t = \alpha(\beta - X_t)dt + \sigma X_t dB_t,$.

    b) $dX_t = \frac{X_t}{t}dt + \sigma t X_t dB_t,$.

    c) $dX_t = X_t^\alpha + \sigma X_t dB_t$. For what values of α the solution of the equation explodes (is equal to ∞ or $-\infty$ for a finite time t).

16.12 Show that the stochastic process

$$e^{\int_0^t c(s)dB_s - \frac{1}{2}\int_0^t c^2(s)ds}$$

is a martingale for any deterministic function $c(t)$. Does the result change if $c(t, \omega)$ is a stochastic process such that the stochastic integral is well defined?

16.13 Give an explicit solution for the mean-reverting Ørnstein–Uhlenbeck SDE

$$dX_t = \alpha(\mu - X_t)dt + \sigma dB_t,$$

with $X_0 = x$.

16.14 [Exponential martingale] Suppose $\theta(t, \omega) = (\theta_1(t, \omega), \theta_2(t, \omega), \ldots, \theta_n(t, \omega))$ is a stochastic process in $\mathbb{R}^n$ such that $\theta_i(t, \omega) \in \nu_T$ for all i. Define the process

$$Z_t = e^{\int_0^t \theta(s,\omega)d\mathbf{B}_s - \frac{1}{2}\int_0^t \theta^2(s,\omega)ds},$$

where $\mathbf{B}_t$ is an n-dimensional Brownian motion and all terms in the integrals are calculated as scalar products, for example,

$$\theta(s, \omega)d\mathbf{B}_s = \sum_{i=1}^n \theta_i(s, \omega)dB_i(s)$$

and

$$\theta^2(s, \omega)ds = \sum_{i=1}^n \theta_i^2(s, \omega).$$

Note that the process Z_t is one dimensional. Use Itô formula to show that

$$dZ_t = Z_t\theta(t, \omega)d\mathbf{B}_t$$

and derive that Z_t is a martingale if and only if $Z_t\theta_i(t, \omega) \in \nu_T$ for all i.

Remark 16.21 *The previous exponential martingale has been extensively studied in finance since it is the basis of the Girsanov theorem which is crucial in this area. A strong sufficient condition for the process Z_t to be a martingale is the Novikov condition:*

$$\mathbf{E}\left[e^{\frac{1}{2}\int_0^T \theta^2(t,\omega)ds}\right] < \infty.$$

See, for example, Karatzas and Shreve (1991) for details.

16.15   Generate and plot 10 Monte Carlo simulations of the process

$$dX_t = 10(20 - X_t)dt + 0.7dB_t,$$

with $X_0 = 20$. Use the Euler scheme for this problem.

16.16   Give the Euler–Milstein approximation scheme for the following SDE:

$$dS_t = \mu_S S_t dt + \sigma S_t^\beta dB_t,$$

where $\beta \in (0, 1]$. Generate 10 paths and plot them for the following parameter values:

$$\mu_S = 0.1, \quad \sigma = 0.7, \quad \beta = \frac{1}{3}, \quad S_0 = 100$$

16.17   Consider the Hull and White (1987) model

$$dS_t = \mu_S S_t dt + \sqrt{Y_t} S_t dB_t$$
$$dY_t = \mu_Y Y_t dt + \xi Y_t dW_t.$$

The volatility process is $\sqrt{Y_t}$. Show that the volatility process has moments

$$\mathbf{E}[\sqrt{Y_t}] = \sqrt{Y_0}\, e^{\frac{1}{2}\mu_Y t - \frac{1}{8}\xi^2 t}$$
$$V[\sqrt{Y_t}] = \sqrt{Y_0}^{\,2}\, e^{\mu_Y t}\left(1 - e^{-\frac{1}{4}\xi^2 t}\right)$$

and study what happens with these moments when $t \to \infty$ depending on the values of the parameters in the model.

16.18   Consider a generalization of the Hull and White (1987) and Wiggins (1987) model:

$$dS_t = \mu_S S_t dt + \sqrt{Y_t} S_t^\beta dB_t$$
$$dY_t = \mu_Y Y_t dt + \xi Y_t dW_t,$$

where $\beta \in (0, 1]$ and B_t and W_t are Brownian motions correlated with correlation coefficient ρ. Create an Euler–Milstein approximating scheme for this model. Then use the following parameter values to create and plot 10 paths of both Y_t and S_t processes:

$$S_0 = 100, \quad Y_0 = 0.5, \quad \mu_S = 0.02, \quad \mu_Y = 0.001$$
$$\xi = 0.2, \quad \beta = 0.3.$$

16.19 Consider the model

$$dS_t = \mu_S S_t dt + \sqrt{Y_t} S_t^\beta dB_t$$
$$dY_t = \mu_Y Y_t dt + \xi Y_t dW_t,$$

with $\beta \in (0, 1]$, and correlated Brownian motions with coefficient ρ. Show that, if we take the parameters in the model as $\mu_y = \alpha^2$ and $\xi = 2\alpha$, $\mu_S = 0$, we can write the model above as

$$dS_t = y_t S_t^\beta dW_t$$
$$dy_t = \alpha y_t dZ_t.$$

This is the SABR (stochastic alpha beta rho) model introduced in Hagan, Kumar, Lesniewski, and Woodward (2002). Calculate the moments of the volatility driving process y_t in this case.

16.20 Consider the Scott (1987), and the later Chesney and Scott (1989) model which introduced for the first time in finance a so-called mean-reverting volatility process. The model is

$$dS_t = \mu_S S_t dt + e^{Y_t} S_t dW_t,$$
$$dY_t = \alpha(\bar{Y} - Y_t)dt + \sigma_Y dZ_t$$

with the difference between the two papers being that in the first paper the two Brownian motions are uncorrelated ($\rho = 0$).

Calculate the dynamics of the volatility term e_t^Y and its mean and variance. Use Milstein scheme and generate and plot 10 paths.

APPENDIX A

APPENDIX: LINEAR ALGEBRA AND SOLVING DIFFERENCE EQUATIONS AND SYSTEMS OF DIFFERENTIAL EQUATIONS

This appendix contains basic notions from undergraduate mathematics. This appendix is no replacement for a Linear Algebra book or a Differential Equations class but we will state the important concepts and results that are used throughout this book. Furthermore, I am going to assume that the reader is familiar with basic notions of matrix algebra such as determinant, rank of a matrix, linearly independent vectors, inverse of a matrix, when such inverse exists, etc. Such notions should be taught in high school. For more details we refer to Hoffman and Kunze (1971) for linear algebra and Nagle et al. (2011) for differential equations and Boyce and DiPrima (2004) for Ordinary Differential Equations also containing the theory on difference equations.

Probability and Stochastic Processes, First Edition. Ionuţ Florescu

A.1 Solving difference equations with constant coefficients

This methodology is given for second order difference equations but higher order equations are solved in a very similar way. Suppose we are given an equation of the form:

$$a_n = Aa_{n-1} + Ba_{n-2},$$

with some boundary conditions.

The idea is to look for solutions of the form $a_n = cy^n$, with c some constant and y needs to be determined. Note that if we have two solutions of this form (say $c_1 y_1^n$ and $c_2 y_2^n$), then any linear combination of them is also a solution. We substitute this proposed form and obtain:

$$y^n = Ay^{n-1} + By^{n-2}.$$

Dividing by y^{n-2} we obtain the characteristic equation:

$$y^2 = Ay + B.$$

Next, we solve this equation and obtain real solutions y_1 and y_2 (if they exist). It may be possible that the characteristic equation does not have solutions in $\mathbb{R}$ in which case the difference equation does not have solutions either. Now we have two cases:

1. If y_1 and y_2 are distinct then the solution is $a_n = Cy_1^n + Dy_2^n$ where C, D are constants that are going to be determined from the initial conditions.

2. If $y_1 = y_2$ the solution is $a_n = Cy_1^n + Dny_1^n$. Again, C and D are determined from the initial conditions.

In the case when the difference equation contains p terms for example:

$$a_n = A_1 a_{n-1} + A_2 a_{n-2} + \cdots + A_p a_{n-p},$$

the procedure is identical even replicating the multiplicity issues.

A.2 Generalized matrix inverse and pseudo-determinant

For square matrices with n rows and n columns the inverse of the matrix A is defined as a matrix A^{-1} such that:

$$AA^{-1} = A^{-1}A = I_n$$

where I_n is the identity matrix dimension n.

We know that a square matrix is invertible if and only if its determinant is not equal to 0. So what do we do when the determinant is equal to 0? This is a problem that appears a lot in practical applications. Specifically, the covariance matrix of a vector $\mathbf{X} = (X_1, X_2, \ldots, X_n)$ is supposed to be positive definite and symmetric.

However, when some of the random variables $X_1, X_2, \ldots, X_n$ are linearly related the covariance matric is degenerate (determinant equals to 0).

Even if the random variables have some linear combination close to 0 this will happen in practice. In fact, in finance in portfolio theory we often deal with hundreds of random variables and most of the time they are related so this is a very common problem. The big issue in finance is that we need to maximize a portfolio and that will involve calculating the variance of the portfolio which will involve calculating the inverse of the matrix.

To deal with this issue mathematicians created the pseudo-determinant and a matrix that kind of behave like the inverse.

The pseudo-determinant of a square matrix A of dimension n is defined as:

$$|A|_+ = \lim_{\alpha \to 0} \frac{|A + \alpha I_n|}{\alpha^{n - rank(A)}},$$

where $|A + \alpha I_n|$ denotes the regular determinant of the matrix $A + \alpha I_n$, and $rank(A)$ denotes the rank of the matrix A, that is the largest dimension of the square matrix made with columns and rows of A (minor) and with nonzero determinant.

For a square matrix the most common concept of generalized inverse is the Moore–Penrose pseudoinverse. For a square matrix A of dimension n, this is defined as a matrix A^+ such that:

$$AA^+A = A$$
$$A^+AA^+ = A^+$$
$$(AA^+)^t = AA^+$$
$$(A^+A)^t = A^+A$$

Most constructions in practice will insure these properties are satisfied. where we used A^t notation for the transpose of the matrix A.

A.3 Connection between systems of differential equations and matrices

The connection is best seen in the chapter 13 on the Markov Processes. In population dynamics (which in fact are modeled as Markov processes) they appear as Kolmogorov Backward or Forward equations. Let us present some notations. We will be working and presenting results about n dimensional systems and matrices. Given a time dependent deterministic matrix:

$$P(t) = \begin{pmatrix} P_{11}(t) & P_{12}(t) & P_{13}(t) & \cdots & P_{1n}(t) \\ P_{21}(t) & P_{22}(t) & P_{23}(t) & \cdots & P_{2n}(t) \\ \vdots & \vdots & \vdots & & \vdots \\ P_{n1}(t) & P_{n2}(t) & P_{n3}(t) & \cdots & P_{nn}(t) \end{pmatrix},$$

We will call the matrix derivative:

$$P'(t) = \frac{dP}{dt}(t) = \begin{pmatrix} P'_{11}(t) & P'_{12}(t) & P'_{13}(t) & \cdots & P'_{1n}(t) \\ P'_{21}(t) & P'_{22}(t) & P'_{23}(t) & \cdots & P'_{2n}(t) \\ \vdots & \vdots & \vdots & & \\ P'_{n1}(t) & P'_{n2}(t) & P'_{n3}(t) & \cdots & P'_{nn}(t) \end{pmatrix},$$

the matrix with each element the derivative. We will denote the integral:

$$\int P(t)dt = \begin{pmatrix} \int P_{11}(t)dt & \int P_{12}(t)dt & \int P_{13}(t)dt & \cdots & \int P_{1n}(t)dt \\ \int P_{21}(t)dt & \int P_{22}(t) & \int P_{23}(t)dt & \cdots & \int P_{2n}(t)dt \\ \vdots & \vdots & \vdots & & \\ \int P_{n1}(t)dt & \int P_{n2}(t)dt & \int P_{n3}(t)dt & \cdots & \int P_{nn}(t)dt \end{pmatrix},$$

So pretty much everything work as normal.

Furthermore, differentiating rules work pretty much as usual:

$$\frac{d}{dt}(CP) = C\frac{dP}{dt}, \qquad C \text{ a constant matrix}$$

$$\frac{d}{dt}(P+Q) = \frac{dP}{dt} + \frac{dQ}{dt}$$

$$\frac{d}{dt}(PQ) = \frac{dP}{dt}Q + P\frac{dQ}{dt}$$

where in the last the order of multiplication is important since matrix multiplication is not commutative.

A.3.1   Writing a system of differential equations in matrix form

Now suppose we have a system of differential equations of first order. such a system can always be written in matrix form. For example:

$$\begin{cases} x'_1 & = 4tx_2 + x_5 \\ x'_2 & = x_1 + x_3 \\ x'_3 & = 3x_1 - 2(1 + \sin t)x_2 + x_4 - 16x_5 + \sin(t^2 + 1) , \\ x'_4 & = x_1 + 2x_2 + \cos t\, x_3 - \cos(t^2 + 1) \\ x'_5 & = 13(t^2 + 1)x_1 + 2t^2 x_2 + x_4 - x_5 \end{cases}$$

where we suppressed writing the dependency of the functions x_i on t for brevity of notation. Now if we denote the vector $\mathbf{x}(t) = (x_1(t), x_2(t), x_3(t), x_4(t), x_5(t))^T$ where ^{T} denote the transposition operation we can write the same system in matrix form as:

$$\mathbf{x}'(t) = P\mathbf{x}(t) + g(t),$$

where the $\mathbf{x}'$ denotes the derivative vector and the matrices are:

$$P = \begin{pmatrix} 0 & 4t & 0 & 0 & 1 \\ 1 & 0 & 1 & 0 & 0 \\ 3 & -2(1+\sin t) & 0 & 1 & -16 \\ 1 & 2 & \cos t & 0 & 0 \\ 13(t^2+1) & 2 & 0 & 1 & -1 \end{pmatrix} \quad g(t) = \begin{pmatrix} 0 \\ 0 \\ \sin(t^2+1) \\ \cos(t^2+1) \\ 0 \end{pmatrix}$$

Finally, any higher order differential equation may be written as a system of differential equations as the next example will show.

Consider the fourth order differential equation:

$$x^{(4)} + 4x^{(3)} + \sin t\, x' + 15x = \cos t$$
$$x(0) = 0, x'(0) = 1, x''(0) = 2, x^{(3)}(0) = 3$$

If we now denote $x_1 = x$, $x_2 = x'$, $x_3 = x''$, and $x_4 = x^{(3)}$, we can write the following system:

$$\begin{cases} x_1' = x_2 \\ x_2' = x_3 \\ x_3' = x_4 \\ x_4' = -15x_1 - \sin t\, x_2 - 4x_4 + \cos t \end{cases} \quad, \quad \mathbf{x}(0) = \begin{pmatrix} 0 \\ 1 \\ 2 \\ 3 \end{pmatrix}$$

and we can further identify the matrix $P(t)$ and the vector $g(t)$.

The system of differential equations is called **homogeneous** if $g(t) = 0$ and **non-homogeneous** otherwise. When the matrix $P(t)$ is a constant we are obtaining a system that can be solved more easily.

Note that $\mathbf{x}(t) = (x_1(t), x_2(t), x_3(t), x_4(t), x_5(t))^T$ is a vector valued function; more precisely, $\mathbf{x} : I \to \mathbb{R}^n$. In most of our applications the domain of the function will be $I = (0, \infty)$, but the theory will be talking about a general interval $I \subseteq \mathbb{R}$.

Theorem A.1 (Existence and Uniqueness) *Suppose $P(t)$ and $g(t)$ are continuous on an open interval that contains the initial time t_0. Then, for any choice of initial value $\mathbf{x}_0$, there exists a unique solution of the initial value problem:*

$$\mathbf{x}'(t) = P\mathbf{x}(t) + g(t), \quad \mathbf{x}(t_0) = x_0,$$

for any $t \in I$.

So the theorem says that the solution exists and is unique. That is reassuring but what is the solution? For that we need more notions.

Definition A.2 (Linear independence of vector functions) *Suppose we have m vector functions: $\mathbf{x}^1, \ldots, \mathbf{x}^m$, that is each $\mathbf{x}^i : I \to \mathbb{R}^n$. The functions are called* *linearly dependent on I if there exist constants $c_1, \ldots, c_m$ at least one nonzero such that:*

$$c\mathbf{x}^1 + \ldots + c_m\mathbf{x}^m = 0,$$

for all the t in the interval I. If such constants do not exist then the vectors are called linearly independent.

Wronskian. *The Wronskian determined by n vector valued functions $\mathbf{x}^i : I \to \mathbb{R}^n$ is the determinant of the matrix which has as columns the vectors $\mathbf{x}^i$. If we denote the elements in the vector $\mathbf{x}^i(t) = (x_1^i(t), x_2^i(t), \ldots, x_n^i(t))^T$ then the Wronskian is:*

$$W[\mathbf{x}^1, \ldots, \mathbf{x}^m](t) = \begin{vmatrix} x_1^1(t) & x_1^2(t) & \ldots & x_1^n(t) \\ x_2^1(t) & x_2^2(t) & \ldots & x_2^n(t) \\ \vdots & \vdots & \vdots \\ x_n^1(t) & x_n^2(t) & \ldots & x_n^n(t) \end{vmatrix},$$

where $|W| = \det(W)$ is the notation for the determinant of the matrix W.

Checking linear independence. *Suppose that we are given n vector valued functions $\mathbf{x}^i : I \to \mathbb{R}^n$. If the Wronskian of these functions is not zero for any $t \in I$ then the n functions are linearly independent on I*

Finally we can now write the form of the solution. You will see the form is quite useless for now.

Theorem A.3 (System solution representation) *Suppose we are given the system of n differential equations:*

$$\mathbf{x}'(t) = P\mathbf{x}(t) + g(t).$$

Let $P(t)$ and $g(t)$ continuous on the interval I.

Assume that by some method we find n linearly independent solutions $\mathbf{x}^1, \mathbf{x}^2, \ldots, \mathbf{x}^n$ of the homogeneous part of the system $(\mathbf{x}'(t) = P\mathbf{x}(t))$. Also, by some other mysterious method we can find one particular solution $\mathbf{x}^p$ of the original system.

Then any solution (read the most general form of the solution) can be expressed as:

$$\mathbf{x}(t) = \mathbf{x}^p(t) + c_1\mathbf{x}^1(t) + c_2\mathbf{x}^2(t) + \ldots + c_n\mathbf{x}^n(t) = \mathbf{x}^p(t) + \mathbf{X}(t)\mathbf{c},$$

where $\mathbf{X}(t) = [\mathbf{x}^1(t), \ldots, \mathbf{x}^m(t)]$, called the fundamental matrix is the matrix which has columns each of the linearly independent vector valued solutions and $\mathbf{c} = (c_1, \ldots, c_n)^T$.

Finding the fundamental solutions (that is the linearly independent functionals) $[\mathbf{x}^1(t), \ldots, \mathbf{x}^m(t)]$ as well as the particular solution $\mathbf{x}^p(t)$ is therefore the big issue. Here is the simple answer to how to find them. If the matrix appearing in the system $P(t) = P$ that is the matrix is constant then we can solve anything by many different methods. For a non-constant matrix $P(t)$ we are usually out of luck (read can't solve) unless the matrix $P(t)$ has a special structure or it can be simplified.

To proceed we need elements from linear algebra.

A.4   Linear Algebra results

A.4.1   Eigenvalues, eigenvectors of a square matrix

Given a constant, $n \times n$ square matrix P, a non-zero $n \times 1$ column vector $\mathbf{u} = (u_1, u_2, \ldots, u_n)^T$ is called an eigenvector of the matrix P corresponding to eigenvalue λ if it satisfies the characteristic equation:

$$P\mathbf{u} = \lambda\mathbf{u}$$

Note that the eigenvectors corresponding to the same λ are not unique. For example, for any existing eigenvector-eigenvalue pair $(\mathbf{u}_0, \lambda_0)$ a multiplication with a constant $c\mathbf{u}_0$ produces a new eigenvector-eigenvalue pair.

To find the eigenvalues-eigenvector pairs of a matrix one writes the characteristic equation above as:

$$(P - \lambda I)\mathbf{u} = \mathbf{0},$$

where I is the $n \times n$ identity matrix. If the squared matrix $P - \lambda I$ is invertible then one multiplies with the inverse above and thus the unique solution is the null solution $\mathbf{u} = \mathbf{0}$. Therefore, the pairs (eigenvalue, eigenvector) (nonzero by definition) may only be obtained when the matrix is singular. Thus, the eigenvalues λ are obtained as solutions of the characteristic equation:

$$\det(P - \lambda I) = 0,$$

(det denotes the determinant of a square matrix). This produces n roots (possibly complex and possibly with multiplicity greater than one) $\lambda_1, \lambda_2 \ldots, \lambda_n$.

Once an eigenvalue is obtained a corresponding eigenvector is obtained by solving a reduced system and imposing a size condition on the eigenvector (for example making the eigenvector unitary $|\mathbf{u}| = \sqrt{u_1^2 + u_2^2 + \ldots + u_n^2} = 1$). That is, for a fixed λ_i the system to be solved is

$$(P - \lambda_i I)\mathbf{u} = \mathbf{0}.$$

By hand this amounts to a simple Gaussian elimination routine, but this of course, is not practical if the dimension $n \geq 5$. The eigenvalues, eigenvectors for higher dimensional situations are obtained using software. For example in R which is the best statistical software (and is free as of the time of this writing) one would just input the matrix as P and use the command `eigen(P)`. This automatically produces the eigenvalues (fast calculation) and corresponding eigenvectors (this may be slow for large dimensional matrices).

These eigenvalues are really useful for example, the trace of a matrix (sum of the elements on the diagonal)

$$Tr(P) = \lambda_1 + \lambda_2 + \cdots + \lambda_n$$

And the determinant:

$$\det(P) = \lambda_1\lambda_2 \ldots \lambda_n$$

But for us they are important as we shall see due to their connection to solving systems of differential equations.

A.4.2 Matrix Exponential Function

Definition A.4 *Suppose P is an $n \times n$ constant matrix. We define the matrix exponential e^{Pt} as the matrix:*

$$e^{Pt} = I + Pt + P^2 \frac{t^2}{2!} + \cdots + P^k \frac{t^k}{k!} + \ldots$$

where P^k is the matrix P multiplied with itself k times and I is the $n \times n$ identity matrix.

Note that due to constant multiplication $e^{Pt} = e^{tP}$. Calculating e^{Pt} is generally complex and we'll come back to this in a minute. First let us present properties of this matrix.

Proposition A.5 (Properties of the matrix exponential function) *Suppose P and Q are $n \times n$ constant matrices and c, t, s are real (or complex) numbers. Then:*

1. $e^{P0} = e^{0} = I.$

2. $e^{P(s+t)} = e^{Ps} e^{Pt}.$

3. $\left(e^{Pt} \right)^{-1} = e^{-Pt}.$

4. *If $PQ = QP$ then $e^{(P+Q)t} = e^{Pt} e^{Qt}.$*

5. $e^{cIt} = e^{ct} I$

6. $\frac{d}{dt}\left(e^{Pt} \right) = P e^{Pt}$

Property 2 and 3 show that the matrix exponential is always invertible. Property 6 says that the matrix exponential function is in fact a solution to the **matrix** differential equation $\mathbf{X}' = P\mathbf{X}$. this is going to be very important for the continuous time Markov chains. But there is more. The columns of this matrix are linearly independent and in fact the columns of the matrix form a fundamental solution set. This is summarized next:

Corollary A.6 *If P is a $n \times n$ constant matrix then the columns of the matrix exponential e^{Pt} form a fundamental set of solutions for the system $\mathbf{x}'(t) = P\mathbf{x}(t)$. Therefore, the fundamental matrix is e^{Pt} and the general solution is $e^{Pt}\mathbf{c}$.*

There still remains the question: how to calculate this matrix exponential. Linear algebra to the rescue.

A.4.3 Relationship between Exponential matrix and Eigenvectors

Lemma A.7 (The spectral decomposition) *If P is a constant matrix which has n linearly independent eigenvectors $\mathbf{u}_1, \mathbf{u}_2, \ldots, \mathbf{u}_n$ and corresponding eigenvalues $\lambda_1, \ldots, \lambda_n$ then the matrix P can be written as:*

$$P = U\Lambda U^{-1}$$

where U is the matrix which has columns the eigenvectors $\mathbf{u}_1, \mathbf{u}_2, \ldots, \mathbf{u}_n$ and Λ is a diagonal matrix (zero outside the main diagonal) with the diagonal elements the corresponding eigenvalues $\lambda_1, \ldots, \lambda_n$ (we denote a diagonal matrix with those elements on the diagonal with $\Lambda = diag(\lambda_1, \ldots, \lambda_n)$).

The lemma above is usually called the diagonalization criterion for a matrix. Note that if the number of linearly independent eigenvectors is less than n the criterion fails. With this criterion we now can calculate the exponential matrix and thus provide a general solution to the homogeneous system of differential equations.

Theorem A.8 *If P is a constant matrix which has n linearly independent (unit norm) eigenvectors $\mathbf{u}_1, \mathbf{u}_2, \ldots, \mathbf{u}_n$ and corresponding eigenvalues $\lambda_1, \ldots, \lambda_n$ then the exponential matrix e^{Pt} is:*

$$e^{Pt} = Ue^{\Lambda t}U^{-1},$$

where $e^{\Lambda t}$ is the diagonal matrix with diagonal elements $e^{\lambda_1 t}, \ldots, e^{\lambda_n t}$

This is very easy to show by following the definition of an exponential matrix and the property that

$$P^k \frac{t^k}{k!} = (U\Lambda U^{-1})\ldots(U\Lambda U^{-1})\frac{t^k}{k!} = U\Lambda^k \frac{t^k}{k!}U^{-1}$$

and using that $I = UU^{-1}$; $\Lambda^k = diag(\lambda_1^k, \ldots, \lambda_n^k)$.

Corollary A.9 *If P is a constant matrix which has n linearly independent (unit norm) eigenvectors $\mathbf{u}_1, \mathbf{u}_2, \ldots, \mathbf{u}_n$ and corresponding eigenvalues $\lambda_1, \ldots, \lambda_n$ then the system of equations:*

$$\mathbf{x}'(t) = P\mathbf{x}(t)$$

has the general solution:

$$x(t) = c_1 e^{\lambda_1 t}\mathbf{u}_1 + c_2 e^{\lambda_2 t}\mathbf{u}_2 + \cdots + c_n e^{\lambda_n t}\mathbf{u}_n.$$

Furthermore, the system is equivalent with solving the uncoupled system:

$$\mathbf{y}'(t) = \Lambda\mathbf{y}(t),$$

where $\mathbf{y}(t) = U^{-1}\mathbf{x}(t)$ and the diagonal matrix $\Lambda = U^{-1}PU$.

The corollary is a consequence of all the results. The second assertion is immediate by multiplying with U^{-1} to the left in the original system.

So the issue is now to find the eigenvectors corresponding to the eigenvalues $\lambda_1, \lambda_2 \ldots, \lambda_n$ and furthermore showing they are linearly independent. We have several cases depending on the nature of the eigenvalues.

A.5   Finding fundamental solution of the homogeneous system

Recall that the eigenvalues λ are solutions to an n order polynomial with real coefficients. Thus we only have three cases to consider.

A.5.1   The case when all the eigenvalues are distinct and real

This is case which is the nicest and the easiest to deal with. We have in this case:

Theorem A.10 *Suppose the $n \times n$ constant matrix P has n distinct eigenvalues $\lambda_1, \lambda_2 \ldots, \lambda_n$. In this case the associated eigenvectors $\mathbf{u}_1, \ldots, \mathbf{u}_n$ are linearly independent and thus the fundamental set of solutions for the homogeneous system is:*

$$\mathbf{X}(t) = [e^{\lambda_1 t}\mathbf{u}_1, e^{\lambda_2 t}\mathbf{u}_2 \ldots, e^{\lambda_n t}\mathbf{u}_n]$$

Not only this but we actually have a criterion when we always obtain real distinct eigenvalues.

Lemma A.11 *If the matrix P is symmetric ($P^T = P$) then all its associated eigenvalues are real and there always exist n linearly independent eigenvectors.*

We shall not prove this lemma or the theorem please refer to Lay (2006) for details.

A.5.2   The case when some of the eigenvalues are complex

Recall that λ are roots of a n degree polynomial with real coefficients. Thus if a root is complex then its conjugate is also a root of the polynomial. So we only need to see what happens when two eigenvalues are complex conjugates. When we have more (say 6 complex conjugate eigenvalues) whatever we do for two is similar for all the three pairs. Suppose now that $\lambda_1 = \alpha + i\beta$ and $\lambda_2 = \alpha - i\beta$ are the two complex conjugate eigenvalues. It is not hard to see that the two corresponding eigenvectors are also conjugate.

Fist off, $\mathbf{u}_1$ the eigenvector corresponding to λ_1 must be complex, otherwise we could not have

$$(P - (\alpha + i\beta)I)\mathbf{u}_1 = \mathbf{0}.$$

Recall that P is real thus the only way the complex part is zero is if $\beta I\mathbf{u} = \mathbf{0}$ and since $\beta \neq \mathbf{0}$ $\mathbf{u}$ must be the zero vector – contradiction.

So we can write $\mathbf{u} = \mathbf{a} + i\mathbf{b}$ and since we know that $(P - (\alpha + i\beta)I)\mathbf{u}_1 = \mathbf{0}$, the conjugate of the expression must be zero as well. But this means that:

$$\overline{(P - (\alpha + i\beta)I)\mathbf{u}_1} = (P - (\alpha - i\beta)I)\overline{\mathbf{u}}_1 = \mathbf{0},$$

thus, the conjugate of $\mathbf{u}_1$ is an eigenvector corresponding to λ_2. We finally can write the fundamental solution in this case too.

Theorem A.12 *Suppose that the matrix P has two complex eigenvalues $\alpha \pm i\beta$ with corresponding eigenvectors $\mathbf{a} \pm i\mathbf{b}$. The the two columns in the fundamental solution* *matrix in Theorem A.10 corresponding to these eigenvalues are replaced with:*

$$e^{\alpha t} \cos \beta t \ \mathbf{a} - e^{\alpha t} \sin \beta t \ \mathbf{b} \quad and \quad e^{\alpha t} \sin \beta t \ \mathbf{a} + e^{\alpha t} \cos \beta t \ \mathbf{b},$$

and this once again forms a linearly independent system.

A.5.3   The case of repeated real eigenvalues

This is the hardest case to deal with. We have to solve for each multiplicity separately to find an eigenvector separately. So for example if λ_i has multiplicity m_i we have to solve m_i systems to find m_i linearly independent eigenvectors. But first an easy case.

Suppose that the matrix P has one eigenvalue say λ_1 with multiplicity n. That is the characteristic polynomial can be written as $\det(P - \lambda I) = c(\lambda - \lambda_1)^n$. Then applying the Cayley-Hamilton theorem which says that a matrix satisfies its own characteristic equation we have:

$$(P - \lambda_1 I)^n = 0.$$

But now note that we can write:

$$e^{Pt} = e^{\lambda_1 I t} e^{(P - \lambda_1 I)t} = e^{\lambda_1 t} I \sum_{k=0}^{\infty} (P - \lambda_1 I)^k \frac{t^k}{k!},$$

and since all the powers in the sum greater than and including n are equal to $\mathbf{0}$ we can calculate the matrix exponential as a discrete sum of powers:

$$e^{Pt} = e^{\lambda_1 t} \left(I + (P - \lambda_1 I)t + \cdots + (P - \lambda_1 I)^{n-1} \frac{t^{n-1}}{(n-1)!} \right)$$

Therefore, the solution is simply $e^{Pt}\mathbf{c}$. This gives the idea on what to do in different multiplicity cases.

Algorithm to find the eigenvectors corresponding to the multiplicity $m_1 > 1$ eigenvalue λ_1   Given a constant matrix P $n \times n$ dimensional assume that its characteristic polynomial has an eigenvalue λ_1 with multiplicity m_1. Then one needs to calculate m_1 linearly independent eigenvectors which are also linearly independent of the rest $(n - m_1)$ eigenvectors. This is how one proceeds.

First, the multiplicity 1. We need to solve the system:

$$(A - \lambda_1 I)\mathbf{u} = 0$$

We chose a particular eigenvector (with say $|\mathbf{u}_1| = 1$), then the corresponding solution is:

$$\mathbf{x}(t) = e^{-\lambda_1 t}\mathbf{u}_1.$$

Then for every multiplicity k from 2 to m_1 one proceeds as follows. The generalized system is solved:

$$(A - \lambda_1 I)^k \mathbf{u} = 0 \tag{A.1}$$

where k is the running multiplicity and a nontrivial solution is found. Once these m_1 vectors are obtained one needs to have them linearly independent. For this purpose a standard Gramm-Schmidt orthogonalization procedure will suffice.

When the vectors $(\mathbf{u}_2, \dots \mathbf{u}_{m_1})$ are obtained we calculate the their corresponding solutions as

$$\mathbf{x}(t) = e^{Pt}\mathbf{u}_k = e^{\lambda_1 It}e^{(P-\lambda_1 I)t}\mathbf{u}_k = e^{\lambda_1 t}\sum_{i=0}^{\infty}(P-\lambda_1 I)^i\frac{t^i}{i!}\mathbf{u}_k$$

$$= e^{\lambda_1 t}\left(\mathbf{u}_k + t(P-\lambda_1 I)\mathbf{u}_k + \cdots + \frac{t^{k-1}}{(k-1)!}(P-\lambda_1 I)^{k-1}\mathbf{u}_k\right)$$

where we have used (A.1), the defining property of each $\mathbf{u}_k$.

A.6   The nonhomogeneous system

Now we know what to do when the system is homogeneous and when P is constant. What about the nonhomogeneous system:

$$\mathbf{x}'(t) = P\mathbf{x}(t) + g(t)$$

or even the more complicated one when P is not constant:

$$\mathbf{x}'(t) = P(t)\mathbf{x}(t) + g(t)?$$

We need to find a particular solution to this system and then using the solution representation given in Theorem A.3 we have solved the nonhomogeneous system. In that theorem we talk about the mysterious ways to find the particular solution. These are not mysterious at all and in fact are the two classical ways of finding particular solutions to differential equations.

A.6.1   The method of undetermined coefficients

This method applies only when P is constant and the vector values functional $g(t)$ contains polynomials, exponential functions or trigonometric functions. We can then try a particular solution of the same form as the entries in $g(t)$.

The basic guide for the 1 dimensional equations is as follows.

- If some entries in $g(t)$ are of the form $P_m(t)e^{\lambda t}$ where $P_m(t)$ is some polynomial of degree m then we need to try a solution of the form:

$$x_p(t) = t^s(A_m t^m + \dots A_1 t + A_0)e^{\lambda t},$$

where $s = 0$ if λ is not an eigenvalue of the associated homogeneous system and s is equal to the multiplicity of the eigenvalue if it is.

- If some entry is $P_m(t)e^{\alpha t}\cos\beta t + Q_l(t)e^{\alpha t}\sin\beta t$, with Q_n another polynomial of degree l then we try:

$$x_p(t) = t^s(A_k t^k + \dots A_1 t + A_0)e^{\alpha t}\cos\beta t + t^s(B_k t^k + \dots B_1 t + B_0)e^{\alpha t}\sin\beta t,$$

where $k = \max(m, l)$ and s is once again the multiplicity of the eigenvalue $\alpha + i\beta$.

I will give some examples because above applies to higher order differential equations with constant coefficients. Although, these are equivalent with systems they differ in the case when the particular solution is among the ones of the homogeneous system.

■ **EXAMPLE A.1**

Suppose we want to solve a system:

$$\mathbf{x}'(t) = P\mathbf{x}(t) + g(t),$$

where $g(t) = (t, 1 + t, t^2)^T$. Note that we can write:

$$g(t) = \begin{pmatrix} 0 \\ 1 \\ 0 \end{pmatrix} + t \begin{pmatrix} 1 \\ 1 \\ 0 \end{pmatrix} + t^2 \begin{pmatrix} 0 \\ 0 \\ 1 \end{pmatrix}$$

We look for a particular solution of the form:

$$x_p(t) = \mathbf{a} + t\,\mathbf{b} + t^2\,\mathbf{c}$$

where $\mathbf{a}, \mathbf{b}, \mathbf{c}$ are constant vectors that need to be determined.

As another example suppose $g = (1, t^2, \sin t)$. In this case one will try:

$$x_p(t) = \mathbf{a} + t\,\mathbf{b} + t^2\,\mathbf{c} + \sin t\,\mathbf{d} + \cos t\,\mathbf{e}$$

But in the case when the resulting $x_p(t)$ is also a solution of the homogeneous system the strategy needs to be modified. For example in the first $g(t)$ example if x_p is a solution we would identify the corresponding eigenvalue and determine its multiplicity. If it has multiplicity say s we would then reattempt to find a particular solution as:

$$x_p(t) = \mathbf{a}_0 + t\,\mathbf{a}_1 + t^2\,\mathbf{a}_2 + \ldots t^{s+2}\,\mathbf{a}_{s+2}$$

Please note that this is different from the one dimensional guide presented above. For more details and examples please consult a standard textbook such as (Nagle et al., 2011, Chapter 9).

A.6.2   The method of variation of parameters

This method is applicable to time varying P matrices as well assuming that we have a way to determine the solution of the associated homogeneous system. More precisely suppose we need to find a particular solution of the system:

$$\mathbf{x}'(t) = P(t)\mathbf{x}(t) + g(t).$$

Suppose that we can calculate the fundamental matrix $\mathbf{X}(t)$ associated with the homogeneous system:

$$\mathbf{x}'(t) = P(t)\mathbf{x}(t)$$

Then a particular solution of the nonhomogeneous system is:

$$\mathbf{x}_p = \mathbf{X}(t) \int \mathbf{X}^{-1}(t) g(t) dt. \tag{A.2}$$

In the formula above $\mathbf{X}^{-1}(t)$ is a notation for the inverse of the matrix $\mathbf{X}(t)$.

A.7   Solving systems when P is non-constant

Suppose we have to solve

$$\mathbf{x}'(t) = P(t)\mathbf{x}(t).$$

We know what to do if we are dealing with a non-homogeneous system and we can solve the homogeneous one above – we just apply variation of parameters formula and we find a particular solution.

The big issue is that there is no method to solve the time varying system above. It can only be solved in particular cases. One such example is when the matrix $P(t)$ is diagonal. In this case the system is uncoupled and we are in fact solving one dimensional differential equations.

A second important strategy is to perform a change of variables to make the matrix constant.

For example a simple case is when $P(t) = Ph(t)$ where P is a constant matrix and $h(t)$ a one dimensional function of t. By using the eigenvalue eigenvector decomposition again one may write:

$$U^{-1}\mathbf{x}'(t) = Dh(t)U^{-1}\mathbf{x}(t),$$

where U is the matrix containing the eigenvectors. By making the change of variables $\mathbf{y}(t) = U^{-1}\mathbf{x}(t)$ we end up with a decoupled system (recall that D is a diagonal matrix and it will be the covariance matrix of $\mathbf{y}(t)$) and thus easier to solve. The original system solution is found as $\mathbf{x}(t) = U\mathbf{y}(t)$.

Bibliography

Bertrand, J. L. F. (1889). *Calcul des probabilités*. Paris: Gauthier-Villars et fils.

Bibby, B. and M. Sørensen (1995). Martingale estimating functions for discretely observed diffusion processes. *Bernoulli 1*, 17–39.

Billingsley, P. (1995). *Probability and measure* (3 ed.). Wiley.

Black, F. and M. Scholes (1973). The valuation of options and corporate liability. *Journal of Political Economy 81*, 637–654.

Blæsild, P. and J. Granfeldt (2002). *Statistics with Applications in Biology and Geology*. CRC Press.

Borodin, A. and P. Salminen (2002). *Handbook of Brownian Motion - Facts and Formulae* (2nd ed.). Berlin: Springer.

Boyce, W. E. and R. C. DiPrima (2004). *Elementary Differential Equations and Boundary Value Problems* (8 ed.). Wiley.

Bucher, C. G. (1988). Adaptive Sampling – An iterative fast Monte Carlo procedure. *Structural Safety 5*, 119–128.

Casella, G. and R. L. Berger (2001). *Statistical Inference* (second ed.). Duxbury Press.

Probability and Stochastic Processes, First Edition. Ionuţ Florescu
© 2015 John Wiley & Sons, Inc. Published 2015 by John Wiley & Sons, Inc.

Cauchy, A. L. (1821). *Analyse algébrique*. Imprimerie Royale.

Chesney, M. and L. Scott (1989). Pricing european currency options: a comparison of the modified black-scholes model and a random variance model. *J. Finan. Quant. Anal. 24*, 267–284.

Chung, K. L. (2000). *A Course in Probability Theory Revised* (2nd ed.). Academic Press.

Cox, J., J. Ingersoll, and S. Ross (1985). A theory of the term structure of interest rates. *Econometrica* (53), 385–407.

Dembo, A. (2008). Lecture notes in probability. available on http://www-stat.stanford.edu/~adembo/.

Doob, J. (1953). *Stochastic Processes*. New York: Wiley.

Fisher, R. A. (1925). Applications of "student's" distribution. *Metron*, 90–104.

Florescu, I. and C. G. Păsărică (2009). A study about the existence of the leverage effect in stochastic volatility models. *Physica A: Statistical Mechanics and its Applications 388*(4), 419 – 432.

Florescu, I. and C. Tudor (2013). *Handbook of probability*. Wiley.

Glasserman, P. (2003). *Monte Carlo Methods in Financial Engineering*. Stochastic Modelling and Applied Probability. Springer.

Good, I. J. (1986). Some statistical applications of poisson's work. *Statistical Science 1*(2), 157–170.

Grimmett, G. and D. Stirzaker (2001). *Probability and Random Processes* (3 ed.). Oxford University Press.

Gross, D. and C. M. Harris (1998). *Fundamentals of Queueing Theory*. Wiley.

Hagan, P., D. Kumar, A. Lesniewski, and D. Woodward (2002). Managing smile risk. *Wilmott Magazine*.

Halmos, P. R. (1998, Jan.). *Naive Set Theory*. Undergraduate Texts in Mathematics. Springer.

Heston, S. L. (1993). A closed-form solution for options with stochastic volatility with applications to bond and currency options. *Review of Financial Studies 6*(2), 327–43.

Hoffman, K. and R. Kunze (1971). *Linear Algebra*. Prentice Hall.

Hull, J. C. and A. D. White (1987, June). The pricing of options on assets with stochastic volatilities. *Journal of Finance 42*(2), 281–300.

Jona-Lasinio, G. (1985). *Some recent applications of stochastic processes in quantum mechanics*, Volume 1159 of *Lecture Notes in Mathematics*, pp. 130–241. Springer Berlin / Heidelberg.

Karatzas, I. and S. E. Shreve (1991). *Brownian Motion and Stochastic Calculus* (second ed.). Springer.

Karlin, S. and H. M. Taylor (1975). *A first course in stochastic processes* (2 ed.). Academic Press.

Kendall, M. G. (2004). *A Course in the Geometry of n Dimensions*. Dover Publications.

Khintchine, A. Y. and P. Lévy (1936). Sur les lois stables. *C. R. Acad. Sci. Paris 202*, 374–376.

Kingman, J. F. C. (1993). *Poisson processes*. Oxford University Press.

Kloeden, P. E. and E. Platen (1992). *Numerical Solution of Stochastic Differential Equations*. Stochastic Modelling and Applied Probability. Springer.

Koponen, I. (1995). Analytic approach to the problem of convergence of truncated lévy flights towards the gaussian stochastic process. *Phys. Rev. E 52*(1), 1197–1199.

Kutner, M., C. Nachtsheim, J. Neter, and W. Li (2004, August). *Applied Linear Statistical Models* (5th ed.). McGraw-Hill/Irwin.

Lay, D. C. (2006). *Linear Algebra and Its Applications* (3 ed.). Reading, Mass: Addison-Wesley.

Lévy, P. (1925). *Calcul des probabilités*. Gauthier-Villars.

Lu, T.-C., Y.-S. Hou, and R.-J. Chen (1996). A parallel poisson generator using parallel prefix. *Computers & Mathematics with Applications 31*(3), 33 – 42.

Lyons, R. and Y. Peres (2010). *Probability on Trees and Networks*. In preparation. Current version available at `http://mypage.iu.edu/~rdlyons/`: Cambridge University Press.

Mansuy, R. (2009, June). The origins of the word "martingale". *Electronic Journal for History of Probability and Statistics 5/1*, 1–10.

Mantegna, R. N. and H. E. Stanley (1994). Stochastic process with ultraslow convergence to a gaussian: The truncated lévy flight. *Phys. Rev. Lett. 73*.

Marsaglia, G. and W. W. Tsang (2000). The ziggurat method for generating random variables. *Journal of Statistical Software 5*, 1–7.

Morgan, J. P., N. R. Chaganty, R. C. Dahiya, and M. J. Doviak (1991). Let's make a deal: The player's dilemma. *American Statistician 45*, 284–287.

Mörters, P. and Y. Peres (2010). *Brownian Motion.* Cambridge University Press.

Mueser, P. R. and D. Granberg (1991). The monty hall dilemma revisited: Understanding the interaction of problem definition and decision making. working paper 99-06, University of Missouri.

Nagle, R. K., E. Saff, and D. Snider (2011, March). *Fundamentals of Differential Equations and Boundary Value Problems* (6 ed.). Addison Wesley.

Nikodym, O. (1930). Sur un généralisation des intégrales de m. j. radon. *Fundamenta Mathematicae 15*, 131–179.

Øksendal, B. (2003). *Stochastic Differential Equations* (5 ed.). Springer Verlag.

Press, W., S. Teukolsky, W. Vetterling, and B. Flannery (1992). *Numerical Recipes in C* (2nd ed.). Cambridge University Press.

Press, W., S. Teukolsky, W. Vetterling, and B. Flannery (2007). *Numerical Recipes in C* (3rd ed.). Cambridge University Press.

Protter, P. E. (2003). *Stochastic Integration and Differential Equations* (second ed.). Springer.

Renshaw, E. (2011). *Stochastic Population Processes: Analysis, Approximations, Simulations.* Oxford Scholarship Online.

Revuz, D. and M. Yor (2004). *Springer.* Continuous Martingales and Brownian Motion.

Rosenthal, J. (2008, September). Monty hall, monty fall, monty crawl. *Math Horizons*, 5–7.

Ross, S. (1995). *Stochastic Processes* (2nd ed.). Wiley.

Royden, H. (1988). *Real Analysis* (3rd ed.). Prentice Hall.

Rubin, D. B. (1998). Using the sir algorithm to simulate posterior distributions. In *Mobile HCI.*

Rubin, H. and B. C. Johnson (2006). Efficient generation of exponential and normal deviates. *Journal of Statistical Computation and Simulation 76*(6), 509–518.

Samorodnitsky, G. and M. S. Taqqu (1994). *Stable non-Gaussian random processes: Stochastic models with infinite variance.* New York: Chapman and Hall.

Scott, L. O. (1987). Option pricing when the variance changes randomly: Theory, estimation, and an application. *The Journal of Financial and Quantitative Analysis 22*(4), 419–438.

Snell, J. L. (1997). A conversation with joe doob. *Statistical Science 12*(4), 301–311.

Stroock, D. W. and S. R. S. Varadhan (1979). *Multidimensional Diffussion Processes.* Springer.

Student (1908, March). The probable error of a mean. *Biometrika 6*(1), 1–25.

Vasicek, O. (1977). An equilibrium characterisation of the term structure. *Journal of Financial Economics* (5), 177–188.

Ville, J. (1939). *Etude critique de la notion de collectif.* Paris: Gauthier-Villars.

Wheeden, R. L. and A. Zygmund (1977). *Measure and Integral: An Introduction to Real Analysis.* New York: Marcel Dekker Inc.

Whiteley, N. and A. Johansen (2011). Recent developments inauxiliary particle filtering. In D. Barber, A. Cemgil, and S. Chiappa (Eds.), *Inference and Learning in Dynamic Models.* Cambridge University Press. in press.

Wichura, M. J. (1988). Algorithm as241: The percentage points of the normal distribution. *Journal of the Royal Statistical Society. Series C (Applied Statistics) 37*(3), 477–484.

Wiggins, J. B. (1987, December). Option values under stochastic volatility: Theory and empirical estimates. *Journal of Financial Economics 19*(2), 351–372.

Williams, D. (1991, February). *Probability with Martingales.* Cambridge University Press.

Zehna, P. W. (1966). Invariance of maximum likelihood estimators. *Ann. Math. Statist. 37*(3), 744.

Index

Probability and Stochastic Processes, First Edition. Ionuţ Florescu
© 2015 John Wiley & Sons, Inc. Published 2015 by John Wiley & Sons, Inc.